Natural Image Statistics

Computational Imaging and Vision

This comprehensive book series embraces state-of-the-art expository works and advanced research monographs on any aspect of this interdisciplinary field.

Topics covered by the series fall in the following four main categories:

- Imaging Systems and Image Processing
- Computer Vision and Image Understanding
- Visualization
- Applications of Imaging Technologies

Only monographs or multi-authored books that have a distinct subject area, that is where each chapter has been invited in order to fulfill this purpose, will be considered for the series.

Volume 39

For other titles published in this series, go to
www.springer.com/series/5754

Natural Image Statistics

A Probabilistic Approach to Early Computational Vision

Written by

Aapo Hyvärinen
University of Helsinki, Finland

Jarmo Hurri
University of Helsinki, Finland

and

Patrik O. Hoyer
University of Helsinki, Finland

Springer

Aapo Hyvärinen
University of Helsinki
Dept. Mathematics & Statistics and
Dept. Computer Science
Finland

Patrik O. Hoyer
University of Helsinki
Dept. Computer Science
Finland

Jarmo Hurri
University of Helsinki
Dept. Computer Science
Finland

ISSN 1381-6446
Computational Imaging and Vision
ISBN 978-1-84996-844-7 e-ISBN 978-1-84882-491-1
Springer Dordrecht Heidelberg London New York

British Library Cataloguing in Publication Data
A catalogue record for this book is available from the British Library

AMS Codes: 62H35, 92B20, 91E30, 68U10, 68T05, 94A08, 92C20

Cover design: WMHDesign Gmbh

Printed on acid-free paper

Springer is part of Springer Science+Business Media (www.springer.com)

Preface

Aims and Scope

This book is both an introductory textbook and a research monograph on modeling the statistical structure of natural images. In very simple terms, "natural images" are photographs of the typical environment where we live. In this book, their statistical structure is described using a number of statistical models whose parameters are estimated from image samples.

Our main motivation for exploring natural image statistics is computational modeling of biological visual systems. A theoretical framework which is gaining more and more support considers the properties of the visual system to be reflections of the statistical structure of natural images because of evolutionary adaptation processes. Another motivation for natural image statistics research is in computer science and engineering, where it helps in development of better image processing and computer vision methods.

While research on natural image statistics has been growing rapidly since the mid-1990s, no attempt has been made to cover the field in a single book, providing a unified view of the different models and approaches. This book attempts to do just that. Furthermore, our aim is to provide an accessible introduction to the field for students in related disciplines.

However, not all aspects of such a large field of study can be completely covered in a single book, so we have had to make some choices. Basically, we concentrate on the neural modeling approaches at the expense of engineering applications. Furthermore, those topics on which the authors themselves have been doing research are inevitably given more emphasis.

Targeted Audience and Prerequisites

The book is targeted for advanced undergraduate students, graduate students and researchers in vision science, computational neuroscience, computer vision, and image processing. It can also be read as an introduction to the area by people with a background in mathematical disciplines (mathematics, statistics, theoretical physics).

Due to the multi-disciplinary nature of the subject, the book has been written so as to be accessible to an audience coming from very different backgrounds such as psychology, computer science, electrical engineering, neurobiology, mathematics, statistics, and physics. Therefore, we have attempted to reduce the prerequisites to a minimum. The main thing needed are basic mathematical skills as taught in introductory university-level mathematics courses. In particular, the reader is assumed to know the basics of

- univariate calculus (e.g. one-dimensional derivatives and integrals)
- linear algebra (e.g. inverse matrix, orthogonality)
- probability and statistics (e.g. expectation, probability density function, variance, covariance)

To help readers with a modest mathematical background, a crash course on linear algebra is offered at Chap. 19, and Chap. 4 reviews probability theory and statistics on a rather elementary level.

No previous knowledge of neuroscience or vision science is necessary for reading this book. All the necessary background on the visual system is given in Chap. 3, and an introduction to some basic image processing methods is given in Chap. 2.

Structure of the Book and Its Use as a Textbook

This book is a hybrid of a monograph and an advanced graduate textbook. It starts with background material which is rather classic, whereas the latter parts of the book consider very recent work with many open problems. The material in the middle is quite recent but relatively established.

The book is divided into the following parts:

Introduction which explains the basic setting and motivation.

Part I which consists of background chapters. This is mainly classic material found in many textbooks in statistics, neuroscience, and signal processing. However, here it has been carefully selected to ensure that the reader has the right background for the main part of the book.

Part II starts the main topic, considering the most basic models for natural image statistics. These models are based on the statistics of linear features, i.e. linear combinations of image pixel values.

Part III considers more sophisticated models of natural image statistics, in which dependencies (interactions) of linear features are considered, which is related to computing non-linear features.

Part IV applies the models already introduced to new kinds of data: color images, stereo images, and image sequences (video). Some new models on the temporal structure of sequences are also introduced.

Part V consists of a concluding chapter. It provides a short overview of the book and discusses open questions as well as alternative approaches to image modeling.

Part VI consists of mathematical chapters which are provided as a kind of an appendix. Chapter 18 is a rather independent chapter on optimization theory. Chapter 19 is background material which the reader is actually supposed to know; it is provided here as a reminder. Chapters 20 and 21 provide sophisticated supplementary mathematical material for readers with such interests.

Dependencies of the parts are rather simple. When the book is used as a textbook, **all readers should start by reading the first seven chapters** in the order they are given (i.e. Introduction, Part I, and Part II except for the last chapter), unless the reader is already familiar with some of the material. After that, it is possible to jump to later chapters in almost any order, except for the following:

- Chapter 10 requires Chap. 9, and Chap. 11 requires Chaps. 9 and 10.
- Chapter 14 requires Sect. 13.1.

Some of the sections are marked with an asterisk *, which means that they are more sophisticated material which can be skipped without interrupting the flow of ideas.

An introductory course on natural image statistics can be simply constructed by going through the first n chapters of the book, where n would typically be between 7 and 17, depending on the amount of time available.

Referencing and Exercises

To keep the text readable and suitable for a textbook, the first 11 chapters do not include references in the main text. References are given in a separate section at the end of the chapter. In the latter chapters, the nature of the material requires that references are given in the text, so the style changes to a more scholarly one. Likewise, mathematical exercises and computer assignments are given for the first 10 chapters.

Code for Reproducing Experiments

For pedagogical purposes as well as to ensure the reproducibility of the experiments, the Matlab™ code for producing most of the experiments in the first 11 chapters, and some in Chap. 13, is distributed on the Internet at

www.naturalimagestatistics.net

This web site will also include other related material.

Acknowledgements

We would like to thank Michael Gutmann, Asun Vicente, and Jussi Lindgren for detailed comments on the manuscript. We have also greatly benefited from discussions with Bruno Olshausen, Eero Simoncelli, Geoffrey Hinton, David Field, Peter Dayan, David Donoho, Pentti Laurinen, Jussi Saarinen, Simo Vanni, and many others. We are also very grateful to Dario Ringach for providing the reverse correlation results in Fig. 3.7. During the writing process, the authors were funded by the University of Helsinki (Department of Computer Science and Department of Mathematics and Statistics), the Helsinki Institute for Information Technology, and the Academy of Finland.

Helsinki

Aapo Hyvärinen
Jarmo Hurri
Patrik Hoyer

Contents

Part VI Appendix: Supplementary Mathematical Tools

Abbreviations

DFT	discrete Fourier transform
FFT	fast Fourier transform
ICA	independent component analysis
ISA	independent subspace analysis
LGN	lateral geniculate nucleus
MAP	maximum a posteriori
MRF	Markov random field
NMF	non-negative matrix factorization
PCA	principal component analysis
RF	receptive field
RGB	red–green–blue
V1	primary visual cortex
V2, V3, ...	other visual cortical areas

Chapter 1
Introduction

1.1 What this Book Is All About

The purpose of this book is to present a general theory of early vision and image processing. The theory is normative, i.e. it says what is the optimal way of doing these things. It is based on construction of statistical models of images combined with Bayesian inference. Bayesian inference shows how we can use prior information on the structure of typical images to greatly improve image analysis, and statistical models are used for learning and storing that prior information.

The theory predicts what kind of features should be computed from the incoming visual stimuli in the visual cortex. The predictions on the primary visual cortex have been largely confirmed by experiments in visual neuroscience. The theory also predicts something about what should happen in higher areas such as V2, which gives new hints for people doing neuroscientific experiments.

Also, the theory can be applied on engineering problems to develop more efficient methods for de-noising, synthesis, reconstruction, compression, and other tasks of image analysis, although we do not go into the details of such applications in this book.

The statistical models presented in this book are quite different from classic statistical models. In fact, they are so sophisticated that many of them have been developed only during the last 10 years, so they are interesting in their own right. The key point in these models is the non-Gaussianity (non-normality) inherent in image data. The basic model presented is independent component analysis, but that is merely a starting point for more sophisticated models.

A preview of what kind of properties these models learn is in Fig. 1.1. The figure shows a number of linear features learned from natural images by a statistical model. Chapters 5–7 will already consider models which learn such linear features. In addition to the features themselves, the results in Fig. 1.1 show another visually striking phenomenon, which is their spatial arrangement, or topography. The results in the figure actually come from a model called Topographic ICA, which is explained in Chap. 11. The spatial arrangement is also related to computation of non-linear, invariant features, which is the topic of Chap. 10. Thus, the result in this figure combines several of the results we develop in this book. All of these properties are similar to those observed in the visual system of the brain.

In the rest of this Introduction, we present the basic problem of image analysis, and an overview of the various ideas discussed in more detail in this book.

A. Hyvärinen, J. Hurri, P.O. Hoyer, *Natural Image Statistics*,
Computational Imaging and Vision 39,

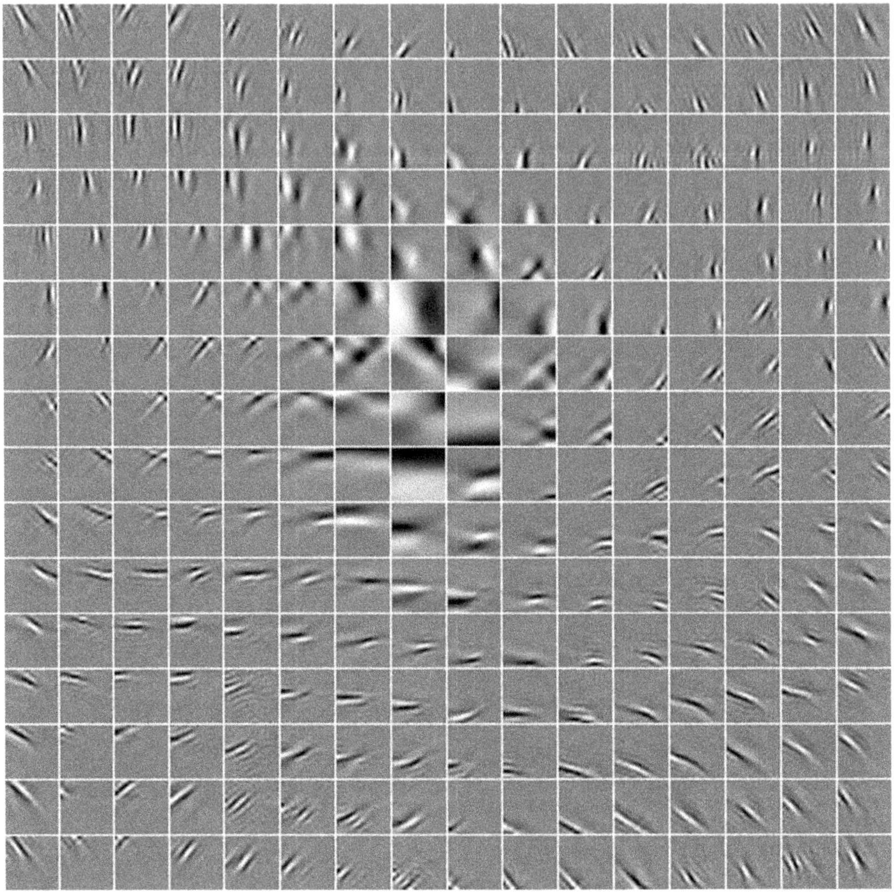

Fig. 1.1 An example of the results we will obtain in this book. Each *small square* in the image is one image feature, grey-scale coded to that middle—*grey* means zero, *white* positive, and *black* negative values. The model has learned local, oriented features which are similar to those computed by cells in the brain. Furthermore, the model uses the statistical dependencies of the features to arrange them on a 2D surface. Such a spatial arrangement can also be observed in the visual cortex. The arrangement is also related to computation of non-linear, invariant features

1.2 What Is Vision?

We can define vision as the process of acquiring knowledge about environmental objects and events by extracting information from the light the objects emit or reflect. The first thing we will need to consider is in what form this information initially is available.

The light emitted and reflected by objects has to be collected and then measured before any information can be extracted from it. Both biological and artificial systems typically perform the first step by projecting light to form a two-dimensional *image*. Although there are, of course, countless differences between the eye and any camera, the image formation process is essentially the same. From the image, the intensity of the light is then measured in a large number of spatial locations, or sampled. In the human eye, this is performed by the photo-receptors, whereas artificial systems employ a variety of technologies. However, all systems share the fundamental idea of converting the light first into a two-dimensional image and then into some kind of signal that represents the intensity of the light at each point in the image.

Although in general the projected images have both temporal and chromatic dimensions, we will be mostly concerned with static, monochrome (grey-scale) images. Such an image can be defined as a scalar function over two dimensions, $I(x, y)$, giving the intensity (luminance) value at every location (x, y) in the image. Although in the general case both quantities (the position (x, y) and the intensity $I(x, y)$) take continuous values, we will focus on the typical case where the image has been sampled at discrete points in space. This means that in our discussion x and y take only integer values, and the image can be fully described by an array containing the intensity values at each sample point.[1] In digital systems, the sampling is typically *rectangular*, i.e. the points where the intensities are sampled form a rectangular array. Although the spatial sampling performed by biological systems is not rectangular or even regular, the effects of the sampling process are not very different.

It is from this kind of image data that vision extracts information. Information about the physical environment is contained in such images, but only *implicitly*. The visual system must somehow transform this implicit information into an explicit form, for example by recognizing the identities of objects in the environment. This is not a simple problem, as the demonstration of the next section attempts to illustrate.

1.3 The Magic of Your Visual System

Vision is an exceptionally difficult computational task. Although this is clear to vision scientists, it might come as a surprise to others. The reason for this is that we are equipped with a truly amazing visual system that performs the task effortlessly and quite reliably in our daily environment. We are simply not aware of the whole computational process going on in our brains, rather we experience only the result of that computation.

To illustrate the difficulties in vision, Fig. 1.2 displays an image in its numerical format (as described in the previous section), where light intensities have been measured and are shown as a function of spatial location. In other words, if you were to color each square with the shade of grey corresponding to the contained number you would see the image in the form we are used to, and it would be easily interpretable. Without looking at the solution just yet, take a minute and try to decipher what the image portrays. You will probably find this extremely difficult.

[1] When images are stored on computers, the entries in the arrays also have to be discretized; this is, however, of less importance in the discussion that follows, and we will assume that this has been done at a high enough resolution so that this step can be ignored.

0	1	2	3	4	5	6	7	8	9

1	3	3	3	3	3	3	2	2	2	2	2	2	2	2	2	2	1	1	3	3	3	3	2	2	1	1	1	1	1	0	2	4	5	3	3	3	3	3	4	4	2	2	3	4	2
2	4	5	5	5	5	4	3	3	4	4	3	3	2	3	3	3	2	3	4	6	5	5	3	4	5	5	5	6	5	3	4	3	3	4	4	4	4	4	5	6	5	4	4	5	3
2	3	4	4	4	4	3	3	4	4	4	4	2	2	3	3	2	3	4	4	5	4	5	6	5	4	4	5	4	3	2	2	2	3	5	5	5	4	4	5	6	6	4	4	4	3
3	4	4	3	3	3	4	4	4	3	2	3	4	3	3	2	2	3	4	5	3	3	4	3	4	4	3	4	3	2	2	2	3	3	4	3	4	5	4	3	4	5	5	4	5	3
3	5	5	3	2	3	3	2	2	2	2	2	2	3	3	3	3	4	5	5	4	3	3	3	3	2	2	3	4	6	5	4	4	5	5	4	4	4	4	5	5	5	6	5	5	3
3	5	4	2	2	2	2	1	1	3	1	2	3	3	4	4	4	3	5	4	3	3	4	3	2	2	3	3	2	3	4	3	3	3	4	5	5	4	3	3	4	5	5	5	5	3
3	5	4	2	1	1	1	1	2	2	2	3	3	3	4	3	3	4	4	3	3	4	4	3	3	3	4	5	5	4	4	4	3	2	3	3	3	5	3	3	3	5	6	5	4	3
4	5	4	3	2	1	0	1	2	1	2	3	2	3	3	2	4	4	3	3	5	4	3	3	2	1	2	4	5	4	3	4	3	2	3	4	4	5	5	3	3	3	4	4	3	1
4	5	3	3	2	1	1	2	3	3	3	2	3	2	2	3	4	4	3	4	5	4	3	2	2	2	3	3	4	5	5	4	4	4	4	4	4	5	5	4	3	4	4	4	3	1
3	5	4	3	2	2	2	3	4	5	4	3	3	2	3	3	3	4	3	4	5	4	3	4	4	4	3	3	3	4	5	5	4	3	3	3	3	4	5	4	4	4	4	5	4	1
3	6	4	4	3	4	5	6	6	6	5	5	4	3	3	4	4	3	4	5	5	4	5	5	5	4	3	3	3	3	4	4	4	3	2	2	3	4	5	4	4	4	3	4	4	1
3	6	6	5	4	5	6	6	7	7	6	5	5	5	4	4	4	5	5	5	5	5	5	5	5	5	4	3	4	4	3	3	3	2	2	2	3	3	3	4	4	4	4	3	4	2
3	5	6	5	4	6	6	7	7	7	7	6	6	6	5	6	6	6	6	6	6	6	6	5	5	5	5	4	3	3	3	3	4	2	3	2	3	3	3	5	5	5	4	4	3	2
2	5	6	5	4	6	7	7	7	7	7	7	7	8	8	7	7	7	7	7	7	7	7	6	6	6	5	4	4	3	3	3	4	3	3	3	3	3	3	3	3	3	4	5	4	2
3	5	6	4	5	7	7	7	7	7	7	8	8	8	8	8	8	8	8	8	7	7	7	6	6	6	5	5	4	4	4	4	3	3	3	3	3	3	2	2	3	3	4	4	5	3
2	5	5	5	6	7	7	7	7	7	8	8	8	9	8	8	8	8	8	8	8	7	7	6	6	6	5	4	4	4	3	2	3	2	3	3	3	3	3	4	4	4	4	5	5	3
3	5	5	5	7	7	7	7	8	8	8	9	8	9	9	9	8	8	8	8	8	8	7	7	6	6	6	5	5	5	4	3	3	2	1	2	2	2	3	4	5	4	4	4	5	3
3	4	4	6	8	7	7	7	8	8	9	8	8	8	9	8	8	8	8	8	8	8	7	7	7	6	6	5	5	5	5	4	4	4	3	2	2	3	3	3	5	5	5	4	5	4
3	5	5	6	8	7	7	7	8	8	9	9	9	9	8	8	8	8	8	8	8	7	7	6	6	6	6	5	5	5	5	4	4	4	4	3	3	3	4	4	5	6	4	4	5	3
3	6	5	6	7	7	7	7	8	8	8	9	9	8	8	8	8	8	8	7	7	7	6	6	6	6	6	6	6	6	5	5	5	4	3	3	4	4	4	4	4	3	4	3	3	2
2	5	5	7	7	7	8	7	8	8	8	8	8	7	7	8	7	7	7	7	7	7	7	6	6	6	6	6	6	6	5	5	5	4	3	3	3	4	5	4	5	5	5	4	4	3
2	3	5	7	7	7	7	7	7	7	8	7	7	7	7	8	7	7	7	7	7	7	6	6	6	6	6	6	5	5	5	5	5	4	4	3	3	3	4	3	4	5	5	6	6	4
3	4	6	7	7	7	7	7	7	8	8	8	8	8	7	8	7	7	8	7	7	7	6	6	6	6	6	6	5	5	5	4	5	4	4	3	3	3	5	5	4	4	5	6	6	4
4	4	6	7	7	7	7	7	7	8	8	8	8	8	8	8	8	8	8	8	8	8	7	7	6	6	6	6	5	5	5	5	4	4	4	4	3	3	3	5	4	4	4	6	6	3
3	4	6	7	7	6	7	7	8	8	8	8	9	9	8	8	8	8	8	8	8	7	6	5	5	5	5	5	5	4	5	5	5	4	4	4	4	3	3	4	4	3	4	4	2	2
2	4	6	6	6	6	6	7	7	6	6	7	7	8	8	8	8	8	8	8	7	6	5	5	5	5	5	5	5	5	4	4	4	4	5	5	4	4	3	4	4	4	3	3	1	1
2	4	6	6	5	6	6	6	6	6	7	7	7	7	8	7	7	7	8	7	6	6	6	6	6	6	6	6	5	5	5	5	5	5	5	5	5	4	4	5	5	5	4	3	2	1
1	4	6	6	5	6	7	7	8	8	8	8	7	7	7	8	7	7	7	6	5	6	8	7	7	6	6	5	5	5	4	4	5	5	5	5	5	5	4	5	5	4	2	2	2	1
2	5	6	6	6	7	7	7	8	7	6	7	7	7	7	8	8	7	6	4	3	7	6	4	5	4	2	2	3	4	4	4	4	5	5	5	4	4	4	4	5	2	1	2	2	1
3	5	6	6	7	6	6	5	4	3	2	2	4	6	7	7	8	7	5	1	2	5	5	5	8	6	2	2	2	1	2	4	4	5	5	4	4	4	4	3	4	3	0	1	2	2
3	5	6	7	7	6	4	2	5	5	3	3	4	4	6	7	7	7	4	1	3	5	5	5	7	6	5	4	4	3	2	3	4	5	5	4	4	4	3	4	4	4	1	0	2	2
1	4	6	7	7	6	5	5	5	7	6	5	4	6	7	6	7	6	4	2	2	5	6	6	6	7	7	6	5	5	4	4	5	5	6	5	4	4	3	4	4	4	2	0	3	2
0	2	6	7	7	6	6	7	7	8	6	5	5	6	6	6	6	6	4	2	2	2	6	7	7	7	7	6	5	4	4	5	6	6	6	5	4	4	3	3	4	3	2	3	4	3
0	2	6	7	7	7	7	7	7	7	7	7	7	7	6	6	6	6	4	3	3	2	4	7	7	7	6	5	5	5	5	6	6	6	6	5	4	3	3	3	4	3	2	5	5	3
1	3	6	6	7	7	7	7	7	7	8	7	6	6	6	6	7	6	5	3	3	3	3	5	6	6	6	6	6	6	6	6	6	6	5	4	4	4	3	4	4	3	4	5	5	2
1	4	5	6	7	7	7	7	7	7	7	6	7	6	6	7	6	6	5	3	3	3	3	4	5	5	5	6	6	6	6	6	6	6	5	4	4	4	3	3	4	2	3	4	4	1
0	2	5	6	7	7	7	7	7	6	6	6	7	7	7	6	6	6	5	4	3	3	3	4	5	6	6	6	6	6	6	6	6	5	5	4	4	4	3	3	4	3	3	4	2	2
0	1	3	6	6	7	7	7	7	7	7	7	7	7	6	6	6	6	5	4	3	2	3	5	6	7	7	6	6	5	5	6	6	5	5	4	3	4	4	4	4	3	3	1	1	2
0	1	2	6	6	6	6	7	7	8	8	8	8	7	6	6	7	6	5	5	4	3	2	4	6	7	7	7	6	6	5	5	5	5	4	4	3	3	4	4	3	2	2	0	1	2
0	0	1	6	6	6	7	7	8	8	8	7	7	7	6	8	8	7	5	4	4	3	3	2	5	7	7	7	6	6	5	5	5	4	4	4	3	3	3	4	3	0	2	1	1	2
0	1	1	5	6	6	6	7	8	8	8	7	7	6	6	8	8	7	5	4	3	3	3	2	5	7	7	7	6	6	6	5	4	4	4	4	3	3	3	3	4	2	3	1	2	2
0	1	1	5	6	6	6	7	7	8	7	7	8	6	5	6	6	6	5	3	2	1	2	2	4	6	7	7	6	6	5	5	4	4	4	4	3	3	3	3	4	4	3	0	2	2
0	1	0	4	6	6	7	7	7	7	7	8	7	6	5	5	5	5	4	1	1	2	3	3	4	5	7	7	6	6	5	5	5	4	4	3	3	3	3	3	4	4	3	1	2	3
0	1	1	3	6	6	7	7	7	7	8	7	7	6	6	6	6	6	5	4	3	3	4	4	4	5	6	7	6	6	6	6	5	5	4	3	3	3	3	3	4	4	3	1	2	3
0	2	4	5	6	7	7	7	7	7	7	7	6	6	6	6	8	7	7	7	6	4	5	4	4	5	7	6	6	6	6	6	5	5	4	3	3	3	3	3	4	3	3	1	2	3
1	5	8	7	6	7	7	7	7	7	7	7	7	6	6	6	7	7	8	7	5	5	4	4	4	4	7	6	5	6	6	6	5	5	4	3	3	3	3	3	4	1	1	1	2	3
2	6	9	8	6	6	7	7	7	7	7	7	5	4	4	5	5	5	5	5	5	3	4	3	2	1	6	7	5	6	6	6	5	4	4	3	3	3	3	3	4	1	0	0	2	2
1	5	8	7	4	6	7	7	7	7	7	6	4	2	2	7	9	9	9	9	9	6	4	2	1	4	6	7	6	7	6	6	5	4	3	3	3	2	3	3	4	2	0	0	2	3
0	2	4	4	2	5	6	7	7	7	7	6	6	6	3	4	5	6	6	5	6	2	1	2	4	5	6	6	7	7	6	5	5	4	3	3	2	2	2	3	4	5	0	0	0	2
0	1	2	1	1	3	6	7	7	7	7	7	7	7	6	5	4	3	3	3	3	4	4	4	4	5	6	6	7	6	5	5	5	4	3	2	2	2	2	3	4	8	1	0	0	0
0	1	1	0	1	2	4	6	7	7	7	7	7	7	6	6	6	7	6	6	6	5	5	4	5	6	6	6	6	5	5	5	4	3	2	2	2	2	2	2	5	9	2	0	0	0
0	1	0	0	1	2	2	5	6	7	6	7	7	7	7	6	7	7	7	7	6	5	5	5	5	6	6	6	6	5	5	5	4	3	2	1	1	1	2	2	6	9	2	0	0	0
0	1	1	0	1	2	2	3	5	6	6	7	7	7	7	7	7	7	7	7	6	6	5	5	6	6	5	6	5	4	4	4	3	2	1	1	1	1	2	2	7	9	2	0	0	0
0	1	0	0	1	2	2	1	3	6	6	7	7	7	7	7	8	8	7	7	7	6	6	6	6	6	5	5	4	3	4	4	2	1	1	1	1	1	2	3	8	7	1	0	0	0
0	1	0	0	1	2	2	2	2	4	6	7	7	7	8	8	8	7	7	7	6	6	6	6	6	6	5	5	4	3	3	2	1	1	1	1	1	1	1	6	9	6	0	0	0	0
0	0	0	0	1	2	2	1	2	3	4	6	7	7	7	7	7	7	7	7	6	5	6	6	6	5	5	4	3	2	2	1	1	1	1	1	1	1	2	8	8	3	0	0	0	0
0	1	0	1	2	3	2	1	1	1	1	4	7	7	6	6	6	7	7	6	6	5	5	5	6	5	4	3	2	2	2	1	1	2	2	1	1	1	6	8	7	1	0	0	0	0
0	1	1	2	5	4	2	1	0	0	0	0	4	6	6	6	7	7	7	7	6	5	5	5	5	4	3	3	2	1	1	1	2	2	2	2	1	4	8	8	5	0	0	0	0	0
0	1	3	5	6	3	1	0	0	0	0	0	1	5	6	7	7	7	7	7	7	6	5	5	4	3	2	2	1	2	2	2	2	2	2	1	3	7	8	7	2	0	0	0	0	0
1	3	4	2	1	0	0	0	0	0	0	0	0	3	6	6	7	7	7	7	6	6	5	5	4	2	2	2	2	3	3	3	2	2	1	3	8	8	8	5	0	0	0	0	0	0
1	1	0	0	0	0	0	0	0	0	0	0	0	2	4	3	4	4	4	4	4	3	3	2	2	1	2	2	2	2	2	2	2	1	1	5	5	5	5	1	0	0	0	0	0	0

Fig. 1.2 An image displayed in numerical format. The shade of grey of each square has been replaced by the corresponding numerical intensity value. What does this mystery image depict?

Fig. 1.3 The image of Fig. 1.2. It is immediately clear that the image shows a male face. Many observers will probably even recognize the specific individual (note that it might help to view the image from relatively far away)

Now, have a look at the solution in Fig. 1.3. It is immediately clear what the image represents! Our visual system performs the task of recognizing the image completely effortlessly. Even though the image at the level of our photo-receptors is represented essentially in the format of Fig. 1.2, our visual system somehow manages to make sense of all this data and figure out the real-world object that caused the image.

In the discussion thus far, we have made a number of drastic simplifications. Among other things, the human retina contains photo-receptors with varying sensitivity to the different wavelengths of light, and we typically view the world through two eyes, not one. Finally, perhaps the most important difference is that we normally perceive dynamic images rather than static ones. Nonetheless, these differences do not change the fact that the optical information is, at the level of photo-receptors, represented in a format analogous to that we showed in Fig. 1.2, and that the task of the visual system is to understand all this data.

Most people would agree that this task initially seems amazingly hard. But after a moment of thought it might seem reasonable to think that perhaps the problem is not so difficult after all? Image intensity edges can be detected by finding oriented segments where small numbers border with large numbers. The detection of such features can be computationally formalized and straightforwardly implemented. Perhaps such oriented segments can be grouped together and subsequently object form be analyzed? Indeed, such computations can be done, and they form the basis of many computer vision algorithms. However, although current computer vision systems work fairly well on synthetic images or on images from highly restricted environments, they still perform quite poorly on images from an unrestricted, natural environment. In fact, perhaps one of the main findings of computer vision research to date has been that the analysis of real-world images is extremely difficult! Even such a basic task as identifying the contours of an object is complicated because often there is no clear image contour along some part of its physical contour, as illustrated in Fig. 1.4.

In light of the difficulties computer vision research has run into, the computational accomplishment of our own visual system seems all the more amazing. We perceive our environment quite accurately almost all the time, and only relatively rarely make perceptual mistakes. Quite clearly, biology has solved the task of every-

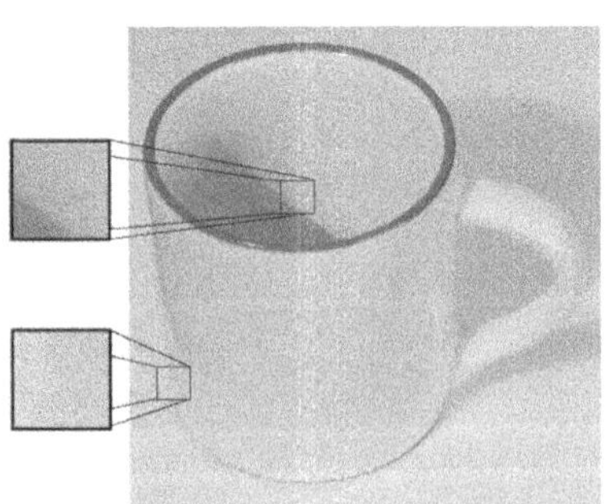

Fig. 1.4 This image of a cup demonstrates that physical contours and image contours are often very different. The physical edge of the cup near the *lower-left corner* of the image yields practically no image contour (as shown by the magnification). On the other hand, the shadow casts a clear image contour where there in fact is no physical edge

day vision in a way that is completely superior to any present-day machine vision system.

This being the case, it is natural that computer vision scientists have tried to draw inspiration from biology. Many systems contain image processing steps that mimic the processing that is known to occur in the early parts of the biological visual system. However, beyond the very early stages, little is actually known about the representations used in the brain. Thus, there is actually not much to guide computer vision research at the present.

On the other hand, it is quite clear that good computational theories of vision would be useful in guiding research on biological vision, by allowing hypothesis-driven experiments. So, it seems that there is a dilemma: computational theory is needed to guide experimental research, and the results of experiments are needed to guide theoretical investigations. The solution, as we see it, is to seek synergy by multi-disciplinary research into the computational basis of vision.

1.4 Importance of Prior Information

1.4.1 Ecological Adaptation Provides Prior Information

A very promising approach for solving the difficult problems in vision is based on adaptation to the statistics of the input. An adaptive representation is one that does not attempt to represent all possible kinds of data; instead, the representation is adapted to a particular kind of data. The advantage is that then the representation can concentrate on those aspects of the data that are useful for further analysis. This is in stark contrast to classic representations (e.g. Fourier analysis) that are fixed based on some general theoretical criteria, and completely ignore what kind of data is being analyzed.

Thus, the visual system is not viewed as a general signal processing machine or a general problem-solving system. Instead, it is acknowledged that it has evolved to solve some very particular problems that form a small subset of all possible problems. For example, the biological visual system needs to recognize faces under different lighting environments, while the people are speaking, possibly with different emotional expressions superimposed; this is definitely an extremely demanding problem. But on the other hand, the visual system does *not* need to recognize a face when it is given in an unconventional format, as in Fig. 1.2.

What distinguished these two representations (numbers vs. a photographic image) from each other is that the latter is *ecologically valid*, i.e. during the evolution of the human species, our ancestors have encountered this problem many times, and it has been important for their survival. The case of an array of numbers does definitely not have any of these two characteristics. Most people would label it as "artificial".

In vision research, more and more emphasis is being laid on the importance of the enormous amount of prior information that the brain has about the structure of the world. A formalization of these concepts has recently been pursued under the heading "Bayesian perception", although the principle goes back to the "maximum likelihood principle" by Helmholtz in the 19th century. Bayesian inference is the natural theory to use when inexact and incomplete information is combined with prior information. Such prior information should presumably be reflected in the whole visual system.

Similar ideas are becoming dominant in computer vision as well. Computer vision systems have been used on many different kinds of images: "ordinary" (i.e. optical) images, satellite images, magnetic resonance images, to name a few. Is it realistic to assume that the same kind of processing would adequately represent all these different kinds of data? Could better results be obtained if one uses methods (e.g. features) that are specific to a given application?

1.4.2 Generative Models and Latent Quantities

The traditional computational approach to vision focuses on how, from the image data I, one can compute quantities of interest called s_i, which we group together in a vector $\mathbf{s}$. These quantities might be, for instance, scalar variables such as the distances to objects, or binary parameters such as signifying if an object belongs to some given categories. In other words, the emphasis is on a function $\mathbf{f}$ that transforms images into world or object information, as in $\mathbf{s} = \mathbf{f}(I)$. This operation might be called image *analysis*.

Several researchers have pointed out that the opposite operation, image *synthesis*, often is simpler. That is, the mapping $\mathbf{g}$ that generates the image given the state of the world

$$I = \mathbf{g}(\mathbf{s}), \tag{1.1}$$

is considerably easier to work with, and more intuitive, than the mapping $\mathbf{f}$. This operation is often called *synthesis*. Moreover, the framework based on a fixed analyzing function $\mathbf{f}$ does not give much room for using prior information. Perhaps, by intelligently choosing the function $\mathbf{f}$, some prior information on the data could be incorporated.

Generative models use (1.1) as a starting point. They attempt to explain observed data by some underlying hidden (latent) causes or factors s_i about which we have only indirect information.

The key point is that the models incorporate a set of *prior probabilities for the latent variables* s_i. That is, it is specified how often different combinations of latent variables occur together. For example, this probability distribution could describe, in the case of a cup, the *typical* shape of a cup. Thus, this probability distribution for the latent variables is what formalizes the prior information on the structure of the world.

This framework is sufficiently flexible to be able to accommodate many different kinds of prior information. It all depends on how we defined the latent variables, and the synthesis function $\mathbf{g}$.

But how does knowing $\mathbf{g}$ help us, one may ask. The answer is that one may then search for the parameters $\hat{\mathbf{s}}$ that produce an image $\hat{I} = \mathbf{g}(\hat{\mathbf{s}})$ which, as well as possible, matches the observed image I. In other words, a combination of latent variables that is the "most likely". Under reasonable assumptions, this might lead to a good approximation of the correct parameters $\mathbf{s}$.

To make all this concrete, consider again the image of the cup in Fig. 1.4. The traditional approach of vision would propose that an early stage extracts local edge information in the image, after which some sort of grouping of these edge pieces would be done. Finally, the evoked edge pattern would be compared with patterns in memory, and recognized as a cup. Meanwhile, analysis of other scene variables, such as lighting direction or scene depth, would proceed in parallel. The analysis-by-synthesis framework, on the other hand, would suggest that our visual system has an unconscious internal model for image generation. Estimates of object identity, lighting direction, and scene depth are all adjusted until a satisfactory match between the observed image and the internally generated image is achieved.

1.4.3 Projection onto the Retina Loses Information

One very important reason why it is natural to formulate vision as inference of latent quantities is that the world is three dimensional whereas the retina is only two-dimensional. Thus, the whole 3D structure of the world is seemingly lost in the eye! Our visual system is so good in reconstructing a three-dimensional perception of the world that we hardly realize that a complicated reconstruction procedure is necessary. Information about the depth of objects and the space between is only implicit in the retinal image.

We do benefit from having two eyes which give slightly different views of the outside world. This helps a bit in solving the problem of depth perception, but it is only part of the story. Even if you close one eye, you can still understand which object is in front of another. Television is also based on the principle that we can quite well reconstruct the 3D structure of the world from a 2D image, especially if the camera (or the observer) is moving.

1.4.4 Bayesian Inference and Priors

The fundamental formalism for modeling how prior information can be used in the visual system is based on what is called Bayesian inference. Bayesian inference refers to statistically estimating the hidden variables $\mathbf{s}$ given an observed image I. In most models, it is impossible (even in theory) to know the precise values of $\mathbf{s}$, so one must be content with a probability density $p(\mathbf{s}|I)$. This is the probability of the latent variables *given* the observed image. By Bayes' rule, which is explained in

Sect. 4.7, this can be calculated as

$$p(\mathbf{s}|I) = \frac{p(I|\mathbf{s})p(\mathbf{s})}{p(I)}. \tag{1.2}$$

To obtain an estimate of the hidden variables, many models simply find the particular $\mathbf{s}$ which maximize this density,

$$\hat{\mathbf{s}} = \arg\max_{\mathbf{s}} p(\mathbf{s}|I). \tag{1.3}$$

Ecological adaptation is now possible by learning the prior probability distribution from a large number of natural images. Learning refers, in general, to the process of constructing a representation of the regularities of data. The dominant theoretical approach to learning in neuroscience and computer science is the probabilistic approach, in which learning is accomplished by statistical estimation: the data is described by a statistical model that contains a number of parameters, and learning consists of finding "good" values for those parameters, based on the input data. In statistical terminology, the input data is a sample that contains observations.

The advantage of formulating adaptation in terms of statistical estimation is very much due to the existence of an extensive theory of statistical theory and inference. Once the statistical model is formulated, the theory of statistical estimation immediately offers a number of tools to estimate the parameters. And after estimation of the parameters, the model can be used in inference according to the Bayesian theory, which again offers a number of well-studied tools that can be readily used.

1.5 Natural Images

1.5.1 The Image Space

How can we apply the concept of prior information about the environment in early vision? "Early" vision refers to the initial parts of visual processing, which are usually formalized as the computation of features, i.e. some relatively simple functions of the image (features will be defined in Sect. 1.8 below). Early vision does not yet accomplish such tasks as object recognition. In this book, we consider early vision only.

The central concept we need here is the image space. Earlier, we described an image representation in which each image is represented as a numerical array containing the intensity values of its picture elements, or *pixels*. To make the following discussion concrete, say that we are dealing with images of a fixed size of 256-by-256 pixels. This gives a total of $65\,536 = 256^2$ pixels in an image. Each image can then be considered as a point in a 65 536-dimensional space, each axis of which specifies the intensity value of one pixel. Conversely, each point in the space specifies one particular image. This space is illustrated in Fig. 1.5.

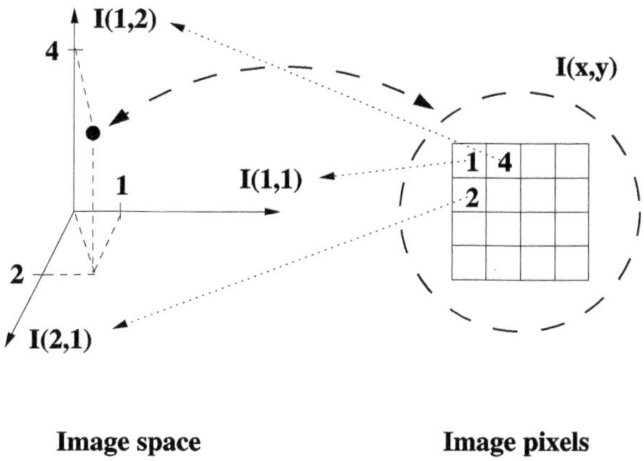

Fig. 1.5 The space representation of images. Images are mapped to points in the space in a one-to-one fashion. Each axis of the image space corresponds to the brightness value of one specific pixel in the image

Next, consider taking an enormous set of images, and plotting each as the corresponding point in our image space. (Of course, plotting a 65 536-dimensional space is not very easy to do on a two-dimensional page, so we will have to be content with making a thought experiment.) An important question is: how would the points be distributed in this space? In other words, what is the probability density function of our images like? The answer, of course, depends on the set of images chosen. Astronomical images have very different properties from holiday snapshots, for example, and the two sets would yield very different clouds of points in our space.

It is this probability density function of the image set in question that we will model in this book.

1.5.2 Definition of Natural Images

In this book, we will be specifically concerned with a particular set of images called *natural images* or images of *natural scenes*. Some images from our data set are shown in Fig. 1.6. This set is supposed to resemble the natural input of the visual system we are investigating. So, what is meant by "natural input"? This is actually not a trivial question at all. The underlying assumption in this line of research is that biological visual systems are, through a complex combination of the effects of evolution and development, adapted to process the kind of sensory input that they receive. Natural images are thus some set that we believe has similar statistical structure to that which the visual system is adapted to.

This poses an obvious problem, at least in the case of human vision. The human visual system has evolved in an environment that is in many ways different from the one most of us experience daily today. It is probably quite safe to say that images of skyscrapers, cars, and other modern entities have not affected our genetic makeup to any significant degree. On the other hand, few people today experience nature as omnipresent as it was tens of thousands of years ago. Thus, the input on the time-scale of evolution has been somewhat different from that on the time-scale

Fig. 1.6 Three representative examples from our set of natural images

Fig. 1.7 Three images drawn randomly from a uniform distribution in the image space. Each pixel is drawn independently from a uniform distribution from black to white

of the individual. Should we then choose images of nature or images from a modern, urban environment to model the "natural input" of our visual system? Most work to date has focused on the former, and this is also our choice in this book. Fortunately, this choice of image set does not have a drastic influence on the results of the analysis: Most image sets collected for the purpose of analyzing natural images give quite similar results in statistical modeling, and these results are usually completely different from what you would get using most artificial, randomly generated data sets.

Returning to our original question, how would *natural images* be distributed in the image space? The important thing to note is that they would not be anything like uniformly distributed in this space. It is easy for us to draw images from a uniform distribution, and they do not look anything like our natural images! Figure 1.7 shows three images randomly drawn from a uniform distribution over the image space. As there is no question that we can easily distinguish these images from natural images (Fig. 1.6), it follows that these are drawn from separate, very different, distributions. In fact, the distribution of natural images is highly non-uniform. This is the same as saying that natural images contain a lot of *redundancy*, an information-theoretic term that we turn to now.

1.6 Redundancy and Information

1.6.1 Information Theory and Image Coding

At this point, we make a short excursion to a subject that may seem, at first sight, to be outside of the scope of statistical modeling: information theory.

The development of the theory of information by Claude Shannon and others is one of the milestones of science. Shannon considered the transmission of a message across a communication channel and developed a mathematical theory that quantified the variables involved (these will be presented in Chap. 8). Because of its generality the theory has found, and continues to find, a growing number of applications in a variety of disciplines.

One of the key ideas in information theory is that the amount of memory needed to store an image is often less than what is needed in a trivial representation (code), where each pixel is stored using a fixed number of bits, such as 8 or 24. This is because some of the memory capacity is essentially consumed by *redundant structure* in the image. The more rigid the structure, the less bits is really needed to code the image. Thus, the contents of any image, indeed any signal, can essentially be divided into information and redundancy. This is depicted in Fig. 1.8.

To make this more concrete, consider the binary image of Fig. 1.9. The image contains a total of $32 \times 22 = 704$ pixels. Thus, the trivial representation (where the color of each pixel is indicated by a '1' or a '0') for this image requires 704 bits. But it is not difficult to imagine that one could compress it into a much smaller number of bits. For example, one could invent a representation that assumes a white background on which black squares (with given positions and sizes) are printed. In such a representation, our image could be coded by simply specifying the top-left corners of the squares ($(5,5)$ and $(19,11)$) and their sizes (8 and 6). This could certainly be coded in less than 704 bits.[2]

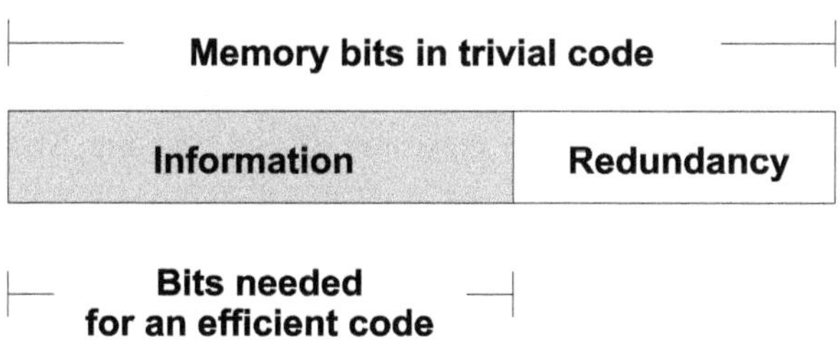

Fig. 1.8 Redundancy in a signal. Some of the memory consumed by the storage of a typical image is normally unnecessary because of redundancy (structure) in the image. If the signal is optimally compressed, stripping it of all redundancy, it can be stored using much less bits

[2]The specification of each square requires three numbers which each could be coded in 5 bits, giving a total of 30 bits for two squares. Additionally, a few bits might be needed to indicate how many squares are coded, assuming that we do not know a priori that there are exactly two squares.

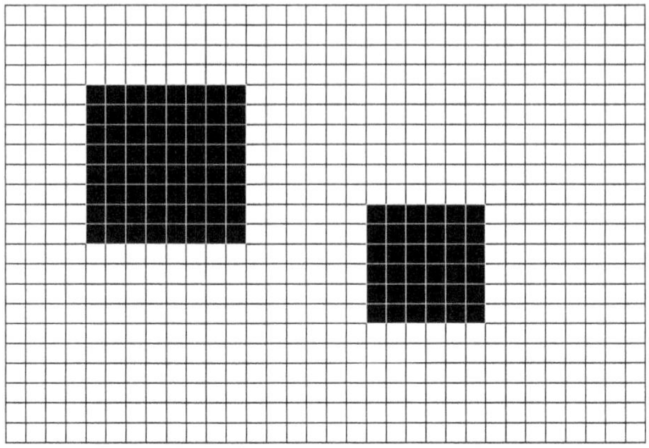

Fig. 1.9 A binary image containing a lot of structure. Images like this can be coded efficiently; see main text for discussion

The important thing to understand is that this kind of representation is good for certain kinds of images (those with a small number of black squares) but not others (that do not have this structure, and thus require a huge amount of squares to be completely represented). Hence, if we are dealing mostly with images of the former kind, and we are using the standard binary coding format, then our representation is highly redundant. By compressing it using our black-squares-on-white representation, we achieve an efficient representation. Although natural images are much more variable than this hypothetical class of images, it is nonetheless true that they also show structure and can be compressed.

Attneave was the first to explicitly point out the redundancy in images in 1954. The above argument is essentially the same as originally given by Attneave, although he considered a 'guessing game' in which subjects guessed the color of pixels in the image. The fact that subjects perform much better than chance proves that the image is predictable, and information theory ensures that predictability is essentially the same thing as redundancy.

Making use of this redundancy of images is essential for vision. But the same statistical structure is, in fact, also crucial for many other tasks involving images. Engineers who seek to find compact digital image formats for storing or transmitting images also need to understand this structure. Image synthesis and noise reduction are other tasks that optimally would make use of this structure. Thus, the analysis of the statistical properties of images has widespread applications indeed, although perhaps understanding vision is the most profound.

1.6.2 Redundancy Reduction and Neural Coding

Following its conception, it did not take long before psychologists and biologists understood that information theory was directly relevant to the tasks of biological systems. Indeed, the sensory input is a signal that carries *information* about the outside world. This information is *communicated* by sensory neurons by means of action potentials.

In Attneave's original article describing the redundancy inherent in images, Attneave suggested that the visual system recodes the inputs to *reduce redundancy*, providing an 'economical description' of the sensory signals. He likened the task of the visual system to that of an engineer who seeks to represent pictures with the smallest possible number of bits. It is easy to see the intuitive appeal of this idea. Consider again the image of Fig. 1.9. Recoding images of this kind using our black-squares-on-white representation, we reduce redundancy and obtain an efficient representation. However, at the same time, we have *discovered the structure* in the signal: we now have the concept of 'squares' which did not exist in the original representation. More generally: to reduce redundancy one must first identify it. Thus, redundancy reduction *requires* discovering structure.

Although he was arguably the first to spell it out explicitly, Attneave was certainly not the only one to have this idea. Around the same time, Barlow, in 1961, provided similar arguments from a more biological/physiological viewpoint. Barlow has also pointed out that the idea, in the form of 'economy of thought', is clearly expressed already in the writings of Mach and Pearson in the 19th century. Nevertheless, with the writings of Attneave and Barlow, the redundancy reduction (or efficient coding) hypothesis was born.

1.7 Statistical Modeling of the Visual System

1.7.1 Connecting Information Theory and Bayesian Inference

Earlier, we emphasized the importance of prior information and Bayesian modeling, but in the preceding section we talked about information theory and coding. This may seem a bit confusing at first sight, but the reason is that the two approaches are very closely related.

Information theory wants to find an economical representation of the data for efficient compression, while Bayesian modeling uses prior information on the data for such purposes as de-noising and recovery of the 3D structure. To accomplish their goals, both of these methods fundamentally need the same thing: *a good model of the statistical distribution of the data.* That is the basic motivation for this book. It leads to a new approach to normative visual modeling as will be discussed next.

1.7.2 Normative vs. Descriptive Modeling of Visual System

In visual neuroscience, the classic theories of receptive field structure[3] can be called *descriptive* in the sense that they give us mathematical tools (such as Fourier and

[3]I.e., the way visual neurons respond to stimulation, see Sect. 3.3.

Gabor analysis, see Chap. 2) that allow us to describe parts of the visual system in terms of a small number of parameters.

However, the question we really want to answer is: *Why* is the visual system built the way neuroscientific measurements show? The basic approach to answer such a question in neuroscience is to assume that the system in question has been *optimized* by evolution to perform a certain function. (This does not mean that the system would be completely determined genetically, because evolution can just have designed mechanisms for self-organization and learning that enable the system to find the optimal form.)

Models based on the assumption of optimality are often called *normative* because they tell how the system *should* behave. Of course, there is no justification to assume that evolution has optimized all parts of the organism; most of them may be far from the optimum, and such an optimum may not even be a well-defined concept.

However, in certain cases, it can be demonstrated that the system is not far from optimal in certain respects. This happens to be the case with the early cortical visual processing system (in particular, the primary visual cortex, see Chap. 3 for a brief description of the visual system). That brain area seems to function largely based on principles of statistical modeling, as will be seen in this book. Thus, there is convincing proof that parts of the system are optimal for statistical inference, and it is this proof that justifies these normative models.

Previous models of the early visual system did not provide satisfactory normative models, they only provided practical descriptive models. Although there were some attempts to develop a normative theory, the predictions were too vague.[4] The statistical approach is the first one to give exact quantitative models of visual processing, and these have been found to provide a good match with neuroscientific measurements.

1.7.3 Toward Predictive Theoretical Neuroscience

Let us mention one more important application of this framework: a mode of modeling where we are able to predict the properties of visual processing beyond the primary visual cortex. Then we obtain quantitative predictions on what kinds of visual processing should take place in areas whose function is not well understood at this point.

Almost all the experimental results in early visual processing have concerned the primary visual cortex, or even earlier areas such as the retina. Likewise, most research in this new framework of modeling natural image statistics has been on very low-level features. However, the methodology of statistical modeling can most probably be extended to many other areas.

---

[4]The main theory attempting to do this is the joint space–frequency localization theory leading to Gabor models; see Sect. 2.4.2. However, this does not provide predictions on how the parameters in Gabor models should be chosen, and what is more serious, it is not really clear why the features should be jointly localized in space and frequency in the first place.

Formulating statistical generative models holds great promise as a framework that will give new testable theories for visual neuroscience, for the following reasons:

- This framework is highly constructive. From just a couple of simple theoretical specifications, natural images lead to the emergence of complex phenomena, e.g. the forms of the receptive fields of simple cells and their spatial organization in Fig. 1.1.
- This framework is, therefore, less subjective than many other modeling approaches. The rigorous theory of statistical estimation makes it rather difficult to insert the theorist's subjective expectations in the model and, therefore, the results are strongly determined by the data, i.e. the objective reality. Thus, the framework can be called data-driven.
- In fact, in statistical generative models, we often see emergence of new kinds of feature detectors—sometimes very different from what was expected when the model was formulated.

So far, experiments in vision research have been based on rather vague, qualitative predictions. (This is even more true for other domains of neuroscience.) However, using the methodology described here, visual neuroscience has the potential of starting a new mode of operation where theoretical developments directly give new *quantitative* hypotheses to be falsified or confirmed in experimental research. Becoming theory-driven would be a real revolution in the way neuroscience is done. In fact, this same development is what gave much of the driving force to exact natural sciences in the 19th and 20th centuries.

1.8 Features and Statistical Models of Natural Images

1.8.1 Image Representations and Features

Most statistical models of natural images are based on computing *features*. The word "feature" is used rather loosely for any function of the image which is to be used in further visual processing. The same word can be used for the output (value) of the function, or the computational operation of which computes that value.

A classic approach to represent an image is a linear weighted sum of features. Let us denote each feature by $A_i(x, y)$, $i = 1, \dots, n$. These features are assumed to be fixed. For each incoming image, the coefficients of each feature in an image are denoted by s_i. Algebraically, we can write:

$$I(x, y) = \sum_{i=1}^{n} A_i(x, y) s_i. \tag{1.4}$$

If we assume for simplicity that the number of features n equals the number of pixels, the system in (1.4) can be inverted. This means that for a given image I, we can find the coefficients s_i that fulfill this equation. In fact, they can be computed linearly as

$$s_i = \sum_{x,y} W_i(x, y) I(x, y) \tag{1.5}$$

for certain inverse weights W. The terminology is not very fixed here, so either A_i, W_i, or s_i can be called a "feature". The W_i can also be called a *feature detector*.

There are many different sets of features that can be used. Classic choices include Fourier functions (gratings), wavelets, Gabor functions, features of discrete cosine transform, and many more. What all these sets have in common is that they attempt to represent all possible images, not just natural images, in a way which is "optimal".

What we want to do is to *learn* these features so that they are adapted to the properties of *natural* images. We do not believe that there could be a single set of features which would be optimal for all kinds of images. Also, we want to use the features to build a statistical model of natural images. The basis for both of these is to consider the statistics of the features s_i.

1.8.2 Statistics of Features

The most fundamental statistical properties of images are captured by the histograms of the outputs s_i of linear feature detectors. Let us denote the output of a single linear feature detector with weights $W(x, y)$ by s:

$$s = \sum_{x,y} W(x, y) I(x, y). \tag{1.6}$$

Now, the point is to look at the statistics of the output when the input of the detector consists of a large number of natural image patches. Natural image patches means small sub-images (windows) taken in random locations in randomly selected natural images. Thus, the feature s is a random variable, and for each input patch we get a realization (observation) of that random variable. (This procedure is explained in more detail in Sect. 4.1.)

Now, we shall illustrate this with real natural image data. Let us consider a couple of simple feature detectors and the histograms of their output when the input consists of natural images. In Fig. 1.10, we show three simple feature detectors. The first is a Dirac detector, which means that all the weights $W(x, y)$ are zero except for one. The second is a simple one-dimensional grating. The third is a basic Gabor edge detector. All three feature detectors have been normalized to unit norm, i.e. $\sum_{x,y} W(x, y)^2 = 1$.

The statistics of the output are contained in the histogram of the outputs. In Fig. 1.11, we show the output histograms for the three different kinds of linear detectors. We can see that the histograms are rather different. In addition to the different shapes, note that their variances are also quite different from each other.

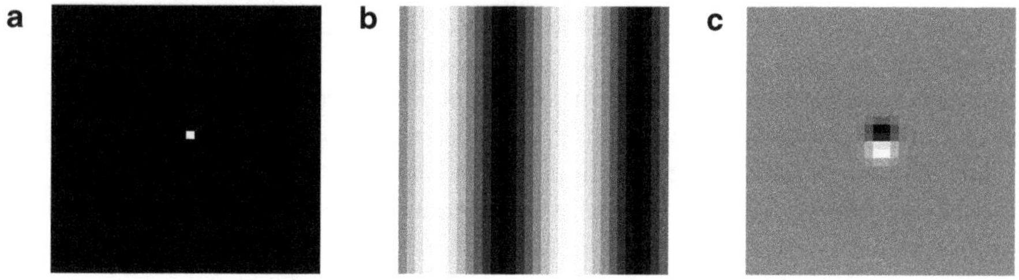

Fig. 1.10 Three basic filters. **a** A Dirac feature, i.e. only one pixel is non-zero. **b** A sinusoidal grating. **c** Gabor edge detector

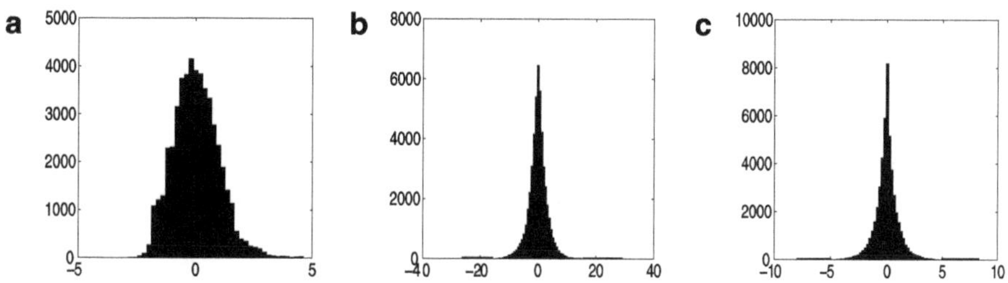

Fig. 1.11 The histograms of the outputs of the filters in Fig. 1.10 when the input is natural images with mean pixel value subtracted. **a** Output of Dirac filter, which is the same as the histogram of the original pixels themselves. **b** Output of grating feature detector. **c** Output of edge detector. Note that the scales of both axes are different in the three plots

Thus, we see that different feature detectors are characterized by different statistics of their outputs for natural image input. This basic observation is the basis for the theory in this book. We can *learn features* from image data by optimizing some statistical properties of the features s_i.

1.8.3 From Features to Statistical Models

The Bayesian goal of building a statistical (prior) model of the data, and learning features based on their output statistics are intimately related. This is because the most practical way of building a statistical model proceeds by using features and building a statistical model for them. The point is that the statistical model for features can be much simpler than the corresponding model for the pixels, so it makes sense to first transform the data into a feature space.

In fact, a large class of model builds *independent* models for each of the features s_i in (1.5). Independence is here to be taken both in an intuitive sense, and in the technical sense of statistical independence. Most models in Part II of this book are based on this idea. Even if the features are not modeled independently, the interactions (dependencies) of the features are usually much simpler than those of the original pixels; such models are considered in Part III of this book.

Thus, we will describe most of the models in this book based on the principle of learning features. Another reason for using this approach is that the most interesting neurophysiological results concern usually the form of the features obtained. In fact, it is very difficult to interpret or visualize a probability distribution given by the model; comparing the distribution with neurophysiological measurements is next to impossible. It is the features which give a simple and intuitive idea of what kind of visual processing these normative models dictate, and they allow a direct comparison with measured properties (receptive fields) of the visual cortex.

1.9 The Statistical–Ecological Approach Recapitulated

This chapter started with describing the difficulty of vision, and ended up proposing one particular solution, which can be called the statistical–ecological approach. The two basic ingredients in this approach are

- Ecology: The visual system is only interested in properties that are important in a real environment. This is related to the concept of *situatedness* in cognitive sciences.
- Statistics: Natural images have regularities. The regularities in the ecologically valid environment could be modeled by different formal frameworks, but statistical modeling seems to be the one that is most relevant.

Thus, we take the following approach to visual modeling:

1. Different sets of features are good for different kinds of data.
2. The images that our eyes receive have certain statistical properties (regularities).
3. The visual system has learned a model of these statistical properties.
4. The model of the statistical properties enables (close to) optimal statistical inference.
5. The model of the statistical properties is reflected in the measurable properties of the visual system (e.g. receptive fields of the neurons).

Most of this book will be concerned on developing different kinds of statistical models for natural images. These statistical models are based on very few theoretical assumptions, while they give rise to detailed quantitative predictions. We will show how these normative models are useful in two respects:

1. They provide predictions that are largely validated by classic neuroscientific measurements. Thus, they provide concise and theoretically well-justified explanations of well-known measurements. This is the evidence that justifies our normative modeling.
2. Moreover, the models lead to new predictions of phenomena which have not yet been observed, thus enabling theory-driven neuroscience. This is the big promise of natural image statistics modeling.

Another application of these models is in computer science and engineering. Such applications will not be considered in detail in this book: We hope we will have convinced the reader of the wide applicability of such methods. See below for references on this topic.

1.10 References

For textbook accounts of computer vision methods, such as edge detection algorithms, see, e.g. Marr (1982), Sonka et al. (1998), Gonzales and Woods (2002); for information theory, see Cover and Thomas (2006). The approach of generative models is presented in, e.g. Grenander (1976–1981), Kersten and Schrater (2002), Mumford (1994), Hinton and Ghahramani (1997).

For short reviews on using natural image statistics for visual modeling, see Field (1999), Simoncelli and Olshausen (2001), Olshausen (2003), Olshausen and Field (2004), Hyvärinen et al. (2005b). For reviews on engineering applications of statistical models, see Simoncelli (2005).

Historical references include Mach (1886), Pearson (1892), Helmholtz (1867), Shannon (1948), Attneave (1954), Barlow (1961). See also Barlow (2001a, 2001b) on a discussion on the history of redundancy reduction.

Part I
Background

Chapter 2
Linear Filters and Frequency Analysis

This chapter reviews some classical image analysis tools: linear filtering, linear bases, frequency analysis, and space-frequency analysis. Some of the basic ideas are illustrated in Fig. 2.1. These basic methods need to be understood before the results of statistical image models can be fully appreciated. The idea of processing of different frequencies is central in the reviewed tools. Therefore, a great deal of the following material is devoted to explaining what a *frequency-based representation* of images is, and why it is relevant in image analysis.

2.1 Linear Filtering

2.1.1 Definition

Linear filtering is a fundamental image-processing method in which a *filter* is applied to an input image to produce an output image.

Figure 2.2 illustrates the way in which the filter and the input image interact to form an output image: the filter is centered at each image location (x, y), and the pixel value of the output image $O(x, y)$ is given by the linear correlation of the filter and the filter-size subarea of the image at coordinate (x, y). (Note that the word "correlation" is used here in a slightly different way than in the statistical context.) Letting $W(x, y)$ denote a filter with size $(2K + 1) \times (2K + 1)$, $I(x, y)$ the input image, and $O(x, y)$ the output image, linear filtering is given by

$$O(x, y) = \sum_{x_*=-K}^{K} \sum_{y_*=-K}^{K} W(x_*, y_*) I(x + x_*, y + y_*). \tag{2.1}$$

An example of linear filtering is shown in Fig. 2.3.

What (2.1) means is that we "slide" the filter over the whole image and compute a weighted sum of the image pixel values, separately at each pixel location.

Visual inspection of a filter alone is usually not sufficient to interpret a filtering operation. This is also the case in the example in Fig. 2.3: what does this filtering operation actually accomplish? For a complete interpretation of a filtering operation a different type of mathematical language is needed. This language utilizes a frequency-based representation of images, as explained in Sect. 2.2 below.

A. Hyvärinen, J. Hurri, P.O. Hoyer, *Natural Image Statistics*,
Computational Imaging and Vision 39,

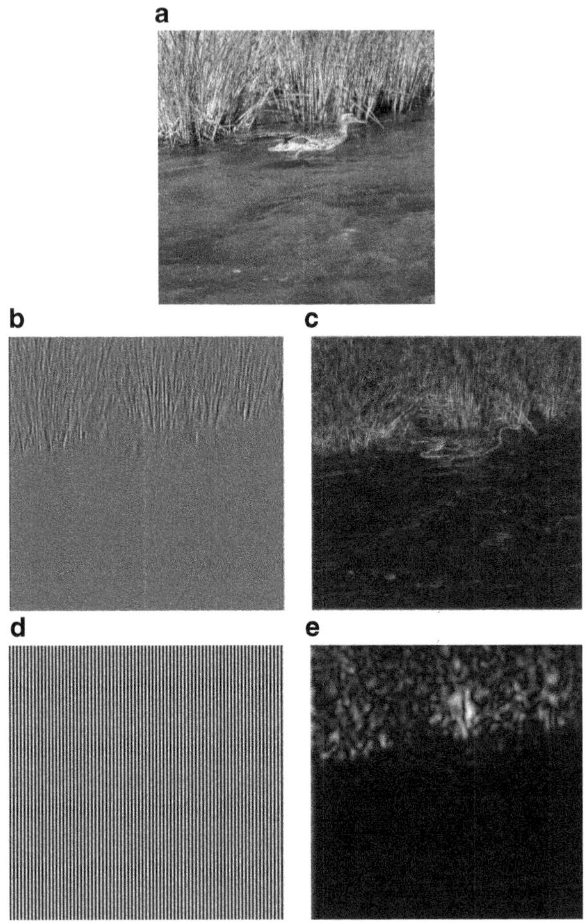

Fig. 2.1 The two classical image analysis tools reviewed in this chapter are linear filtering (**b**–**c**) and space-frequency analysis (**d**–**e**). **a** An input image. **b** An example output of linear filtering of **a**; in this case, the filter has retained medium-scaled vertical structures in the image. A more complete description of what a linear filtering operation does is provided by the frequency representation (Sect. 2.2). **c** An example of how the outputs of several linear filters can be combined in image analysis. In this case, the outputs of four filters have been processed non-linearly and added together to form an edge image: in the image, *lighter areas* correspond to image locations with a luminance edge. This kind of a result could be used, for example, as a starting point to locate objects of a certain shape. **d**–**e** An example of space-frequency analysis, where the main idea is to analyze the magnitude of a frequency **d** at different locations. The end result **e** reflects the magnitude of this frequency at different points in the input image **a**. From the point of view of image analysis, this result suggests that the *upper part* of the image is of different texture than the *lower part*

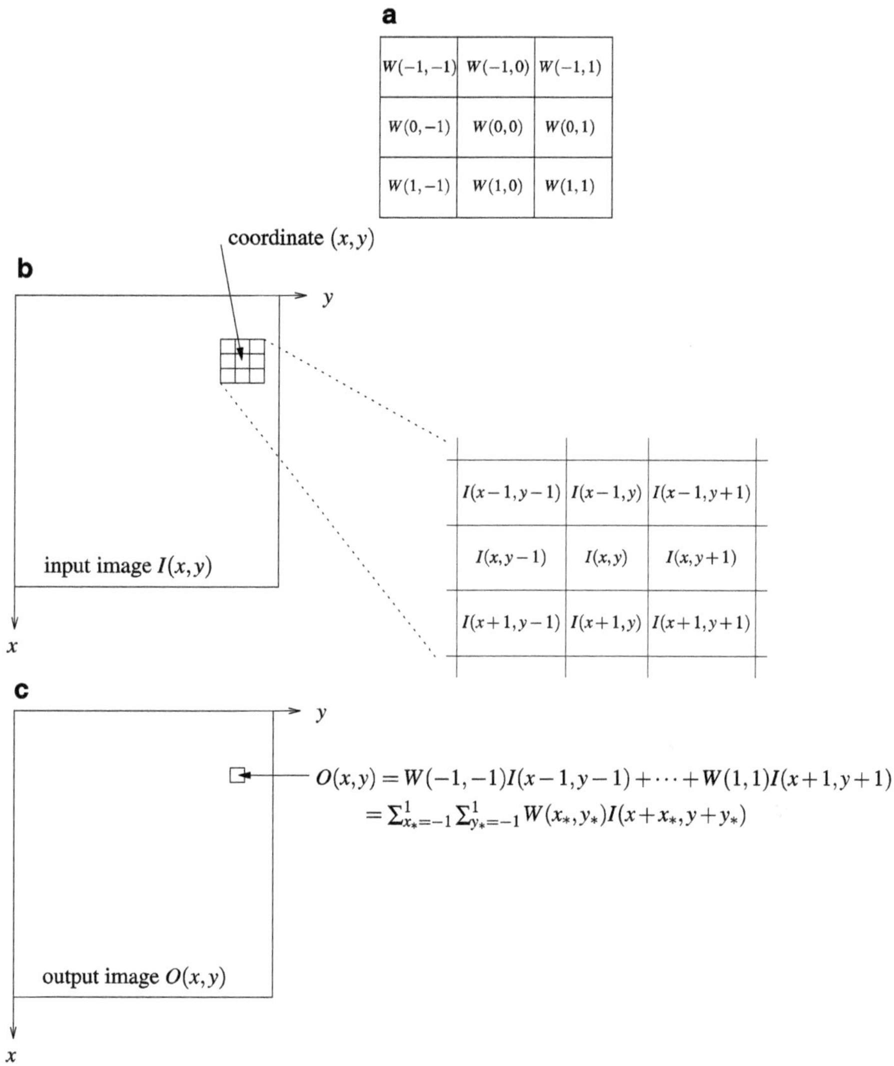

Fig. 2.2 Linear filtering is an operation that involves a filter (denoted here by $W(x, y)$) an input image (here $I(x, y)$) and yields an output image (here $O(x, y)$). The pixel value of the output image at location (x, y), that is, $O(x, y)$, is given by the linear correlation of the filter $W(x, y)$ and a filter-size subarea of the input image $I(x, y)$ centered at coordinate (x, y). **a** A 3×3 linear filter (template) $W(x, y)$. **b** An image $I(x, y)$ and a 3×3 subarea of the image centered at location (x, y). **c** The output pixel value $O(x, y)$ is obtained by taking the pixel-wise multiplication of the filter **a** and image subarea **b**, and summing this product over both x- and y-dimensions. Mathematically, $O(x, y) = \sum_{x_*} \sum_{y_*} W(x_*, y_*) I(x + x_*, y + y_*)$

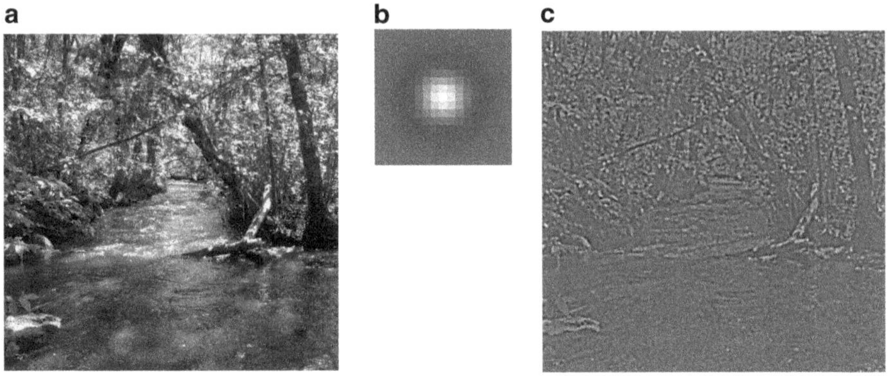

Fig. 2.3 An example of linear filtering. **a** An input image. **b** A filter. **c** An output image

2.1.2 Impulse Response and Convolution

The impulse response $H(x, y)$ is the response of a filter to an impulse

$$\delta(x, y) = \begin{cases} 1, & \text{if } x = 0 \text{ and } y = 0, \\ 0, & \text{otherwise,} \end{cases} \tag{2.2}$$

that is, to an image in which a single pixel is "on" (equal to 1) and the others are "off" (equal to 0). The impulse response characterizes the system just as well as the original filter coefficients $W(x, y)$. In fact, in frequency-based analysis of linear filtering, rather than filtering with a filter, it is customary to work with impulse responses and an operation called *convolution*. This is because the frequency-modifying properties of the linear filter can be read out directly from the frequency-based representation of the impulse response, as will be seen below. (In general, this holds for any linear shift-invariant system, which are defined in Sect. 20.1.)

Based on the definition of filtering in (2.1), it is not difficult to see that

$$H(x, y) = W(-x, -y) \tag{2.3}$$

Thus, the impulse response $H(x, y)$ is a "mirror image" of the filter weights $W(x, y)$: the relationship is simply that of a 180° rotation around the center of the filter. This is due to the change of signs of x_* and y_*; the impulse response is equal to one only if $x + x_* = 0$, which implies that only at points $x_* = -x$ we have one and elsewhere the impulse response is zero. For a filter that is symmetric with respect to this rotation, the impulse response is identical to the filter.

The convolution of two images I_1 and I_2 is defined as

$$I_1(x, y) * I_2(x, y) = \sum_{x_*=-\infty}^{\infty} \sum_{y_*=-\infty}^{\infty} I_1(x - x_*, y - y_*) I_2(x_*, y_*) \tag{2.4}$$

The only real difference to the definition of a filtering operation in (2.1) is that we have minus signs instead of plus signs. Note that convolution is symmetric in the sense that we can change the order of I_1 and I_2, since by making the change in summation index $x'_* = x - x_*$ and $y'_* = y - y_*$ we get the same formula with the roles of I_1 and I_2 interchanged (this is left as an exercise).

Therefore, we can express the filtering operation using the impulse response (which is considered just another image here) and the convolution simply as

$$O(x, y) = I(x, y) * H(x, y) \tag{2.5}$$

which is a slight modification of the original definition in (2.1). Introduction of this formula may seem like splitting hairs, but the point is that convolution is a well-known mathematical operation with interesting properties, and the impulse response is an important concept as well, so this formula shows how filtering can be interpreted using these concepts.

2.2 Frequency-Based Representation

2.2.1 Motivation

Frequency-based representation is a very useful tool in the analysis of image-processing systems. In particular, a frequency-based representation can be used to interpret what happens during linear filtering: it describes linear filtering as modification of strengths (amplitudes) and spatial locations (phases) of frequencies (sinusoidal components) that form the input image. As an example and sneak preview, Figs. 2.8a–d on page 36 show how the filtering operation of Fig. 2.3 can be interpreted as attenuation of low and high frequencies, which can be seen in the output image as disappearance of large- and fine-scale structures or, alternatively, preservation of medium-scale structures. This interpretation can be read out from Fig. 2.8d, which shows the frequency amplification map for this filter: this map, which is called the *amplitude response* of the filter, shows that both low frequencies (in the middle of the figure) and high frequencies (far from the middle) are attenuated; in the map, higher grey-scale value indicates larger amplitude response.

In what follows, we will first describe the frequency-based representation, and then demonstrate its special role in the analysis and design of linear filters.

2.2.2 Representation in One and Two Dimensions

Figure 2.4 illustrates the main idea of the frequency-based representation in the case of one-dimensional data. In the usual (spatial) representation (Fig. 2.4a), a signal is represented by a set of numbers at each point $x = 0, \ldots, N - 1$; in this example, $N = 7$. Therefore, to reconstruct the signal in Fig. 2.4a, we need 7 numbers. In the

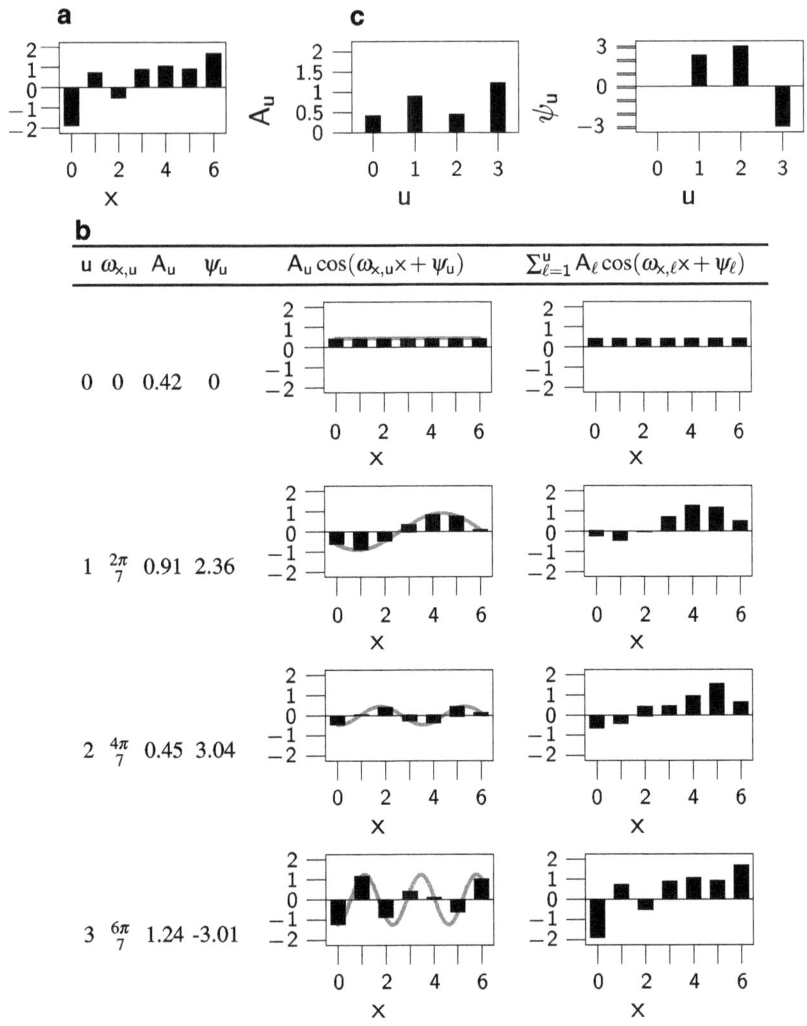

Fig. 2.4 In frequency-based representation, a signal is represented as amplitudes and phases of sinusoidal components. **a** A signal. **b** A table showing how the signal in **a** can be constructed from sinusoidal components. In the table, the number of sinusoidal components runs from 1 to 4 (frequency index u runs from 0 to 3), and the *rightmost column* shows the cumulative sum of the sinusoidal components with frequencies $\omega_{x,u}$ having amplitudes A_u and phases ϕ_u. In the *fifth column*, the *grey continuous lines* show the continuous frequency components from which the discrete versions have been sampled. Here, the frequency components are added in increasing frequency, that is, $\omega_{x,u_1} > \omega_{x,u_2}$ if $u_1 > u_2$. **c** A frequency-based representation for the signal in **a**: the signal is represented by the set of frequency amplitudes A_u, which is also called the amplitude spectrum (on the *left*), and the set of frequency phases ψ_u (on the *right*) of the corresponding frequencies $\omega_{x,u}$, $u = 0, \ldots, 3$. Note that the phase of the constant component $u = 0$ corresponding to frequency $\omega_{x,0} = 0$ is irrelevant; thus 7 numbers are needed in the frequency-based representation of the signal, just as in the usual representation **a**

frequency-based representation of this signal, we also use 7 numbers to describe the contents of the signal, but in this case the numbers have a totally different meaning: they are the *amplitudes* and *phases* of sinusoidal components, that is, parameters A and ψ of signals of the form $A\cos(\omega x + \psi)$, where ω is the frequency parameter; see Fig. 2.4c.

The theory of the *discrete Fourier transform* (treated in detail in Chap. 20) states that *any* signal of length 7 can be represented by the four amplitudes and the three phases of the four frequencies; the phase of the constant signal corresponding to $\omega = 0$ is irrelevant because the constant signal does not change when it is shifted spatially. For a family of signals of given length N, the set of frequencies ω_u, $u = 0, \ldots, U-1$, employed in the representation is fixed; in our example, these frequencies are listed in the second column of the table in Fig. 2.4b. Overall, the frequency-based representation is given by the sum

$$I(x) = \sum_{u=0}^{U-1} A_u \cos(\omega_u x + \psi_u), \tag{2.6}$$

where ω_u are the frequencies and A_u their amplitudes and ψ_u their phases.

In the case of images—that is, two-dimensional data—the sinusoidal components are of the form

$$A\cos(\omega_x x + \omega_y y + \psi), \tag{2.7}$$

where ω_x is the frequency in the x-direction and ω_y in the y-direction. In order to grasp the properties of such a component, let us define vector $\omega = (\omega_x, \omega_y)$, and denote the dot-product by $\langle . \rangle$. Then the component (2.7) can be written as

$$\begin{aligned} A\cos(\omega_x x + \omega_y y + \psi) &= A\cos\big(\langle (x,y), \omega\rangle + \psi\big) \\ &= A\cos\Bigg(\underbrace{\|\omega\|}_{\text{“frequency”}} \underbrace{\left\langle (x,y), \frac{\omega}{\|\omega\|}\right\rangle}_{\text{projection}} + \psi\Bigg), \end{aligned} \tag{2.8}$$

which shows that computation of the argument of the cosine function can be interpreted as a projection of coordinate vector (x, y) onto the direction of the vector ω, followed by a scaling with frequency $\|\omega\|$. Figure 2.5 illustrates this dependency of the frequency and the direction of the sinusoidal component on ω_x and ω_y.

Figure 2.5 also illustrates why it is necessary to consider both positive and negative values of either ω_x or ω_y: otherwise, it is not possible to represent all directions in the (x, y)-plane. However, there is a certain redundancy in this representation. For example, the frequency pairs $\omega = (\omega_x, \omega_y)$ and $-\omega = (-\omega_x, -\omega_y)$ represent sinusoidal components that have the same direction and frequency, because ω and $-\omega$ have the same direction and length. So, it can be seen that any half of the (ω_x, ω_y)-plane suffices to represent all directions. In practice, it has become customary to use

$\omega = (\omega_x, \omega_y)$	$\|\omega\|$	$\frac{\omega}{\|\omega\|}$	$I(x,y)$
$(\frac{1}{10}, \frac{1}{10})$	$\frac{\sqrt{2}}{10} \approx 0.14$	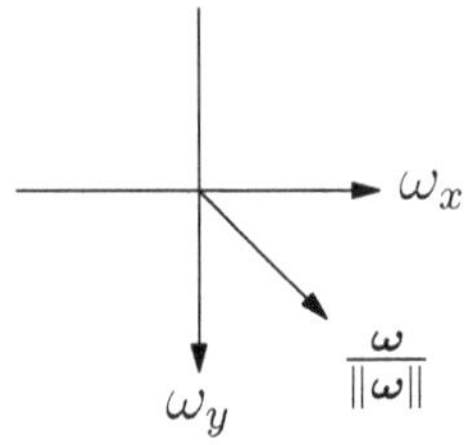	
$(\frac{1}{10}, -\frac{1}{10})$	$\frac{\sqrt{2}}{10} \approx 0.14$	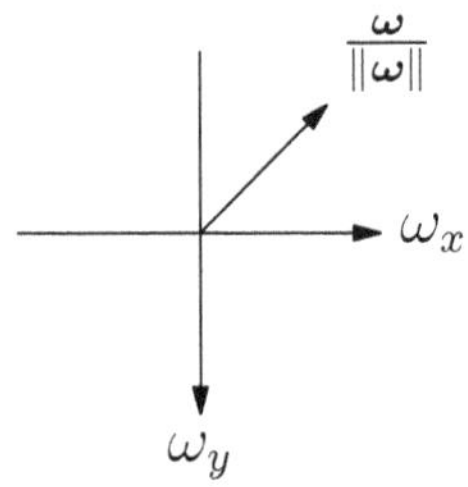	
$(\frac{6}{10}, \frac{3}{10})$	$\frac{3\sqrt{5}}{10} \approx 0.67$	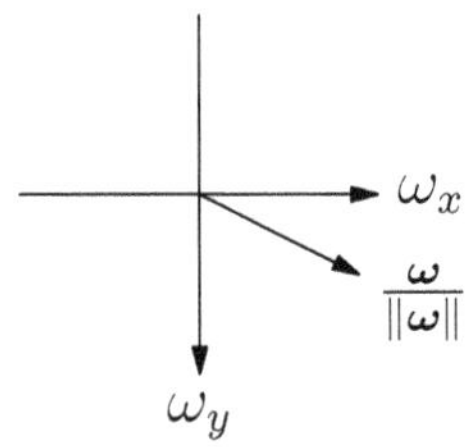	

Fig. 2.5 In an equation $I(x, y) = \cos(\omega_x x + \omega_y y)$ of a two-dimensional sinusoidal image, the frequencies ω_x and ω_y determine the direction and the frequency of the component. More specifically, if $\omega = (\omega_x, \omega_y)$, then the length of ω determines the frequency of the component, and the direction of ω determines the direction of the component. This is illustrated here for three different (ω_x, ω_y) pairs; the sinusoidal components are of size 128×128 pixels. Notice that in the plots in the *third column*, ω_y runs from *top* to *bottom* because of the convention that in images the y-axis runs in this direction. See (2.8) on page 31 for details

a redundant frequency representation which employs the whole (ω_x, ω_y)-plane, that is, negative and positive parts of *both* the ω_x- and ω_y-axis.[1]

Figure 2.6 shows an example of the resulting frequency representation. Again, the theory of discrete Fourier transform (see Chap. 20) states that *any* image of size 3×3 pixels can be represented by five amplitudes and four phases, a total of

[1]In fact, because cosine is an even function—that is, $\cos(-\alpha) = \cos(\alpha)$—the frequency components corresponding to these two frequency pairs are identical when $A\cos[\langle(x, y), \omega\rangle + \psi] = A\cos[-(\langle(x, y), \omega\rangle + \psi)] = A\cos[\langle(x, y), (-\omega)\rangle + (-\psi)]$, that is, when their amplitudes are the same and their phases are negatives or each other. Therefore, when employing the whole (ω_x, ω_y)-plane, the amplitude of a frequency component are customarily split evenly among the frequency pairs ω and $-\omega$ with phases ψ and $-\psi$.

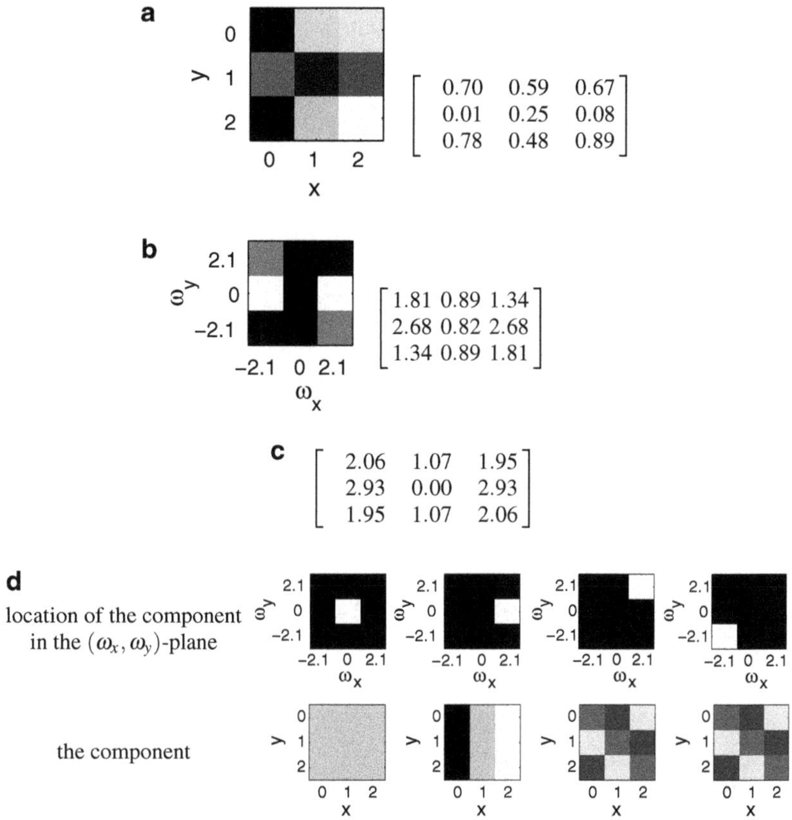

Fig. 2.6 An example of a two-dimensional frequency representation. **a** The grey-scale (*left*) and numerical (*right*) representation of an image of size 3 × 3 pixels. **b** Amplitude information of the frequency representation of the image in **a**: the grey-scale (*left*) and numerical (*right*) representation of the amplitudes of the different frequencies. Notice the symmetries/redundancies: the amplitude of frequency ω is the same as the amplitude of frequency $-\omega$. **c** Phase information of the frequency representation of the image in **a**; the axis of this representation are the same as in **b**. Notice the symmetries/redundancies: the phase of ω is the negative of the phase of $-\omega$. **d** Four examples of the actual sinusoidal components that make up the image in **a** in the frequency representation. In each column, the *first row* shows the location of the component in the (ω_x, ω_y)-plane, while the *second row* shows the actual component. The leftmost component is the constant component corresponding to $\omega = (0, 0)$. The second component is a horizontal frequency component. Because of the symmetry in the frequency representation, the third and the fourth components are identical. Notice that the second component (the horizontal frequency component) is stronger than the other components, which can also be seen in the amplitude representation in **b**

9 numbers as in the usual spatial representation; the phase of the constant signal corresponding to $\omega = (0, 0)$ is irrelevant as before.

Note on Terminology The square of the amplitude A^2 is also called the *Fourier energy* or *power*, and when it is computed for many different frequencies, we get what is called the *power spectrum*. The power spectrum is a classic way of characterizing which frequencies are present and with which "strengths" in a given signal or image. If the original amplitudes are used, we talk about the amplitude spectrum (Fig. 2.4c). It should be noted that the terminology of frequency-based analysis is not very well standardized, so other sources may use different terminology.

2.2.3 Frequency-Based Representation and Linear Filtering

Sinusoidals also play a special role in the analysis and design of linear filters. What makes sinusoidals special is the fact that when a sinusoidal is input into a linear filter, the response is a sinusoidal *with the same frequency*, but possibly different amplitude and phase. Figure 2.7 illustrates this phenomenon. Furthermore, both the

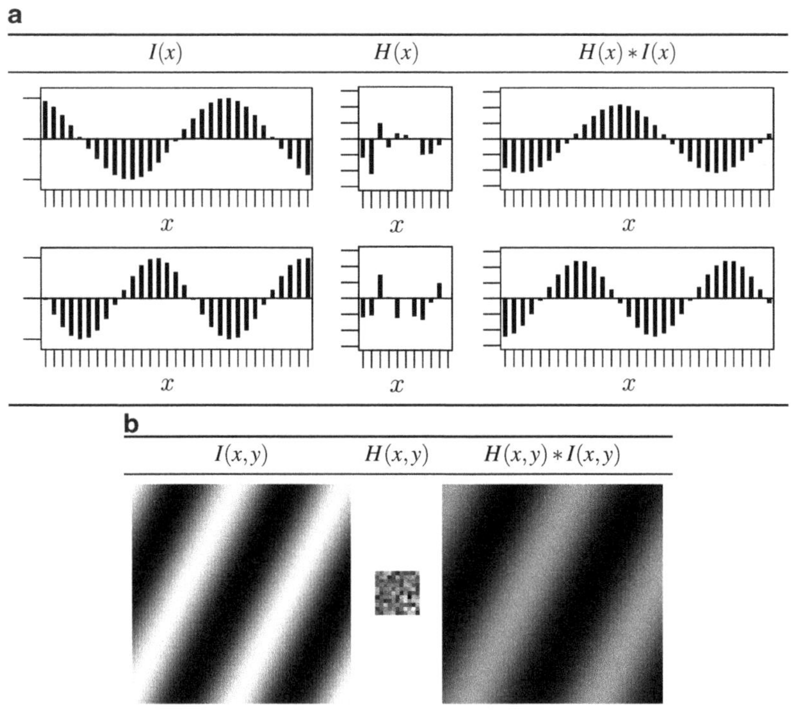

Fig. 2.7 Sinusoidal components play a special role in the analysis and design of linear systems, because if a sinusoidal is input into a linear filter, the output is a sinusoidal with the same frequency but possibly different amplitude and phase. **a** Two examples of this phenomenon are shown here in the one-dimensional case, one on each row. The *left column* shows the input signal, which is here assumed to be a sinusoidal of infinite duration. The *middle column* shows a randomly selected impulse response, and the *column on the right* the response of this linear system to the sinusoidal. Notice the different scale in the output signals, which is related to the amplitude change taking place in the filtering. **b** An illustration of the two-dimensional case, with a 64×64-pixel input *on the left*, a random 11×11-pixel impulse response *in the middle*, and the output *on the right*

amplification factor of the amplitude and the shift in the phase depend only on the frequency of the input, and not its amplitude or phase (see Chap. 20 for a detailed discussion).

For a linear filter with impulse response $H(x, y)$, let $|\tilde{H}(\omega_x, \omega_y)|$ denote the amplitude magnification factor of the system for horizontal frequency ω_x and vertical frequency ω_y, and $\angle\tilde{H}(\omega_x, \omega_y)$ denote the phase shift of the filter. Then if the input signal has frequency-based representation

$$I(x, y) = \sum_{\omega_x} \sum_{\omega_y} A_{\omega_x,\omega_y} \cos(\omega_x x + \omega_y y + \psi_{\omega_x,\omega_y}), \tag{2.9}$$

where the sum over ω_x and ω_y is here and below taken over both positive and negative frequencies, the response of the linear filter has the following frequency-based representation

$$\begin{aligned} O(x, y) &= H(x, y) * I(x, y) \\ &= \sum_{\omega_x} \sum_{\omega_y} \underbrace{|\tilde{H}(\omega_x, \omega_y)| A_{\omega_x,\omega_y}}_{\text{amplitude}} \cos\Big(\omega_x x + \omega_y y + \underbrace{\psi_{\omega_x,\omega_y} + \angle\tilde{H}(\omega_x, \omega_y)}_{\text{phase}}\Big). \end{aligned} \tag{2.10}$$

The amplitude magnification factor $|\tilde{H}(\omega_x, \omega_y)|$ is called the *amplitude response* of the linear system, while the phase shift $\angle\tilde{H}(\omega_x, \omega_y)$ is called the *phase response*. The way these quantities are determined for a linear filter is described shortly below; for now, let us just assume that they are available.

Figure 2.8 shows an example of the insight offered by the frequency representation of linear filtering (2.10). The example shows how a linear filter can be analyzed or designed by its amplitude response (the phase response is zero for all frequencies in this example). Notice that while relating the forms of the filters themselves (Figs. 2.8b and f) to the end result of the filtering is very difficult, describing what the filter does is straightforward once the frequency-based representation (Figs. 2.8d and e) is available.

How can the amplitude and phase responses of a linear system be determined? Consider a situation where we feed into the system a signal which contains a mixture of all frequencies with unit amplitudes and zero phases:

$$I(x, y) = \sum_{\omega_x} \sum_{\omega_y} \cos(\omega_x x + \omega_y y). \tag{2.11}$$

Then applying (2.10), the frequency-based representation of the output is

$$\begin{aligned} O(x, y) &= H(x, y) * I(x, y) \\ &= \sum_{\omega_x} \sum_{\omega_y} \underbrace{|\tilde{H}(\omega_x, \omega_y)|}_{\text{amplitude}} \cos\Big(\omega_x x + \omega_y y + \underbrace{\angle\tilde{H}(\omega_x, \omega_y)}_{\text{phase}}\Big). \end{aligned} \tag{2.12}$$

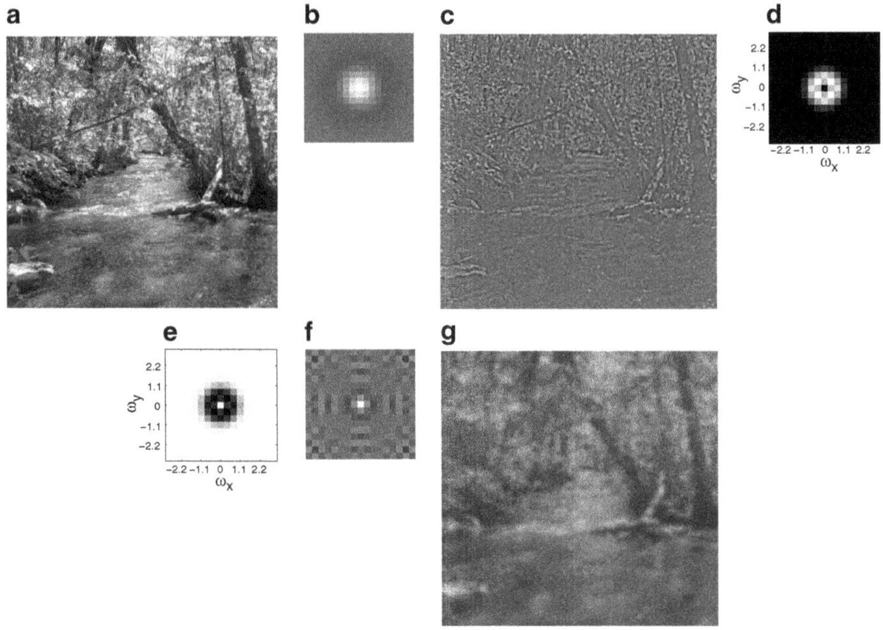

Fig. 2.8 An example of the usefulness of frequency-based representation in the analysis and design of linear filters. **a** An input image. **b** A filter of size 17×17 pixels. **c** The output of linear filtering of image **a** with filter **b**. **d** The amplitude response of the filter in **b**; in this representation dark pixels indicate amplitude response values close to zero and bright pixels values close to one. The amplitude response shows that the filter *attenuates low and high frequencies*, that is, frequencies which are either close to the origin $(\omega_u, \omega_v) = (0, 0)$ or far away from it. This can be verified in **c**, where medium-scaled structures have been preserved in the image, while details and large-scale grey-scale variations are no longer visible. The phase response of the filter is zero for all frequencies. **e** Assume that we want to design a filter that has a reverse effect than the filter shown in **b**: our new filter attenuates medium frequencies. The filter can be designed by specifying its amplitude and phase response. The amplitude response is shown here, the phase response is zero for all frequencies. **f** The filter corresponding to the frequency-based representation **e**. **g** The result obtained when the filter **f** is applied to the image in **a**. The results is as expected: the filter preserves details and large-scale grey-scale variations, while medium-scale variations are no longer visible. Notice that just by examining the filters themselves (**b** and **f**) it is difficult to say what the filters do, while this becomes straightforward once the frequency-based representations (**d** and **e**) are available

In other words, the amplitude and phase responses of the linear system can be read from the frequency-based representation of the output $O(x, y)$. What remains to be determined is what kind of a signal the signal $I(x, y)$ in (2.11) is. The theory of the Fourier transform states that the image obtained when all frequencies with identical amplitudes and zero phases are added together is an impulse. Figure 2.9 illustrates this when image size is 3×3 pixels. To summarize, when we feed an impulse into a linear filter,

- from the point of view of frequency-based description, we are giving the system an input with equal amplitudes of all frequencies at phase zero

k	0	1	2	3	4	5	6	7	8
$\omega_{x,k}$	$\frac{2\pi}{3}$	$-\frac{2\pi}{3}$	$-\frac{2\pi}{3}$	0	0	0	$\frac{2\pi}{3}$	$\frac{2\pi}{3}$	$\frac{2\pi}{3}$
$\omega_{y,k}$	$\frac{2\pi}{3}$	0	$\frac{2\pi}{3}$	$-\frac{2\pi}{3}$	0	$\frac{2\pi}{3}$	$-\frac{2\pi}{3}$	0	$\frac{2\pi}{3}$

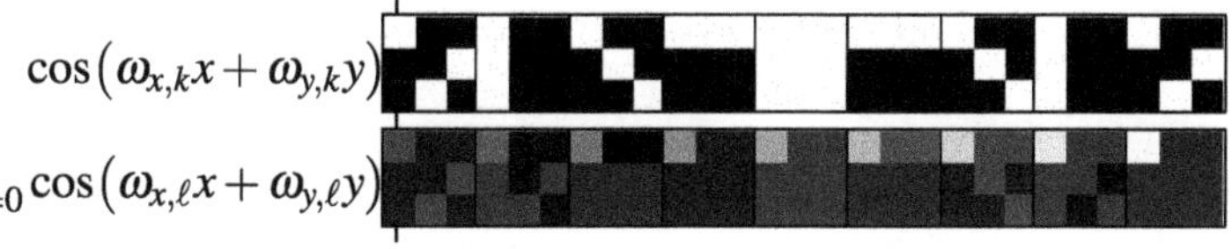

Fig. 2.9 An image containing all frequencies with unit amplitudes and zero phases is an impulse. Here, the different frequency components are being added from left to right; the right-most image is an impulse response

- the linear system modifies the amplitudes and phases of these frequencies according to the amplitude and phase response of the system
- the amplitude and phase response properties can be easily read out from the impulse response, since the amplitudes of the input were equal and phases were all zero.

In other words, the *amplitude and phase responses of a linear filter are obtained from the frequency-based representation of the impulse response: the amplitude responses are the amplitudes of the frequencies, and the phase responses are the phases of the frequencies.* An example of this principle is shown in Fig. 2.8: the amplitude response images (d) and (e) are in fact the amplitudes of the frequency-based representations of the impulse responses (b) and (f).

2.2.4 Computation and Mathematical Details

Above, we have outlined the nature of the frequency-based representation in the one- and two-dimensional case, and the usefulness of this representation in the design and analysis of linear systems. The material presented so far should therefore provide the reader with the knowledge needed to understand what the frequency-based representation is, and why it is used.

However, a number of questions have been left unanswered in the text above, including the following:

- What exactly are the values of the frequencies ω used in the frequency-based representation?
- How is the frequency-based representation computed?
- What guarantees that a frequency-based representation exists?

The set of mathematical tools used to define and analyze frequency-based representations are part of mathematics called *Fourier analysis*. In particular, *Fourier transforms* are used to convert data and impulse responses to and from frequency-based representation. There are different types of Fourier transforms for different purposes: continuous/discrete and finite/infinite data. When working with digital images, the most important Fourier transform is the *discrete Fourier transform (DFT)*, which is particularly suited for representation of discrete and finite data in computers. The basics of DFT are described in Chap. 20. The computational implementation of DFT is usually through a particular algorithm called Fast Fourier Transform, or FFT.

The DFT has fairly abstract mathematical properties, because complex numbers are employed in the transform. The results of the DFT are, however, quite easily understood in terms of the frequency-based representation: for example, Fig. 2.8d was computed by taking the DFT of the filter in Fig. 2.8b, and then showing the magnitudes of the complex numbers of the result of the DFT.

A working knowledge of the frequency-based representation is not needed in reading this book: it is sufficient to understand what the frequency-based representation is and why it is used. If you are interested in working with frequency-based representations, then studying the DFT is critical, because the DFT has some counterintuitive properties that must be known when working with results of transforms; for example, the DFT assumes that the data (signal or image) is periodic, which causes periodicity effects when filtering is done in the frequency-based representation.

2.3 Representation Using Linear Basis

Now we consider a general and widely-used framework for image representation: a linear basis. We will also see how frequency-based representation can be seen as a special case in this framework.

2.3.1 Basic Idea

A traditional way to represent grey-scale images is the pixel-based representation, where the value of each pixel denotes the grey-scale value at that particular location in the image. For many different purposes, more convenient representations of images can be devised.

Consider the three 3×2 images f_1, f_2 and I_3 shown in Fig. 2.10. The traditional pixel-based representations of the images are

$$f_1 = \begin{bmatrix} 4\,0\,0 \\ 4\,0\,0 \end{bmatrix},$$

$$f_2 = \begin{bmatrix} 0\,0\,4 \\ 0\,0\,4 \end{bmatrix},$$

Fig. 2.10 Three different 3×2 images consisting of vertical lines: **a** f_1; **b** f_2; **c** I_3. Here, *black* denotes a grey-scale value of zero, *light grey* a grey-scale value of 4, and the *darker grey* a grey-scale value of 2

$$I_3 = \begin{bmatrix} 2\,0\,2 \\ 2\,0\,2 \end{bmatrix}.$$

These images consist of vertical lines with different grey-scale values. A compact way to describe a set of images containing only vertical lines is to define the following *basis images*

$$B_1 = \begin{bmatrix} 1\,0\,0 \\ 1\,0\,0 \end{bmatrix},$$

$$B_2 = \begin{bmatrix} 0\,1\,0 \\ 0\,1\,0 \end{bmatrix},$$

$$B_3 = \begin{bmatrix} 0\,0\,1 \\ 0\,0\,1 \end{bmatrix},$$

and then represent each image as a *weighted sum* of these basis images. For example, $f_1 = 4B_1$, $f_2 = 4B_3$ and $I_3 = 2B_1 + 2B_3$. Such a representation could convey more information about the inherent structure in these images. Also, if we had a very large set of such images, consisting of only vertical lines, and were interested in compressing our data, we could store the basis images, and then for each image just save the three coefficients of the basis images.

This simple example utilized a special property of the images f_1, f_2, and I_3, that is, the fact that in this case each image contains only vertical lines. Note that not every possible 3×2 image can be represented as a weighted sum of the basis images B_1, B_2, and B_3 (one example is an image with a single non-zero pixel at any image location). This kind of a basis is called an *under-complete* basis. Usually the word *basis* is used to refer to a *complete basis*, a basis which—when used in the form of a weighted sum—can be used to represent any image. For the set of 3×2 images, one example of a complete basis is the set of 6 images with a single one at exactly one image location (x, y). This basis is typically associated with the traditional pixel-based image representation: each pixel value denotes the coefficient of the corresponding basis image.

A particularly important case is an *orthogonal* basis. Then the coefficients in the basis simply equal the dot-products with the basis vectors. For more on bases and orthogonality, see Sect. 19.6.

The use of different bases in the representation of images has several important areas of application in image processing. Two of these were mentioned already: the description and analysis of the structure of images, and image compression. Other important applications are in the domain of image processing systems: different image representations are central in the analysis of these systems (including the analysis of the visual system), design of such systems, and in their efficient implementation.

2.3.2 Frequency-Based Representation as a Basis

Now, we consider how frequency-based representation can be rephrased as finding the coefficients in a basis. Consider the situation where we want to analyze, in a given signal, the amplitude A and phase ψ of a sinusoidal component

$$A\cos(\omega x + \psi). \tag{2.13}$$

The key point here is that instead of determining A and ψ directly, we can determine the coefficients C and S of a cosine and sine signal with (centered) phase zero:

$$C\cos(\omega x) + S\sin(\omega x). \tag{2.14}$$

To show this, we will demonstrate that there is a one-to-one correspondence between signals of the form (2.13) and of the form (2.14). First, a given sinusoidal of the form (2.13) can be represented in form (2.14) as follows:

$$\begin{aligned} A\cos(\omega x + \psi) &= A\big(\cos(\omega x)\cos\psi - \sin(\omega x)\sin\psi\big) \\ &= \underbrace{A\cos\psi}_{=C}\cos(\omega x) + \underbrace{A(-\sin\psi)}_{=S}\sin(\omega x) \\ &= C\cos(\omega x) + S\sin(\omega x). \end{aligned} \tag{2.15}$$

Conversely, if we are given a signal in form (2.14), the derivation (2.15) can be reversed (this is left as an exercise), so that we get

$$A = \sqrt{C^2 + S^2}, \tag{2.16}$$

$$\psi = -\operatorname{atan}\frac{S}{C}. \tag{2.17}$$

Thus, to analyze the amplitude and phase of frequency ω in a given signal, it suffices to find coefficients C and S in equation $C\cos(\omega x) + S\sin(\omega x)$; after the coefficients have been computed, (2.16) and (2.17) can be used to compute the amplitude and phase. In particular, the square of the amplitude ("Fourier power" or "Fourier energy") is obtained as the sum of squares of the coefficients.

The formula in (2.16) is also very interesting from a neural modelling viewpoint because it shows how to compute the amplitude using quite simple operations, since computation of the coefficients in a linear basis is a linear operation (at least if the basis is complete). Computation of Fourier energy in a given frequency thus requires two linear operations, followed by squaring, and summing of the squares. As we will see in Chap. 3, something similar to linear Fourier operators seems to be computed in the early parts of visual processing, which makes computation of Fourier energy rather straightforward.

How are the coefficients C and S then computed? The key is orthogonality. The signals $\cos(\omega x)$ and $\sin(\omega x)$ are orthogonal, at least approximately. So, the coefficients C and S are simply obtained as the dot-products of the signal with the basis vectors given by the cos and sin functions.

In fact, Discrete Fourier Transform can be viewed as defining a basis with such cos and sin functions with many different frequencies, and those frequencies are carefully chosen so that the sinusoidals are exactly orthogonal. Then the coefficients for *all* the sin and cos functions, in the different frequencies, can be computed as the dot-products

$$\sum_x I(x)\cos(\omega x) \quad \text{and} \quad \sum_x I(x)\sin(\omega x). \tag{2.18}$$

The idea generalizes to two dimensions (images) in the same way as frequency-based analysis was shown to generalize above. More details on the DFT can be found in Chap. 20.

2.4 Space-Frequency Analysis

2.4.1 Introduction

The frequency representation utilizes global sinusoidal gratings, that is, components which span the whole image (see, e.g. Fig. 2.5). This is particularly useful for the analysis and design of linear filters. However, because of the global nature of sinusoidals, the frequency representation tells us nothing about the spatial relationship of different frequencies in an image. This is illustrated in Figs. 2.11a and b, which show an image and the amplitudes of its frequency representation. The upper part of the image contains grass, which tends to have a more vertical orientation and sharper structure than the lower part of the image. The amplitudes of the frequency representation (Fig. 2.11b) show that many horizontal or near-horizontal frequencies—that is, frequencies in the vicinity of axis $\omega_y = 0$—have a relatively large amplitude, even at fairly high frequencies. (Notice that structures with vertical lines correspond to *horizontal* frequencies.) From the amplitude spectrum there is, however, no way to tell the spatial location of these frequencies.

The spatial locations of different frequencies can contain important information about the image. In this example, if we are able to locate those areas which tend to have more horizontal frequencies, we can use that information, for example, to

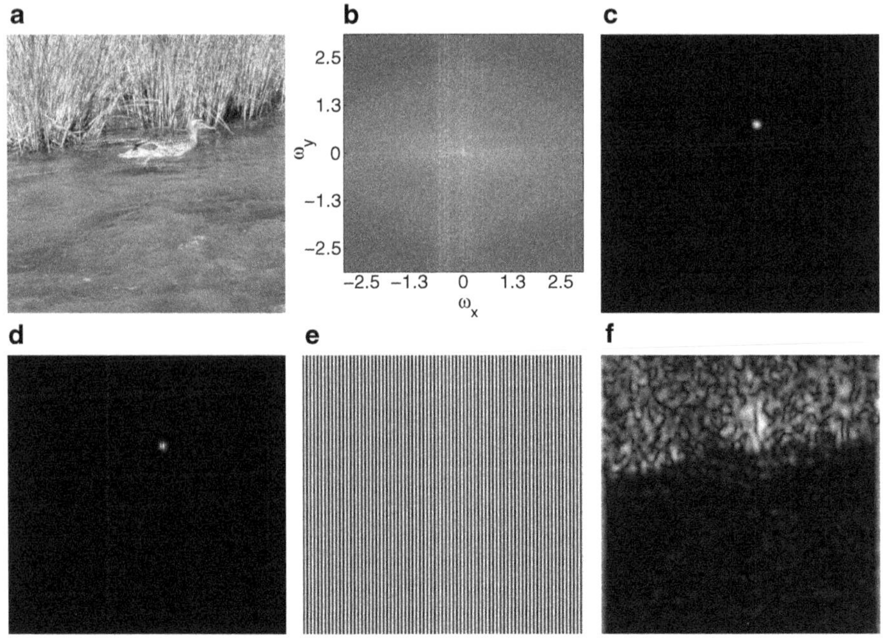

Fig. 2.11 The main idea in space-frequency analysis is to consider the amplitudes/phases of different frequencies *at different locations in an image*. **a** An image. Notice how different areas of the image differ in their frequency contents. **b** The standard frequency representation: the amplitudes of the different frequencies that make up the frequency representation of the image in **a**. Note that while this representation suggests that fairly high horizontal frequencies are present in the image, it does not convey information about the spatial location of different frequencies. For purposes of visibility, the amplitudes of different frequencies have been transformed with logarithmic mapping before display. **c** A spatially localized non-negative windowing function. **d** A localized area of the image can be obtained by multiplying the image **a** with the windowing function **c**. **e** In this example, the amplitude (strength) of this horizontal frequency is analyzed at each point in the image. **f** Applying the weighting scheme **c**–**d** at each point in the image **a**, and then analyzing the amplitude of the frequency **e** at each of these points results in this spatial map of the amplitude of the frequency. As can be seen, the frequency tends to have larger amplitudes in the upper part of the image, as can be expected

facilitate the identification of the grass area in the image. How can the spatial locations of these frequencies be found? A straightforward way to do this is to analyze the frequency contents of *limited spatial areas*. Figures 2.11c–f illustrate this idea. The original image (Fig. 2.11a) is multiplied with a non-negative *windowing function* (Fig. 2.11c) to examine the frequency contents of a spatially localized image area (Fig. 2.11d). For example, for the horizontal frequency shown in Fig. 2.11e, a spatial map of the amplitude of this frequency is shown in Fig. 2.11f; the map has been obtained by applying the weighting scheme (Figs. 2.11c and d) at *every point* of the image, and analyzing the amplitude of the frequency in the localized image area. Now, we see that in Fig. 2.11f that the different areas (grass vs. water) are clearly separated by this space-frequency analysis.

In the case of space-frequency analysis, the computational operations underlying the analysis are of great interest to us. This is because some important results obtained with statistical modelling of natural images can be interpreted as performing space-frequency analysis, so the way the analysis is computed needs to be understood to appreciate these results. Because of this connection, we need to delve a little deeper into the mathematics.

Before going into the details, we first state the main result: space-frequency analysis can be done by the method illustrated in Fig. 2.12: by filtering the image with two different localized sinusoidal filters, and computing the amplitudes and phases (in this example only the amplitudes) from the outputs of these two filters. The following section explains the mathematics behind this method.

2.4.2 Space-Frequency Analysis and Gabor Filters

Consider a situation where we want to analyze the local frequency contents of an image: we want to compute the amplitude and phase for each location (x_0, y_0). Alternatively, we could compute a set of coefficients $C(x_0, y_0)$ and $S(x_0, y_0)$ as in Sect. 2.3.2, which now are functions of the location. The analysis is made local by applying a weighting function, say $W(x, y)$, centered at (x_0, y_0) before the analysis.

We can simply modify the analysis in Sect. 2.3.2 by centering the signal f around the point (x_0, y_0) and weighting it by w before the analysis. Thus, we get a formula for the coefficient C at a given point:

$$C(x_0, y_0) \approx \sum_x \sum_y I(x, y) W(x - x_0, y - y_0) \cos\big(\omega_x (x - x_0) + \omega_y (y - y_0)\big) \tag{2.19}$$

and similarly for $S(x_0, y_0)$. Note that the order in which we multiply the three images (image f, weighting image w, sinusoidal cos) inside the sum is irrelevant. Therefore, it does not make any difference whether we apply the weighting to the image $I(x, y)$ or to the sinusoidal $\cos(\omega_x (x - x_0) + \omega_y (y - y_0))$. If we define a new weighting function

$$W_2(x, y) = W(x, y) \cos(\omega_x x + \omega_y y), \tag{2.20}$$

(2.19) becomes

$$C(x_0, y_0) \approx \sum_x \sum_y I(x, y) W_2(x - x_0, y - y_0). \tag{2.21}$$

Equations (2.20) and (2.21) show that computation of coefficients $C(x_0, y_0)$ can be approximated by filtering the image with a filter that is the product of a sinusoidal $\cos(\omega_x x + \omega_y y)$ and the weighting window. Similar analysis applies to coefficients $S(x_0, y_0)$, except that in that case the sinusoidal is $\sin(\omega_x x + \omega_y y)$. Because the magnitude of the weighting function $W(x, y)$ typically falls off quite fast from the

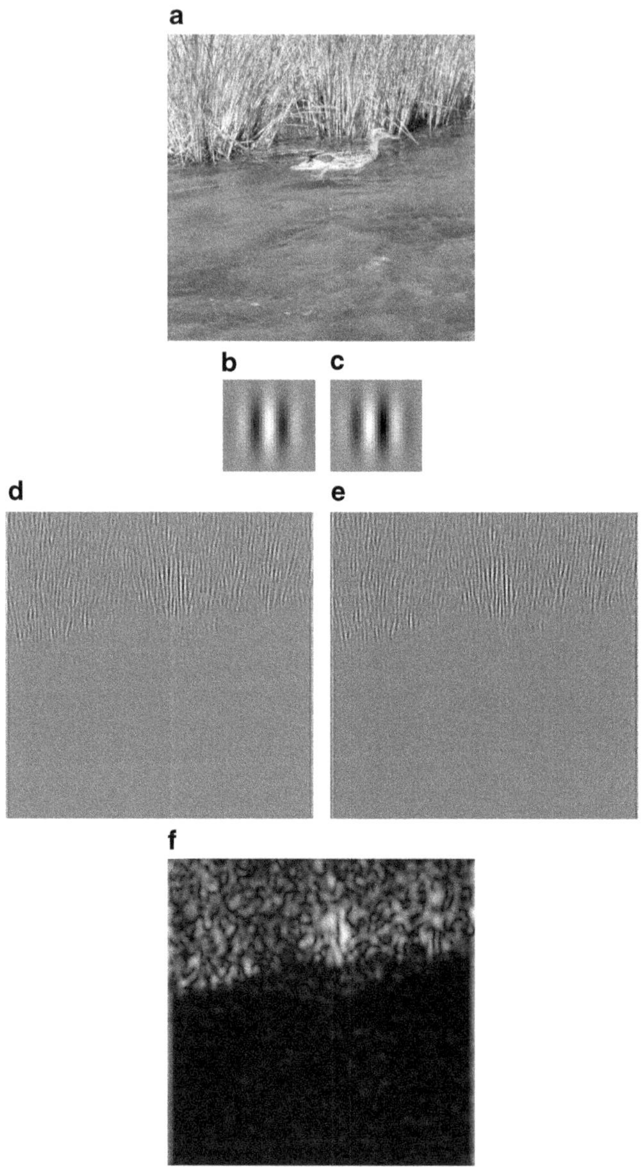

Fig. 2.12 An example showing how the spatial map of amplitudes of Fig. 2.11f on page 42 was computed. **a** The analyzed image. **b** The spatial filter (called a Gabor filter) obtained by multiplying $\cos(\omega_x x + \omega_y y)$ with the weighting window $W(x, y)$; the filter has been truncated to a size of 33×33 pixels. **c** The spatial filter obtained by multiplying $\sin(\omega_x x + \omega_y y)$ with the weighting window $W(x, y)$. **d** Coefficients $C(x_0, y_0)$, obtained by filtering image **a** with filter **b**. **e** Coefficients $S(x_0, y_0)$, obtained by filtering image **a** with filter **c**. **f** The spatial amplitude map $A(x_0, y_0) = \sqrt{C(x_0, y_0)^2 + S(x_0, y_0)^2}$

origin $(x, y) = (0, 0)$, computational savings can be obtained by truncating the filter to zero for (x, y) far away from the origin.

To summarize the result in an example, Fig. 2.12 shows how the amplitude map of Fig. 2.11f on page 42 was computed: $C(x_0, y_0)$ and $S(x_0, y_0)$ were computed by filtering the image with weighted versions of $\cos(\omega_x x + \omega_y y)$ and $\sin(\omega_x x + \omega_y y)$, respectively, and the amplitude map $A(x_0, y_0)$ was obtained by $A(x_0, y_0) = \sqrt{C(x_0, y_0)^2 + S(x_0, y_0)^2}$.

The two filters used in the computation of $C(x_0, y_0)$ and $S(x_0, y_0)$ are often called a *quadrature-phase pair*. This is because $\sin(x + \frac{\pi}{2}) = \cos(x)$, so the two filters are $W(x, y)\cos(\omega_x x + \omega_y y)$ and $W(x, y)\cos(\omega_x x + \omega_y y + \frac{\pi}{2})$, that is, they are otherwise identical expect for a a phase difference of one quarter of a whole cycle: $\frac{2\pi}{4} = \frac{\pi}{2}$.

When the weighting function is a Gaussian window, which in the one-dimensional case is of the form

$$W_\sigma(x) = \frac{1}{d} e^{-\frac{x^2}{\sigma^2}}, \tag{2.22}$$

the resulting filter is called a *Gabor filter*; parameter σ determines the width of the window, and the scaling constant d is typically chosen so that $\sum_x W_\sigma(x) = 1$. Overall, a one-dimensional Gabor function

$$W_{\sigma,\omega,\psi}(x) = W_\sigma(x)\cos(\omega x + \psi) \tag{2.23}$$

has three parameters, width σ, frequency ω and phase ψ. One-dimensional Gabor functions are illustrated in Fig. 2.13.

In the two-dimensional case, a Gabor filter has a few additional parameters that control the two-dimensional shape and orientation of the filter. When the sinusoidal in the filter has vertical orientation the filter is given by the following equation

$$W_{\sigma_x,\sigma_y,\omega,\psi}(x, y) = \frac{1}{d} e^{-\left(\frac{x^2}{\sigma_x^2} + \frac{y^2}{\sigma_y^2}\right)} \cos(\omega x + \psi); \tag{2.24}$$

here, σ_x and σ_y control the width of the weighting window in the x- and y-directions, respectively. A Gabor-filter with orientation α can be obtained by rotating the original (x, y) coordinate system by $-\alpha$ to yield a new coordinate system (x_*, y_*) (this rotation is equivalent to the rotation of the filter itself by α). The equations that relate the two coordinate systems are

$$x = x_* \cos\alpha + y_* \sin\alpha, \tag{2.25}$$

$$y = -x_* \sin\alpha + y_* \cos\alpha. \tag{2.26}$$

Substituting (2.25) and (2.26) into (2.24) gives the final form (not shown here).

Examples of two-dimensional Gabor functions were already given in Figs. 2.12b–c; two more will be given in the next section, Fig. 2.15.

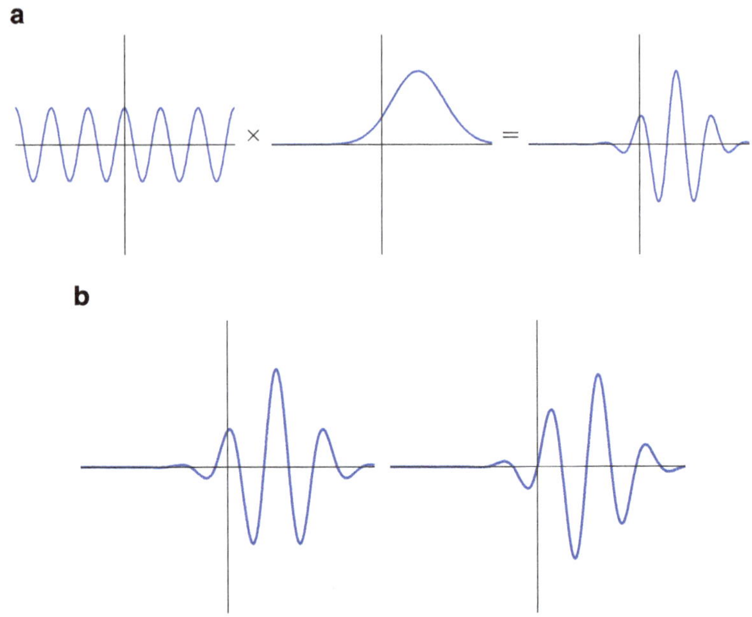

Fig. 2.13 Illustration of one-dimensional Gabor functions. **a** Construction of the function by multiplication of the envelope with a sinusoidal function. **b** Two Gabor functions in quadrature phase.

2.4.3 Spatial Localization vs. Spectral Accuracy

Above, we have outlined a scheme where the frequency contents of an image at a certain point is analyzed by first multiplying the image with a localized weighting window, and then analyzing the frequency contents of the weighted image. How accurate is this procedure, that is, how well can it capture the localized frequency structure?

The answer is that there is a trade-off between spatial localization and spectral (frequency) accuracy because the use of a weighting window changes the spectral contents. Figure 2.14 illustrates this phenomenon by showing how the results of space-frequency analysis of a pure sinusoidal $I(x) = \cos(\frac{1}{2}x)$ depend on the degree of spatial localization. The mathematical theory behind this phenomenon goes under the name *time-bandwidth product theorem* in signal processing, or *uncertainty principle* in physics. These theories state that there is a lower bound on the product of the spread of the energy in the spatial domain and the frequency domain. See the References at the end of this chapter for more details.

With images, the extra dimensions introduce another factor: uncertainty about orientation. This parameter behaves just like frequency and location in the sense that if we want to have a filter which is very localized in orientation, we have to give up localization in the other parameters. This is illustrated in Fig. 2.15, in which we see that a basic Gabor which is very localized in space (a) responds to a wider range of different orientations than the Gabor in (b). The Gabor in (b) has has been

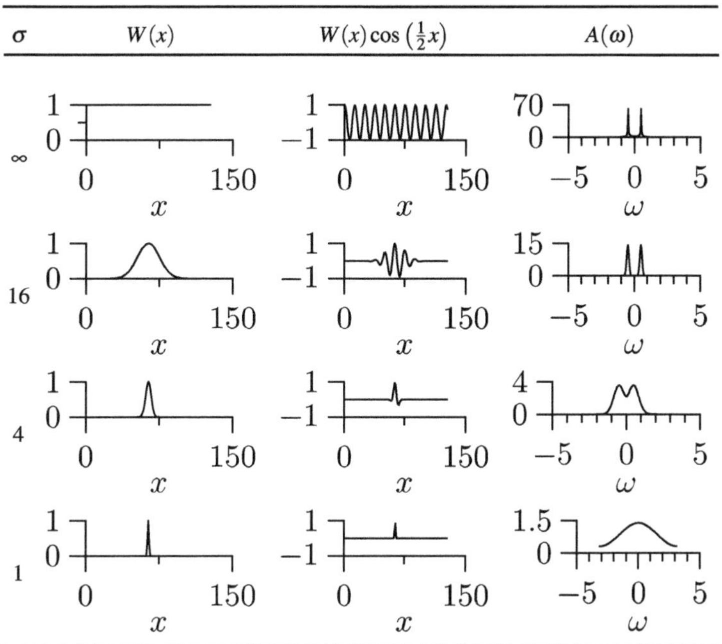

Fig. 2.14 In space-frequency analysis, there is a trade-off between spatial localization and spectral accuracy. This example shows how the use of a weighting window $W(x)$ (*second column*) changes the spectral (frequency) contents of a pure sinusoidal $I(x) = \cos(\frac{1}{2}x)$, $x = 0, \ldots, 127$. The *rightmost column* shows the amplitude spectrum of the localized signal $W(x)I(x)$, which in turn is plotted in the *third column*. With no spatial localization (window width $\sigma = \infty$; *top row*), the amplitude spectrum $A(\omega)$ shows a clear peak at the location $\omega = \frac{1}{2}$. As window width σ decreases, spatial localization increases, but accuracy in the spectral domain decreases: the two peaks in $A(\omega)$ spread out and eventually fuse so that $A(\omega)$ is a unimodal function when $\sigma = 1$

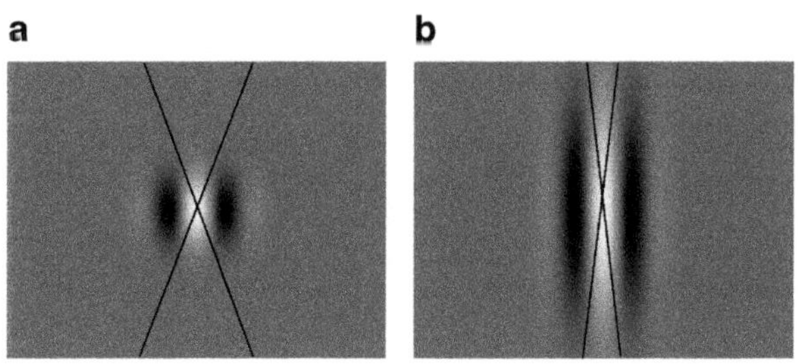

Fig. 2.15 Uncertainty in two dimensions. Compared with a "basic" Gabor function in **a**, the Gabor in **b** is very localized in orientation, that is, it responds to only a small range of different orientations. The *thin black lines* show the orientations of *thin lines* which still fall on the white (positive) area of the function: a line which is more oblique will partly fall on the black (negative) areas, and thus the response of the filter will the reduced. We can see that in **b**, the range of orientations producing very large responses (falling on the white area only) is much smaller than in **a**. This illustrates that in order to make the basic Gabor function in **a** more localized in orientation, it is necessary to make it longer, and thus to reduce its spatial localization

designed to respond only to a small range of orientations, which was only possible by making it more extended in space.

2.5 References

Most of the material covered in this chapter can be found in most image-processing textbooks. A classic choice is (Gonzales and Woods 2002), which does not, however, consider space-frequency analysis. A large number of textbooks explain time-frequency analysis, which is the one-dimensional counterpart of space-frequency analysis, for example (Cohen 1995). Related material can also be found in textbooks on wavelets, which are a closely related method (see Section 17.3.2 for a very short introduction), for example (Vetterli and Kovačević 1995; Strang and Nguyen 1996).

2.6 Exercises

Mathematical Exercises

1. Show that convolution is a symmetric operation.
2. Show (2.3).
3. Prove (2.17). Hint: Find two different values for x so that you get the two equations

$$A \cos \psi = C, \tag{2.27}$$

$$-A \sin \psi = S. \tag{2.28}$$

   Now, solve for A and ψ as follows. First, take the squares of both sides of both (2.27) and (2.28) and sum the two resulting equations. Recall the sum of squares of a sine and a cosine function. Second, divide both sides of (2.27) and (2.28) with each other.

Computer Assignments

The computer assignments in this book are designed to be made with Matlab™. Most of them will work on Matlab clones such as Octave. We will assume that you know the basics of Matlab.

1. The command meshgrid is very useful for image processing. It creates two-dimensional coordinates, just like a command such as [5:.01:5] creates a one-dimensional coordinate. Give the command [X,Y]=meshgrid([-5:0.1:5]); and plot the matrices X and Y using imagesc.

2. Create Fourier gratings by sin(X), sin(Y), sin(X+Y). Plot the gratings.
3. Create a Gabor function using these X and Y, simply by plugging in those matrices in the formula in (2.24). Try out different values for the parameters until you get a function which looks like the one in Fig. 2.12b.
4. Change the roles of X and Y to get a Gabor filter in a different orientation.
5. Try out a Gabor function of a different orientation by plugging in X+Y instead of X and X-Y instead of Y.
6. Linear filtering is easily done with the function conv2 (the "2" means two-dimensional convolution, i.e. images). Take any image, import it to Matlab, and convolve it with the three Gabor functions obtained above.

Chapter 3
Outline of the Visual System

In this chapter, we review very briefly the structure of the human visual system. This exposition contains a large number of terms which are likely to be new for readers who are not familiar with neuroscience. Only a few of them are needed later in this book; they are given in italics for emphasis.

3.1 Neurons and Firing Rates

Neurons The main information processing workload of the brain is carried by nerve cells, or *neurons*. Estimates of the number of neurons in the brain typically vary between 10^{10} and 10^{11}. What distinguishes neurons from other cells are their special information-processing capabilities. A neuron receives signals from other neurons, processes them, and sends the result of that processing to other neurons. A schematic diagram of a neuron is shown in Fig. 3.1, while a more realistic picture is given in Fig. 3.2.

Axons How can such tiny cells send signals to other cells which may be far away? Each neuron has one very long formation called an *axon* which connects it to other cells. Axons can be many centimeters or even a couple of meters long, so they can reach from one place in the brain to almost any other. Axons have a sophisticated biochemical machinery to transmit signals over such relatively long distances. The machinery is based on a phenomenon called *action potential*.

Action Potentials An action potential is a very short (1 ms) electrical impulse traveling via the axon of the neuron. Action potentials are illustrated in Fig. 3.3. Due to their typical shape, action potential are also called *spikes*. Action potentials are fundamental to the information processing in neurons; they constitute the signals by which the brain receives, analyzes, and conveys information.

Action potentials are all-or-none, in the sense that they always have the same strength (a potential of about 100 mV) and shape. Thus, a key principle in brain function is that the meaning of a spike is not determined by what the spike is like (because they are all the same), but rather, *where* it is, i.e. which axon is it traveling along, or equivalently, which neuron sent it. (Of course, the meaning also depends on *when* the spike was fired.)

Signal Reception and Processing At the receiving end, action potentials are input to neurons via shorter formations called dendrites. Typically, an axon has many branches, and each of them connects to a dendrite of another neuron. Thus, the axon

A. Hyvärinen, J. Hurri, P.O. Hoyer, *Natural Image Statistics*,
Computational Imaging and Vision 39,

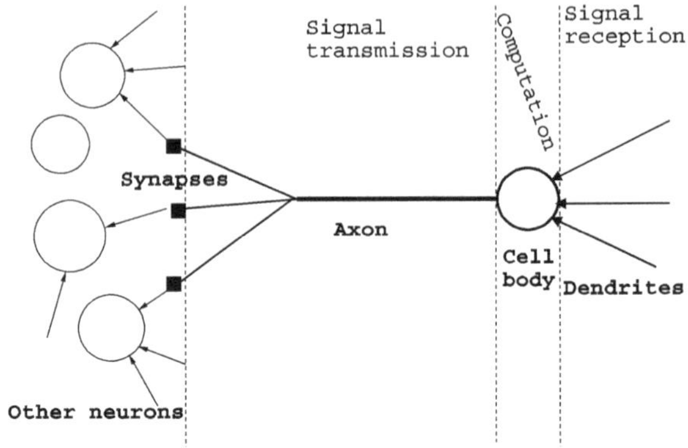

Fig. 3.1 A schematic diagram of information-processing in a neuron. Flow of information is from right to left

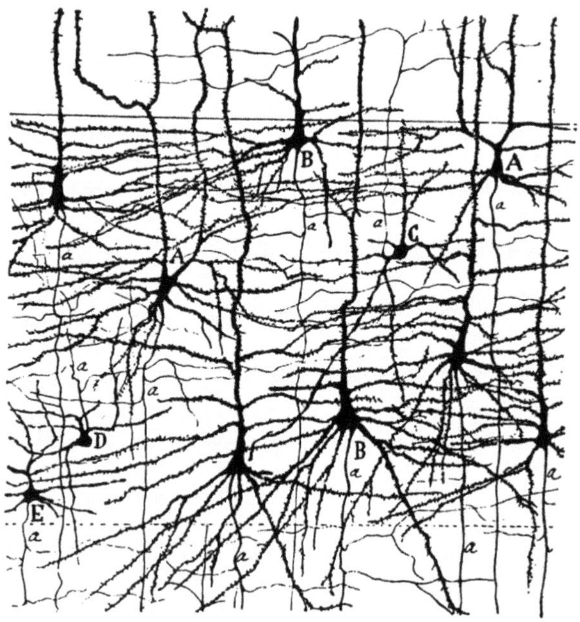

Fig. 3.2 Neurons (*thick bodies*, some lettered), each with one axon (*thicker line* going up) for sending out the signal, and many dendrites (*thinner lines*) for receiving signals. Drawing by Santiago Ramón y Cajal in 1900

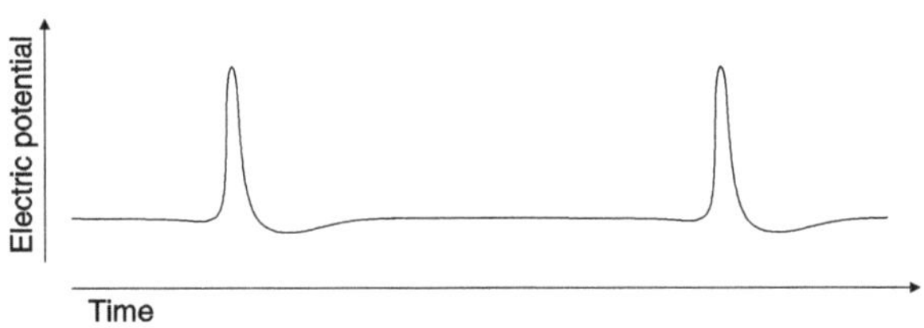

Fig. 3.3 An action potential is a wave of electrical potential which travels along the axon. It travels quite fast, and is very short both in time and its spatial length (along the axon). The figure shows the potentials in different parts of the axon soon after the neuron has emitted two action potentials

could be thought of as output wires along which the output signal of the neuron is sent to other neurons; dendrites are input wires which receive the signal from other neurons. The site where an axon meets a dendrite is called a synapse. The main cell body, or soma, is often thought of as the main "processor" which does the actual computation. However, an important part of the computation is already done in the dendrites.

Firing Rate The output of the neuron consists of a sequence of spikes emitted (a "spike train"). To fully describe such a sequence, one should record the time intervals between each successive spike. To simplify the situation, most research in visual neuroscience has concentrated on the neurons' *firing rates*, i.e. the number of spikes "fired" (emitted) by a neuron per second. This gives a single scalar quantity which characterizes the activity of the cell. Since it is these action potentials which are transmitted to other cells, the firing rate can also be viewed as the "result" of the computation performed by the neuron, in other words, its output.

Actually, most visual neurons are emitting spikes all the time, but with a relatively low frequency (of the order of 1 Hz). The "normal" firing rate of the neuron when there is no specific stimulation is called the *spontaneous firing rate*. When the firing rate is increased from the spontaneous one, the neuron is said to be active or activated.

Computation by the Neuron How is information processed, i.e. how are the incoming signals integrated in the soma to form the out-coming signal? This question is extremely complex and we can only give an extremely simplified exposition here.

A fundamental principle of neural computation is that the reception of a spike at a dendrite can either excite (increase the firing rate) of the receiving neuron, or inhibit it (decrease the firing rate), depending on the neuron from which the signal came. Furthermore, depending on the dendrite and the synapse, some incoming signals have a stronger tendency to excite or inhibit the neuron. Thus, a neuron can be thought of as an elementary pattern-matching device: its firing rates is large when it receives input from those neurons which excite it (strongly), and no input from those neurons which inhibit it. A basic mathematical model for such an action is to consider the firing rate as a linear combination of incoming signals; we will consider linear models below.

Thinking in terms of the original visual stimuli, it is often thought that a neuron is active when the input contains a feature for which the neuron is specialized—but this is a very gross simplification. Thus, for example, a hypothetical "grandmother cell" is one that only fires when the brain perceives, or perhaps thinks of, the grandmother. Next, we will consider what are the actual response properties of neurons in the visual system.

3.2 From the Eye to the Cortex

Figure 3.4 illustrates the earliest stages of the main visual pathway. Light enters the eye, reaching the *retina*. The retina is a curved, thin sheet of brain tissue that

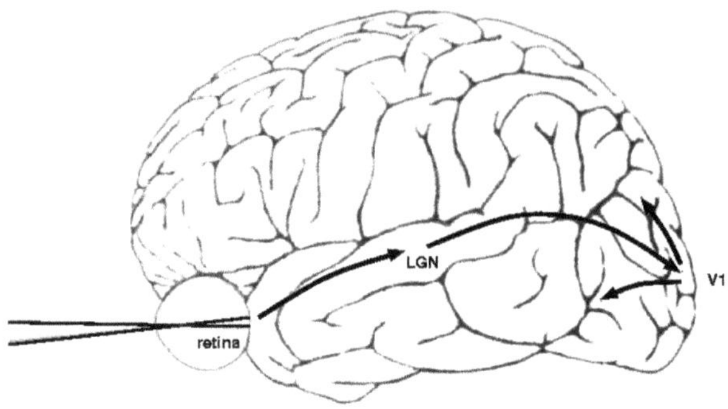

Fig. 3.4 The main visual pathway in the human brain

grows out into the eye to provide the starting point for neural processing of visual signals. The retina is covered by a more than a hundred million photo-receptors, which convert the light into an electric signal, i.e. neural activity.

From the photo-receptors, the signal is transmitted through a couple of neural layers. The last of the retinal processing layer consists of ganglion cells, which send the output of the retina (in form of action potentials) away from the eye using their very long axons. The axons of the ganglion cells form the optic nerve. The optic nerve transmits the visual signals to the lateral geniculate nucleus (LGN) of the thalamus. The thalamus is a structure in the middle of the brain through which most sensory signals pass on their way from the sensory organs to the main sensory processing areas in the brain.

From the LGN, the signal goes to various other destinations, the most important being the visual cortex at the back of the head, where most of the visual processing is performed. Cortex, or cerebral cortex to be more precise, means here the surface of the two cerebral hemispheres, also called the "grey matter". Most of the neurons associated with sensory or cognitive processing are located in the cortex. The rest of the cerebral cortex consists mainly of axons connecting cortical neurons with each other, or the "white matter".

The visual cortex contains some $1/5$ of the total cortical area in humans, which reflects the importance of visual processing to us. It consists of a number of distinct areas. The *primary visual cortex*, or V1 for short, is the area to which most of the retinal output first arrives. It is the most widely-studied visual area, and also the main focus in this book.

3.3 Linear Models of Visual Neurons

3.3.1 Responses to Visual Stimulation

How to make sense of the bewildering network of neurons processing visual information in the brain? Much of visual neuroscience has been concerned with measuring the firing rates of cells as a function of some properties of a visual input. For

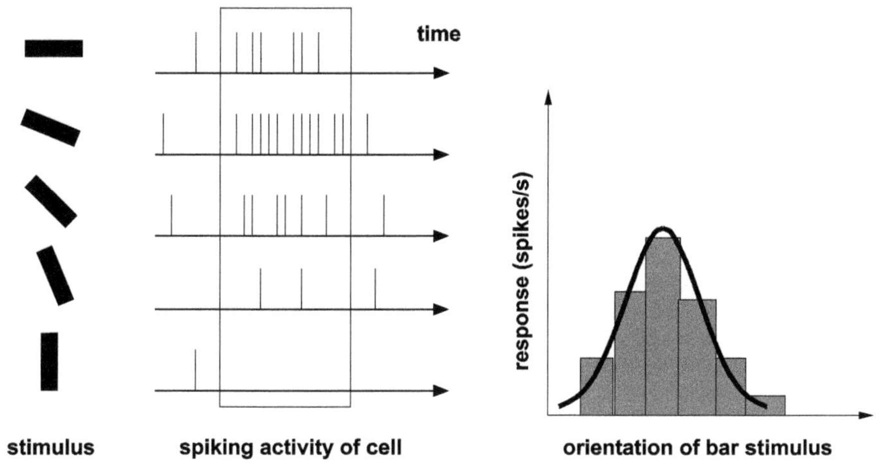

Fig. 3.5 A caricature of a typical experiment. A *dark bar* on a *white background* is flashed onto the screen, and action potentials are recorded from a neuron. Varying the orientation of the bar yields varying responses. Counting the number of spikes elicited within a fixed time window following the stimulus, and plotting these counts as a function of bar orientation, one can construct a mathematical model of the response of the neuron

example, an experiment might run as follows: An image is suddenly projected onto a (previously blank) screen that an animal is watching, and the number of spikes fired by some recorded cell in the next second are counted. By systematically changing some properties of the stimulus and monitoring the elicited response, one can make a quantitative model of the response of the neuron. An example is shown in Fig. 3.5. Such a model mathematically describes the response (firing rate) r_j of a neuron as a function of the stimulus $I(x, y)$.

In the early visual system, the response of a typical neuron depends only on the intensity pattern of a very small part of the visual field. This area, where light increments or decrements can elicit increased firing rates, is called the (classical) *receptive field* (RF) of the neuron. More generally, the concept also refers to the particular light pattern that yields the maximum response.

So, what kind of light patterns actually elicit the strongest responses? This of course varies from neuron to neuron. One thing that most cells have in common is that they don't respond to a static image which consists of a uniform surface. They respond to stimuli in which there is some change, either temporally or spatially; such change is called *contrast* in vision science.

The retinal ganglion cells as well as cells in the lateral geniculate nucleus typically have circular center-surround receptive field structure: Some neurons are excited by light in a small circular area of the visual field, but inhibited by light in a surrounding annulus. Other cells show the opposite effect, responding maximally to light that fills the surround but not the center. This is depicted in Fig. 3.6a.

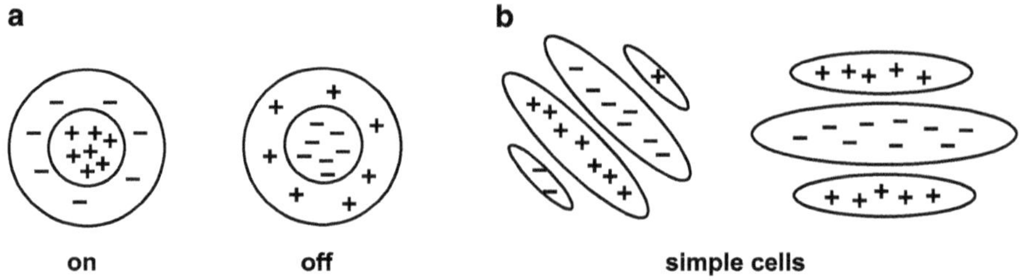

Fig. 3.6 Typical classical receptive fields of neurons early in the visual pathway. Plus signs denote regions of the visual field where light causes excitation, minuses regions where light inhibits responses. **a** Retinal ganglion and LGN neurons typically exhibit center-surround receptive field organization, in one of two arrangements. **b** The majority of simple cells in V1, on the other hand, have oriented receptive fields

3.3.2 Simple Cells and Linear Models

The cells that we are modeling are mainly in the primary visual cortex (V1). Cells in V1 have more interesting receptive fields than those in the retina or LGN. The so-called *simple cells* typically have adjacent elongated (instead of concentric circular) regions of excitation and inhibition. This means that these cells respond maximally to *oriented* image structure. This is illustrated in Fig. 3.6b.

Linear models are the ubiquitous workhorses of science and engineering. They are also the simplest successful neuron models of the visual system. A linear model for a visual neuron[1] means that the response of a neuron is modeled by a weighted sum of the image intensities, as in

$$r_j = \sum_{x,y} W_j(x, y) I(x, y) + r_0, \tag{3.1}$$

where $W_j(x, y)$ contains the pattern of excitation and inhibition for light for the neuron j in question. The constant r_0 is the spontaneous firing rate. We can define the spontaneous firing rate to be the baseline (zero) by subtracting it from the firing rate:

$$\tilde{r}_j = r_j - r_0, \tag{3.2}$$

which will be done in all that follows.

[1]Note that there are two different kinds of models one could develop for a visual neuron. First, one can model the output (firing rate) as a function of the *input stimulus*, which is what we do here. Alternatively, one could model the output as a function of the *direct inputs* to the cell, i.e. the rates of action potentials received in its dendrites. This latter approach is more general because it can be applied to any neuron in the brain. However, it is not usually used in vision research because it does not tell us much about the function of the visual system unless we already know the response properties of those neurons whose firing rates are input to the neuron via dendrites, and just finding those cells whose axons connect to a given neuron is technically very difficult.

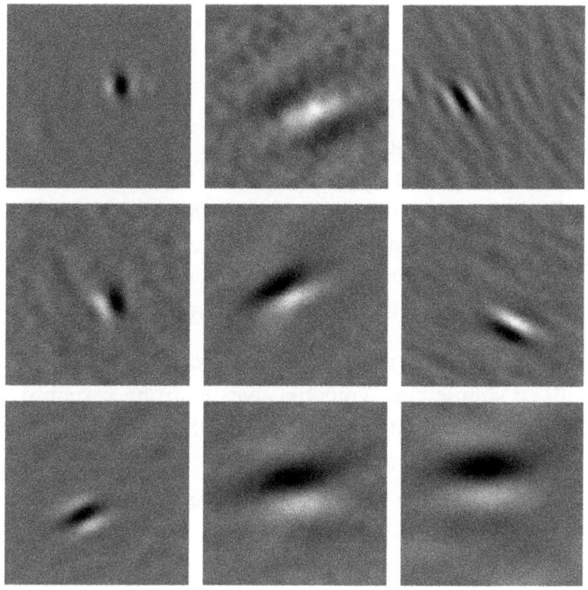

Fig. 3.7 Receptive fields of simple cells estimated by reverse correlation based on single-cell recordings in a macaque monkey. Courtesy of Dario Ringach, UCLA

Linear receptive-field models can be estimated from visual neurons by employing a method called reverse correlation. In this method, a linear receptive field is estimated so that the mean square error between the estimated r_j in (3.1), and the actual firing rate is minimized, where the mean is taken over a large set of visual stimuli. The name "reverse correlation" comes from the fact that the general solution to this problem involves the computation of the time-correlation of stimulus and firing rate. However, the solution is simplified when temporally and spatially uncorrelated ("white noise", see Sect. 4.6.4) sequences are used as visual stimuli—in this case, the optimal W_j is obtained by computing an average stimulus over those stimuli which elicited a spike. Examples of estimated receptive fields are shown in Fig. 3.7.

3.3.3 Gabor Models and Selectivities of Simple Cells

How can we describe the receptive field of simple cells in mathematical terms? Typically, this is based on modeling the receptive fields by Gabor functions, reviewed in Sect. 2.4.2. A Gabor function consists of an oscillatory sinusoidal function which generates the alternation between the excitatory and inhibitory ("white/black") areas, and a Gaussian "envelope" function which determines the spatial size of the receptive field. In fact, when comparing the receptive field in Fig. 3.7 with the Gabor functions in Fig. 2.12b–c, it seems obvious that Gabor functions provide a reasonable model for the receptive fields.

Using a Gabor function, the receptive field is reduced to a small number of parameters:

- Orientation of the oscillation.
- Frequency of oscillation.
- Phase of the oscillation.
- Width of the envelope (in the direction of the oscillation).
- Length of the envelope (in the direction orthogonal to the oscillation). The ratio of the length to the width is called the aspect ratio.
- The location in the image (on the retina).

These parameters are enough to describe the basic selectivity properties of simple cells: a simple cell typically gives a strong response when the input consists of a Gabor function with approximately the right ("preferred") values for all, or at least most, of these parameters (the width and the length of the envelope are not critical for all simple cells). Thus, we say that simple cells are selective for frequency, orientation, phase, and location.

In principle, one could simply try to find a Gabor function which gives the best fit to the receptive field estimated by reverse correlation. In practice, however, more direct methods are often used since reverse correlation is rather laborious. Typically, what is computed are *tuning curves* for some of these parameters. This was illustrated in Fig. 3.5. Typical stimuli include two-dimensional Fourier gratings (see Fig. 2.5) and simple, possibly short, lines or bars. Examples of such analyses will be seen in Chaps. 6 and 10.

3.3.4 Frequency Channels

The selectivity of simple cells (as well as many other cells) to frequency is related to the concept of "frequency channels" which is widely used in vision science. The idea is that in the early visual processing (something like V1), information of different frequencies is processed independently. Justification for talking about different channels is abundant in research on V1. In fact, the very point in using Gabor models is to model the selectivity of simple cells to a particular frequency range.

Furthermore, a number of psychological experiments point to such a division of early processing. For example, in Fig. 3.8, the information in the high- and low-frequency parts are quite different, yet observes have no difficulty in processing (reading) them separately. This figure also illustrates the practical meaning of frequency selectivity: some of the cells in V1 respond to the "yes" letters but do not respond to the "no" letters, while for other cells, the responses are the other way round. (The responses depend, however, on viewing distance: stimuli which are low-frequency when viewed from a close distance will be high-frequency when viewed from far away.)

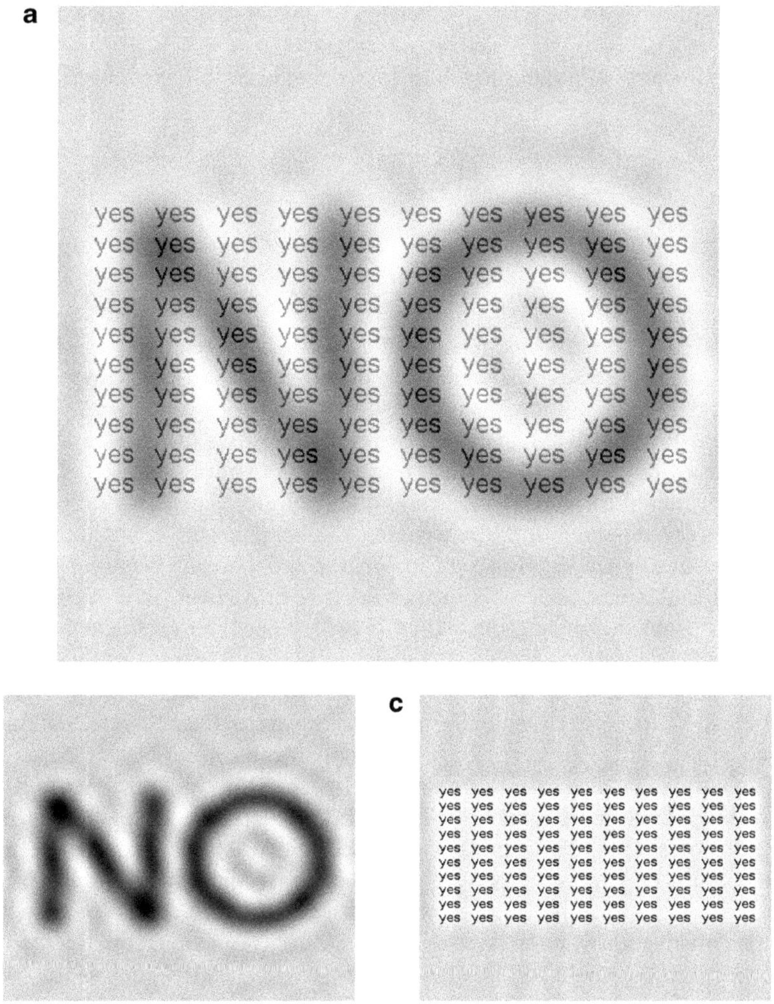

Fig. 3.8 A figure with independent (contradictory?) information in different frequency channels. **a** The original figure, **b** low-frequency part of figure in **a**, obtained by taking the Fourier transform and setting to zero all high-frequency components (whose distance from zero is larger than a certain threshold), **c** high-frequency part of figure in **a**. The sum of the figures in **b** and **c** equals **a**

3.4 Non-linear Models of Visual Neurons

3.4.1 Non-linearities in Simple-Cell Responses

Linear models are widely used in modeling visual neurons, but they are definitely a rough approximation of the reality. Real neurons exhibit different kinds of non-linear behavior. The most basic non-linearities can be handled by adding a simple scalar non-linearity to the model, which leads to what is simply called a *linear/non-linear* model.

In the linear/non-linear model, a linear stage is followed by a static non-linearity f:

$$\tilde{r}_j = f\left(\sum_{x,y} W_j(x,y)I(x,y)\right). \tag{3.3}$$

A special case of the linear/non-linear model is *half-wave rectification*, defined by

$$f(\alpha) = \max\{0, \alpha\}. \tag{3.4}$$

One reason for using this model is that if a neuron has a relatively low spontaneous firing rate, the firing rates predicted by the linear model may then tend to be negative. The firing rate, by definition, cannot be negative.

We must distinguish here between two cases. Negative firing rates are, of course, impossible by definition. In contrast, it is possible to have positive firing rates that are smaller than the spontaneous firing rate; they give a negative $\tilde{r}_j$ in (3.2). Such firing rates correspond to the sum term in (3.1) being negative, but not so large that the r_j becomes negative. However, in V1, the spontaneous firing rate tends to be rather low, and the models easily predict negative firing rates for cortical cells. (This is less of a problem for ganglion and LGN cells, since their spontaneous firing rates are relatively high.)

Thus, half-wave rectification offers one way to interpret the purely linear model in (3.1) in a more physiologically plausible way: the linear model combines the outputs of two half-wave rectified (non-negative) cells with reversed polarities into a single output r_j—one cell corresponds to linear RF W_j and the other to RF $-W_j$.

The linear/non-linear model is flexible and can accommodate a number of other properties of simple cell responses as well. First, when the linear model predicts small outputs, i.e. the stimulus is weak, no output (increase in firing rate) is actually observed in simple cells. In other words, it seems there is a *threshold* which the stimulus must attain to elicit any response. This phenomenon, combined with half-wave rectification, could be modeled by using a non-linearity such as

$$f(\alpha) = \max(0, \alpha - c) \tag{3.5}$$

where c is a constant that gives the threshold.

Second, due to biological properties, neurons have a maximum firing rate. When the stimulus intensity is increased above a certain limit, no change in the cell response is observed, a phenomenon called *saturation*. This is in contradiction with the linear model, which has no maximum response: if you multiply the input stimulus by, say, 1 000 000, the output of the neuron increases by the same factor. To take this property into account, we need to use a non-linearity that saturates as well, i.e. has a maximum value. Combining the three non-linear properties listed here leads us to a linear/non-linear model with the non-linearity

$$f(\alpha) = \min\big(d, \max(0, \alpha - c)\big), \tag{3.6}$$

where d is the maximum response. Figure 3.9 shows the form of this function.

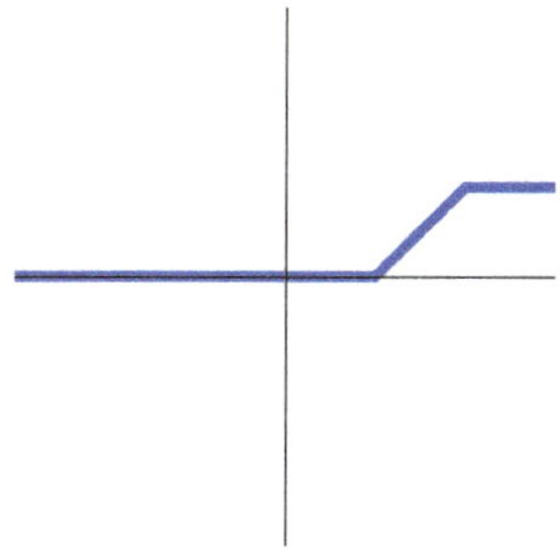

Fig. 3.9 The non-linear function in (3.6)

Alternatively, we could use a smooth function with the same kind of behavior, such as

$$f(\alpha) = d\frac{\alpha^2}{c' + \alpha^2}, \tag{3.7}$$

where c' is another constant that is related to the threshold c.

3.4.2 Complex Cells and Energy Models

Although linear/non-linear models are useful in modeling many cells, there are also neurons in V1 called *complex cells* for which these models are completely inadequate. These cells do not show any clear spatial zones of excitation or inhibition. Complex cells respond, just like simple cells, selectively to bars and edges at a particular location and of a particular orientation; they are, however, relatively invariant to the spatial phase of the stimulus. An example of this is that reversing the contrast polarity (e.g. from white bar to black bar) of the stimulus does not markedly alter the response of a typical complex cell.

The responses of complex cells have often been modeled by the classical 'energy model'. (The term 'energy' simply denotes the squaring operation.) In such a model (see Fig. 3.10), we have

$$r_j = \left(\sum_{x,y} W_{j_1}(x, y)I(x, y)\right)^2 + \left(\sum_{x,y} W_{j_2}(x, y)I(x, y)\right)^2,$$

where $W_{j_1}(x, y)$ and $W_{j_2}(x, y)$ are quadrature-phase Gabor functions, i.e. they have a phase-shift of 90 degrees, one being odd-symmetric and the other being even-symmetric. It is often assumed that V1 complex cells pool the responses of simple cells, in which case the linear responses in the above equation are outputs of simple cells.

The justification for this model is that since the two linear filters are Gabors in quadrature-phase, the model is computing the local Fourier "energy" in a particular range of frequencies and orientations, see (2.16). This provides a model of a cell which is selective for frequency and orientation, and is also spatially localized, but

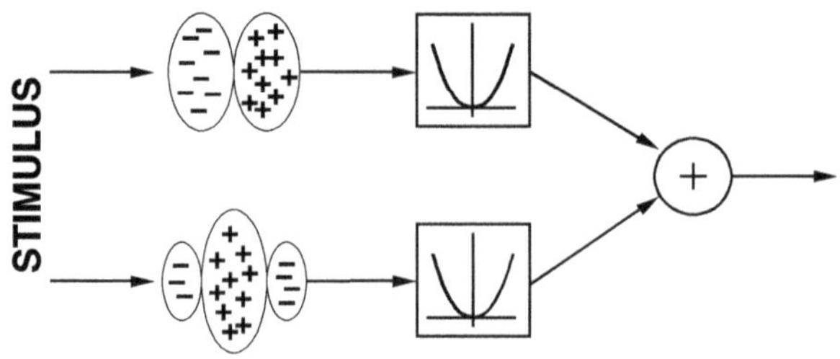

Fig. 3.10 The classic energy model for complex cells. The response of a complex cell is modeled by linearly filtering with quadrature-phase Gabor filters (Gabor functions whose sinusoidal components have a 90 degrees phase difference), taking squares, and summing. Note that this is purely a mathematical description of the response and should not be directly interpreted as a hierarchical model summing simple cell responses

does not care about the phase of the input. In other words it is *phase-invariant*. (This will be discussed in more detail in Chap. 10.)

The problem of negative responses considered earlier suggests a simple modification of the model, where each linear RF again corresponds to two simple cells. The output of a linear RF is divided to the positive and negative parts and half-wave rectified. In this case, the half-wave rectified outputs are further squared so that they compute the squaring operation of the energy model. In addition, complex cells saturate just as simple cells, so it makes sense to add a saturating non-linearity to the model as well.

3.5 Interactions between Visual Neurons

In the preceding models, V1 cells are considered completely independent units: each of them just takes its input and computes its output. However, different kinds interactions between the cells have been observed.

The principal kind of interaction seems to be *inhibition*: when a cell j is active, the responses of another cell i is reduced from what they would be without that cell j being active. To be more precise, let us consider two simple cells whose receptive fields W_i and W_j are orthogonal (for more on orthogonality, see Chap. 19). Take, for example, two cells in the same location, one with vertical and the other with horizontal orientation. Take any stimulus I_0 which excites the cell W_j. For example, we could take a stimulus which is equal to the receptive field W_j itself. Now, we add another stimulus pattern, say I_1, to I_0. This simply means that we add the intensities pixel-by-pixel, showing the following stimulus to the retina:

$$I(x, y) = I_0(x, y) + I_1(x, y). \tag{3.8}$$

The added stimulus I_1 is often called a mask or a pedestal.

The point is that by choosing I_1 suitably, we can demonstrate a phenomenon which is probably due to interaction between the two cells. Specifically, let us

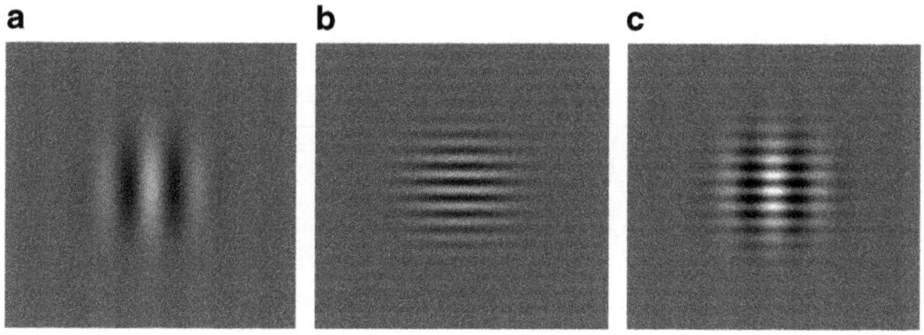

Fig. 3.11 Interaction between different simple cells. **a** Original stimulus I_0 of a simple cell, chosen here as equal the receptive field of W_j. **b** Masking pattern I_1 which is orthogonal to I_0. **c** Compound stimulus I. The response to I is smaller than the response to I_0 although the linear models predicts that the responses should be equal

choose a stimulus which is equal to the receptive field of cell i: $I_1 = W_i$. This is maximally excitatory for the cell i, but it is orthogonal to the receptive field of cell j. With this kind of stimuli, the typical empirical observation is that the cell j has a lower firing rate for the compound stimulus $I = I_0 + I_1$ than for I_0 alone. This inhibition cannot be explained by the linear models (or the linear/non-linear models). The mask I_1 should have no effect on the linear filter stage, because the mask is orthogonal to the receptive field W_j. So, to incorporate this phenomenon in our models, we must include some interaction between the linear filters: The outputs of some model cells must reduce the outputs of others. (It is not completely clear whether this empirical phenomenon is really due to interaction between the cells, but that is a widely-held view, so it makes sense to adopt it in our models.)

This phenomenon is typically called "contrast gain control". The idea is that when there is more contrast in the image (due to the addition of the mask), the system adjusts its responses to be generally weaker. It is thought to be necessary because of the saturating non-linearity in the cells and the drastic changes in illumination conditions observed in the real world. For example, the cells would be responding with the maximum value most of the time in bright daylight (or a brightly lit part of the visual scene), and they would be responding hardly at all in a dim environment (or a dimly lit part of the scene). Gain control mechanisms alleviate this problem by normalizing the variation of luminance over different scenes, or different parts of the same scene. For more on this point, see Sect. 9.5.[2]

This leads us to one of the most accurate currently known simple-cell models, in terms of predictive power, the divisive normalization model. Let $W_1, \ldots, W_K$ denote the receptive fields of those cells whose receptive fields are approximately in the same location, and σ a scalar parameter. In the divisive normalization model,

[2] In fact, different kinds of gain control mechanisms seem to be operating in different parts of the visual system. In the retina, such mechanisms normalize the general luminance level of the inputs, hence the name "luminance gain control". Contrast gain control seems to be done after that initial gain control. The removal of the mean grey-scale value (DC component) that we do in later chapters can be thought to represent a simple luminance gain control mechanism.

the output of the cell corresponding to RF W_j is given by

$$r_j = \frac{f(\sum_{x,y} W_j(x,y)I(x,y))}{\sum_{i=1}^{K} f(\sum_{x,y} W_i(x,y)I(x,y)) + \sigma^2}, \tag{3.9}$$

where f is again a static non-linearity, such as the half-wave rectification followed by squaring. This divisive normalization model provides a simple account of contrast gain control mechanisms. In addition, it also automatically accounts for such simple-cell non-linearities as response saturation and threshold. In fact, if the input stimulus is such that it only excites cell j, and the linear responses in the denominator are all zero expect for the one corresponding to cell j, the model reduces to the linear/non-linear model in Sect. 3.4.1. If we further define f to be the square function, we get the non-linearity in (3.7).

3.6 Topographic Organization

A further interesting point is how the receptive fields of neighboring cells are related. In the retina, the receptive fields of retinal ganglion cells are necessarily linked to the physical position of the cells. This is due to the fact that the visual field is mapped in an orderly fashion to the retina. Thus, neighboring retinal ganglion cells respond to neighboring areas of the visual field. However, there is nothing to guarantee the existence of a similar organization further up the visual pathway.

But the fact of the matter is that, just like in the retina, neighboring neurons in the LGN and in V1 tend to have receptive fields covering neighboring areas of the visual field. This phenomenon is called *retinotopy*. Yet this is only one of several types of organization. In V1, the orientation of receptive fields also tends to shift gradually along the surface of the cortex. In fact, neurons are often approximately organized according to several functional parameters (such as location, frequency, orientation) simultaneously. This kind of *topographic organization* also exists in many other visual areas.

Topographical representations are not restricted to cortical areas devoted to vision, but are present in various forms throughout the brain. Examples include the tonotopic map (frequency-based organization) in the primary auditory cortex and the complete body map for the sense of touch. In fact, one might be pressed to find a brain area that would not exhibit any sort of topography.

3.7 Processing after the Primary Visual Cortex

From V1, the visual signals are sent to other areas, such as V2, V4, and V5, called *extra-striate* as another name for V1 is the "striate cortex". The function of some of these areas (mainly V5, which analyzes motion) is relatively well understood, but the function of most of them is not really understood at all. For example, it is assumed that V2 is the next stage in the visual processing, but the differences in the features computed in V1 and V2 are not really known. V4 has been variously described as being selective to long contours, corners, crosses, circles, "non-Cartesian" gratings, color, or temporal changes (see the references section below). Another problem is that the extra-striate cortex may be quite different in humans and monkeys (not to mention other experimental animals), so results from animal experiments may not generalize to humans.

3.8 References

Among general introductions to the visual system, see, e.g. Palmer (1999). A most interesting review of the state of modeling of the visual cortex, with extensive references to experiments, is in Carandini et al. (2005).

For a textbook account of reverse correlation, see e.g. Dayan and Abbott (2001); reviews are Ringach and Shapley (2004), Simoncelli et al. (2004). Classic application of reverse correlation for estimating simple cell receptive fields is Jones and Palmer (1987a, 1987b), Jones et al. (1987). For spatiotemporal extensions, see DeAngelis et al. (1993a, 1993b). LGN responses are estimated, e.g. in Cai et al. (1997), and retinal ones, e.g. in Davis and Naka (1980).

The non-linearities in neuron responses are measured in Anzai et al. (1999b), Ringach and Malone (2007); theoretical studies include Hansel and van Vreeswijk (2002), Miller and Troyer (2002). These studies concentrate on the "thresholding" part of the non-linearity, ignoring saturation. Reverse correlation in the presence of non-linearities is considered in Nykamp and Ringach (2002).

A review on contrast gain control can be found in Carandini (2004). The divisive normalization model is considered in Heeger (1992), Carandini et al. (1997, 1999). More on the interactions can be found in Albright and Stoner (2002). For review of the topographic organization in different parts of the cortex, see Mountcastle (1997).

A discussion on our ignorance of V2 function can be found in Boynton and Hedgé (2004). Proposed selectivities in V4 include long contours (Pollen et al. 2002), corners and related features (Pasupathy and Connor 1999, 2001), crosses, circles, and other non-Cartesian gratings (Gallant et al. 1993; Wilkinson et al. 2000), as well as temporal changes (Gardner et al. 2005). An alternative viewpoint is that the processing might be quite similar in most extra-striate areas, the main difference being the spatial scale (Hegdé and Essen 2007). A model of V5 is proposed in Simoncelli and Heeger (1998).

Basic historical references on the visual cortex include Hubel and Wiesel (1962, 1963, 1968, 1977).

3.9 Exercises

Mathematical Exercises

1. Show that the addition of a mask which is orthogonal to the receptive field, as in Sect. 3.5 should not change the output of the cell in the linear model.

2. What is the justification for using the same letter d for the constants in (3.6) and (3.7)?

Computer Assignments

1. Plot function in (3.7) and compare with the function in (3.6).
2. Receptive fields in the ganglion cells and the LGN are often modeled as a "difference-of-Gaussians" model in which $W(x, y)$ is defined as

$$\begin{aligned} &\exp\left(-\frac{1}{2\sigma_1^2}\left[(x-x_0)^2+(y-y_0)^2\right]\right) \\ &\quad - a\exp\left(-\frac{1}{2\sigma_2^2}\left[(x-x_0)^2+(y-y_0)^2\right]\right). \end{aligned} \tag{3.10}$$

   Plot the receptive fields for some choices of the parameters. Find some parameter values that reproduce a center-surround receptive field.

Chapter 4
Multivariate Probability and Statistics

This chapter provides the theoretical background in probability theory and statistical estimation needed in this book. This is not meant as a first introduction to probability, however, and the reader is supposed to be familiar with the basics of probability theory. The main emphasis here is on the extension of the basic notions to *multidimensional* data.

4.1 Natural Images Patches as Random Vectors

To put the theory on a concrete basis, we shall first discuss the fundamental idea on how natural image patches can be considered as a random vector.

A random vector is a vector whose elements are random variables. Randomness can be defined in different ways. In probability theory, it is usually formalized by assuming that the value of the random variable or vector depends on some other variable ("state of the world") whose value we do not know. So, the randomness is due to our ignorance.

In this book, an image patch I is typically modeled as a random vector, whose obtained values (called "observations") are the numerical grey-scale values of pixels in a patch (window) of a natural image. A patch simply means a small sub-image, such as the two depicted in Fig. 1.4. We use small patches because whole images have too large dimensions for existing computers (we must be able to perform complicated computations on a large number of such images or patches). A typical patch size we use in this book is 32×32 pixels.

To get one observation of the random vector in question, we randomly select one of the images in the image set we have, and then randomly select the location of the image patch. The randomness of the values in the vector stems from the fact that the patch is taken in some "random" position in a "randomly" selected image from our database. The "random" position and image selection are based on a random number generator implemented in a computer.

It may be weird to call an image patch a "vector" as it is two-dimensional and could also be called a matrix. However, for the purposes of most of this book, a two-dimensional image patch has to be treated like a one-dimensional vector, or a point in the image space, as was illustrated in Fig. 1.5. (A point and a vector are basically the same thing in this context.) This is because observed data are typically considered to be such vectors in statistics, and matrices are used for something quite different (that is, to represent linear transformations, see Sect. 19.3 for details). In practical calculations, one often has to transform the image patches into one-dimensional vectors, i.e. "vectorize" them. Such a transformation can be done

A. Hyvärinen, J. Hurri, P.O. Hoyer, *Natural Image Statistics*,
Computational Imaging and Vision 39,

in many different ways, for example by scanning the numerical values in the matrix row-by-row; the statistical analysis of the vector is not at all influenced by the choice of that transformation. In most of the chapters in this book, it is assumed that such a transformation has been made.

On a more theoretical level, the random vector can also represent the whole set of natural images, i.e. each observation is one natural image, in which case the database is infinitely large and does not exist in reality.

4.2 Multivariate Probability Distributions

4.2.1 Notation and Motivation

In this chapter, we will denote random variables by $z_1, z_2, \ldots, z_n$ and $s_1, s_2, \ldots, s_n$ for some number n. Taken together, the random variables $z_1, z_2, \ldots, z_n$ form an n-dimensional random vector which we denote by $\mathbf{z}$:

$$\mathbf{z} = \begin{pmatrix} z_1 \\ z_2 \\ \vdots \\ z_n \end{pmatrix}. \tag{4.1}$$

Likewise, the variables $s_1, s_2, \ldots, s_n$ can be collected to a random vector, denoted by $\mathbf{s}$.

Although we will be considering general random vectors, in order to make things concrete, you can think of each z_i as the grey-scale value of a pixel in the image patch. In the simple case of two variables, z_1 and z_2, this means that you take samples of two adjacent pixels (say, one just to the right of the other). A scatter plot of such a pixel pair is given in Fig. 4.1. However, this is by no means the only thing the variables can represent; in most chapters of this book, we will also consider various kinds of features which are random variables as well.

The fundamental goal of the models in this book is to describe the probability distribution of the random vector of natural image patches. So, we need to next consider the concept of a probability density function.

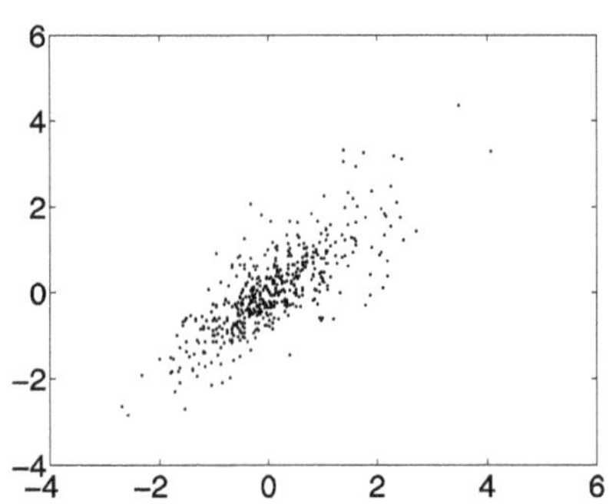

Fig. 4.1 Scatter plot of the grey-scale values of two neighboring pixels. The *horizontal axis* gives the value of the pixel on the *left*, and the *vertical axis* gives the value of the pixel on the *right*. Each *dot* corresponds to one observed pair of pixels

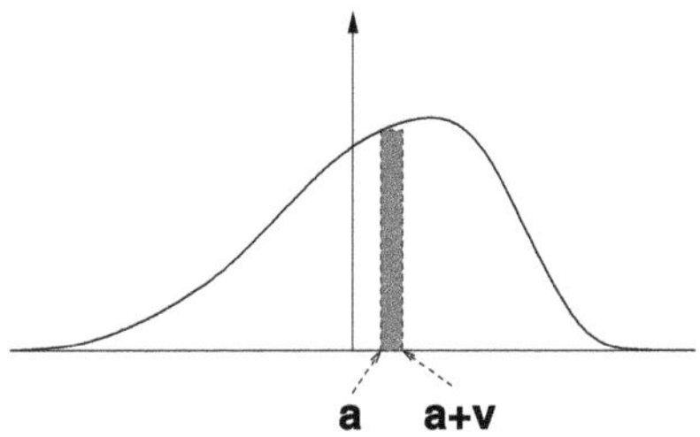

Fig. 4.2 The pdf of a random variable at a point a gives the probability that the random variable takes a value in a small interval $[a, a+v]$, divided by the length of that interval, i.e. v. In other words, the *shaded area*, equal to $p(a)v$, gives the probability that the variable takes a value in that interval

4.2.2 Probability Density Function

A probability distribution of a random vector such as $\mathbf{z}$ is usually represented using a *probability density function* (pdf). The pdf at a point in the n-dimensional space is denoted by $p_{\mathbf{z}}$.

The definition of the pdf of a multidimensional random vector is a simple generalization of the definition of the pdf of a random variable in one dimension. Let us first recall that definition. Denote by z a random variable. The idea is that we take a small number v, and look at the probability that z takes a value in the interval $[a, a+v]$ for any given a. Then we divide that probability by v, and that is the value of the probability density function at the point a. That is,

$$p_z(a) = \frac{P(z \text{ is in } [a, a+v])}{v}. \tag{4.2}$$

This principle is illustrated in Fig. 4.2. Rigorously speaking, we should take the limit of an infinitely small v in this definition.

This principle is simple to generalize to the case of an n-dimensional random vector. The value of the pdf function at a point, say $\mathbf{a} = (a_1, a_2, \ldots, a_n)$, gives the probability that an observation of $\mathbf{z}$ belongs to a small neighborhood of the point $\mathbf{a}$, divided by the volume of the neighborhood. Computing the probability that the values of each z_i are between the values of a_i and $a_i + v$, we obtain

$$p_{\mathbf{z}}(\mathbf{a}) = \frac{P(z_i \text{ is in } [a_i, a_i + v] \text{ for all } i)}{v^n} \tag{4.3}$$

where v^n is the volume of the n-dimensional cube whose edges all have length v. Again, rigorously speaking, this equation is true only in the limit of infinitely small v.

A most important property of a pdf is that it is normalized: its integral is equal to one

$$\int p_{\mathbf{z}}(\mathbf{a})\, d\mathbf{a} = 1. \tag{4.4}$$

This constraint means that you cannot just take any non-negative function and say that it is a pdf: you have to normalize the function by dividing it by its integral. (Calculating such an integral can actually be quite difficult and sometimes leads to serious problems, as discussed in Chap. 21).

For notational simplicity, we often omit the subscript $\mathbf{z}$. We often also write $p(\mathbf{z})$ which means the value of $p_{\mathbf{z}}$ at the point $\mathbf{z}$. This simplified notation is rather ambiguous because now $\mathbf{z}$ is used as an ordinary vector (like $\mathbf{a}$ above) instead of a random vector. However, often it can be used without any confusion.

Example 1 The most classic probability density function for two variables is the Gaussian, or normal, distribution. Let us first recall the one-dimensional Gaussian distribution, which in the basic case is given by

$$p(z) = \frac{1}{\sqrt{2\pi}} \exp\left(-\frac{1}{2}z^2\right). \tag{4.5}$$

It is plotted in Fig. 4.3b. This is the "standardized" version (mean is zero and variance is one), as explained below. The most basic case of a two-dimensional Gaussian distribution is obtained by taking this one-dimensional pdf separately for each variables, and multiplying them together. (The meaning of such multiplication is that the variables are independent, as will be explained below.) Thus, the pdf is given by

$$p(z_1, z_2) = \frac{1}{2\pi} \exp\left(-\frac{1}{2}(z_1^2 + z_2^2)\right). \tag{4.6}$$

A scatter plot of the distribution is shown in Fig. 4.3a. The two-dimensional pdf itself is plotted in Fig. 4.3c.

Example 2 Let us next consider the following two-dimensional pdf:

$$p(z_1, z_2) = \begin{cases} 1, & \text{if } |z_1| + |z_2| < 1, \\ 0, & \text{otherwise.} \end{cases} \tag{4.7}$$

This means that the data is uniformly distributed inside a square which has been rotated 45 degrees. A scatter plot of data from this distribution is shown in Fig. 4.4a.

4.3 Marginal and Joint Probabilities

Consider the random vector $\mathbf{z}$ whose pdf is denoted by $p_{\mathbf{z}}$. It is important to make a clear distinction between the *joint* pdf and the *marginal* pdf's. The joint pdf is just what we called pdf above. The marginal pdf's are what you might call the "individual" pdf's of z_i, i.e. the pdf's of those variables, $p_{z_1}(z_1), p_{z_2}(z_2), \ldots$ when we just consider one of the variables and ignore the existence of the other variables.

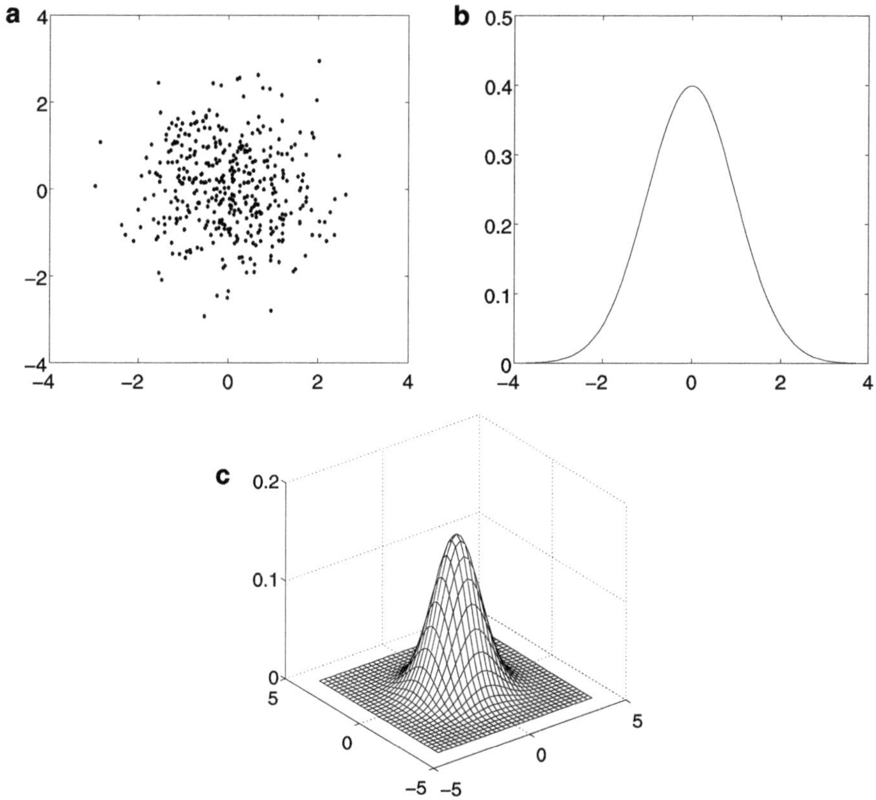

Fig. 4.3 **a** Scatter plot of the two-dimensional Gaussian distribution in (4.6). **b** The one-dimensional standardized Gaussian pdf. As explained in Sect. 4.3, it is also the marginal distribution of one of the variables in **a**, and furthermore, turns out to be equal to the conditional distribution of one variable given the other variable. **c** The probability density function of the two-dimensional Gaussian distribution

There is actually a simple connection between marginal and joint pdf's. We can obtain a marginal pdf by integrating the joint pdf over one of the variables. This is sometimes called "integrating out". Consider for simplicity the case where we only have two variables, z_1 and z_2. Then the marginal pdf of z_1 is obtained by

$$p_{z_1}(z_1) = \int p_{\mathbf{z}}(z_1, z_2)\, dz_2. \tag{4.8}$$

This is a continuous-space version of the intuitive idea that for a given value of z_1, we "count" how many observations we have with that value, going through all the possible values of z_2.[1] (In this continuous-valued case, no observed values of z_1 are

[1]Note again that the notation in (4.8) is sloppy, because now z_1 in the parentheses, both on the left and the right-hand side, stands for any value z_1 might obtain, although the same notation is used

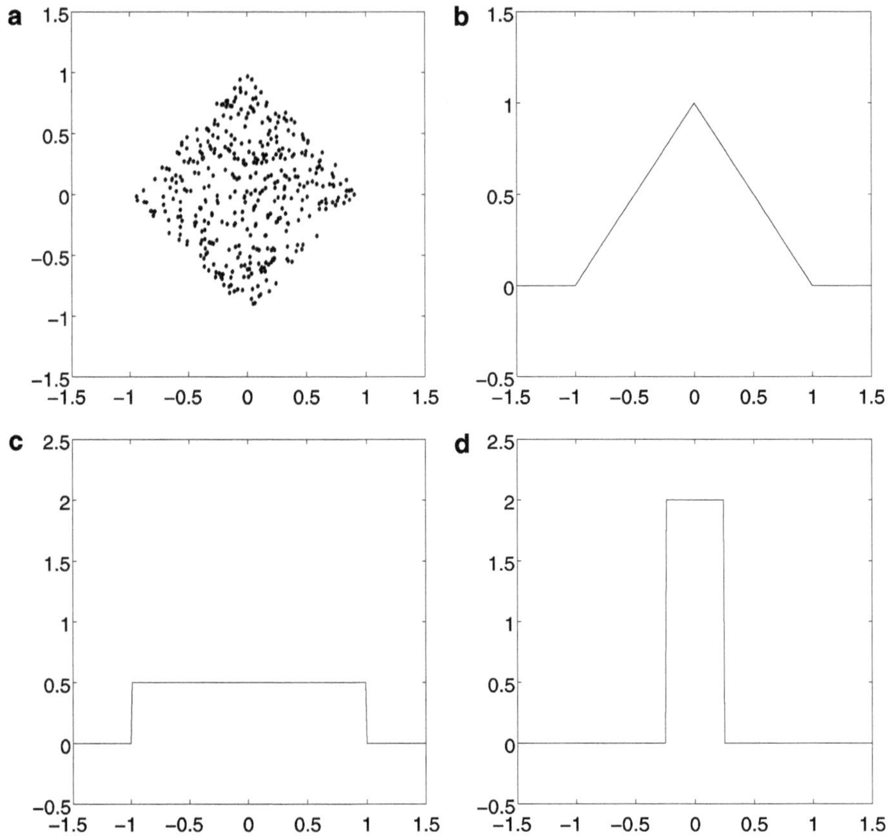

Fig. 4.4 **a** Scatter plot of data obtained from the pdf in (4.7). **b** Marginal pdf of one of the variables in **a**. **c** Conditional pdf of z_2 given $z_1 = 0$. **d** Conditional pdf of z_2 given $z_1 = 0.75$

likely to be exactly equal to the specified value, but we can use the idea of a small interval centered around that value as in the definition of the pdf above.)

Example 3 In the case of the Gaussian distribution in (4.6), we have

$$p(z_1) = \int p(z_1, z_2)\, dz_2 = \int \frac{1}{2\pi} \exp\left(-\frac{1}{2}(z_1^2 + z_2^2)\right) dz_2$$

---

for the random quantity itself. A more rigorous notation would be something like:

$$p_{z_1}(v_1) = \int p_{\mathbf{z}}(v_1, v_2)\, dv_2 \tag{4.9}$$

where we have used two new variables, v_1 to denote the point where we want to evaluate the marginal density, and v_2 which is the integration variable. However, in practice we often do not want to introduce new variable names in order to keep things simple, so we use the notation in (4.8).

$$= \frac{1}{\sqrt{2\pi}} \exp\left(-\frac{1}{2}z_1^2\right) \int \frac{1}{\sqrt{2\pi}} \exp\left(-\frac{1}{2}z_2^2\right) dz_2. \tag{4.10}$$

Here, we used the fact that the pdf is factorizable since $\exp(a+b) = \exp(a)\exp(b)$. In the last integral, we recognize the pdf of the one-dimensional Gaussian distribution of zero mean and unit variance given in (4.5). Thus, that integral is one, because the integral of any pdf is equal to one. This means that the marginal distribution $p(z_1)$ is just the classic one-dimensional standardized Gaussian pdf.

Example 4 Going back to our example in (4.7), we can calculate the marginal pdf of z_1 to equal

$$p_{z_1}(z_1) = \begin{cases} 1 - |z_1|, & \text{if } |z_1| < 1, \\ 0, & \text{otherwise} \end{cases} \tag{4.11}$$

which is plotted in Fig. 4.4b, and shows the fact that there is more "stuff" near the origin, and no observation can have an absolute value larger than one. Due to symmetry, the marginal pdf of z_2 has exactly the same form.

4.4 Conditional Probabilities

Another important concept is the *conditional* pdf of z_2 given z_1. This means the pdf of z_2 when we have observed the value of z_1. Let us denote the observed value of z_1 by a. Then conditional pdf is basically obtained by just fixing the value of z_1 to a in the pdf, which gives $p_{\mathbf{z}}(a, z_2)$. However, this is not enough because a pdf must have an integral equal to one. Therefore, we must normalize $p_{\mathbf{z}}(a, z_2)$ by dividing it by its integral. Thus, we obtain the conditional pdf, denoted by $p(z_2 \mid z_1 = a)$ as

$$p(z_2 \mid z_1 = a) = \frac{p_{\mathbf{z}}(a, z_2)}{\int p_{\mathbf{z}}(a, z_2) dz_2}. \tag{4.12}$$

Note that the integral in the denominator equals the marginal pdf of z_1 at point a, so we can also write

$$p(z_2 \mid z_1 = a) = \frac{p_{\mathbf{z}}(a, z_2)}{p_{z_1}(a)}. \tag{4.13}$$

Again, for notational simplicity, we can omit the subscripts and just write

$$p(z_2 \mid z_1 = a) = \frac{p(a, z_2)}{p(a)} \tag{4.14}$$

or, we can even avoid introducing the new quantity a and write

$$p(z_2 \mid z_1) = \frac{p(z_1, z_2)}{p(z_1)}. \tag{4.15}$$

Example 5 For the Gaussian density in (4.6), the computation of the conditional pdf is quite simple, if we use the same factorization as in (4.10):

$$p(z_2|z_1) = \frac{p(z_1, z_2)}{p(z_1)} = \frac{\frac{1}{\sqrt{2\pi}}\exp(-\frac{1}{2}z_1^2)\frac{1}{\sqrt{2\pi}}\exp(-\frac{1}{2}z_2^2)}{\frac{1}{\sqrt{2\pi}}\exp(-\frac{1}{2}z_1^2)}$$
$$= \frac{1}{\sqrt{2\pi}}\exp\left(-\frac{1}{2}z_2^2\right) \tag{4.16}$$

which turns out to be the same as the marginal distribution of z_2. (This kind of situation where $p(z_2|z_1) = p(z_2)$ is related to independence as discussed in Sect. 4.5 below.)

Example 6 In our example pdf in (4.7), the conditional pdf changes quite a lot as a function of the value a of z_1. If z_1 is zero (i.e. $a = 0$), the conditional pdf of z_2 is the uniform density in the interval $[-1, 1]$. In contrast, if z_1 is close to 1 (or -1), the values that can be taken by z_2 are quite small. Simply fixing $z_1 = a$ in the pdf, we have

$$p(a, z_2) = \begin{cases} 1, & \text{if } |z_2| < 1 - |a|, \\ 0, & \text{otherwise} \end{cases} \tag{4.17}$$

which can be easily integrated:

$$\int p(a, z_2)\, dz_2 = 2\big(1 - |a|\big). \tag{4.18}$$

(This is just the length of the segment in which z_2 is allowed to take values.) So, we get

$$p(z_2|z_1) = \begin{cases} \frac{1}{2-2|z_1|}, & \text{if } |z_2| < 1 - |z_1|, \\ 0, & \text{otherwise} \end{cases} \tag{4.19}$$

where we have replaced a by z_1. This pdf is plotted for $z_1 = 0$ and $z_1 = 0.75$ in Figs. 4.4a and b, respectively.

Generalization to Many Dimensions The concepts of marginal and conditional pdf's extend naturally to the case where we have n random variables instead of just two. The point is that instead of two random variables, z_1 and z_2, we can have two random vectors, say $\mathbf{z}_1$ and $\mathbf{z}_2$, and use exactly the same formulas as for the two random variables. So, starting with a random vector $\mathbf{z}$, we take some of its variables and put them in the vector $\mathbf{z}_1$, and leave the rest in the vector $\mathbf{z}_2$

$$\mathbf{z} = \begin{pmatrix} \mathbf{z}_1 \\ \mathbf{z}_2 \end{pmatrix}. \tag{4.20}$$

Now, the marginal pdf of $\mathbf{z}_1$ is obtained by the same integral formula as above:

$$p_{\mathbf{z}_1}(\mathbf{z}_1) = \int p_{\mathbf{z}}(\mathbf{z}_1, \mathbf{z}_2)\, d\mathbf{z}_2 \tag{4.21}$$

and, likewise, the conditional pdf of $\mathbf{z}_2$ given $\mathbf{z}_1$ is given by:

$$p(\mathbf{z}_2 \mid \mathbf{z}_1) = \frac{p(\mathbf{z}_1, \mathbf{z}_2)}{p(\mathbf{z}_1)}. \tag{4.22}$$

Both of these are, naturally, multidimensional pdf's.

Discrete-Valued Variables For the sake of completeness, let us note that these formulas are also valid for random variables with discrete values; then the integrals are simply replaced by sums. For example, for the conditional probabilities, we simply have

$$P(z_2 \mid z_1) = \frac{P_{\mathbf{z}}(z_1, z_2)}{P_{z_1}(z_1)} \tag{4.23}$$

where marginal probability of z_1 can be computed as

$$P_{z_1}(z_1) = \sum_{z_2} P_{\mathbf{z}}(z_1, z_2). \tag{4.24}$$

4.5 Independence

Let us consider two random variables, z_1 and z_2. Basically, the variables z_1 and z_2 are said to be statistically independent if information on the value taken by z_1 does not give any information on the value of z_2, and vice versa.

As an example, let us consider again the grey-scale values of two neighboring pixels. As in Fig. 4.1, we go through many different locations in an image in random order, and take the grey-scale values of the pixels as the observed values of the two random variables. These random variables will *not* be independent. One of the basic statistical properties of natural images is that two neighboring pixels are dependent. Intuitively, it is clear that two neighboring pixels tend to have very similar grey-scale values: If one of them is black, then the other one is black with a high probability, so they do give information on each other. This is seen in the oblique shape (having an angle of 45 degrees) of the data "cloud" in Fig. 4.1. Actually, the grey-scale values are correlated, which is a special form of dependence as we will see below.

The idea that z_1 gives no information on z_2 can be intuitively expressed using conditional probabilities: the conditional probability $p(z_2 \mid z_1)$ should be just the same as $p(z_2)$:

$$p(z_2 \mid z_1) = p(z_2) \tag{4.25}$$

for any observed value a of z_1. This implies

$$\frac{p(z_1, z_2)}{p(z_1)} = p(z_2) \tag{4.26}$$

or

$$p(z_1, z_2) = p(z_1)p(z_2) \tag{4.27}$$

for any values of z_1 and z_2. Equation (4.27) is usually taken as the definition of independence because it is mathematically so simple. It simply says that the joint pdf must be a product of the marginal pdf's. The joint pdf is then called factorizable.

The definition is easily generalized to n variables $z_1, z_2, \ldots, z_n$, in which case it is

$$p(z_1, z_2, \ldots, z_n) = p(z_1)p(z_2)\ldots p(z_n). \tag{4.28}$$

Example 7 For the Gaussian distribution in (4.6) and Fig. 4.3, we have

$$p(z_1, z_2) = \frac{1}{\sqrt{2\pi}} \exp\left(-\frac{1}{2}z_1^2\right) \times \frac{1}{\sqrt{2\pi}} \exp\left(-\frac{1}{2}z_2^2\right). \tag{4.29}$$

So, we have factorized the joint pdf as the product of two pdf's, each of which depends on only one of the variables. Thus, z_1 and z_2 are independent. This can also be seen in the form of the conditional pdf in (4.16), which does not depend on the conditioning variable at all.

Example 8 For our second pdf in (4.7), we computed the conditional pdf $p(z_2|z_1)$ in (4.19). This is clearly not the same as the marginal pdf in (4.11); it depends on z_1. So, the variables are not independent. (See the discussion just before (4.17) for an intuitive explanation of the dependencies.)

Example 9 Consider the uniform distribution on a square:

$$p(z_1, z_2) = \begin{cases} \frac{1}{12}, & \text{if } |z_1| \leq \sqrt{3} \text{ and } |z_2| \leq \sqrt{3}, \\ 0, & \text{otherwise.} \end{cases} \tag{4.30}$$

A scatter plot from this distribution is shown in Fig. 4.5. Now, z_1 and z_2 are independent because the pdf can be expressed as the product of the marginal distributions, which are

$$p(z_1) = \begin{cases} \frac{1}{2\sqrt{3}}, & \text{if } |z_1| \leq \sqrt{3}, \\ 0, & \text{otherwise} \end{cases} \tag{4.31}$$

and the same for z_2.

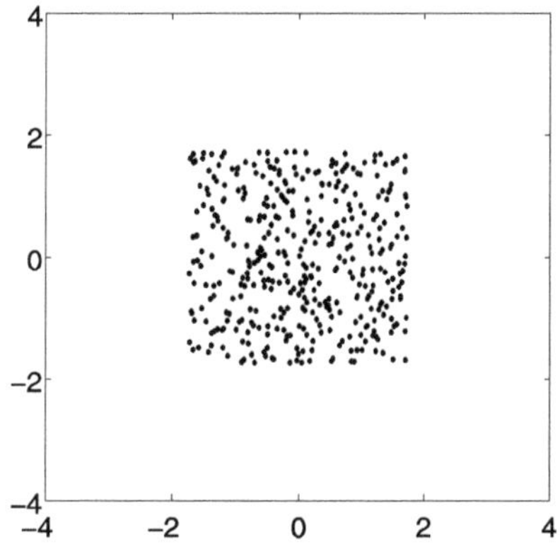

Fig. 4.5 A scatter plot of the two-dimensional uniform distribution in (4.30)

4.6 Expectation and Covariance

4.6.1 Expectation

The expectation of a random vector, or its "mean" value is, in theory, obtained by the same kind of integral as for a single random variable

$$E\{\mathbf{z}\} = \int p_{\mathbf{z}}(\mathbf{z})\mathbf{z}\,d\mathbf{z}. \tag{4.32}$$

In practice, the expectation can be computed by taking the expectation of each variable separately, completely ignoring the existence of the other variables

$$E\{\mathbf{z}\} = \begin{pmatrix} E\{z_1\} \\ E\{z_2\} \\ \vdots \\ E\{z_n\} \end{pmatrix} = \begin{pmatrix} \int p_{z_1}(z_1)z_1\,dz_1 \\ \int p_{z_2}(z_2)z_2\,dz_2 \\ \vdots \\ \int p_{z_n}(z_n)z_n\,dz_n \end{pmatrix}. \tag{4.33}$$

The expectation of any transformation $\mathbf{g}$, whether one- or multi-dimensional, can be computed as:

$$E\{\mathbf{g}(\mathbf{z})\} = \int p_{\mathbf{z}}(\mathbf{z})\mathbf{g}(\mathbf{z})\,d\mathbf{z}. \tag{4.34}$$

The expectation is a linear operation, which means

$$E\{a\mathbf{z} + b\mathbf{s}\} = aE\{\mathbf{z}\} + bE\{\mathbf{s}\} \tag{4.35}$$

for any constants a and b. In fact, this generalizes to any multiplication by a matrix $\mathbf{M}$:

$$E\{\mathbf{M}\mathbf{z}\} = \mathbf{M}E\{\mathbf{z}\}. \tag{4.36}$$

4.6.2 Variance and Covariance in One Dimension

The variance of a random variable is defined as

$$\mathrm{var}(z_1) = E\{z_1^2\} - (E\{z_1\})^2. \tag{4.37}$$

This can also be written $\mathrm{var}(z_1) = E\{(z_1 - E\{z_1\})^2\}$, which more clearly shows how variance measures average deviation from the mean value.

When we have more than one random variable, it is useful to analyze the *covariance*:

$$\mathrm{cov}(z_1, z_2) = E\{z_1 z_2\} - E\{z_1\}E\{z_2\} \tag{4.38}$$

which measures how well we can predict the value of one of the variables using a simple linear predictor, as will be seen below.

The covariance is often normalized to yield the correlation coefficient

$$\mathrm{corr}(z_1, z_2) = \frac{\mathrm{cov}(z_1, z_2)}{\sqrt{\mathrm{var}(z_1)\,\mathrm{var}(z_2)}} \tag{4.39}$$

which is invariant to the scaling of the variables, i.e. it is not changed if one or both of the variables is multiplied by a constant.

If the covariance is zero, which is equivalent to saying that the correlation coefficient is zero, the variables are said to be *uncorrelated.*

4.6.3 Covariance Matrix

The variances and covariances of the elements of a random vector $\mathbf{z}$ are often collected to a *covariance matrix* whose i, jth element is simply the covariance of z_i and z_j:

$$\mathbf{C}(\mathbf{z}) = \begin{pmatrix} \mathrm{cov}(z_1, z_1) & \mathrm{cov}(z_1, z_2) & \dots & \mathrm{cov}(z_1, z_n) \\ \mathrm{cov}(z_2, z_1) & \mathrm{cov}(z_2, z_2) & \dots & \mathrm{cov}(z_2, z_n) \\ \vdots & & \ddots & \vdots \\ \mathrm{cov}(z_n, z_1) & \mathrm{cov}(z_n, z_2) & \dots & \mathrm{cov}(z_n, z_n) \end{pmatrix}. \tag{4.40}$$

Note that the covariance of z_i with itself is the same as the variance of z_i. So, the diagonal of the covariance matrix gives the variances. The covariance matrix is basically a generalization of variance to random vectors: in many cases, when moving from a single random variable to random vectors, the covariance matrix takes the place of variance.

In matrix notation, the covariance matrix is simply obtained as a generalization of the one-dimensional definitions in (4.38) and (4.37) as

$$\mathbf{C}(\mathbf{z}) = E\{\mathbf{z}\mathbf{z}^T\} - E\{\mathbf{z}\}E\{\mathbf{z}\}^T \tag{4.41}$$

where taking the transposes in the correct places is essential. In most of this book, we will be dealing with random variables whose means are zero, in which case the second term in (4.41) is zero.

If the variables are uncorrelated, the covariance matrix is diagonal. If they are all further standardized to unit variance, the covariance matrix equals the identity matrix.

The covariance matrix is the basis for the analysis of natural images in the next chapter. However, in many further chapters, the covariance matrix is not enough, and we need further concepts, such as independence, so we need to understand the connection between these concepts.

4.6.4 Independence and Covariances

A most important property of *independent* random variables z_1 and z_2 is that the expectation of any product of a function of z_1 and a function of z_2 is equal to the product of the expectations:

$$E\{g_1(z_1)g_2(z_2)\} = E\{g_1(z_1)\}E\{g_2(z_2)\} \tag{4.42}$$

for any functions g_1 and g_2. This implies that *independent variables are uncorrelated*, since we can take $g_1(z) = g_2(z) = z$, in which case equation (4.42) simply says that the covariance is zero.

Example 10 In the standardized Gaussian distribution in (4.6), the means of both z_1 and z_2 are zero, and their variances are equal to one (we will not try to prove this here). Actually, the word "standardized" means exactly that the means and variances have been standardized in this way. The covariance $\text{cov}(z_1, z_2)$ equals zero, because the variables are independent, and thus uncorrelated.

Example 11 What would be the covariance of z_1 and z_2 in our example pdf in (4.7)? First, we have to compute the means. Without computing any integrals, we can actually see that $E\{z_1\} = E\{z_2\} = 0$ because of symmetry: both variables are symmetric with respect to the origin, so their means are zero. This can be justified as follows: take a new variable $y = -z_1$. Because of symmetry of the pdf with respect to zero, the change of sign has no effect and the pdf of y is just the same as the pdf of z_1. Thus, we have

$$E\{y\} = E\{-z_1\} = -E\{z_1\} = E\{z_1\} \tag{4.43}$$

which implies that $E\{z_1\} = 0$. Actually, the covariance is zero because of the same kind of symmetry with respect to zero. Namely, we have $\text{cov}(y, z_2) = \text{cov}(z_1, z_2)$ because again, the change of sign has no effect and the joint pdf of y, z_2 is just the same as the pdf of z_1, z_2. This means $\text{cov}(y, z_2) = E\{(-z_1)z_2\} = -E\{z_1 z_2\} = E\{z_1 z_2\}$. This obviously implies that the covariance is zero. The covariance matrix of the vector $\mathbf{z}$ is thus diagonal (we don't bother to compute the diagonal elements, which are the variances).

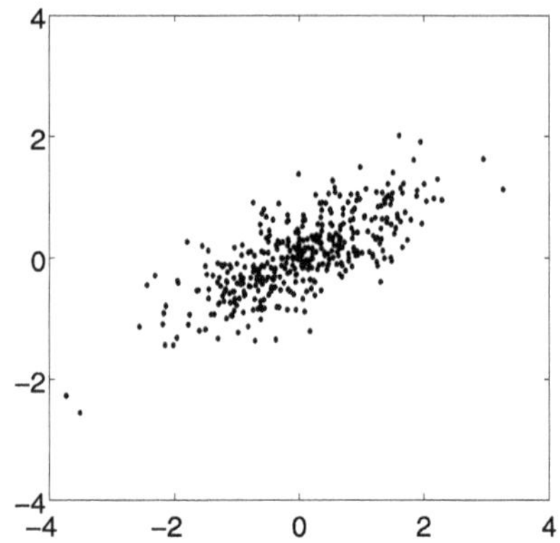

Fig. 4.6 A scatter plot of the distribution created by the dependence relation in (4.44)

Example 12 Let's have a look at a more classical example of covariances. Assume that z_1 has mean equal to zero and variance equal to one. Assume that n (a "noise" variable) is independent from z_1. Let us consider a variable z_2 which is a linear function of x, with noise added:

$$z_2 = az_1 + n. \tag{4.44}$$

What is the covariance of the two variables? We can calculate

$$\begin{aligned}\mathrm{cov}(z_1, z_2) &= E\{z_1(az_1 + n)\} + 0 \times E\{z_2\} = aE\{z_1^2\} + E\{z_1 n\} \\ &= a + E\{z_1\}E\{n\} = a + 0 \times E\{z_1\} = a.\end{aligned} \tag{4.45}$$

Here, we have the equality $E\{z_1 n\} = E\{z_1\}E\{n\}$ because of the uncorrelatedness of z_1 and n, which is implied by their independence. A scatter plot of such data, created for parameter a set at 0.5 and with noise variance $\mathrm{var}(n) = 1$, is shown in Fig. 4.6. The covariance matrix of the vector $\mathbf{z} = (z_1, z_2)$ is equal to

$$\mathbf{C}(\mathbf{z}) = \begin{pmatrix} 1 & a \\ a & 1 \end{pmatrix}. \tag{4.46}$$

Example 13 *White noise* refers to a collection of random variables which are independent and have the same distribution. (In some sources, only uncorrelatedness is required, not independence, but in this book the definition of white noise includes independence.) Depending on the context, the variables could be the value of noise at different time points $n(t)$, or at different pixels $N(x, y)$. In the first case, white noise in the system is independent at different time points; in the latter, noise at different pixels is independent. When modeling physical noise, which can be found in most measurement devices, it is often realistic and mathematically simple to assume that the noise is white.

4.7 Bayesian Inference

Bayesian inference is a framework that has recently been increasingly applied to model such phenomena as perception and intelligence. There are two viewpoints on what Bayesian inference is.

1. Bayesian inference attempts to infer underlying causes when we observe their effects.
2. Bayesian inference uses prior information on parameters in order to estimate them better.

Both of these goals can be accomplished by using the celebrated Bayes' formula, which we will now explain.

4.7.1 Motivating Example

Let us start with a classic example. Assume that we have a test for a rare genetic disorder. The test is relatively reliable but not perfect. For a patient with the disorder, the probability of a positive test result is 99%, whereas for a patient without the disorder, the probability of a positive test is only 2%. Let us denote the test result by t and the disorder by d. A positive test result is expressed as $t = 1$ and a negative one as $t = 0$. Likewise, $d = 1$ means that the patient really has the disorder, whereas $d = 0$ means the patients doesn't. Then the specifications we just gave can be expressed as the following conditional probabilities:

$$P(t = 1 \mid d = 1) = 0.99, \tag{4.47}$$

$$P(t = 1 \mid d = 0) = 0.02. \tag{4.48}$$

Because probabilities sum to one, we immediately find the following probabilities as well:

$$P(t = 0 \mid d = 1) = 0.01, \tag{4.49}$$

$$P(t = 0 \mid d = 0) = 0.98. \tag{4.50}$$

Now, the question we want to answer is: *Given a positive test result, what is the probability that the patient has the disorder?* Knowing this probability is, of course, quite important when applying this medical test. Basically, we then want to compute the conditional probability of the form $p(d = 1 \mid t = 1)$. The order of the variables in this conditional probability is reversed from the formulas above. This is because the formulas above gave us the observable effects given the causes, but now we want to know the causes, given observations of their effects.

To find that probability, let's try to use the definition in (4.23)

$$P(d = 1 \mid t = 1) = \frac{P(d = 1, t = 1)}{P(t = 1)} \tag{4.51}$$

which presents us with two problems: We know neither the denominator nor the numerator. To get further, let's assume we know the marginal distribution $P(d)$. Then we can easily find the numerator by using the definition of conditional probability

$$P(d=1,t=1)=P(t=1\mid d=1)P(d=1) \tag{4.52}$$

and after some heavy thinking, we see that we can also compute the denominator in (4.51) by using the formula for marginal probability:

$$P(t=1)=P(d=1,t=1)+P(d=0,t=1) \tag{4.53}$$

which can be computed once we know the joint probabilities by (4.52) and its corresponding version with $d=0$. Thus, in the end, we have

$$P(d=1\mid t=1)=\frac{P(t=1\mid d=1)P(d=1)}{P(t=1\mid d=1)P(d=1)+P(t=1\mid d=0)P(d=0)}. \tag{4.54}$$

So, we see that the key to inferring the causes from observed effects is to know the marginal distribution of the causes, in this case $P(d)$. This distribution is also called the *prior* distribution of d, because it incorporates our knowledge of the cause d prior to any observations. For example, let's assume 0.1% of the patients given this test have the genetic disorder. Then before the test our best guess is that a given patient has the disorder with the probability of 0.001. However, after making the test, we have more information on the patient, and that information is given by the conditional distribution $P(d\mid t=1)$ which we are trying to compute. This distribution, which incorporates both our prior knowledge on d and the observation of t, is called the *posterior* probability.

To see a rather surprising phenomenon, let us plug in the value $P(d=1)=0.001$ as the prior probability of disorder in (4.54). Then we can calculate

$$P(d=1\mid t=1)=\frac{0.99\times 0.001}{0.99\times 0.001+0.02\times(1-0.001)}\approx 0.05. \tag{4.55}$$

Thus, even after a positive test result, the probability that the patient has the disorder is approximately 5%. Many people find this quite surprising, because they would have guessed that the probability is something like 99%, as the test gives the right result in 99% of the cases.

This posterior probability depends very much on the prior probability. Assume that half the tested patients actually have the disorder, $P(d=1)=0.5$. Then the posterior probability is 99%. This prior actually gives us no information because the chances are 50–50, and the 99% accuracy of the test is directly seen in the posterior probability.

Thus, in cases where the prior assigns very different probabilities to different causes, Bayesian inference shows that the posterior probabilities of the causes can be very different from what one might expect by just looking at the effects.

4.7.2 Bayes' Rule

The logic of the previous section was actually the proof of the celebrated Bayes' rule. In the general case, we consider a continuous-valued random vector $\mathbf{s}$ that gives the causes and $\mathbf{z}$ that gives the observed effects. The Bayes' rule then takes the form

$$p(\mathbf{s} \mid \mathbf{z}) = \frac{p(\mathbf{z} \mid \mathbf{s}) p_{\mathbf{s}}(\mathbf{s})}{\int p(\mathbf{z} \mid \mathbf{s}) p_{\mathbf{s}}(\mathbf{s}) \, d\mathbf{s}} \tag{4.56}$$

which is completely analogous to (4.54) and can be derived in the same way. This is the Bayes' rule, in one of its formulations. It gives the posterior distribution of $\mathbf{s}$ based on its prior distribution $p(\mathbf{s})$ and the conditional probabilities $p(\mathbf{z}|\mathbf{s})$. Note that instead of random variables, we can directly use vectors in the formula without changing anything.

To explicitly show what is random and what is observed in Bayes rule, we should rewrite it as

$$p(\mathbf{s} = \mathbf{b} \mid \mathbf{z} = \mathbf{a}) = \frac{p_{\mathbf{z}|\mathbf{s}}(\mathbf{a} \mid \mathbf{b}) p_{\mathbf{s}}(\mathbf{b})}{\int p_{\mathbf{z}|\mathbf{s}}(\mathbf{a} \mid \mathbf{u}) p_{\mathbf{s}}(\mathbf{u}) \, d\mathbf{u}} \tag{4.57}$$

where $p_{\mathbf{z}|\mathbf{s}}(\mathbf{a}|\mathbf{b})$ is the conditional probability $p(\mathbf{z} = \mathbf{a}|\mathbf{s} = \mathbf{b})$, $\mathbf{a}$ is the observed value of $\mathbf{z}$, and $\mathbf{b}$ is a possible value of $\mathbf{s}$. This form is, of course, much more difficult to read than (4.56).

In theoretical treatment, Bayes rule can sometimes be simplified because the denominator is actually equal to $p(\mathbf{z})$, which gives

$$p(\mathbf{s} \mid \mathbf{z}) = \frac{p(\mathbf{z} \mid \mathbf{s}) p_{\mathbf{s}}(\mathbf{s})}{p_{\mathbf{z}}(\mathbf{z})}. \tag{4.58}$$

However, in practice, we usually have to use the form in (4.56) because we do not know how to directly compute $p_{\mathbf{z}}$.

The prior $p_{\mathbf{s}}$ contains the prior information on the random variable $\mathbf{s}$. The conditional probabilities $p(\mathbf{z} \mid \mathbf{s})$ show the connection between the observed quantity $\mathbf{z}$ (the "effect") and the underlying variable $\mathbf{s}$ (the "cause").

Where do we get the prior distribution $p(\mathbf{s})$? In some cases, $p(\mathbf{s})$ can be estimated, because we might be able to observe the original $\mathbf{s}$. In the medical example above, the prior distribution $p(d)$ can be estimated if some of the patients are subjected to additional tests which are much more accurate (thus usually much more expensive) so that we really know for sure how many of the patients have the disorder. In other cases, the prior might be formulated more subjectively, based on the opinion of an expert.

4.7.3 Non-informative Priors

Sometimes, we have no information on the prior probabilities $p_{\mathbf{s}}$. Then we should use a non-informative prior that expresses this fact. In the case of discrete variables, a non-informative prior is one that assigns the same probability to all the possible values of $\mathbf{s}$ (e.g. 50% probability of a patient to have the disorder or not).

In the case of continuous-valued priors defined in the whole real line $[-\infty, \infty]$, the situation is a bit more complicated. If we take a "flat" pdf that is constant, $p(\mathbf{s}) = c$, it cannot be a real pdf because the integral of such a pdf is infinite (or zero if $c = 0$). Such a prior is called improper. Still, they can often be used in Bayesian inference even though the non-integrability may pose some theoretical problems.

What happens in the Bayes rule if we take such a flat, non-informative prior? We get

$$p(\mathbf{s} \mid \mathbf{z}) = \frac{p(\mathbf{z} \mid \mathbf{s})c}{\int p(\mathbf{z} \mid \mathbf{s})c\, d\mathbf{s}} = \frac{p(\mathbf{z} \mid \mathbf{s})}{\int p(\mathbf{z} \mid \mathbf{s})\, d\mathbf{s}}. \tag{4.59}$$

The denominator does not depend on $\mathbf{s}$ (this is always true in Bayes' rule), so we see that $p(\mathbf{s} \mid \mathbf{z})$ is basically the same of $p(\mathbf{z} \mid \mathbf{s})$; it is just rescaled so that the integral is equal to one. What this shows is that if we have no information on the prior probabilities, the probabilities of effects given the causes are simply proportional to the probabilities of causes given the effects. However, if the prior $p(\mathbf{s})$ is far from flat, these two probabilities can be very different from each other, as the example above showed in the case where the disorder is rare.

4.7.4 Bayesian Inference as an Incremental Learning Process

The transformation from the prior probability $p(\mathbf{s})$ to $p(\mathbf{s} \mid \mathbf{z})$ can be compared to an incremental (on-line) learning process where a biological organism receives more and more information in an uncertain environment.

In the beginning, the organism's belief about the value of a quantity $\mathbf{s}$ is the prior probability $p(\mathbf{s})$. Here, we assume that the organism performs probabilistic inference: the organism does not "think" that it knows the value of $\mathbf{s}$ with certainty; rather, it just assigns probabilities to different values $\mathbf{s}$ might take. This does not mean that we assume the organism is highly intelligent and knows Bayes' rule. Rather, we assume that the neural networks in the nervous system of the organism have evolved to perform something similar.

Then the organism receives information via sensory organs or similar means. A statistical formulation of "incoming information" is that the organism observes the value of a random variable $\mathbf{z}_1$. Now, the belief of the organism is expressed by the posterior pdf $p(\mathbf{s} \mid \mathbf{z}_1)$. This pdf gives the probabilities that the organism assigns to different values of $\mathbf{s}$.

Next, assume that the organism observes another piece of information, say $\mathbf{z}_2$. Then the organism's belief is changed to $p(\mathbf{s} \mid \mathbf{z}_1, \mathbf{z}_2)$

$$p(\mathbf{s} \mid \mathbf{z}_1, \mathbf{z}_2) = \frac{p(\mathbf{z}_1, \mathbf{z}_2 \mid \mathbf{s})p(\mathbf{s})}{p(\mathbf{z}_1, \mathbf{z}_2)}. \tag{4.60}$$

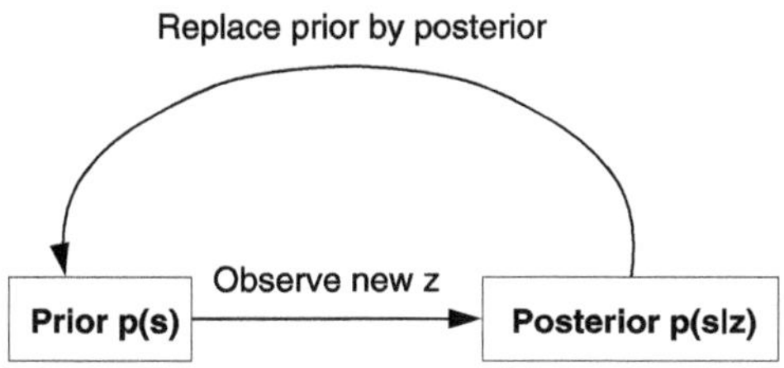

Fig. 4.7 Computation of posterior as an incremental learning process. Given the current prior, the organism observes the input $\mathbf{z}$, and computes the posterior $p(\mathbf{s}|\mathbf{z})$. The prior is then replaced by this new posterior, which is used as the prior in the future

Assume further that $\mathbf{z}_2$ is independent from $\mathbf{z}_1$ given $\mathbf{s}$, which means $p(\mathbf{z}_1, \mathbf{z}_2 \mid \mathbf{s}) = p(\mathbf{z}_1 \mid \mathbf{s})p(\mathbf{z}_2 \mid \mathbf{s})$ (see Sect. 4.5 for more on independence). Then the posterior becomes

$$p(\mathbf{s} \mid \mathbf{z}_1, \mathbf{z}_2) = \frac{p(\mathbf{z}_1 \mid \mathbf{s})p(\mathbf{z}_2 \mid \mathbf{s})p(\mathbf{s})}{p(\mathbf{z}_1)p(\mathbf{z}_2)} = \frac{p(\mathbf{z}_2 \mid \mathbf{s})}{p(\mathbf{z}_2)} \frac{p(\mathbf{z}_1 \mid \mathbf{s})p(\mathbf{s})}{p(\mathbf{z}_1)}. \tag{4.61}$$

Now, the expression $p(\mathbf{z}_1 \mid \mathbf{s})p(\mathbf{s})/p(\mathbf{z}_1)$ is nothing but the posterior $p(\mathbf{s} \mid \mathbf{z}_1)$ that the organism computed previously. So, we have

$$p(\mathbf{s} \mid \mathbf{z}_1, \mathbf{z}_2) = \frac{p(\mathbf{z}_2 \mid \mathbf{s})p(\mathbf{s} \mid \mathbf{z}_1)}{p(\mathbf{z}_2)}. \tag{4.62}$$

The right-hand side is just like the Bayes' rule applied on $\mathbf{s}$ and $\mathbf{z}_2$ but instead of the prior $p(\mathbf{s})$ it has $p(\mathbf{s} \mid \mathbf{z}_1)$. Thus, the new posterior (after observing $\mathbf{z}_2$) is computed *as if the previous posterior were a prior*.

This points out an incremental learning interpretation of Bayes rule. When the organism observes new information (new random variables), it updates its belief about the world by the Bayes rule, where the current belief is taken as the prior, and the new belief is computed as the posterior. This is illustrated in Fig. 4.7

Such learning can happen on different time scales. It could be that $\mathbf{s}$ is a very slowly changing parameter, say, the length of the arms (or tentacles) of the organism. In that case, the organism can collect a large number of observations over time, and the belief would change very slowly. The first "prior" belief that the organism may have had before collection of any data, eventually loses its significance (see next section).

On the other hand, $\mathbf{s}$ could be a quantity that has to be computed instantaneously, say, the probability that the animal in front of you is trying to eat you. Then only a few observed quantities (given by the current visual input) are available. Such inference can then be heavily influenced by the prior information that the organism has at the moment of encountering the animal. For example, if the animal is small and cute, the prior probability is small, and even if the animal seems to behaves in an aggressive way, you will probably infer that it is not going to try to eat you.

4.8 Parameter Estimation and Likelihood

4.8.1 Models, Estimation, and Samples

A *statistical model* describes the pdf of the observed random vector using a number of parameters. The parameters typically have an intuitive interpretation; for example, in this book, they often define image features. A model is basically a conditional density of the observed data variable, $p(z \mid \alpha)$, where α is the parameter. The parameter could be a multidimensional vector as well. Different values of the parameter imply different distributions for the data, which is why this can be thought of as a conditional density.

For example, consider the one-dimensional Gaussian pdf

$$p(z \mid \alpha) = \frac{1}{\sqrt{2\pi}} \exp\left(-\frac{1}{2}(z-\alpha)^2\right). \tag{4.63}$$

Here, the parameter α has an intuitive interpretation as the mean of the distribution. Given α, the observed data variable z then takes values around α, with variance equal to one.

Typically, we have a large number of observations of the random variable z, which might come from measuring some phenomenon n times, and these observations are independent. The set of observations is called a *sample* in statistics.[2] So, we want to use all the observations to better estimate the parameters. For example, in the model in (4.63), it is obviously not a very good idea to estimate the mean of the distribution based on just a single observation.

Estimation has a very boring mathematical definition, but basically it means that we want to find a reasonable approximation of the value of the parameter based on the observations in the sample. A method (a formula or an algorithm) that estimates α is called an estimator. The value given by the estimator for a particular sample is called an estimate. Both are usually denoted by a hat: $\hat{\alpha}$.

Assume we now have a sample of n observations. Let us denote the observed values by $z(1), z(2), \dots, z(n)$. Because the observations are independent, the joint probability is simply obtained by multiplying the probabilities of the observations, so we have

$$p\big(z(1), z(2), \dots, z(n) \mid \alpha\big) = p\big(z(1) \mid \alpha\big) \times p\big(z(2) \mid \alpha\big) \times \cdots \times p\big(z(n) \mid \alpha\big). \tag{4.64}$$

This conditional density is called the *likelihood*. It is often simpler to consider the logarithm, which transforms products into sums. If we take the logarithm, we have

---

[2] In signal processing, sampling refers the process of reducing a continuous signal to a discrete signal. For example, an image $I(x, y)$ with continuous-valued coordinates x and y is reduced to a finite-dimensional vector in which the coordinates x and y take only a limited number of values (e.g. as on a rectangular grid). These two meanings of the word "sample" need to be distinguished.

the log-likelihood as

$$\log p\big(z(1), z(2), \dots, z(n) \mid \alpha\big) = \log p\big(z(1) \mid \alpha\big) + \log p\big(z(2) \mid \alpha\big) + \cdots + \log p\big(z(n) \mid \alpha\big). \tag{4.65}$$

4.8.2 Maximum Likelihood and Maximum a Posteriori

The question is then, How can we estimate α? In a Bayesian interpretation, we can consider the parameters as "causes" in Bayes' rule, and the observed data are the effects. Then the estimation of the parameters means that we compute the posterior pdf of the parameters using Bayes rule:

$$p(\alpha \mid z(1), \dots, z(n)) = \frac{p(z(1), \dots, z(n) \mid \alpha) p(\alpha)}{p(z(1), \dots, z(n))}. \tag{4.66}$$

In estimating parameters of the model, one usually takes a flat prior, i.e. $p(\alpha) = c$. Moreover, the term $p(z(1), \dots, z(n))$ does not depend on α, it is just for normalization, so we don't need to care about its value. Thus, we see that

$$p\big(\alpha \mid z(1), \dots, z(n)\big) = p\big(z(1), z(2), \dots, z(n) \mid \alpha\big) \times \text{const.} \tag{4.67}$$

the posterior of the parameters is proportional to the likelihood in the case of a flat prior.

Usually, we want a single value as an estimate. Thus, we have to somehow summarize the posterior distribution $p(\alpha \mid z(1), \dots, z(n))$, which is a function of α. The most widespread solution is to use the value of α that gives the highest value of the posterior pdf. Such estimation is called *maximum a posteriori* (MAP) estimation.

In the case of a flat prior, the maximum of the posterior distribution is obtained at the same point as the maximum of the likelihood, because likelihood is then proportional to the posterior. Thus, the estimation is then called the *maximum likelihood estimator*. If the prior is not flat, the maximum a posteriori estimator may be quite different from the maximum likelihood estimator.

The maximum likelihood estimator has another intuitive interpretation: it gives *the parameter value that gives the highest probability for the observed data.* This interpretation is slightly different from the Bayesian interpretation that we used above.

Sometimes the maximum likelihood estimator can be computed by a simple algebraic formula, but in most cases, the maximization has to be done numerically. For a brief introduction to optimization methods, see Chap. 18.

Example 14 In the case of the model in (4.63), we have

$$\log p(z \mid \alpha) = -\frac{1}{2}(z - \alpha)^2 + \text{const.} \tag{4.68}$$

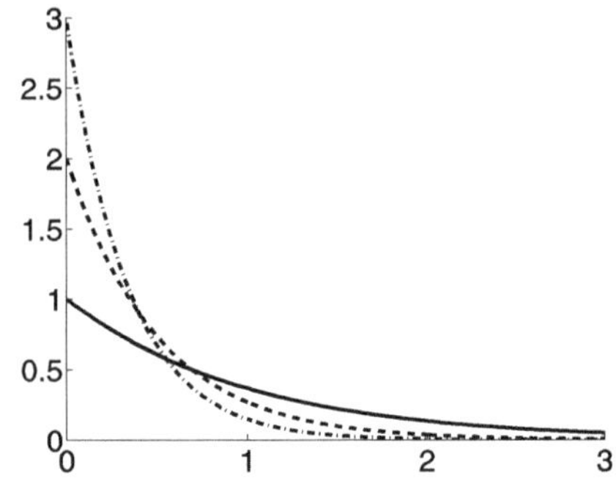

Fig. 4.8 The exponential pdf in (4.71) plotted for three different values of α, which is equal 1, 2, or 3. The value of α is equal to the value of the pdf at zero

where the constant is not important because it does not depend on α. So, we have for a sample

$$\log p\big(z(1), z(2), \ldots, z(n) \mid \alpha\big) = -\frac{1}{2}\sum_{i=1}^{n}\big(z(i)-\alpha\big)^2 + \text{const.} \tag{4.69}$$

It can be shown (this is left as an exercise) that this is maximized by

$$\hat{\alpha} = \frac{1}{n}\sum_{i=1}^{n} z(i). \tag{4.70}$$

Thus, the maximum likelihood estimator is given by the average of the observed values. This is not a trivial result: in some other models, the maximum likelihood estimator of such a location parameter is given by the median.

Example 15 Here's an example of maximum likelihood estimation with a less obvious result. Consider the exponential distribution

$$p(z|\alpha) = \alpha \exp(-\alpha z) \tag{4.71}$$

where z is constrained to be positive. The parameter α determines how likely large values are and what the mean is. Some examples of this pdf are shown in Fig. 4.8. The log-pdf is given by

$$\log p(z|\alpha) = \log \alpha - \alpha z \tag{4.72}$$

so the log-likelihood for a sample equals

$$\log p\big(z(1), z(2), \ldots, z(n) \mid \alpha\big) = n \log \alpha - \alpha \sum_{i=1}^{n} z(i). \tag{4.73}$$

To solve for the α which maximizes the likelihood, we take the derivative of this with respect to α and find the point where it is zero. This gives

$$\frac{n}{\alpha} - \sum_{i=1}^{n} z(i) = 0 \tag{4.74}$$

from which we obtain

$$\hat{\alpha} = \frac{1}{\frac{1}{n}\sum_{i=1}^{n} z(i)}. \tag{4.75}$$

So, the estimate is the reciprocal of the mean of the z in the sample.

4.8.3 *Prior and Large Samples*

If the prior is not flat, we have the log-posterior

$$\log p(\alpha \mid z(1), z(2), \ldots, z(n)) = \log p(\alpha) + \log p(z(1) \mid \alpha) + \log p(z(2) \mid \alpha) + \cdots$$
$$+ \log p(z(n) \mid \alpha) + \text{const.} \tag{4.76}$$

which usually needs to be maximized numerically.

Looking at (4.76), we see an interesting phenomenon: when n grows large, the prior loses its significance. There are more and more terms in the likelihood part, and, eventually, they will completely determine the posterior because the single prior term will not have any influence anymore. In other words, when we have a very large sample, the data outweighs the prior information. This phenomenon is related to the learning interpretation we discussed above: the organism eventually learns so much from the incoming data that the prior belief it had in the very beginning is simply forgotten.

4.9 References

Most of the material in this chapter is very classic. Most of it can be found in basic textbooks to probability theory, while Sect. 4.8 can be found in introductory textbooks to the theory of statistics. A textbook covering both areas is Papoulis and Pillai (2001). Some textbooks on probabilistic machine learning also cover all this material, in particular Mackay (2003), Bishop (2006).

4.10 Exercises

Mathematical Exercises

1. Show that a conditional pdf as defined in (4.15) is properly normalized, i.e. its integral is always equal to one.
2. Compute the mean and variance of a random variable distributed uniformly in the interval $[a, b]$ $(b > a)$.
3. Consider n scalar random variables x_i, $i = 1, 2, \ldots, n$, having, respectively, the variances $\sigma_{x_i}^2$. Show that if the random variables x_i are all uncorrelated, the vari-

ance of their sum equals the sum of their variances

$$\sigma_y^2 = \sum_{i=1}^{n} \sigma_{x_i}^2. \tag{4.77}$$

4. Assume the random vector $\mathbf{x}$ has uncorrelated variables, all with unit variance. Show that the covariance matrix equals the identity matrix.
5. Take a linear transformation of $\mathbf{x}$ in the preceding exercise: $\mathbf{y} = \mathbf{Mx}$ for some matrix $\mathbf{M}$. Show that the covariance matrix of $\mathbf{y}$ equals $\mathbf{MM}^T$.
6. Show that the maximum likelihood estimator of the mean of a Gaussian distribution equals the sample average, i.e. (4.70).
7. Next we consider estimation of the variance parameter in a Gaussian distribution. We have the pdf

$$p(z \mid \sigma) = \frac{1}{\sqrt{2\pi}\sigma} \exp\left(-\frac{z^2}{2\sigma^2}\right). \tag{4.78}$$

   Formulate the likelihood and the log-likelihood, given a sample $z(1), \ldots, z(n)$. Then find the maximum likelihood estimator for σ.

Computer Assignments

1. Generate 1000 samples of 100 independent observations of a Gaussian variable of zero mean and unit variance (e.g. with Matlab's randn function). That is, you generate a matrix of size 1000×100 whose all elements are all independent Gaussian observations.
   a. Compute the average of each sample. This is the maximum likelihood estimator of the mean for that sample.
   b. Plot a histogram of the 1000 sample averages.
   c. Repeat all the above, increasing the sample size to 1000 and to 10 000.
   d. Compare the three histograms. What is changing?
2. Generate a sample of 10 000 observations of a two-dimensional random vector $\mathbf{x}$ with independent standardized Gaussian variables. Put each observation in a column and each random variable in a row, i.e. you have a $2 \times 10\,000$ matrix, denote it by $\mathbf{X}$.
   a. Compute the covariance matrix of this sample of $\mathbf{x}$, e.g. by using the cov function in Matlab. Note that the transpose convention in Matlab is different from what we use here, so you have to apply the cov function of the transpose of $\mathbf{X}$. Compare the result with the theoretical covariance matrix (what is its value?)
   b. Multiply $\mathbf{x}$ (or in practice, $\mathbf{X}$) from the left with the matrix

$$\mathbf{A} = \begin{pmatrix} 2 & 3 \\ 0 & 1 \end{pmatrix}. \tag{4.79}$$

   Compute the covariance matrix of $\mathbf{Ax}$. Compare with $\mathbf{AA}^T$.

Part II
Statistics of Linear Features

Chapter 5
Principal Components and Whitening

The most classical method of analyzing the statistical structure of multi-dimensional random data is principal component analysis (PCA), which is also called the Karhunen–Loève transformation, or the Hotelling transformation. In this chapter, we will consider the application of PCA to natural images. It will be found that it is not a successful model in terms of modeling the visual system. However, PCA provides the basis for all subsequent models. In fact, before applying the more successful models described in the following chapters, PCA is often applied as a preprocessing of the data. So, the investigation of the statistical structure of natural images must start with PCA.

Before introducing PCA, however, we will consider a very simple and fundamental concept: the DC component.

5.1 DC Component or Mean Grey-Scale Value

To begin with, we consider a simple property of an image patch: its DC component. The DC component refers to the mean grey-scale value of the pixels in an image or an image patch.[1] It is often assumed that the DC component does not contain interesting information. Therefore, it is often removed from the image before any further processing to simplify the analysis. Removing the DC component thus means that we preprocess each image (in practice, image patch) as follows

$$\tilde{I}(x,y) = I(x,y) - \frac{1}{m} \sum_{x',y'} I(x',y') \tag{5.1}$$

where m is the number of pixels. All subsequent computations would then use $\tilde{I}$.

In Sect. 1.8, we looked at the outputs of some simple feature detectors when the input is natural images. Let us see what the effect of DC removal is on the statistics of these features; the features are depicted in Fig. 1.10 on page 19. Let us denote the output of a linear feature detector with weights $W_i(x,y)$ by s:

$$s_i = \sum_{x,y} W_i(x,y) I(x,y). \tag{5.2}$$

[1] The name "DC" comes from a rather unrelated context in electrical engineering, in which it originally meant "direct current" as opposed to "alternating current". The expression has become rather synonymous with "constant" in electrical engineering.

A. Hyvärinen, J. Hurri, P.O. Hoyer, *Natural Image Statistics*,
Computational Imaging and Vision 39,

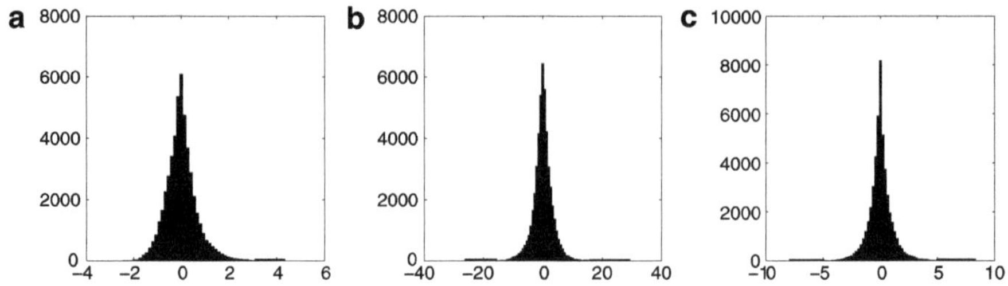

Fig. 5.1 Effect of DC removal. These are histograms of the outputs of the filters in Fig. 1.10 when the output is natural images whose DC component has been removed. *Left*: output of Dirac filter, which is the same as the histogram of the original pixels themselves. *Center*: output of grating feature detector. *Right*: output of edge detector. The scales of the axes are different from those in Fig. 1.10

The ensuing histograms of the s_i, for the three detectors when input with natural images, and after DC removal, are shown in Fig. 5.1. Comparing with Fig. 1.11 on page 19, we can see that the first histogram changes radically, where as the latter two do not. This is because the latter two filters were not affected by the DC component in the first place, which is because the sum of their weight was approximately zero: $\sum_{x,y} W(x, y) = 0$. Actually, the three histograms are now more similar to each other: the main difference is in the scale. However, they are by no means identical, as will be seen in the analyses of this book.

The effect of DC component removal depends on the size of the image patch. Here, the patches were relatively small, so the removal had a large effect on the statistics. In contrast, removing the DC component from whole images has little effect on the statistics.

In the rest of this book, we will assume that the DC component has been removed unless otherwise mentioned. Removing the DC component also means that the mean of any s is zero; this is intuitively rather obvious but needs some assumptions to be shown rigorously (see Exercises). Thus, in what follows, we shall assume that the mean of any feature s is zero.

Some examples of natural image patches with DC component removed are shown in Fig. 5.2. This is the kind of data analyzed in almost all of the rest of this book.

5.2 Principal Component Analysis

5.2.1 A Basic Dependency of Pixels in Natural Images

The point in PCA is to analyze the dependencies of the pixel grey-scale values $I(x, y)$ and $I(x', y')$ for two different pixel coordinate pairs (x, y) and (x', y'). More specifically, PCA considers the second-order structure of natural images, i.e. the variances and covariances of pixel values $I(x, y)$.

If the pixel values were all uncorrelated, PCA would have nothing to analyze. Even a rudimentary analysis of natural images shows, however, that the pixel values

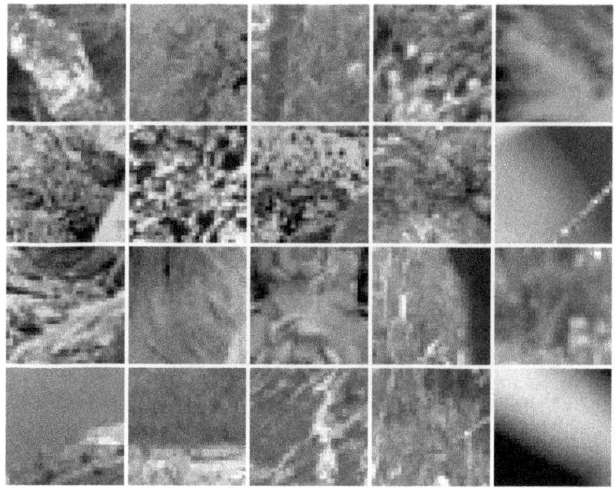

Fig. 5.2 Some natural image patches, with DC component removed

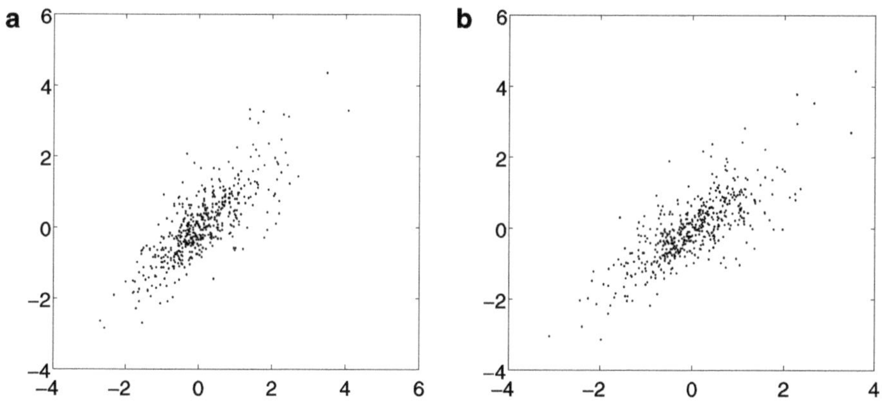

Fig. 5.3 Scatter plot of the grey-scale values of two neighboring pixels. **a** Original pixel values. The values have been scaled so that the mean is zero and the variance one. **b** Pixel values after removal of DC component in a 32 × 32 patch

are far from independent. It is intuitively rather clear that natural images are typically *smooth* in the sense that quite often, the pixel values are very similar in two near-by pixels. This can be easily demonstrated by a scatter plot of the pixel values for two neighboring pixels sampled from natural images. This is shown in Fig. 5.3. The scatter plot shows that the pixels are correlated. In fact, we can compute the correlation coefficient (equation (4.39)), and it turns out to be approximately equal to 0.9.

Actually, we can easily compute the correlation coefficients of a single pixel with all near-by pixels. Such a plot is shown in grey-scale in Fig. 5.4, both without removal of DC component (in a) and with DC removal (in b). We see that the correlation coefficients (and thus, the covariances) fall off with increasing distance. These two plots, with or without DC removal, look rather similar because the plots use

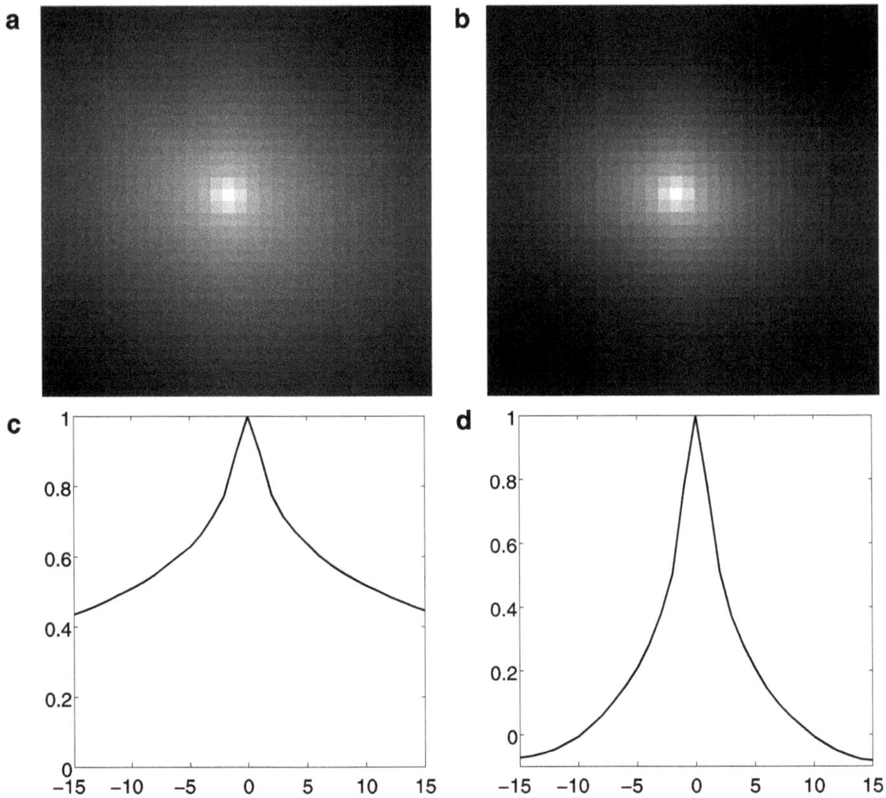

Fig. 5.4 The correlation coefficients of a pixel (in the middle) with all other pixels. **a** For original pixels. *Black* is small positive *white* is one. **b** After removing DC component. The scale is different from **a**: *black* is now negative and *white* is plus one. **c** A cross-section of **a** in 1D to show the actual values. **d** A cross-section of **b** in 1D

different scales; the actual values are quite different. We can take one-dimensional cross-sections to see the actual values. They are shown in Fig. 5.4c and d. We see that without DC removal, all the correlation coefficients are strongly positive. Removing the DC components reduces the correlations to some extent, and introduces negative correlations.

5.2.2 Learning One Feature by Maximization of Variance

5.2.2.1 Principal Component as Variance-Maximizing Feature

The covariances found in natural images pixel values can be analyzed by PCA. In PCA, the point is to find linear features that explain most of the variance of the data.

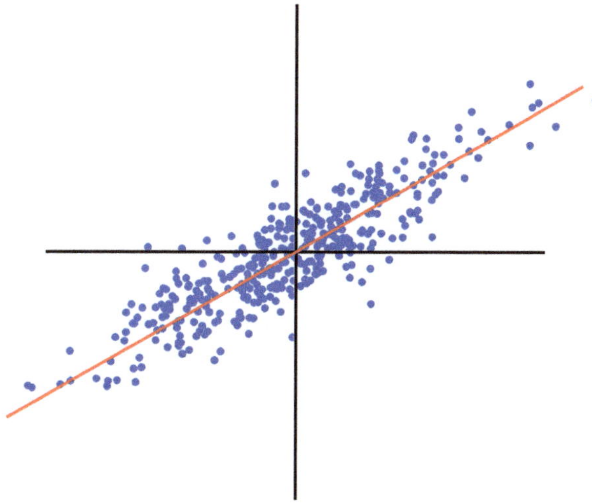

Fig. 5.5 Illustration of PCA. The principal component of this (artificial) two-dimensional data is the oblique axis plotted. Projection on the principal axis explains more of the variance of the data than projection on any other axis

It is natural to start the definition of PCA by looking at the definition of the first principal component. We consider the variance of the output:

$$\text{var}(s) = E\{s^2\} - \left(E\{s\}\right)^2 = E\{s^2\} \tag{5.3}$$

where the latter equality is true because s has zero mean.

Principal components are features s that contain (or "explain") as much of the variance of the input data as possible. It turns out that the amount of variance explained is directly related to the variance of the feature, as will be discussed in Sect. 5.3.1 below. Thus, the first principal component is defined as the feature, or linear combination of the pixel values, which has the maximum variance. Finding a feature with maximum variance can also be considered interesting in its own right. The idea is to find the "main axis" of the data cloud, which is illustrated in Fig. 5.5.

Some constraint on the weights W, which we call the *principal component weights*, must be imposed as well. If no constraint were imposed, the maximum of the variance would be attained when W becomes infinitely large (and the minimum would be attained when all the $W(x,y)$ are zero). In fact, just multiplying all the weights in W by a factor of two, we would get a variance that is four times as large, and by dividing all the coefficients by two the variance decreases to one quarter.

A natural thing to do is to constrain the norm of W:

$$\|W\| = \sqrt{\sum_{x,y} W(x,y)^2}. \tag{5.4}$$

For simplicity, we constrain the norm to be equal to one, but any other value would give the same results.

Fig. 5.6 The feature detectors giving the first principal component of image windows of size 32 × 32, computed for ten different randomly sampled datasets taken from natural images. The feature detector is grey-scale-coded so that the grey-scale value of a pixel gives the value of the coefficient at that pixel. *Grey* pixels mean zero coefficients, *light-grey* or *white* pixels are positive, and *dark-grey* or *black* are negative

5.2.2.2 Learning One Feature from Natural Images

What is then the feature detector that maximizes the variance of the output, given natural image input, and under the constraint that the norm of the detector weights equals one? We can find the solution by taking a random sample of image patches. Let us denote by T the total number of patches used, and by I_t each patch, where t is an index that goes from 1 to T. Then, the expectation of s^2 can be approximated by the average over this sample. Thus, we maximize

$$\frac{1}{T}\sum_{t=1}^{T}\left(\sum_{x,y} W(x,y)I_t(x,y)\right)^2 \tag{5.5}$$

with respect to the weights in $W(x, y)$, while constraining the values of $W(x, y)$ so that the norm in (5.4) is equal to one. The computation of the solution is discussed in Sect. 5.2.4.

Typical solutions for natural images are shown in Fig. 5.6. The feature detector is an object of the same size and shape as an image patch, so it can be plotted as an image patch itself. To test whether the principal component weights are stable, we computed it ten times for different image samples. It can be seen that the component is quite stable.

5.2.3 Learning Many Features by PCA

5.2.3.1 Defining Many Principal Components

One of the central problems with PCA is that it basically gives only one well-defined feature. It cannot be extended to the learning of many features in a very meaningful way. However, if we want to model visual processing by PCA, it would be absurd to compute just a single feature which would then be supposed to analyze the whole visual scene.

Definition Typically, the way to obtain many principal components is by a "deflation" approach: After estimating the first principal component, we want to find the feature of maximum variance *under the constraint* that the new feature must be *orthogonal* to the first one (i.e. the dot-product is zero, as in (5.8)). This will then be called the second principal component. This procedure can be repeated to obtain as many components as there are dimensions in the data space. To put this formally, assume that we have estimated k principal components, given by the weight vectors $W_1, W_2, \ldots, W_k$. Then the $k+1$-th principal component weight vector is defined by

$$\max_W \operatorname{var}\left(\sum_{x,y} I(x,y)W(x,y)\right) \tag{5.6}$$

under the constraints

$$\|W\| = \sqrt{\sum_{x,y} W(x,y)^2} = 1, \tag{5.7}$$

$$\sum_{x,y} W_j(x,y)W(x,y) = 0 \quad \text{for all } j = 1, \ldots, k. \tag{5.8}$$

An interesting property is that any two principal components are uncorrelated, and not only orthogonal. In fact, we could change the constraint in the definition to uncorrelatedness, and the principal components would be the same.

Critique of the Definition This classic definition of many principal components is rather unsatisfactory, however. There is no really good justification for thinking that the second principal component corresponds to something interesting: it is not a feature that maximizes any property in itself. It only maximizes the variance not explained by the first principal component.

Moreover, the solution is not quite well defined, since for natural images, there are many principal components that give practically the same variance. After the first few principal components, the differences of the variances of different directions get smaller and smaller. This is a serious problem for the following reason. If two principal components, say s_i and s_j, have the same variance, then any linear combination $q_1 s_i + q_2 s_j$ has the same variance as well,[2] and the weights $q_1 W_i(x,y) + q_2 W_j(x,y)$ fulfill the constraints of unit norm and orthogonality, if we normalize the coefficients q_1 and q_2 so that $q_1^2 + q_2^2 = 1$. So, not only we cannot say what is the order of the components, but actually there is an infinite number of different components from which we cannot choose the "right" one.

In practice, the variances of the principal components are not exactly equal due to random fluctuations, but this theoretical result means that the principal components are highly dependent on those random fluctuations. In the particular sample of natural images that we are using, the maximum variance (orthogonal to previous components) can be obtained by any of these linear combinations. Thus, we cannot really say what the 100th principal component is, for example, because the result we get from computing it depends so much on random sampling effects. This will be demonstrated in the experiments that follow.

---

[2]This is due to the fact that the principal components are uncorrelated; see Sect. 5.8.1.

Fig. 5.7 The 320 first principal components weights W_i of image patches of size 32×32. The order of decreasing variances is left to right on each row, and top to bottom

5.2.3.2 All Principal Components of Natural Images

The first 320 principal components of natural images patches are shown in Fig. 5.7, while Fig. 5.8 shows the variances of the principal components. For lack of space, we don't show all the components, but it is obvious from the figure what they look like. As can be seen, the first couple of features seem quite meaningful: they are oriented, something like very low-frequency edge detectors. However, most of the

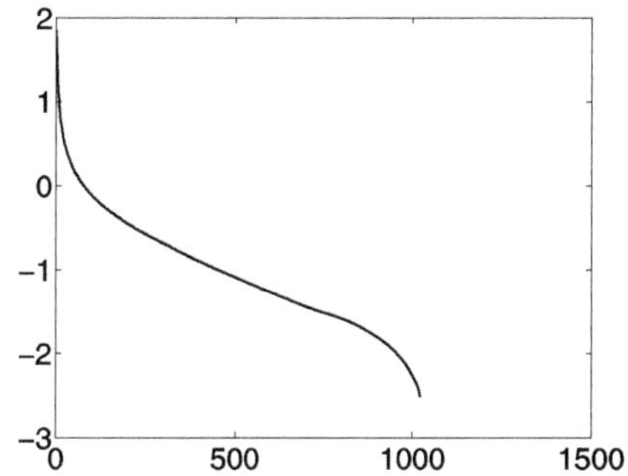

Fig. 5.8 The logarithms of the variances of the principal components for natural image patches, the first of which were shown in Fig. 5.7

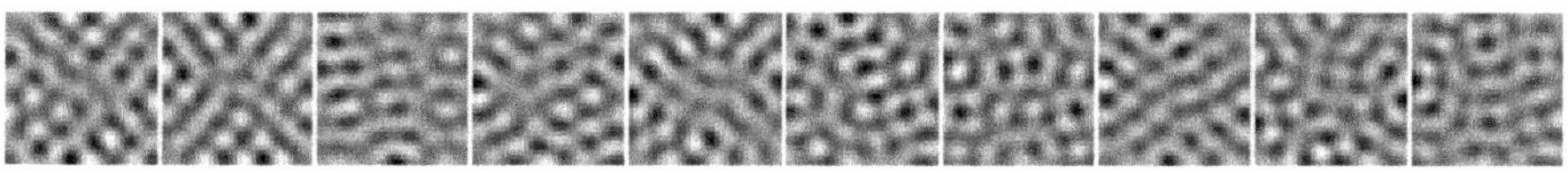

Fig. 5.9 Ten different estimations of the 100th principal component of image windows of size 32×32. The random image sample was different in each run

features given by PCA do not seem to be very interesting. In fact, after the first, say, 50 features, the rest seem to be just garbage. They are localized in frequency as they clearly are very high-frequency features. However, they do not seem to have any meaningful spatial structure. For example, they are not oriented.

In fact, most of the features do not seem to be really well defined for the reason explained in the previous section: the variances are too similar for different features. For example, some of the possible 100th principal components, for different random sets of natural image patches, are shown in Fig. 5.9. The random changes in the component are obvious.

5.2.4 Computational Implementation of PCA

In practice, numerical solution of the optimization problem which defines the principal components is rather simple and based on what is called the *eigenvalue decomposition*. We will not go into the mathematical details here; they can be found in Sect. 5.8.1. Briefly, the computation is based on the following principles

1. The variance of any linear feature as in (5.2) can be computed if we just know the variances and covariances of the image pixels.
2. We can collect the variances and covariances of image pixels in a single matrix, called the *covariance matrix*, as explained in Sect. 4.6.3. Each entry in the matrix then gives the covariance between two pixels—variance is simply covariance of a pixel with itself.
3. Any sufficiently sophisticated software for scientific computation is able to compute the eigenvalue decomposition of that matrix. (However, the amount of com-

putation needed grows fast with the size of the image patch, so the patch size cannot be too large.)

4. As a result of the eigenvalue decomposition we get two things. First, the *eigenvectors*, which give the W_i which are the principal component weights. Second, the *eigenvalues*, which give the variances of the principal components s_i. So, we only need to order the eigenvectors in the order of descending eigenvalues, and we have computed the whole PCA decomposition.

5.2.5 The Implications of Translation-Invariance

Many of the properties of the PCA of natural images are due a particular property of the covariance matrix of natural images. Namely, the covariance for natural images is translation-invariant, i.e. it depends only on the distance

$$\operatorname{cov}\big(I(x, y), I(x', y')\big) = f\big((x - x')^2 + (y - y')^2\big) \tag{5.9}$$

for some function f. After all, the covariance of two neighboring pixels is not likely to be any different depending on whether they are on the left or the right side of the image. (This form of translation-invariance should not be confused with the invariances of complex cells, discussed in Sect. 3.4.2 and Chap. 10.)

The principal component weights $W_i(x, y)$ for a covariance matrix of this form can be shown to always have a very particular form: they are sinusoids:

$$W_i(x, y) = \sin(ax + by + c) \tag{5.10}$$

for some constants a, b, c (the scaling is arbitrary so you could also multiply W_i with a constant d). See Sect. 5.8.2 below for a detailed mathematical analysis which proves this.

The constants a and b determine the frequency of the oscillation. For example, the first principal components have lower frequencies than the later ones. They also determine the orientation of the oscillation. It can be seen in Fig. 5.7 that some of the oscillations are oblique while others are vertical and horizontal. Some have oscillations in one orientation only, while others form a kind of checkerboard pattern. The constant c determines the phase.

The variances associated with the principal components thus tell how strongly the frequency of the sinusoid is present in the data, which is closely related to computing the power spectrum of the data.

Because of random effects in the sampling of image patches and computation of the covariance matrix, the estimated feature weights are not exactly sinusoids. Of course, just the finite resolution of the images makes them different from real sinusoidal functions.

Since the principal component weights are sinusoids, they actually perform some kind of Fourier analysis. If you apply the obtained W_i as feature detectors on an image patch, that will be related to a discrete Fourier transform of the image patch.

In particular, this can be interpreted as computing the coefficients of the patch in the basis given by sinusoidal functions, as discussed in Sect. 2.3.2.

An alternative viewpoint is that you could also consider the computation of the principal components as doing a Fourier analysis of the covariance matrix of the image patch; this interesting connection will be considered in Sect. 5.6.

5.3 PCA as a Preprocessing Tool

So far, we have presented PCA as a method for learning features, which is the classic approach to PCA. However, we saw that the results were rather disappointing in the sense that the features were not interesting as neurophysiological models, and they were not even well defined.

However, PCA is not a useless model. It accomplishes several useful *preprocessing* tasks, which will be discussed in this section.

5.3.1 Dimension Reduction by PCA

One task where PCA is very useful is in reducing the dimension of the data so that the maximum amount of the variance is preserved.

Consider the following general problem that also occurs in many other areas than image processing. We have a very large number, say m, of random variables $x_1, \ldots, x_m$. Computations that use all the variables would be too burdensome. We want to reduce the dimension of the data by linearly transforming the variables into a smaller number, say n, of variables that we denote by $z_1, \ldots, z_n$:

$$z_i = \sum_{j=1}^{m} w_{ij} x_j, \quad \text{for all } i = 1, \ldots, n. \tag{5.11}$$

The number of new variables n might be only 10% or 1% of the original number m. We want to find the new variables so that they preserve as much information on the original data as possible. This "preservation of information" has to be exactly defined. The most wide-spread definition is to look at the squared error that we get when we try to reconstruct the original data using the z_i. That is, we reconstruct x_j as a linear transformation $\sum_i a_{ji} z_i$, minimizing the average error

$$E\left\{\sum_j \left(x_j - \sum_i a_{ji} z_i\right)^2\right\} = E\left\{\left\|\mathbf{x} - \sum_i \mathbf{a}_i z_i\right\|^2\right\} \tag{5.12}$$

where the a_{ji} are also determined so that they minimize this error. For simplicity, let us consider only transformations for which the transforming weights are orthogonal and have unit norm:

$$\sum_j w_{ij}^2 = 1, \quad \text{for all } i, \tag{5.13}$$

$$\sum_j w_{ij} w_{kj} = 0, \quad \text{for all } i \neq k. \tag{5.14}$$

What is the best way of doing this dimension reduction? The solution is to take as the z_i the n first principal components! (A basic version of this result is shown in the exercises.) Furthermore, the optimal reconstruction weight vectors $\mathbf{a}_i$ in (5.12) are given by the very same principal components weights which compute the z_i.

The solution is not uniquely defined, though, because any orthogonal transformation of the z_i is just as good. This is understandable because any such transformation of the z_i contains just the same information: we can make the inverse transformation to get the z_i from the transformed ones.

As discussed above, the features given by PCA suffer from the problem of not being uniquely defined. This problem is much less serious in the case of dimension reduction. What matters in the dimension reduction context is not so much the actual components themselves, but the *subspace* which they span. The *principal subspace* means the set of all possible linear combinations of the n first principal components. It corresponds to taking all possible linear combinations of the principal component weight vectors W_i associated with the n principal components. As pointed out above, if two principal components s_i and s_j have the same variance, any linear combination $q_1 s_i + q_2 s_j$ has the same variance for $q_1^2 + q_2^2 = 1$. This is not a problem here, however, since such a linear combination still belongs to the same subspace as the two principal components s_i and s_j. Thus, it does not matter if we consider the components s_i and s_j, or two components of the form $q_1 s_i + q_2 s_j$ and $r_1 s_i + r_2 s_j$ where the coefficients r_1 and r_2 give a different linear combination than the q_1 and q_2.

So, the n-dimensional principal subspace is usually uniquely defined even if some principal components have equal variances. Of course, it may happen that the nth and the $(n+1)$-th principal components have equal variances, and that we cannot decide which one to include in the subspace. But the effect on the whole subspace is usually quite small and can be ignored in practice.

Returning to the case of image data, we can rephrase this result by saying that it is the *set* of features defined by the n first principal components and their linear combinations that is (relatively) well defined, and not the features themselves.

5.3.2 Whitening by PCA

5.3.2.1 Whitening as Normalized Decorrelation

Another task for which PCA is quite useful is whitening. Whitening is an important preprocessing method where the image pixels are transformed to a set of new variables $s_1, \ldots, s_n$ so that the s_i are uncorrelated and have unit variance:

$$E\{s_i s_j\} = \begin{cases} 0 & \text{if } i \neq j, \\ 1 & \text{if } i = j. \end{cases} \tag{5.15}$$

(It is assumed that all the variables have zero mean.) It is also said that the resulting vector $(s_1, \ldots, s_n)$ is white.

In addition to the principal components weights being orthogonal, the principal components themselves are uncorrelated, as will be shown in more detail in Sect. 5.8.1. So, after PCA, the only thing we need to do to get whitened data is to normalize the variances of the principal components by dividing them by their standard deviations. Denoting the principal components by y_i, this means we compute

$$s_i = \frac{y_i}{\sqrt{\text{var}(y_i)}} \tag{5.16}$$

to get whitened components s_i. Whitening is a useful preprocessing method that will be used later in this book. The intuitive idea is that it completely removes the second-order information of the data. "Second-order" means here correlations and variances. So, it allows us to concentrate on properties that are not dependent on covariances, such as sparseness in the next chapter.

Whitening by PCA is illustrated in Fig. 5.10.

5.3.2.2 Whitening Transformations and Orthogonality

It must be noted that there are many whitening transformations. In fact, if the random variables s_i, $i = 1, \ldots, n$ are white, then any *orthogonal transformation* of those variables is also white (the proof is left as an exercise). Often, whitening is based on PCA because PCA is a well-known method that can be computed very fast, but it must be kept in mind that PCA is just one among the many whitening transformations. Yet, PCA is a unique method because it allows us to combine three different preprocessing methods into one: dimension reduction, whitening, and anti-aliasing (which will be discussed in the next section).

In later chapters, we will often use the fact that the connection between orthogonality and uncorrelatedness is even stronger for whitened data. In fact, if we compute two linear components $\sum_i v_i s_i$ and $\sum_i w_i s_i$ from white data, they are uncorrelated *only* if the two vectors $\mathbf{v}$ and $\mathbf{w}$ (which contain the entries v_i and w_i, respectively) are orthogonal.

In general, we have the following theoretical result. For white data, multiplication by a square matrix gives white components if *and only if* the matrix is orthogonal. Thus, when we have computed one particular whitening transformation, we also know that *only* orthogonal transformations of the transformed data can be white.

Note here the tricky point in terminology: a matrix if called orthogonal if its columns, or equivalently its rows, are orthogonal, *and* the norms of its columns are all equal to one. To emphasize this, some authors call an orthogonal matrix ortho*normal*. We stick to the word "orthogonal" in this book.

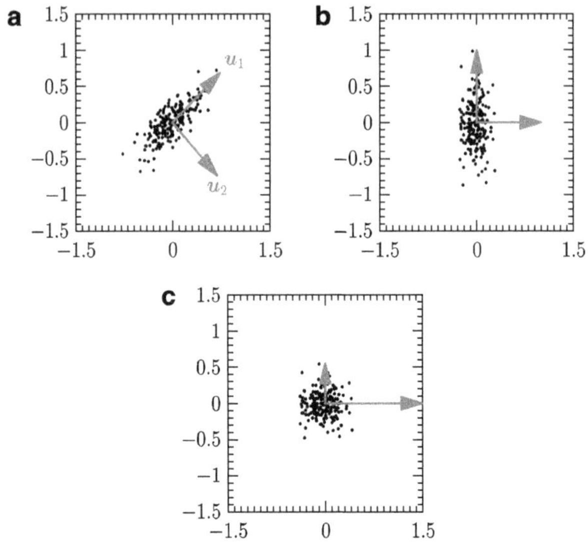

Fig. 5.10 Illustration of PCA and whitening. **a** The original data "cloud". The *arrows* show the principal components. The first one points in the direction of the largest variance in the data, and the second in the remaining orthogonal direction. **b** When the data is transformed to the principal components, i.e. the principal components are taken as the new coordinates, the variation in the data is aligned with those new axes, which is because the principal components are uncorrelated. **c** When the principal components are further normalized to unit variance, the data cloud has equal variance in all directions, which means it has been whitened. The change in the lengths of the arrows reflects this normalization; the larger the variance, the shorter the arrow

5.3.3 Anti-aliasing by PCA

PCA also helps combat the problem of *aliasing*, which refers to a class of problems due to the sampling of the data at a limited resolution—in our case the limited number of pixels used to represent an image. Sampling of the image loses information; this is obvious since we only know the image through its values at a finite number of pixels. However, sampling can also introduce less obvious distortions in the data. Here, we consider two important ones, and show how PCA can help.

5.3.3.1 Oblique Gratings Can Have Higher Frequencies

One problem is that in the case of the rectangular sampling grid, oblique higher frequencies are overrepresented in the data, because the grid is able to represent oblique oscillations which have a higher frequency than either the vertical or horizontal oscillations of the highest possible frequency.

This is because we can have an image which takes the form of a checkerboard as illustrated in Fig. 5.11a. If you draw long oblique lines along the white and black squares, the distance between such lines is equal to $\sqrt{1/2}$ as can be calculated by

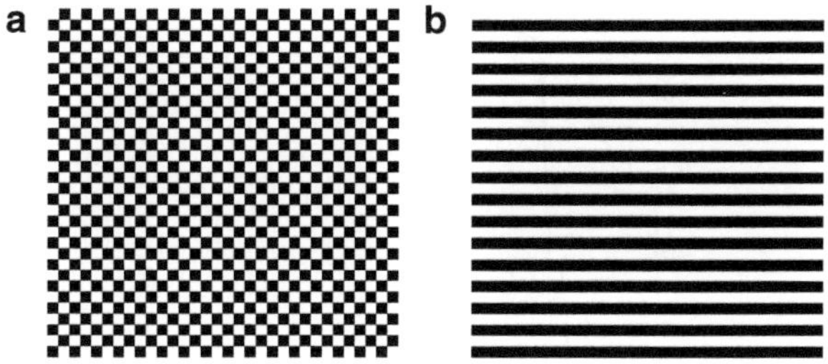

Fig. 5.11 Effects of sampling (limited number of pixels) on very high-frequency gratings. **a** A sinusoidal grating which has a very high frequency in the oblique orientation. The cycle of the oscillation has a length of $2\sqrt{1/2} = \sqrt{2}$ which is shorter than the smallest possible cycle length (equal to two) in the vertical and horizontal orientations. **b** A sinusoidal grating which has the Nyquist frequency. Although it is supposed to be sinusoidal, due to the limited sampling (i.e. limited number of pixels), it is really a block grating

basic trigonometry. This is smaller than one, which is the shortest half-cycle (half the length of an oscillation) we can have in the vertical and horizontal orientation. (It corresponds to the Nyquist frequency as discussed in the next subsection, and illustrated in Fig. 5.11b.)

In the Fourier transform, this lack of symmetry is seen in the fact that the area of possible 2-D frequencies is of the form of a square, instead of a circle as would be natural for data which is the same in all orientations ("rotation-invariant", as will be discussed in Sect. 5.7). Filtering out the highest oblique frequencies is thus a meaningful preprocessing step to avoid any artefacts due to this aliasing phenomenon. (Note that we are here talking about the rectangular form of the sampling grid, i.e. the relation of the pixels center-points to each other. This is not at all related to the shape of the sampling window, i.e. the shape of the patch.)

It turns out that we can simply filter out the oblique frequencies by PCA. With natural images, the last principal components are those that correspond to the highest oblique frequencies. Thus, simple dimension reduction by PCA alleviates this problem.

5.3.3.2 Highest Frequencies Can Have only Two Different Phases

Another problem is that at the highest frequencies, we cannot have sinusoidal gratings with different phases. Let us consider the highest possible frequency, called in Fourier theory the Nyquist frequency, which means that there is one cycle for every two pixels; see Fig. 5.11b). What happens when you change the phase of the grating a little bit, i.e. shift the "sinusoidal" grating a bit? Actually, almost nothing happens: the grating does not shift at all because due to the limited resolution given by the pixel size, it is impossible to represent a small shift in the grating. (The grey-scale values will be changed; they depend on the match between the sampling lattice and the underlying sinusoidal they try to represent.) The sampled image really changes only when the phase is changed so much that the best approximation is to flip all the pixels from white to black and vice versa.

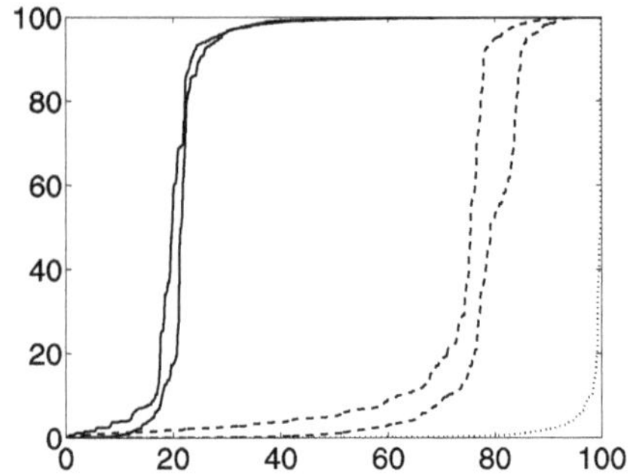

Fig. 5.12 The percentage of different frequencies present in the data as a function of PCA dimension reduction. *Horizontal axis*: Percentage of dimensions retained by PCA. *Vertical axis*: Percentage of energy of a given grating retained. *Solid lines*: Gratings of half Nyquist frequency (vertical and oblique) (wanted). *Dotted line* (see *lower right-hand corner*): checkerboard pattern (unwanted). *Dashed lines*: Gratings of Nyquist frequency (vertical and oblique) (unwanted)

Thus, a grating sampled at the Nyquist frequency can really have only two different phases which can be distinguished. This means that many concepts depending on the phase of the grating, such as the phase tuning curve (Sect. 6.4) or phase-invariance (Chap. 10) are rather meaningless on the Nyquist frequency. So, we would like to low-pass filter the image to be able to analyze such phenomena without the distorting effect of a limited resolution.

Again, we can alleviate this problem by PCA dimension reduction because it amounts to discarding the highest frequencies.

5.3.3.3 Dimension Selection to Avoid Aliasing

So, we would like to do PCA so that we get rid of the checkerboard patterns as well as everything in the Nyquist frequency. On the other hand, we don't want to get rid of any lower frequencies.

To investigate the dimension reduction needed, we computed what amount of checkerboard and Nyquist gratings is present in the data as a function of dimension after PCA. We also computed this for gratings that had half the Nyquist frequency (i.e. a cycle was four pixels), which is a reasonable candidate for the highest frequency patterns that we want to retain.

The results are shown in Fig. 5.12. We can see in the figure that to get rid of checkerboard patterns, not much dimension reduction is necessary: 10% or so seems to be enough.[3] To get rid of the Nyquist frequency, at least 30% seems to be necessary. And if we look at how much we can reduce the dimension without losing any

[3]Note that this may be an underestimate: van Hateren proposed that 30% may be needed (van Hateren and van der Schaaf 1998). This is not important in the following because we will anyway reduce at least 30% for other reasons that will be explained next.

information on the lowest frequencies that we are really interested in, it seems we can easily reduce even 60%–70% of the dimensions.

Thus, the exact number of dimensions is not easy to determine because we don't have a very clear criterion. Nevertheless, a reduction of at least 30% seems to be necessary to avoid the artifacts, and even 60%–70% can be recommended. In the experiments in this book, we usually reduce dimension by 75%.

5.4 Canonical Preprocessing Used in This Book

Now, we have arrived at a preprocessing method that we call "canonical preprocessing" because it is used almost everywhere in this book. Canonical preprocessing means:

1. Remove the DC component as in (5.1).
2. Compute the principal components of the image patches.
3. Retain only the n first principal components and discard the rest. The number n is typically chosen as 25% of the original dimension.
4. Divide the principal components by their standard deviations as in (5.16) to get whitened data.

Here, we see two important (and interrelated) reasons for doing whitening by PCA instead of some other whitening method. We can reduce dimension to combat aliasing and to reduce computational load with hardly any extra computational effort.

Notation The end-product of this preprocessing is an n-dimensional vector for each image patch. The preprocessed vector will be denoted by $\mathbf{z}$, and its elements by $z_1, \ldots, z_n$, when considered as a random vector and random variables. Observations of the random vector will be denoted by $\mathbf{z}_1, \mathbf{z}_2, \ldots$, or more often with the subscript t as in $\mathbf{z}_t$. In the rest of this book, we will often use such canonically preprocessed data. Likewise, observed image patches will be denoted by I, and their individual pixels by $I(x, y)$, when these are considered a random vector and random variables, respectively, and their observations will be denoted by I_t and $I_t(x, y)$.

5.5 Gaussianity as the Basis for PCA

5.5.1 The Probability Model Related to PCA

In PCA and whitening, it is assumed that the only interesting aspect of the data variables $x_1, \ldots, x_n$ is variances and covariances. This is the case with Gaussian data, where the probability density function equals

$$p(x_1, \ldots, x_n) = \frac{1}{(2\pi)^{n/2}|\det \mathbf{C}|^{-1/2}} \exp\left(-\frac{1}{2}\sum_{i,j} x_i x_j [\mathbf{C}^{-1}]_{ij}\right) \qquad (5.17)$$

where $\mathbf{C}$ is the covariance matrix of the data, $\mathbf{C}^{-1}$ is its inverse, and $[\mathbf{C}^{-1}]_{ij}$ is the i, jth element of the inverse. Thus, the probability distribution is completely characterized by the covariances (as always, the means are assumed zero here).

These covariance-based methods are thus perfectly sufficient if the distribution of the data is Gaussian. However, the distribution of image data is typically very far from Gaussian. Methods based on the Gaussian distribution thus neglect some of the most important aspects of image data, as will be seen in the next chapter.

Using the Gaussian distribution, we can also interpret PCA as a statistical model. After all, one of the motivations behind estimation of statistical models for natural images was that we would like to use them in Bayesian inference. For that, it is not enough to just have a set of features. We also need to understand how we can compute the prior probability density function $p(x_1, \ldots, x_n)$ for any given image patch.

The solution is actually quite trivial: we just plug in the covariance of the data in (5.17). There is actually no need to go through the trouble of computing PCA in order to get a probabilistic model! The assumption of Gaussianity is what gives us this simple solution.

In later chapters, we will see the importance of the assumption of Gaussianity, as we will consider models which do not make this assumption.

5.5.2 PCA as a Generative Model

A more challenging question is how to interpret PCA as a generative model, i.e. a model which describes a process which "generated" the data. There is a large literature on such modeling, which is typically called *factor analysis*. PCA is considered a special case, perhaps the simplest one, of factor analytic models. The point is that we can express data as a linear transformation of the principal components

$$I(x, y) = \sum_i W_i(x, y) s_i. \tag{5.18}$$

What we have done here is simply to invert the transformation from the data to the principal components, so that the data is a function of the principal components and not vice versa. This is very simple because the vectors W_i are orthogonal: then the inverse of the system (matrix) they form is just the same matrix transposed, as discussed in Sect. 19.7. Therefore, the feature vectors in this generative model are just the same as the feature detector weights that we computed with PCA.

Now, we define the distribution of the s_i as follows:

1. The distribution of each s_i is Gaussian with variance equal to the variance of the ith principal component.
2. The s_i are statistically independent from each other.

This gives us, using (5.18), a proper generative model of the data. That is, the data can be seen as a function of the "hidden" variables that are now given by the principal components s_i.

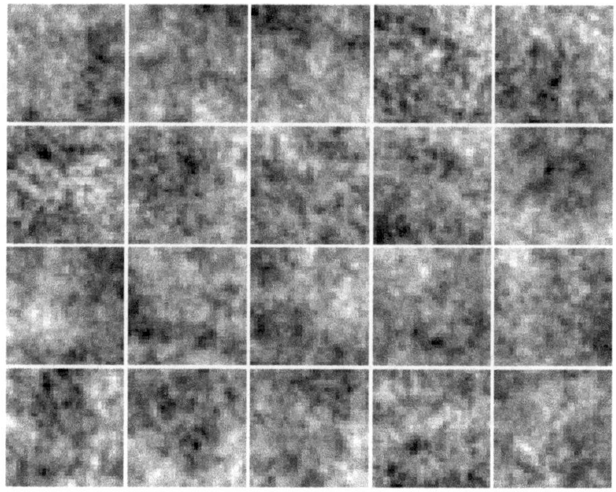

Fig. 5.13 Image synthesis using PCA. 20 patches were randomly generated using the PCA model whose parameters were estimated from natural images. Compare with real natural image patches in Fig. 5.2

5.5.3 Image Synthesis Results

Once we have a generative model, we can do an interesting experiment to test our model: We can generate image patches from our model, and see what they look like. Such results are shown in Fig. 5.13. What we see is that the PCA model captures the general smoothness of the images. The smoothness comes from the fact that the first principal components correspond to feature vectors which change very smoothly. Other structure is not easy to see in these results. The results can be compared with real natural images patches shown in Fig. 5.2 on page 95; they clearly have a more sophisticated structure, visible even in these small patches.

5.6 Power Spectrum of Natural Images

An alternative way of analyzing the covariance structure of images is through Fourier analysis. The covariances and the frequency-based properties are related via the Wiener–Khinchin theorem. We begin by considering the power spectra of natural images and then show the connection.

5.6.1 The $1/f$ Fourier Amplitude or $1/f^2$ Power Spectrum

The fundamental result on frequency-based representation of natural images is that the power spectrum of natural images typically falls off inversely proportional to the square of the frequency. Since the power spectrum is the square of the Fourier amplitude (spectrum), this means that the Fourier amplitude falls off as a function

Fig. 5.14 Two natural images used in the experiments

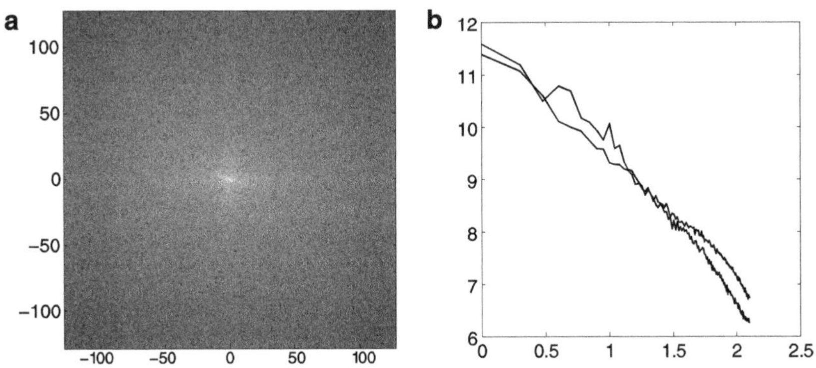

Fig. 5.15 Power spectrum or Fourier amplitude of natural images. **a** The logarithm of two-dimensional power spectrum of natural image in Fig. 5.14a. **b** The average over orientations of one-dimensional cross-sections of the power spectrum of the two images in Fig. 5.14. Only the positive part is shown since this is symmetric with respect to the origin. This is a log–log plot where a logarithm of base 10 has been taken of both the frequency (*horizontal axis*) and the power (*vertical axis*) in order to better show the $1/f^2$ behavior, which corresponds to a linear dependency with slope of -2

c/f where c is some constant and f is the frequency. It is usually more convenient to plot the logarithms. For the logarithm, this means

$$\text{Log Fourier amplitude} = -\log f + \text{const.} \tag{5.19}$$

or

$$\text{Log power spectrum} = -2\log f + \text{const.} \tag{5.20}$$

for some constant which is the logarithm of the constant c.

Figure 5.15a shows the logarithm of the power spectrum of the natural image in Fig. 5.14a. What we can see in this 2D plot is just that the spectrum is smaller for higher frequencies. To actually see how it falls off, we have to look at one-dimensional cross-sections of the power spectrum, so that we average over all orientations. This is how we get Fig. 5.15b, in which we have also taken the logarithm of the frequency as in (5.20). This plot partly verifies our result: it is largely linear with a slope close to minus two, as expected. (Actually, more thorough investigations have found that the log-power spectrum may, in fact, change a bit slower than $1/f^2$, with a exponent closer to 1.8 or 1.9.) In addition, the power spectra are very similar for the two images in Fig. 5.15.

A large literature in physics and other fields has considered the significance of such a behavior of the power spectrum. Many other kinds of data have the same kind of spectra. An important reason for this is that if the data is scale-invariant, or self-similar, i.e. it is similar whether you zoom in or out, the power spectrum is necessarily something like proportional to $1/f^2$; see References section below for some relevant work.

5.6.2 Connection between Power Spectrum and Covariances

What is then the connection between the power spectrum of an image, and the covariances between pixels we have been computing in this chapter? To this end, we need a theorem from the theory of stochastic processes (we will not rigorously define what stochastic processes are because that is not necessary for the purposes of this book). The celebrated Wiener–Khinchin theorem states that for a stochastic process, *the average power spectrum is the Fourier transform of the autocorrelation function.*

The theorem talks about the "autocorrelation function". This is the terminology of stochastic processes, which we have not used in this chapter: we simply considered different pixels as different random variables. The "autocorrelation function" means simply the correlations of variables (i.e. pixel values) as a function of the horizontal and vertical distances between them. Thus, the autocorrelation function is a matrix constructed as follows. First, take one row of the covariance matrix, say the one corresponding to the pixel at (x_0, y_0). To avoid border effects, let's take (x_0, y_0) which is in the middle of the image patch. Then convert this vector back to the shape of the image patch. Thus, we have a matrix $\mathbf{C}(x_0, y_0)$

$$\begin{pmatrix} \mathrm{cov}(I(x_0, y_0), I(1,1)) & \dots & \mathrm{cov}(I(x_0, y_0), I(1,n)) \\ \vdots & & \\ \mathrm{cov}(I(x_0, y_0), I(n,1)) & \dots & \mathrm{cov}(I(x_0, y_0), I(n,n)) \end{pmatrix} \tag{5.21}$$

which has the same size $m \times m$ as the image patch. This matrix is nothing else than what was already estimated from natural images and plotted in Fig. 5.4.

Actually, it is obvious that this matrix essentially contains all the information in the covariance matrix. As discussed in Sect. 5.2.5, it is commonly assumed that image patches are translation-invariant in the sense that the covariances actually only depend on the distance between two pixels, and not on where in the patch the pixels happen to be. (This may not hold for whole images, where the upper half may depict sky more often and lower parts, but it certainly holds for small image patches.) Thus, to analyze the covariance structure of images, all we really need is a matrix like in (5.21).

What the Wiener–Khinchin theorem now says is that when we take the Fourier transformation of $\mathbf{C}(x_0, y_0)$, just as if this matrix were an image patch, the Fourier amplitudes equal the *average* power spectrum of the original image patches. (Due to the special symmetry properties of covariances, the phases in the Fourier transform of $\mathbf{C}(x_0, y_0)$ are all zero, so the amplitudes are also equal to the coefficients of the cos functions.)

Thus, we can see the connection between the $1/f$ Fourier amplitude of natural images and the covariances of the pixels structure. The average $1/f$ Fourier amplitude or the $1/f^2$ power spectrum of single images implies that the Fourier transform of $\mathbf{C}(x_0, y_0)$ also falls of as $1/f^2$. Now, since the features obtained from PCA are not very different from those used in a discrete Fourier transform (sine and cosine functions), and the squares of the coefficients in that basis are the variances of the principal components, we see that the variances of the principal components fall off as $1/f^2$ as a function of frequency. (This cannot be seen in the variance plot in Fig. 5.8 because that plot does not give the variances as a function of frequency.)

Another implication of the Wiener–Khinchin theorem is that it shows how *considering the power spectrum of images alone is related to using Gaussian model.* Since the average power spectrum contains essentially the same information as the covariance matrix, and using covariances only is equivalent to using a Gaussian model, we see that considering the average power spectrum alone is essentially equivalent to modeling the data with a Gaussian pdf as in Sect. 5.5. Since the power spectrum does not contain information about phase, using the phase structure is thus related to using the non-Gaussian aspects of the data, which will be considered in the next chapters.

5.6.3 Relative Importance of Amplitude and Phase

When considering frequency-based representations of natural images, the following question naturally arises: Which is more important, phase or amplitude (power)—or are they equally important? Most researchers agree that the phase information is more important for the perceptual system than the amplitude structure. This view is justified by experiments in which we take the phase structure from one image and the power structure from another, and determine whether the image is more similar to one of the natural images. What this means is that we take the Fourier transform (using the Discrete Fourier Transform) of the two images, and isolate the phase and amplitude from the two transforms. Then we compute the inverse of the Fourier transform from the combination of the phase from the first image and the amplitude from the second; this gives us a new image. We also create another image with the inverse Fourier transform using the phase from the second image and the amplitude from the first.

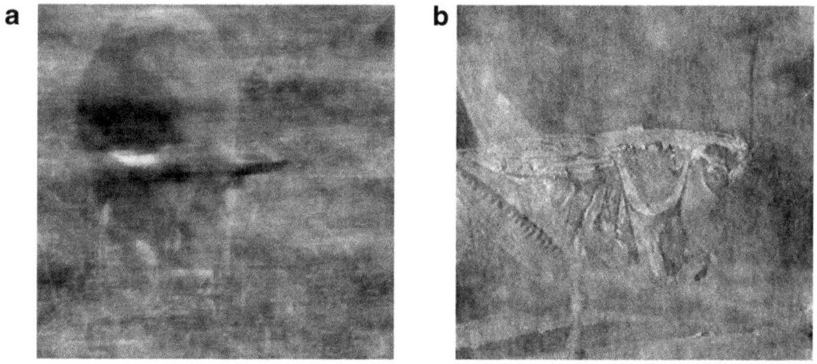

Fig. 5.16 Relative importance of phase and power/amplitude information in natural images. **a** Image which has the Fourier phases of the image in Fig. 5.14a, and the Fourier amplitudes of the image in Fig. 5.14b. **b** Image which has the phases of the image in Fig. 5.14b, and the amplitude structure of the image in Fig. 5.14a. In both cases the images are perceptually more similar to the image from which the phase structure was taken, which indicates that the visual system is more sensitive to the phase structure of natural images

Results of such an experiment are shown in Fig. 5.16. In both cases, the image "looks" more like the image from which the phase structure was taken, although in (a) this is not very strongly so. This may be natural if one looks at the Fourier amplitudes of the images: since they are both rather similar (showing the typical $1/f$ fall-off), they cannot provide much information about what the image really depicts. If all natural images really have amplitude spectra which approximately show the $1/f$ shape, the power spectrum cannot provide much information on any natural image, and thus the phase information has to be the key to identifying the contents in the images.

Thus, one can conclude that since PCA concentrates only on information in the power spectrum, and the power spectrum does not contain a lot of perceptually important information, one cannot expect PCA and related methods to yield too much useful information about the visual system. Indeed, this provides an explanation for the rather disappointing performance of PCA in learning features from natural images as seen in Fig. 5.7—the performance is disappointing, at least, if we want to model receptive fields in V1. In the next chapter, we will see that using information not contained in the covariances gives an interesting model of simple cell receptive fields.

5.7 Anisotropy in Natural Images

The concept of anisotropy refers to the fact that natural images are not completely rotationally invariant (which would be called isotropy). In other words, the statistical structure is not the same in all orientations: if you rotate an image, the statistics change.

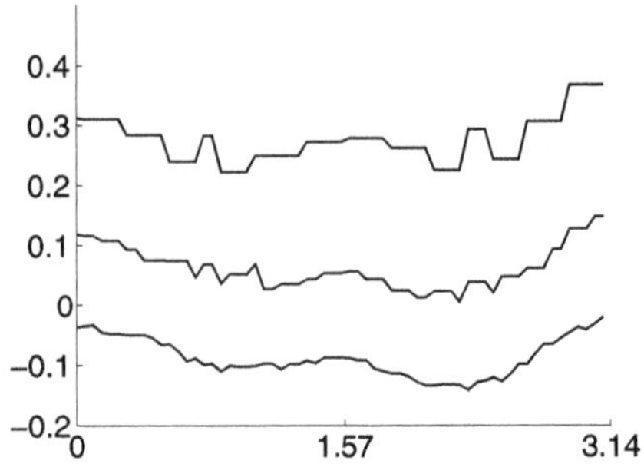

Fig. 5.17 Anisotropy, i.e. lack of rotational invariance, in natural image statistics. We took the correlation coefficients in Fig. 5.4 and plotted them on three circles with different radii (the maximum radius allowed by the patch size, and that multiplied by one half and one quarter). For each of the radii, the plot shows that the correlations are maximized for the orientations of 0 or π, which mean horizontal orientation: the pixels are on the same horizontal line. The vertical orientation $\pi/2$ shows another maximum which is less pronounced

This may come as a surprise after looking at the correlation coefficients in Fig. 5.4d, in which it seems that the correlation is simply a function of the distance: the closer to each other the two pixels are, the stronger their correlation; and the orientation does not seem to have any effect. In fact, isotropy is not a bad first approximation, but a closer analysis reveals some dependencies on orientation.

Figure 5.17 show the results of such an analysis. We have taken the correlation coefficients computed in Fig. 5.4, and analyzed how they depend on the orientation of the line segment connecting the two pixels. An orientation of 0 (or π) means that the two pixels have the same y coordinate; orientation of $\pi/2$ means that they have the same x coordinate. Other values mean that the pixels have an oblique relationship to each other. Figure 5.17 shows that the correlations are the very strongest if the pixels have the same y coordinate, that is, they are on the same horizontal line. The correlations are also elevated if the pixels have the same x coordinate.

In fact, we already saw in Fig. 5.6 that the first principal component is, consistently, a low-frequency horizontal edge. This is in line with the dominance of horizontal correlations. If the images are exactly isotropic, horizontal edges and vertical edges would have exactly the same variance, and the first principal component would not be well defined at all; this would be reflected in Fig. 5.6 so that we would get edges with different random orientations.

Thus, we have discovered a form of anisotropy in natural image statistics. It will be seen in different forms in all the later models and analyses as well.

5.8 Mathematics of Principal Component Analysis*

This section is dedicated to a more sophisticated mathematical analysis of PCA and whitening. It can be skipped by a reader not interested in mathematical details.

5.8.1 Eigenvalue Decomposition of the Covariance Matrix

The "second-order" structure of data is completely described by the covariance matrix, defined in Sect. 4.6.3. In our case with the x and y coordinates of the patches, we can write:

$$C(x, y; x', y') = E\{I(x, y)I(x', y')\}. \tag{5.22}$$

The point is that the covariance of *any* two linear features can be computed by

$$\begin{aligned}
&E\left\{\left[\sum_{x,y} W_1(x, y)I(x, y)\right]\left[\sum_{x,y} W_2(x, y)I(x, y)\right]\right\} \\
&= E\left\{\left[\sum_{xyx'y'} W_1(x, y)I(x, y)W_2(x', y')I(x', y')\right]\right\} \\
&= \sum_{xyx'y'} W_1(x, y)W_2(x', y')E\{I(x, y)I(x', y')\} \\
&= \sum_{xyx'y'} W_1(x, y)W_2(x', y')C(x, y; x', y')
\end{aligned} \tag{5.23}$$

which reduces to a something which can be computed using the covariance matrix. The second-order structure is thus conveniently represented by a single matrix, which enables us to use classic methods of linear algebra to analyze the second-order structure.

To go into detail, we change the notation so that the whole image is in one vector, $\mathbf{x}$, so that each pixel is one element in the vector. This can be accomplished, for example, by scanning the image row by row, as was explained in Sect. 4.1. This simplifies the notation enormously.

Now, considering any linear combination $\mathbf{w}^T\mathbf{x} = \sum_i w_i x_i$ we can compute its variance simply by:

$$E\{(\mathbf{w}^T\mathbf{x})^2\} = E\{(\mathbf{w}^T\mathbf{x})(\mathbf{x}^T\mathbf{w})\} = E\{\mathbf{w}^T(\mathbf{x}\mathbf{x}^T)\mathbf{w}\} = \mathbf{w}^T E\{\mathbf{x}\mathbf{x}^T\}\mathbf{w} = \mathbf{w}^T\mathbf{C}\mathbf{w} \tag{5.24}$$

where we denote the covariance matrix by $\mathbf{C} = E\{\mathbf{x}\mathbf{x}^T\}$. So, the basic PCA problem can be formulated as

$$\max_{\mathbf{w}:\|\mathbf{w}\|=1} \mathbf{w}^T\mathbf{C}\mathbf{w}. \tag{5.25}$$

A basic concept in linear algebra is the eigenvalue decomposition. The starting point is that $\mathbf{C}$ is a symmetric matrix, because $\text{cov}(x_i, x_j) = \text{cov}(x_j, x_i)$. In linear algebra, it is shown that any symmetric matrix can be expressed as a product of the form:

$$\mathbf{C} = \mathbf{U}\mathbf{D}\mathbf{U}^T \tag{5.26}$$

where $\mathbf{U}$ is an orthogonal matrix, and $\mathbf{D} = \text{diag}(\lambda_1, \ldots, \lambda_m)$ is diagonal. The columns of $\mathbf{U}$ are called the *eigenvectors*, and the λ_i are called the *eigenvalues*. Many efficient algorithms exist for computing the eigenvalue decomposition of a matrix.

Now, we can solve PCA easily. Lets us make the change of variables $\mathbf{v} = \mathbf{U}^T\mathbf{w}$. Then we have

$$\mathbf{w}^T\mathbf{C}\mathbf{w} = \mathbf{w}^T\mathbf{U}\mathbf{D}\mathbf{U}^T\mathbf{w} = \mathbf{v}^T\mathbf{D}\mathbf{v} = \sum_i v_i^2 \lambda_i. \tag{5.27}$$

Because $\mathbf{U}$ is orthogonal, $\|\mathbf{v}\| = \|\mathbf{w}\|$, so the constraint is the same for $\mathbf{v}$ as it was for $\mathbf{w}$. Let us make the further change of variables to $m_i = v_i^2$. The constraint of unit norm of $\mathbf{v}$ is now equivalent to the constraints that the sum of the m_i must equal one (they must also be positive because they are squares). Then the problem is transformed to

$$\max_{m_i \geq 0, \sum m_i = 1} \sum_i m_i \lambda_i. \tag{5.28}$$

It is rather obvious that the maximum is found when the m_i corresponding to the largest λ_i is one and the others are zero. Let us denote by i^* the index of the maximum eigenvalue. Going back to the $\mathbf{w}$, this corresponds to $\mathbf{w}$ begin equal to the i^*th eigenvector, that is, the i^*th column of $\mathbf{U}$. Thus, we see how the first principal component is easily computed by the eigenvalue decomposition.

Since the eigenvectors of a symmetric matrix are orthogonal, finding the second principal component means maximizing the variance so that v_{i^*} is kept zero. This is actually equivalent to making the new $\mathbf{w}$ orthogonal to the first eigenvector. Thus, in terms of m_i, we have exactly the same optimization problem, but with the extra constraint that $m_{i^*} = 0$. Obviously, the optimum is obtained when $\mathbf{w}$ is equal to the eigenvector corresponding to the *second* largest eigenvalue. This logic applies to the kth principal component.

Thus, all the principal components can be found by ordering the eigenvectors $\mathbf{u}_i$, $i = 1, \ldots, m$ in $\mathbf{U}$ so that the corresponding eigenvalues are in decreasing order. Let us assume that $\mathbf{U}$ is ordered so. Then the ith principal component s_i is equal to

$$s_i = \mathbf{u}_i^T\mathbf{x}. \tag{5.29}$$

Note that it can be proven that the λ_i are all non-negative for a covariance matrix.

Using the eigenvalue decomposition, we can prove some interesting properties of PCA. First, the principal components are *uncorrelated*, because for the vector of the principal components

$$\mathbf{s} = \mathbf{U}^T\mathbf{x} \tag{5.30}$$

we have

$$\begin{aligned} E\{\mathbf{s}\mathbf{s}^T\} &= E\{\mathbf{U}^T\mathbf{x}\mathbf{x}^T\mathbf{U}\} = \mathbf{U}^T E\{\mathbf{x}\mathbf{x}^T\}\mathbf{U} = \mathbf{U}^T(\mathbf{U}\mathbf{D}\mathbf{U}^T)\mathbf{U} \\ &= (\mathbf{U}^T\mathbf{U})\mathbf{D}(\mathbf{U}^T\mathbf{U}) = \mathbf{D} \end{aligned} \tag{5.31}$$

because of the orthogonality of **U**. Thus, the covariance matrix is diagonal, which shows that the principal components are uncorrelated.

Moreover, we see that the variances of the principal components are equal to the λ_i. Thus, to obtain variables that are white, that is, uncorrelated and have unit variance, it is enough to divide each principal component by the square root of the corresponding eigenvalue. This proves that $\text{diag}(1/\sqrt{\lambda_1}, \ldots, 1/\sqrt{\lambda_m})\mathbf{U}^T$ is a *whitening* matrix for **x**.

This relation also has an important implication for the uniqueness of PCA. If two of the eigenvalues are equal, then the variance of those principal components are equal. Then the principal components are not well-defined anymore because we can make a *rotation* of those principal components without affecting their variances. This is because if z_i and z_{i+1} have the same variance, then linear combinations such as $\sqrt{1/2}z_i + \sqrt{1/2}z_{i+1}$ and $\sqrt{1/2}z_i - \sqrt{1/2}z_{i+1}$ have the same variance as well; all the constraints (unit variance and orthogonality) are still fulfilled, so these are equally valid principal components. In fact, in linear algebra, it is well known that the eigenvalue decomposition is uniquely defined only when the eigenvalues are all distinct.

5.8.2 Eigenvectors and Translation-Invariance

Using the eigenvalue decomposition, we can show why the principal components of a typical image covariance matrix are sinusoids as stated in Sect. 5.2.5. This is because of their property of being translation-invariant, i.e. the covariance depends only on the distance as in (5.9). For simplicity, let us consider a one-dimensional covariance matrix $c(x - x')$. The function c is even-symmetric with respect to zero, i.e. $c(-u) = c(u)$. By a simple change of variable $z = x - x'$, we have

$$\sum_x \text{cov}(x, x') \sin(x + \alpha) = \sum_x c(x - x') \sin(x + \alpha) = \sum_z c(z) \sin(z + x' + \alpha). \tag{5.32}$$

Using the property that $\sin(a + b) = \sin a \cos b + \cos a \sin b$, we have

$$\begin{aligned}\sum_z c(z) \sin(z + x' + \alpha) &= \sum_z c(z)\big(\sin(z)\cos(x' + \alpha) + \cos(z)\sin(x' + \alpha)\big) \\ &= \left[\sum_z c(z)\sin(z)\right]\cos(x' + \alpha) \\ &\quad + \left[\sum_z c(z)\cos(z)\right]\sin(x' + \alpha).\end{aligned} \tag{5.33}$$

Finally, because $c(z)$ is even-symmetric and sin is odd-symmetric, the first sum in brackets is zero. So, we have

$$\sum_x \mathrm{cov}(x,x')\sin(x+\alpha) = \left[\sum_z c(z)\cos(z)\right]\sin(x'+\alpha) \tag{5.34}$$

which shows that the sinusoid is an eigenvector of the covariance matrix, with eigenvalue $\sum_z c(z)\cos(z)$. The parameter α gives the phase of the sinusoid; this formula shows that α can have any value, so sinusoids of any phase are eigenvectors.

This proof can be extended to sinusoids of different frequencies β: they all are eigenvalues with eigenvalues that depend on how strongly the frequency is present in the data: $\sum_z c(z)\cos(\beta z)$.

In the two-dimensional case, we have $\mathrm{cov}(I(x,y), I(x',y')) = c((x-x')^2 + (y-y')^2)$ and with $\xi = x - x'$ and $\eta = y - y'$ we have

$$\begin{aligned}
&\sum_{x,y} c\big((x-x')^2 + (y-y')^2\big)\sin(ax+by+c)\\
&\quad = \sum_{\xi,\eta} c(\xi,\eta)\sin(a\xi + b\eta + ax' + by' + c)\\
&\quad = \sum_{\xi,\eta} c(\xi,\eta)\big[\sin(a\xi+b\eta)\cos(ax'+by'+c)\\
&\qquad + \cos(a\xi+b\eta)\sin(ax'+by'+c)\big]\\
&\quad = 0 + \left[\sum_{\xi,\eta} c(\xi,\eta)\cos(a\xi+b\eta)\right]\sin(ax'+by'+c)
\end{aligned} \tag{5.35}$$

which shows likewise that sinusoids of the form $\sin(ax'+by'+c)$ are eigenvectors.

5.9 Decorrelation Models of Retina and LGN *

In this section, we consider some further methods for whitening and decorrelation of natural images, and the application of such methods as models of processing in the retina and the LGN. This material can be skipped without interrupting the flow of ideas.

5.9.1 Whitening and Redundancy Reduction

The starting point here is the redundancy reduction hypothesis, discussed in Chap. 1. In its original form, this theory states that the early visual system tries to reduce the redundancy in its input. As we have seen in this chapter, image pixel data is highly correlated, so a first approach to reduce the redundancy would be to *decorrelate* image data, i.e. to transform it into uncorrelated components.

One way to decorrelate image data is to whiten it with a spatial filter. In a visual system, this filtering would correspond to a set of neurons, with identical spatial receptive fields, spaced suitably in a lattice. The outputs of the neurons would then be uncorrelated, and if we were to look at the outputs of the whole set of neurons as an image (or a set of images for multiple input images), these output images would on the average have a flat power spectrum.

The whitening theory has led to well known if rather controversial models of the computational underpinnings of the retina and the lateral geniculate nucleus (LGN). In this section, we will discuss spatial whitening and spatial receptive fields according to this line of thought; the case of temporal whitening and temporal receptive fields will be discussed in detail in Sect. 16.3.2 (page 332).

The basic idea is that whitening alone could explain the center-surround structure of the receptive fields of ganglion cells in the retina, as well as those in the LGN. Indeed, certain spatial whitening filters are very similar to ganglion RFs, as we will see below. However, such a proposal is problematic because there are many completely different ways of whitening the image input, and it is not clear why this particular method should be used. Nevertheless, this theory is interesting because of its simplicity and because it sheds light on certain fundamental properties of the covariance structure of images.

There are at least two ways to derive the whitening operation in question. The first is to compute it directly from the covariance matrix of image patches sampled from the data; this will lead to a set of receptive fields, but with a suitable constraint the RFs will be identical except for different center locations, as we will see below. We will call this *patch-based whitening*. The second way is to specify a whitening filter in the frequency domain, which will give us additional insight and control over the process. This we will call *filter-based whitening*.

5.9.2 Patch-Based Decorrelation

Our first approach to spatial whitening is based on the PCA whitening introduced above in Sect. 5.3.2 (page 104). The data transformation is illustrated in the two-dimensional case in Fig. 5.18.

Here, we will use the matrix notation because it directly shows some important properties of the representation we construct. Here, we denote by $\mathbf{U}$ the matrix with the vectors defining the principal components as its columns

$$\mathbf{U} = (\mathbf{u}_1, \mathbf{u}_2, \ldots, \mathbf{u}_k). \tag{5.36}$$

Let $\mathbf{x}$ denote the data vector. Because the vectors $\mathbf{u}$ are orthogonal, each of the principal components $y_k, k = 1, \ldots, K$, of the data vector $\mathbf{x}$ can be computed simply by taking the dot product between the data vector and the kth PCA vector:

$$y_k = \mathbf{u}_k^T \mathbf{x}, \quad k = 1, \ldots, K. \tag{5.37}$$

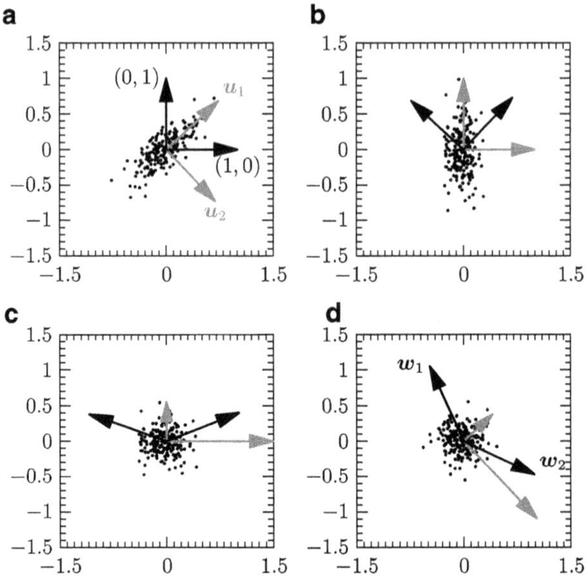

Fig. 5.18 An illustration of the whitening procedure that is used to derive a set of whitening filters $\mathbf{w}_k$, $k = 1, \ldots, K$ (here, we take $K = 2$). The procedure utilizes the PCA basis vectors $\mathbf{u}_k$, $k = 1, \ldots, K$. **a** The original generated data points and the PCA basis vectors $\mathbf{u}_1$ and $\mathbf{u}_2$ (*grey*) and the unit vectors $(1, 0)$ and $(0, 1)$ (*black*). **b** The data points are first rotated so that the new axes match the PCA basis vectors. **c** The data points are then scaled along the axes so that the data have the same variance along both axes. This also makes the two dimensions of the data uncorrelated, so the end result is a whitened data set. (For purposes of visualization of the data points and the vectors, in this illustration this variance is smaller than 1, while in whitening it is 1; this difference corresponds to an overall scaling of the data.) **d** Finally, the data points are rotated back to the original orientation. Note that the data are already white after the second transformation in **c**, and the last transformation is one of infinitely many possible rotations that keep the data white; in this method, it is the one that inverts the rotation done by PCA. Mathematically, the three transformations in **b**–**d** can in fact be combined into a single linear transformation because each transformation is linear; the combined operation can be done by computing the dot products between the original data points and the vectors $\mathbf{w}_1$ and $\mathbf{w}_2$ which are the result of applying the three transformations to the original unit vectors. See text for details

Defining the vector $\mathbf{y} = (y_1, y_2, \ldots, y_K)^T$, (5.37) for all $k = 1, \ldots, K$ can be expressed in a single matrix equation

$$\mathbf{y} = \mathbf{U}^T \mathbf{x}. \tag{5.38}$$

In our two-dimensional illustration, the result of this transformation is shown in Fig. 5.18b.

As in Sect. 5.3.2, we next whiten the data by dividing the principal components with their standard deviations. Thus, we obtain whitened components s_k

$$s_k = \frac{y_k}{\sqrt{\operatorname{var}(y_k)}}, \quad k = 1, \ldots, K, \tag{5.39}$$

This is shown for our illustrative example in Fig. 5.18c.

Again, we can express these K equations by a single matrix equation. Define a vector $\mathbf{s} = (s_1, s_2, \ldots, s_K)^{\mathrm{T}}$, and let $\mathbf{\Lambda}$ denote a diagonal matrix with the inverses of the square roots of the variances on its diagonal:

$$\mathbf{\Lambda} = \begin{bmatrix} \frac{1}{\sqrt{\mathrm{var}(y_1)}} & 0 & \cdots & 0 \\ 0 & \frac{1}{\sqrt{\mathrm{var}(y_2)}} & \cdots & 0 \\ \vdots & \vdots & \ddots & \vdots \\ 0 & 0 & \cdots & \frac{1}{\sqrt{\mathrm{var}(y_K)}} \end{bmatrix}; \tag{5.40}$$

then

$$\mathbf{s} = \mathbf{\Lambda y} = \mathbf{\Lambda U}^{\mathrm{T}}\mathbf{x}. \tag{5.41}$$

So far, we have just expressed PCA whitening as a matrix formulation. Now, we will make a new operation. Among the infinitely many whitening matrices, we choose the one which is given by inverting the PCA computation given by $\mathbf{U}$. In a sense, we go "back from the principal components to the original coordinates" (Fig. 5.18d). Denoting by $\mathbf{z}$ the final component computed, this is defined by the following matrix equation:

$$\mathbf{z} = \mathbf{Us} = \mathbf{U\Lambda U}^{\mathrm{T}}\mathbf{x}. \tag{5.42}$$

The computation presented in (5.43) consist of three linear (matrix) transformations in a cascade. The theory of linear transformation states that a cascade of consecutive linear transformations is simply another linear transformation, and that this combined transformation—which we will here denote by $\mathbf{W}$—can be obtained as the matrix product of the individual transformations:

$$\mathbf{z} = \underbrace{\mathbf{U\Lambda U}^{\mathrm{T}}}_{=\mathbf{W}}\mathbf{x} = \mathbf{Wx}. \tag{5.43}$$

When written as k scalar equations, (5.43) shows that the components of vector $\mathbf{z} = [z_1\ z_2\ \ldots\ z_K]^{\mathrm{T}}$ can be obtained as a dot product between the data vector $\mathbf{x}$ and the kth row of matrix $\mathbf{W} = [\mathbf{w}_1\ \mathbf{w}_2\ \ldots \mathbf{w}_K]^{\mathrm{T}}$:

$$z_k = \mathbf{w}_k^{\mathrm{T}}\mathbf{x}, \quad k = 1, \ldots, K. \tag{5.44}$$

The vectors $\mathbf{w}_k$, $k = 1, \ldots, K$, are of great interest to us, since they are filters which map the input $\mathbf{x}$ to the whitened data. In other words, the vectors $\mathbf{w}_k$, $k = 1, \ldots, K$, can be interpreted as receptive fields. The receptive fields $\mathbf{w}_k$, $k = 1, \ldots, K$, can be obtained simply by computing the matrix product in (5.43).

Matrix Square Root As an aside, we mention an interesting mathematical interpretation of the matrix $\mathbf{W}$. The matrix $\mathbf{W}$ is called the *inverse matrix square root* of the covariance matrix $\mathbf{C}$, and denoted by $\mathbf{C}^{-1/2}$. In other words, the inverse $\mathbf{W}^{-1}$

is called the square root of $\mathbf{C}$, and denoted by $\mathbf{C}^{1/2}$. The reason is that if we multiply $\mathbf{W}^{-1}$ with itself, we get $\mathbf{C}$. This is because, first, $(\mathbf{U}\boldsymbol{\Lambda}\mathbf{U}^T)^{-1} = \mathbf{U}\boldsymbol{\Lambda}^{-1}\mathbf{U}^T$, and second, we can calculate

$$\mathbf{W}^{-1}\mathbf{W}^{-1} = (\mathbf{U}\boldsymbol{\Lambda}^{-1}\mathbf{U}^T)(\mathbf{U}\boldsymbol{\Lambda}^{-1}\mathbf{U}^T) = \mathbf{U}\boldsymbol{\Lambda}^{-1}(\mathbf{U}^T\mathbf{U})\boldsymbol{\Lambda}^{-1}\mathbf{U}^T = \mathbf{U}\boldsymbol{\Lambda}^{-2}\mathbf{U}^T. \quad (5.45)$$

The matrix $\boldsymbol{\Lambda}^{-2}$ is simply a diagonal matrix with the variances in its diagonal, so the result is nothing else than the eigenvalue decomposition of the covariance matrix as in (5.26).

Symmetric Whitening Matrix Another interesting mathematical property of the whitening matrix $\mathbf{W}$ in (5.43) is that it is symmetric, which can be shown as

$$\mathbf{W}^T = (\mathbf{U}\boldsymbol{\Lambda}\mathbf{U}^T)^T = (\mathbf{U}^T)^T\boldsymbol{\Lambda}^T\mathbf{U}^T = \mathbf{U}\boldsymbol{\Lambda}\mathbf{U}^T = \mathbf{W}. \quad (5.46)$$

In fact, it is the only symmetric whitening matrix.

Application to Natural Images When the previous procedure is applied to natural image data, interesting receptive fields emerge. Figure 5.19a shows the resulting whitening filters (rows/columns of $\mathbf{W}$); a closeup of one of the filters is shown in Fig. 5.19b. As can be seen, the whitening principle results in the emergence of filters which have center-surround structure. All of the filters are identical, so processing image patches with such filters is analogous to filtering them with the spatial filter shown in Fig. 5.19b.

As pointed out several times above, whitening can be done in infinitely many different ways: if $\mathbf{W}$ is a whitening transformation, so is any orthogonal transformation of $\mathbf{W}$. Here, the whitening solution in (5.43) has been selected so that it results in center-surround-type filters. This is a general property that we will bump into time and again below: the whitening principle does constrain the form of the emerging filters, but additional assumptions are needed before the results can have a meaningful interpretation.

Note that the theory results in a single receptive field structure, while in the retina, there are receptive fields with differing spatial properties—in particular scale (frequency)—in the retina and the LGN. This is another limitation of the whitening principle, and additional assumptions are needed to produce a range of differing filters.

5.9.3 Filter-Based Decorrelation

Now, we reformulate this theory in a filter-based framework. Then the theory postulates that the amplitude response properties (see Sect. 2.2.3, page 34) of retinal and LGN receptive fields follow from the following two assumptions:

1. The linear filters are whitening natural image data.
2. With the constraint that noise is not amplified unduly.

a

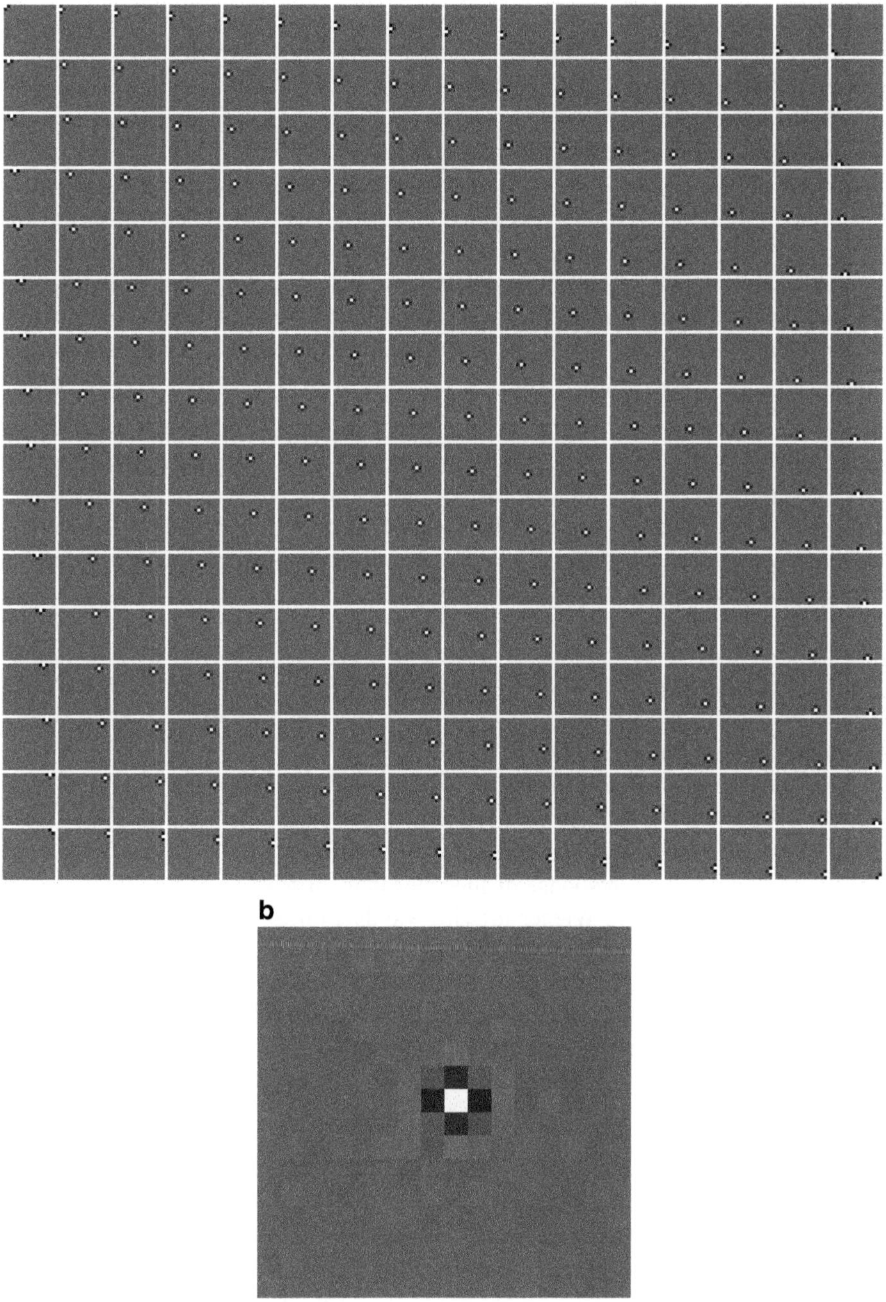

b

Fig. 5.19 The application of the whitening principle results in the emergence of a set of center-surround filters from natural image data. **a** The set of filters (rows/columns of whitening matrix **W**) obtained from image data. **b** A closeup of one of the filters; the other filters are identical except for spatial location

Additional assumptions are needed to derive the phase response in order to specify the filter completely. This is equivalent to the observation made above in the general case of whitening: there are infinitely many whitening transformations. Here, the phases are defined by specifying that the energy of the filter should be concentrated in either time or space, which in the spatial case can be loosely interpreted to mean that the spatial RFs should be as localized as possible.

The amplitude response of the filter will be derived in two parts: the first part is the whitening filter, and the second part suppresses noise. The filter will be derived in the frequency domain, and thereafter converted to the spatial domain by inverse Fourier transform. It is often assumed that the statistics of image data do not depend on spatial orientation; we make the same assumption here and study the orientation-independent spatial frequency ω_{s}. Conversion from the usual two-dimensional frequencies ω_x and ω_y to spatial frequency ω_{s} is given by $\omega_{\mathrm{s}} = \sqrt{\omega_x^2 + \omega_y^2}$. Let $R_{\mathrm{i}}(\omega_{\mathrm{s}})$ denote the average power spectrum in natural images. We know that for uncorrelated/whitened data, the average power spectrum should be a constant (flat). Because the average power spectrum of the filtered data is the product of the average power spectrum of the original data and the squared amplitude response of the whitening filter, which we denote by $|V(\omega_{\mathrm{s}})|^2$; this means that the amplitude response of a whitening filter can be specified by

$$\left|V(\omega_{\mathrm{s}})\right| = \frac{1}{\sqrt{R_{\mathrm{i}}(\omega_{\mathrm{s}})}}, \tag{5.47}$$

since then $|V(\omega_{\mathrm{s}})|^2 R_{\mathrm{i}}(\omega_{\mathrm{s}}) = 1$.

Real measurement data contains noise. Assume that the noise, whose average power spectrum is $R_{\mathrm{n}}(\omega_{\mathrm{s}})$, is additive and uncorrelated with the original image data, whose average power spectrum is $R_{\mathrm{o}}(\omega_{\mathrm{s}})$; then $R_{\mathrm{i}}(\omega_{\mathrm{s}}) = R_{\mathrm{o}}(\omega_{\mathrm{s}}) + R_{\mathrm{n}}(\omega_{\mathrm{s}})$. To derive the amplitude response of the filter that suppresses noise, one can use a *Wiener filtering* approach. Wiener filtering yields a linear filter that can be used to compensate for the presence of additive noise: the resulting filter optimally restores the original signal in the least mean square sense. The derivation of the Wiener filter in the frequency space is somewhat involved, and we will skip it here; see Dong and Atick (1995b). The resulting response properties of the filter are fairly intuitive: the amplitude response $|F(\omega_{\mathrm{s}})|$ of the Wiener filter is given by

$$\left|F(\omega_{\mathrm{s}})\right| = \frac{R_{\mathrm{i}}(\omega_{\mathrm{s}}) - R_{\mathrm{n}}(\omega_{\mathrm{s}})}{R_{\mathrm{i}}(\omega_{\mathrm{s}})}. \tag{5.48}$$

Notice that if there are frequencies that contain no noise—that is, $R_{\mathrm{n}}(\omega_{\mathrm{s}}) = 0$—the amplitude response is simply 1, and that higher noise power leads to decreased amplitude response.

The overall amplitude response of the filter $|W(\omega_{\mathrm{s}})|$ is obtained by cascading the whitening and the noise-suppressive filters ((5.47) and (5.48)). Because this cascading corresponds to multiplication in the frequency domain, the amplitude response of the resulting filter is

$$|W(\omega_s)| = |V(\omega_s)||F(\omega_s)| = \frac{1}{\sqrt{R_i(\omega_s)}} \frac{R_i(\omega_s) - R_n(\omega_s)}{R_i(\omega_s)}. \tag{5.49}$$

In practice, $R_i(\omega_s)$ can be estimated directly from image data for each ω_s, or one can use the parametric form derived in Sect. 5.6 (page 111). (Negative values given by this formula (5.49) have to be truncated to zero.) Noise is assumed to be spatially uncorrelated, implying a constant (flat) power spectrum, and to have power equal to data power at a certain characteristic frequency, denoted by $\omega_{s,c}$, so that

$$R_n(\omega_s) = \frac{R_i(\omega_{s,c})}{2} \quad \text{for all } \omega_s. \tag{5.50}$$

In order to fully specify the resulting filter, we have to define its phase response. Here, we simply set the phase response to zero for all frequencies:

$$\angle W(\omega_s) = 0 \quad \text{for all } \omega_s. \tag{5.51}$$

With the phases of all frequencies at zero, the energy of the filter is highly concentrated around the spatial origin, yielding a highly spatially localized filter. After the amplitude and the phase responses have been defined, the spatial filter itself can be obtained by taking the inverse two-dimensional Fourier transform.

The filter properties that result from the application of (5.49), (5.50) and (5.51) are illustrated in Fig. 5.20 for characteristic frequency value $\omega_{s,c} = 0.3$ cycles per pixel. For this experiment, 100 000 image windows of size 16×16 pixels were sampled from natural images.[4] The average power spectrum of these images was then computed; the average of this spectrum over all spatial orientations is shown in Fig. 5.20a. The squared amplitude response of the whitening filter, obtained from (5.49), is shown in Fig. 5.20b. The power spectrum of the filtered data is shown in Fig. 5.20c; it is approximately flat at lower frequencies and drops off sharply at high frequencies because of the higher relative noise power at high frequencies. The resulting filter is shown in Fig. 5.20d; for comparison, a measured spatial receptive field of an LGN neuron is shown in Fig. 5.20e.

Thus, the center-surround receptive-field structure, found in the retina and the LGN, emerges from this computational model and natural image data. However, we made several assumptions above—such as the spacing of the receptive fields—and obtained as a result only a single filter instead of a range of filters in different scales and locations. In Sect. 16.3.2 (page 332), we will see that in the temporal domain, similar principles lead to the emergence of temporal RF properties of these neurons.

---

[4] Here, we did not use our ordinary data set but that of van Hateren and van der Schaaf (1998).

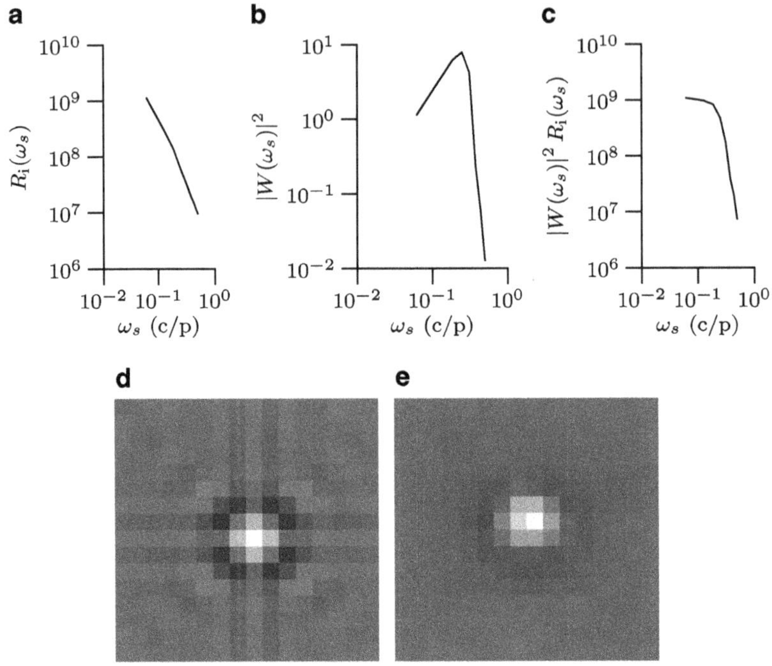

Fig. 5.20 The application of the whitening principle, combined with noise reduction and zero phase response, leads to the emergence of center-surround filters from natural image data. **a** The power spectrum $R_i(\omega_s)$ of natural image data. **b** The squared amplitude response of a whitening filter which suppresses noise: this curve follows the inverse of the data power spectrum at low frequencies, but then drops off quickly at high frequencies, because the proportion of noise is larger at high frequencies. **c** The power spectrum of the resulting (filtered) data, showing approximately flat (white) power at low frequencies, and dropping off sharply at high frequencies. **d** The resulting filter which has been obtained from the amplitude response in **b** and by specifying a zero phase response for all frequencies; see text for details. **e** For comparison, the spatial receptive field of an LGN neuron

5.10 Concluding Remarks and References

This chapter considered models of natural images which were based on analyzing the covariances of the image pixels. The classic model is principal component analysis, in which variance of a linear feature detector is maximized. PCA fails to yield interesting feature detectors if the goal is to model visual cells in brain. However, it is an important model historically and conceptually, and also provides the basis for the preprocessing we use later in this book: dimension reduction combined with whitening. In the next chapter, we will consider a different kind of learning criterion which does yield features which are interesting for visual modeling.

Most of the work on second-order statistics of images is based on the (approximate) $1/f^2$ property of the power spectrum. This was investigated early in Field (1987), Burton and Moorehead (1987), Tolhurst et al. (1992), Ruderman and Bialek

(1994a), van der Schaaf and van Hateren (1996). It has been proposed to explain certain scaling phenomena in the visual cortex, such as the orientation bandwidth (van der Schaaf and van Hateren 1996) and the relative sensitivity of cells tuned to different frequencies (Field 1987). Early work on PCA of images include Sanger (1989), Hancock et al. (1992). The $1/f$ property is closely related to the study of self-similar stochastic processes (Embrechts and Maejima 2000) which has a very long history (Mandelbrot and van Ness 1968). The study of self-critical systems (Bak et al. 1987) may also have some connection. A model of how such self-similarities come about as a result of composing an image of different "objects" is proposed in Ruderman (1997).

A recent paper with a very useful discussion and review of the psychophysical importance of the Fourier powers vs. phases is Wichmann et al. (2006); see also Hansen and Hess (2007).

Another line of research proposes that whitening explains retinal ganglion receptive fields (Atick and Redlich 1992). (An extension of this theory explains LGN receptive field by considering temporal correlations as well (Dan et al. 1996a); see also Chap. 16.) For uniformity of presentation, we follow the mathematical theory of Dong and Atick (1995b) both here in the spatial case and in the temporal case in Sect. 16.3.2. As argued above, the proposal is problematic because there are many ways of whitening data. A possible solution to the problem is to consider energy consumption or wiring length; see Chap. 11 for this concept, as was done in Vincent and Baddeley (2003), Vincent et al. (2005).

The anisotropy of pixel correlations has been used to explain some anisotropic properties in visual psychophysics in Baddeley and Hancock (1991).

An attempt to characterize the proportion of information explained by the covariance structure in natural images can be found in Chandler and Field (2007).

5.11 Exercises

Mathematical Exercises

1. Show that if the expectations of the grey-scale values of the pixels are the same for all x, y:

$$E\{I(x,y)\} = E\{I(x',y')\} \quad \text{for any } x, y, x', y' \tag{5.52}$$

then removing the DC component implies than the expectation of $\tilde{I}(x,y)$ is zero for any x, y.
2. Show that if $\sum_{x,y} W_{x,y} = 0$, the removal of the DC component has no effect on the output of the features detector.
3. Show that if the vector $(y_1, \dots, y_n)^T$ is white, any orthogonal transformation of that vector is white as well.

4. To get used to matrix notation:
   a. The covariance matrix of the vector $\mathbf{x} = (x_1, \ldots, x_n)^T$ is defined as the matrix $\mathbf{C}$ with elements $c_{ij} = \text{cov}(x_i, x_j)$. Under what condition do we have $\mathbf{C} = E\{\mathbf{x}\mathbf{x}^T\}$?
   b. Show that the covariance matrix of $\mathbf{y} = \mathbf{Mx}$ equals $\mathbf{MCM}^T$.
5. Denote by $\mathbf{w}$ a vector which reduces the dimension of $\mathbf{x}$ to one as $z = \sum_i w_i x_i$. Now, we will show that taking the first principal component is the optimal way of reducing dimension if the optimality criterion is least-squares error. This means that we reconstruct the original data as a linear transformation of z as:

$$J(\mathbf{W}) = E\left\{\sum_j (x_j - w_j z)^2\right\}. \tag{5.53}$$

   a. Show that J is equal to

$$\sum_j w_j^2 \sum_{i,i'} w_i w_{i'} \text{cov}(x_i, x_{i'}) - 2\sum_j w_j \sum_i w_i \text{cov}(x_j, x_i) + \sum_j \text{var}(x_j). \tag{5.54}$$

   b. Using this expression for J, show that the $\mathbf{w}$ which minimizes J under the constraint $\|\mathbf{w}\| = 1$ is the first principal component of $\mathbf{x}$.

Computer Assignments

1. Take some images from the web. Take a large sample of extremely small patches of the images, so that the patch contains just two neighboring pixels. Convert the pixels to grey-scale if necessary. Make a scatter plot of the pixels. What can you see? Compute the correlation coefficient of the pixel values.
2. Using the same patches, convert them into two new variables: the sum of the grey-scale values and their difference. Do the scatter plot and computer the correlation coefficient.
3. Using the same images, take a sample of 1000 patches of the form of 1×10 pixels. Compute the covariance matrix. Plot the covariance matrix (because the patches are one-dimensional, you can easily plot this two-dimensional matrix).
4. The same as above, but remove the DC component of the patch. How does this change the covariance matrix?
5. The same as above, but with only 50 patches sampled from the images. How are the results changed and why?
6. *Take the sample of 1000 one-dimensional patches computed above. Compute the eigenvalue decomposition of the covariance matrix. Plot the principal component weights $W_i(x)$.

Chapter 6
Sparse Coding and Simple Cells

In the preceding chapter, we saw how features can be learned by PCA of natural images. This is a classic method of utilizing the second-order information of statistical data. However, the features it gave were not very interesting from a neural modeling viewpoint, which motivates us to find better models. In fact, it is clear that the second-order structure of natural images is scratching the surface of the statistical structure of natural images. Look at the outputs of the feature detectors of Fig. 1.10, for example. We can see that the outputs of different kinds of filters differ from each other in other ways than just variance: the output of the Gabor filter has a histogram that has a strong peak at zero, whereas this is not the case for the histogram of pixel values. This difference is captured in a property called *sparseness*. It turns out that a more interesting model is indeed obtained if we look at the sparseness of the output s instead of the variance as in PCA.

6.1 Definition of Sparseness

Sparseness means that the random variable is most of the time very close to zero, and only occasionally gets clearly non-zero values. One often says that the random variable is "active" only rarely.

It is very important to distinguish sparseness from small variance. When we say "very close to zero", this is relative to the general deviation of the random variable from zero, i.e. relative to its variance and standard deviation. Thus, "very close to zero" would mean something like "an absolute value that is smaller than 0.1 times the standard deviation".

To say that a random variable is sparse needs a baseline of comparison. Here, it is the Gaussian (normal) distribution; a random variable is sparse if it is active more rarely compared to a Gaussian random variable of the same variance (and zero mean). Figure 6.1 shows a sample of a sparse random variable, compared to the Gaussian random variable of the same variance. Another way of looking at sparseness is to consider the probability density function (pdf). The property of being most of the time very close to zero is closely related to the property that the pdf has a peak at zero. Since the variable must have some deviation from zero (variance was normalized to unity), the peak at zero must be compensated by a relatively large probability mass at large values; a phenomenon often called "heavy tails". In between these two extremes, the pdf takes relatively small values, compared to the Gaussian pdf. This is illustrated in Fig. 6.2.[1]

[1] Here, we consider the case of symmetric distributions only. It is possible to talk about the sparseness of non-symmetric distributions as well. For example, if the random variable only obtains

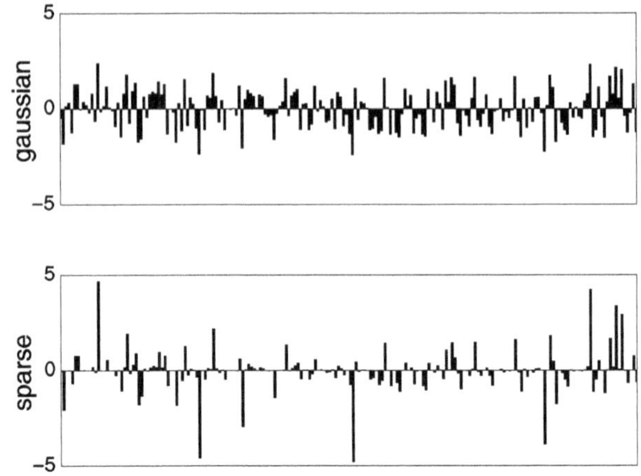

Fig. 6.1 Illustration of sparseness. Random samples of a Gaussian variable (*top*) and a sparse variable (*bottom*). The sparse variable is practically zero most of the time, occasionally taking very large values. Note that the variables have the same variance, and that these are not time series but just observations of random variables

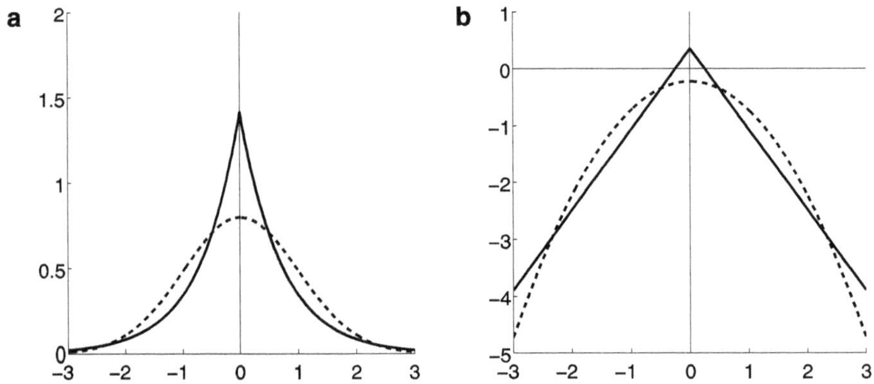

Fig. 6.2 Illustration of a typical sparse probability density. The sparse density function, called Laplacian, is given by the *solid curve* (see (7.18) in the next chapter for an exact formula). For comparison, the density of the absolute value of a Gaussian random variable of the same variance is given by the *dash-dotted curve*. **a** The probability density functions, **b** their logarithms

6.2 Learning One Feature by Maximization of Sparseness

To begin with, we consider the problem of learning a single feature based on maximization of sparseness. As explained in Sect. 1.8, learning features is a simple approach to building statistical models. Similar to the case of PCA, we consider one linear feature s computed using weights $W(x, y)$ as

$$s = \sum_{x,y} W(x, y) I(x, y). \tag{6.1}$$

---

non-negative values, the same idea of being very close to zero most of the time is still valid and is reflected in a peak on the right side of the origin. See Sect. 13.2.3 for more information. However, most distributions found in this book are symmetric.

While a single feature is not very useful for vision, this approach shows the basic principles in a simplified setting. Another way in which we simplify the problem is by postponing the formulation of a proper statistical model. Thus, we do not really *estimate* the feature in this section, but rather learn it by some intuitively justified statistical criteria. In Sect. 6.3, we show how to learn many features, and in Chap. 7 we show how to formulate a proper statistical model and learn the features by estimating it.

6.2.1 Measuring Sparseness: General Framework

To be able to find features that maximize sparseness, we have to develop statistical criteria for measuring sparseness. When measuring sparseness, we can first normalize s to unit variance, which is simple to do by dividing s by its standard deviation. This simplifies the formulation of the measures.

A simple way to approach the problem is to look at the expectation of some function of s, a linear feature of the data. If the function is the square function, we are just measuring variance (which we just normalized to be equal to one), so we have to use something else. Since we know that the variance is equal to unity, we can consider the square function as a baseline and look at the expectations of the form

$$E\{h(s^2)\} \tag{6.2}$$

where h is some *nonlinear* function.

How should the function h be chosen so that the formula in (6.2) measures sparseness? The starting point is the observation that sparse variables have a lot of data (probability mass) around zero because of the peak at zero, as well as a lot of data very far from zero because of heavy tails. Thus, we have two different approaches to measuring sparseness. We can choose h so that it emphasizes values that are close to zero, or values that are much larger than one. However, it may not be necessary to explicitly measure both of them because the constraint of unit variance means that if there is a peak at zero, there has to be something like heavy tails to make the variance equal to unity, and vice versa.

6.2.2 Measuring Sparseness Using Kurtosis

A simple function that measures sparseness with emphasis on large values (heavy tails) is the quadratic function

$$h_1(u) = (u - 1)^2. \tag{6.3}$$

(We denote by u the argument of h, to emphasize that it is not a function of s directly. Typically, $u = s^2$.) Algebraic simplifications show that then the sparseness measure is equal to

$$E\{h_1(s^2)\} = E\{(s^2-1)^2\} = E\{s^4 - 2s^2 + 1\} = E\{s^4\} - 1 \tag{6.4}$$

where the last equality holds because of the unit variance constraint. Thus, this measure of sparseness is basically the same as the fourth moment; subtraction of the constant (one) is largely irrelevant since it just shifts the measurement scale.

Using the fourth moment is closely related to the classic statistic called *kurtosis*

$$\text{kurt}(s) = E\{s^4\} - 3(E\{s^2\})^2. \tag{6.5}$$

If the variance is normalized to 1, kurtosis is in fact the same as the fourth moment minus a constant (three). This constant is chosen so that kurtosis is zero for a Gaussian random variable (this is left as an exercise). If kurtosis is positive, the variable is called leptokurtic (or super-Gaussian); this is a simple operational definition of sparseness.

However, kurtosis is not a very good measure of sparseness for our purposes. The basic problem with kurtosis is its sensitivity to outliers. An "outlier" is a data point that is very far from the mean, possibly due to an error in the data collection process. Consider, for example, a data set that has 1000 scalar values and has been normalized to unit variance. Assume that one of the values is equal to 10. Then kurtosis is necessarily equal to at least $10^4/1000 - 3 = 7$. A kurtosis of 7 is usually considered a sign of strong sparseness. But here it was due to a single value, and not representative of the whole data set at all!

Thus, kurtosis is a very unreliable measure of sparseness. This is due to the fact that h_1 puts much more weight on heavy tails than on values close to zero (it grows infinitely when going far from zero). It is, therefore, useful to consider other measures of sparseness, i.e. other non-linear functions h.

6.2.3 *Measuring Sparseness Using Convex Functions of Square*

Convexity and Sparseness Many valid measures can be found by considering functions h that are *convex*.[2] Convexity means that a line segment that connects two points on the graph is always above the graph of the function, as illustrated in Fig. 6.3. Algebraically, this can be expressed as follows:

$$h\big(\alpha x_1 + (1-\alpha)x_2\big) < \alpha h(x_1) + (1-\alpha)h(x_2) \tag{6.6}$$

for any $0 < \alpha < 1$. It can be shown that this is true if the second derivative of h is positive for all x (except perhaps in single points).

Why is convexity enough to yield a valid measure of sparseness? The reason is that the expectation of a convex function has a large value if the data is concentrated in the extremes, in this case near zero and very far from zero. Any points between

[2]The convexity we consider here is usually called "strict" convexity in mathematical literature.

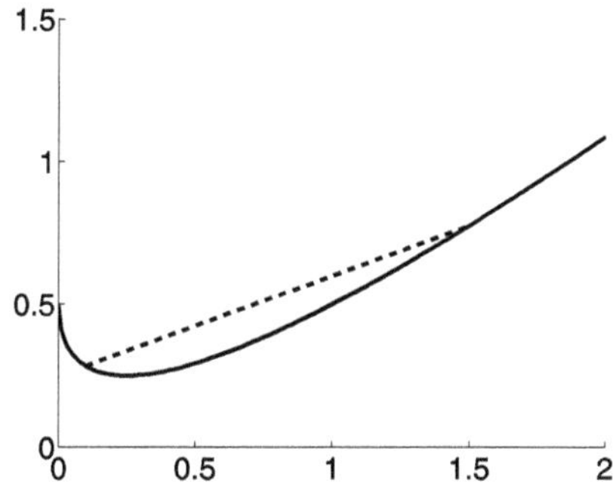

Fig. 6.3 Illustration of convexity. The plotted function is $y = -\sqrt{x} + x + \frac{1}{2}$, which from the viewpoint of measurement of sparseness is equivalent to just the negative square root, as explained in the text. The segment (*dashed line*) connecting two points on its graph is above the graph; actually, this is always the case

the extremes decrease the expectation of the convex h due to the fundamental equation (6.6), where x_1 and x_2 correspond to the extremes, and $\alpha x_1 + (1-\alpha)x_2$ is a point in between.

The function h_1 in (6.3) is one example of a convex function, but below we will propose better ones.

An Example Distribution To illustrate this phenomenon, consider a simple case where s takes only three values:

$$P(s = -\sqrt{5}) = 0.1, \qquad P(s = \sqrt{5}) = 0.1, \qquad P(s = 0) = 0.8. \tag{6.7}$$

This distribution has zero mean, unit variance, and is quite sparse. The square s^2 takes the values 0 and 5, which can be considered very large in the sense that it is rare for a random variable to take values that are $\sqrt{5}$ times the standard deviation, and 0 is, of course an extremely small absolute value. Now, let us move some of the probability mass from 0 to 1, and to preserve unit variance, make the largest value smaller. We define

$$P(s = -2) = 0.1, \qquad P(s = 2) = 0.1, \qquad P(s = 0) = 0.6, \tag{6.8}$$

$$P(s = -1) = 0.1, \qquad P(s = 1) = 0.1. \tag{6.9}$$

We can now compute the value of the measure $E\{h(s^2)\}$ for the new distribution and compare it with the value obtained for the original distribution, based on the definition of convexity:

$$\begin{aligned} &0.2h(4) + 0.2h(1) + 0.6h(0) \\ &\quad = 0.2h(0.8 \times 5 + 0.2 \times 0) + 0.2h(0.2 \times 5 + 0.8 \times 0) + 0.6h(0) \\ &\quad < 0.2 \times \big(0.8h(5) + 0.2h(0)\big) + 0.2 \times \big(0.2h(5) + 0.8h(0)\big) + 0.6h(0) \\ &\quad = 0.2h(5) + 0.8h(0) \end{aligned} \tag{6.10}$$

where the inequality is due to the definition of convexity in (6.6). Now, the last expression is the value of the sparseness measure in the original distribution. Thus, we see that the convexity of h makes the sparseness measure smaller when probability mass is taken away from the extremes. This is true for any convex function.

Suitable Convex Functions A simple convex function—which will be found to be very suitable for our purposes—is given by the negative square root:

$$h_2(u) = -\sqrt{u}. \tag{6.11}$$

This function is actually equivalent to the one in Fig. 6.3, because the addition of $u = s^2$ just adds a constant 1 to the measure: Adding a linear term to the sparseness measure h has no effect because it only adds a constant due to the constraint of unit variance. Adding a linear term has no effect on convexity either (which is left as an exercise). This linear term and the constant term were just added to the function Fig. 6.3 to illustrate the fact that it puts more weight on values near zero and far from zero, but the weight for values far from zero do not grow too fast.

The validity of h_2 as a sparseness measure is easy to see from Fig. 6.3, which shows how the measure gives large values if the data is either around zero, or takes very large values. In contrast to h_1, or kurtosis, it does not suffer from sensitivity to outliers because it is equivalent to using the square root which grows very slowly when going away from zero. Moreover, h_2 also emphasizes the concentration around zero because it has a peak at zero itself.[3]

Another point to consider is that the function $h_2(s^2)$ is actually equal to the negative of the absolute value function $-|s|$. It is not differentiable at zero, because its slope abruptly changes from -1 to $+1$. This may cause practical problems, for example, in the optimization algorithms that will be used to maximize sparseness. Thus, it is often useful to take a smoother function, such as

$$h_3(u) = -\log\cosh\sqrt{u} \tag{6.12}$$

which is as a function of s

$$h_3(s^2) = -\log\cosh s. \tag{6.13}$$

The relevant functions and their derivatives are plotted in Fig. 6.4. Note that the point is to have a function h that is a convex function as a function of the square $u = s^2$ as in (6.12). When expressed as a function of s as in (6.13), the function need not be convex anymore.

Alternatively, one could modify h_2 as

$$h_{2b} = -\sqrt{u + \epsilon} \tag{6.14}$$

[3] One could also argue that h_2 does not give a large value for large values of s at all, but only for s very close to zero, because the function h_2 has a peak at zero. This is a complicated point because we can add a linear function to h_2 as pointed out above. In any case, it is certain that h_2 puts much more weight on values of u very close to zero.

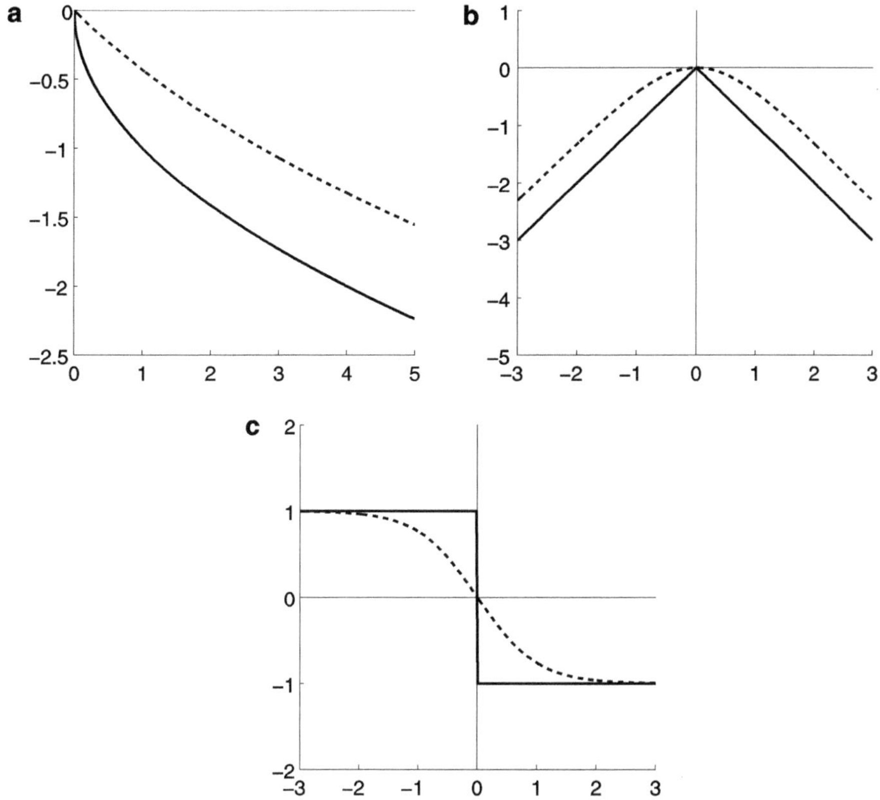

Fig. 6.4 Illustration of the log cosh function and its comparison with the absolute value function. **a** The function h_2 in (6.11) is given in *solid curve*. The function h_3 in (6.12) is given as a *dash-dotted curve*. **b** The same functions h_2 and h_3 are given as function of s (and not its square). **c** The derivatives of the functions in **b**

where ϵ is a small constant. This is another smoother version of the square root function. It has the benefit of being simpler than h_3 when we consider h as a function of u; in contrast, h_3 tends to be simpler when considered a function of s.

There are many different convex function that one might choose, so the question arises whether there is an optimal one that we should use. In fact, estimation theory as described in detail in Chap. 7 shows that the optimal measure of sparseness is basically given by choosing

$$h_{\text{opt}}(s^2) = \log p_s(s) \tag{6.15}$$

where p_s is the probability density function of s. The function h_2 is typically not a bad approximation of this optimal function for natural images. Often, the logarithm of a pdf has an even stronger singularity (peak) at zero than what h_2 has. Thus, to avoid the singularity, it may be better to use something more similar to h_2 or h_3. This will be considered in more detail in Sect. 7.7.2.

Summary To recapitulate, finding linear feature detectors of maximum sparseness can be done by finding a maximum of

$$E\left\{h\left(\left[\sum_{x,y} W(x,y)I(x,y)\right]^2\right)\right\} \tag{6.16}$$

with respect to W, constraining W so that

$$E\left\{\left[\sum_{x,y} W(x,y)I(x,y)\right]^2\right\} = 1, \tag{6.17}$$

where the function h is typically chosen as in (6.12).

Usually, there are many local maxima of the objective function (see Sect. 18.3 for the concept of global and local maxima). Each of the local maxima gives a different feature.

6.2.4 The Case of Canonically Preprocessed Data

In practice, we use data that has been preprocessed by the canonical way described in Sect. 5.4. That is, the dimension of the data has been reduced by PCA to reduce computational load and to get rid of the aliasing artifacts, and the data has been whitened to simplify the correlation structure. Denoting the canonically preprocessed data by z_i, $i = 1, \ldots, n$ the maximization then takes the form

$$E\left\{h\left(\left[\sum_{i=1}^{n} v_i z_i\right]^2\right)\right\} \tag{6.18}$$

with respect to the weights v_i which are constrained so that

$$\|\mathbf{v}\|^2 = \sum_i v_i^2 = 1. \tag{6.19}$$

6.2.5 One Feature Learned from Natural Images

Consider again the three distributions in Fig. 5.1. All of them look quite sparse in the sense that the histograms (which are just estimates of the pdf's) have a peak at zero. It is not obvious what kind of features are maximally sparse. However, optimizing a sparseness measure we can find well-defined features.

Figure 6.5 shows the weights W_i obtained by finding a local maximum of sparseness, using the sparseness measure h_3 and canonically preprocessed data. It turns

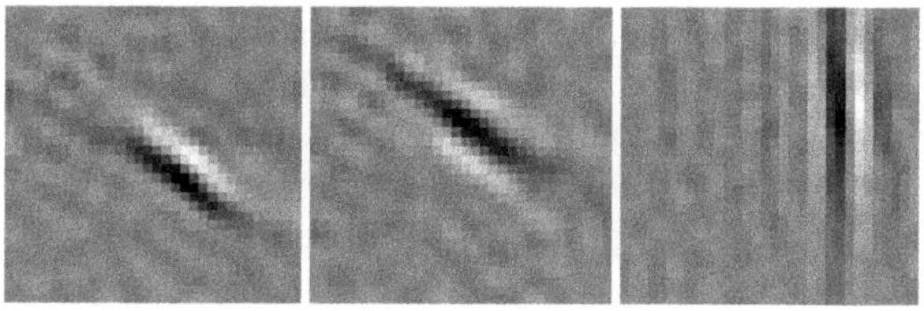

Fig. 6.5 Three weight vectors found by maximization of sparseness in natural images. The maximization was started in three different points which each gave one vector corresponding to one local maximum of sparseness

out that features similar to Gabor functions and simple-cell receptive fields are characterized by maximum sparseness. The features that are local maxima of sparseness, they turn out to have the three basic localization properties: they are localized in the (x, y)-space, localized in frequency (i.e. they are band-pass), and localized in orientation space (i.e. they are oriented).

Note that in contrast to variance, sparseness has many local maxima. Most local maxima (almost all, in fact) are localized in space, frequency, and orientation. The sparsenesses of different local maxima are often not very different from each other. In fact, if you consider a feature detectors whose weights are given by the Gabor functions which are but otherwise similar but are in two different locations, it is natural to assume that the sparsenesses of the two features must be equal, since the properties of natural images should be the same in all locations. The fact that sparseness has many local maxima forms the basis for learning many features.

6.3 Learning Many Features by Maximization of Sparseness

A single feature is certainly not enough: Any vision system needs many features to represent different aspects of an image. Since sparseness is locally maximized by many different features, we could, in principle, just find many different local maxima—for example, by running an optimization algorithm starting from many different random initial conditions. Such a method would not be very reliable, however, because the algorithm could find the same maxima many times.

A better method of learning many features is to find many local maxima that fulfill some given constraint. Typically, one of two options is used. First, we could constrain the detector weights W_i to be orthogonal to each other, just as in PCA. Second, we could constraint the different s_i to be uncorrelated. We choose here the latter because it is a natural consequence of the generative-model approach that will be explained in Chap. 7.

Actually, these two methods are not that different after all, because if the data is whitened as part of canonical preprocessing (see Sect. 5.4), orthogonality and uncorrelatedness are, in fact, the same thing, as was discussed in Sect. 5.3.2.2. This is one of the utilities in canonical preprocessing. Thus, decorrelation is equivalent to *orthogonalization*, which is a classic operation in matrix computations.

Note that there is no order that would be intrinsically defined between the features. This is in contrast to PCA, where the definition automatically leads to the order of the first, second, etc. principal component. One can order the obtained components according to their sparseness, but such an ordering is not as important as in the case of PCA.

6.3.1 Deflationary Decorrelation

There are basically two approaches that one can use in constraining the different feature detectors to have uncorrelated outputs. The first one is called deflation, and proceeds by learning the features one-by-one. First, one learns the first feature. Then one learns a second feature under the constraint that its output must be uncorrelated from the output of the first one, then a third feature whose output must be uncorrelated from the two first ones, and so on, always constraining the new feature to be uncorrelated from the previously found ones. In algorithmic form, this deflationary approach can be described as follows:

1. Set $k = 1$.
2. Find a vector W that maximizes the sparseness:

$$E\left\{h\left(\left[\sum_{x,y} W(x,y)I(x,y)\right]^2\right)\right\} \tag{6.20}$$

under the constraints of unit variance of *deflationary* decorrelation:

$$E\left\{\left(\sum_{x,y} W(x,y)I(x,y)\right)^2\right\} = 1, \tag{6.21}$$

$$E\left\{\sum_{x,y} W(x,y)I(x,y)\sum_{x,y} W_i(x,y)I(x,y)\right\} = 0 \quad \text{for all } 1 \leq i < k. \tag{6.22}$$

3. Store this vector in W_k and increment k by one.
4. If k does not equal the dimension of the space, go back to step 2.

The deflationary approach is easy to understand. However, it is not recommended because of some drawbacks. Basically, in the deflationary approach those features that are found in the beginning are privileged over others. They can do the optimization in the whole space whereas the last vectors (k close to the dimension of the space) have very little space where to optimize. This leads to the gradual deterioration of the features: the latter ones are often rather poor because their form is so severely restricted. In other words, the random errors (due to limited sample size), as well as numerical errors (due to inexact optimization) in the first feature weights propagate to latter weights, and produce new errors in them. A further problem is that the method is not very principled; in fact, the more principled approach to sparse coding discussed in the next chapter leads to the following method, symmetric decorrelation.

6.3.2 Symmetric Decorrelation

It would be more natural and efficient to use a method in which all the features are learned on an equal footing. This is achieved in what is called the symmetric approach. In the symmetric approach, we maximize the *sum* of the sparsenesses of the outputs. In this maximization, the outputs of all units are constrained to be uncorrelated. Thus, no filters are privileged. Using the measures of sparseness defined above, this leads to an optimization problem of the following form:

$$\text{Maximize} \quad \sum_{i=1}^{n} E\left\{h\left(\left[\sum_{x,y} W_i(x,y)I(x,y)\right]^2\right)\right\} \tag{6.23}$$

under the constraints of unit variance and *symmetric* decorrelation:

$$E\left\{\left(\sum_{x,y} W_i(x,y)I(x,y)\right)^2\right\} = 1 \quad \text{for all } i, \tag{6.24}$$

$$E\left\{\sum_{x,y} W_i(x,y)I(x,y)\sum_{x,y} W_j(x,y)I(x,y)\right\} = 0 \quad \text{for all } i \neq j. \tag{6.25}$$

This approach can also be motivated by considering that we are actually maximizing the sparseness of a representation instead of sparsenesses of the features; these concepts will be discussed next.

Whichever method of decorrelation is used, this approach limits the number of features that we can learn to the dimensionality of the data. For canonically preprocessed data, this is the dimensionality chosen in the PCA stage. This is because the features are constrained orthogonal in the whitened space, and there can be at most n orthogonal vectors in an n-dimensional space. Some methods are able to learn more features than this; they will be treated later in Sect. 13.1.

6.3.3 Sparseness of Feature vs. Sparseness of Representation

When considering a group of features, sparseness has two distinct aspects. First, we can look at the distribution of a single feature s when the input consists of many natural images $I_t, t = 1, \ldots, T$, as we did above—this is what we call the sparseness of features (or "lifetime sparseness"). The second aspect is to look at the distribution of the features s_i over the index $i = 1, \ldots, n$, for a single input image I—this is what we call the sparseness of the representation (or "population sparseness").

Sparseness of a representation means that a given image is represented by only a small number of active (clearly non-zero) features. This was, in fact, one of the main motivations of looking for sparse features in the first place, and it has been considered the defining feature of sparse coding, i.e. a sparse representation.

A sparse representation can be compared to a vocabulary in a spoken language. A vocabulary typically consists of tens of thousands of words. Yet, to describe a single event or a single object, we only need to choose a few words. Thus, most of the words are not active in the representation of a single event. In the same way, a sparse representation consists of a large number of potential features; yet, to describe a single input image, only a small subset of them are activated.

This kind of reduction of active elements must be clearly distinguished from dimension reduction techniques such as principal component analysis (PCA). In PCA, we choose once and for all a small set of features that are used for representing all the input patches. The number of these principal features is smaller than the dimension of the original data, which is why this is called dimension reduction. In a sparse representation, the active features are different from patch to patch, and the total number of features in the representation need not be smaller than the number of dimensions in the original data—in fact, it can even be larger.

What is then the connection between these two concepts of sparseness? Basically, we could measure the sparseness of the representation of a given image using the same measures as we used for the sparseness of the features. Thus, *for a single image* I_t, the sparseness of the representation given by the image filters W_i, $i = 1, \ldots, n$ can be measured as:

$$\sum_{i=1}^{n} h\left(\left[\sum_{x,y} W_i(x,y) I_t(x,y)\right]^2\right). \tag{6.26}$$

For this measure to be justified in the same way as we justified it above, it must be assumed that for the single image, the following two normalization conditions hold:

1. The mean of the features is zero.
2. The mean of the square of the features equals one (or any other constant).

While these conditions do not often hold exactly for a single image, they typically are approximately true for large sets of features. In particular, if the features are statistically independent and identically distributed (see Sect. 4.5), the conditions will be approximately fulfilled by the law of large numbers—the basic statistical law that says that the average of independent observations tends to the expectation.

Now, let us assume that we have observed T image patches $I_t(x, y)$, $t = 1, \ldots, T$, and let us simply take the sum of the sparsenesses of each image computed as in (6.26) above. This gives

$$\sum_{t=1}^{T} \sum_{i=1}^{n} h\left(\left[\sum_{x,y} W_i(x,y) I_t(x,y)\right]^2\right). \tag{6.27}$$

Rearranging the summations, we see that this is equal to

$$\sum_{i=1}^{n}\sum_{t=1}^{T} h\left(\left[\sum_{x,y} W_i(x,y)I_t(x,y)\right]^2\right). \tag{6.28}$$

The expression in (6.28) is the sum of the sparsenesses of the features. The expression in (6.27) is the sum of the sparsenesses of representations. Thus, we see that these two measures are equal. However, for this equality to be meaningful, it must be assumed that the normalization conditions given above hold as well. Above, we argued that they are approximately fulfilled if the features are approximately independent.

So, we can conclude that sparseness of features and sparseness of representation give approximately the same function to maximize, hence the same feature set. The functions are closer to equal when the feature sets are large and the features are statistically independent and have identical distributions. However, the measures might be different if the normalization conditions above are far from true.[4]

6.4 Sparse Coding Features for Natural Images

6.4.1 Full Set of Features

Now, we are ready to learn a whole set of features from natural images. We sampled randomly 50 000 image patches of 32×32 pixels, and applied canonical preprocessing to them, reducing the dimension to 256, which meant retaining 25% of the dimensions. We used the log cosh function, i.e. h_3 in (6.12), and symmetric decorrelation. The actual optimization was done using a special algorithm called FastICA, described in Sect. 18.7.

The obtained results are shown in Fig. 6.6. Again, the feature detector weights are coded so that the grey-scale value of a pixel means the value of the coefficient at that pixel. Grey pixels mean zero coefficients.

Visually, one can see that these feature detectors have interesting properties. First, they are localized in space: most of the coefficients are practically zero outside of a small receptive field. The feature detectors are also oriented. Furthermore, they are multiscale in the sense that most of them seem to be coding for small things whereas a few are coding for large things (in fact, so large that they do not fit in the window, so that the detectors are not completely localized).

---

[4]Here's a counterexample in which the sparseness of features is zero but the sparseness of representation is high. Consider ten independent Gaussian features with zero mean. Assume nine have a very small variance, and one of them has a very large variance. Each of the features, considered separately, is Gaussian, and thus not sparse. However, for each image, the feature distribution has nine values close to zero and one which is typically very large, and therefore the distribution is sparse. The key here is that the features have different variances, which violates the normalization conditions.

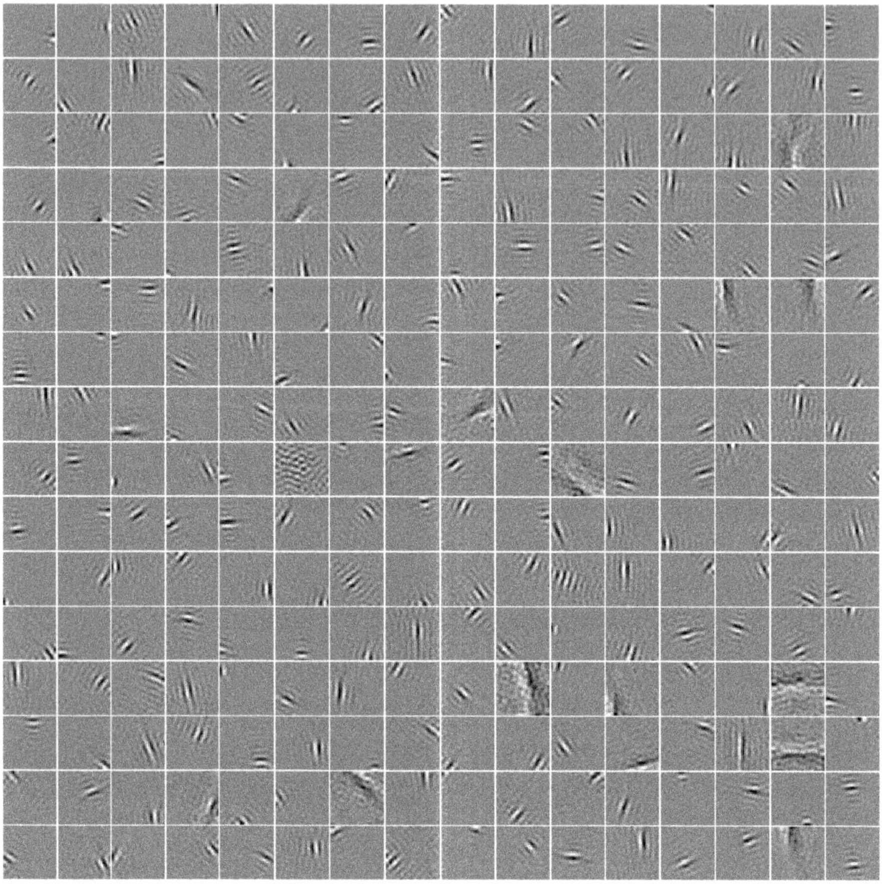

Fig. 6.6 The whole set of symmetrically orthogonalized feature vectors W_i maximizing sparsity, learned from natural images

6.4.2 Analysis of Tuning Properties

We can analyze the feature detectors W_i further by looking at the responses when gratings, i.e. sinusoidal functions, are input to them. In other words, we create artificial images which are two-dimensional sinusoids, and compute the outputs s_i. We consider sinusoidal functions of the form

$$f_o(x, y) = \sin\big(2\pi\alpha\big(\sin(\theta)x + \cos(\theta)y\big)\big), \tag{6.29}$$

$$f_e(x, y) = \cos\big(2\pi\alpha\big(\sin(\theta)x + \cos(\theta)y\big)\big). \tag{6.30}$$

These are sinusoidal gratings where θ gives the orientation (angle) of the oscillation, the x axis corresponding to $\theta = 0$. The parameter α gives the frequency. The two functions give two oscillations in different phases; more precisely, they are in quadrature-phase, i.e. a 90 degrees phase difference.

Now, we compute these functions for a large number of orientations and frequencies. We normalize the obtained functions to unit norm. Then we compute the dot-products of the W_i with each of the gratings. We can then compute the optimal orientation and frequency by finding the α and θ that maximize the sum of the squares of the two dot-products corresponding to the sin and cos functions. (We take the sum of squares because we do not want the phase of the W_i to have influence on this computation.)

This is actually almost the same as computing the 2-D power spectrum for all orientations and frequencies. We could do similar computations using the Discrete (or Fast) Fourier Transform as well, but we prefer here this direct computation for two reasons. First, we see the concrete meaning of the power spectrum in these computations. Second, we can compute the gratings for many more combinations of orientations and frequencies than is possible by the DFT.

In neurophysiology, this kind of analysis is usually done using drifting gratings. In other words, the gratings move on the screen in the direction of their oscillation. The maximum response of the cell for a drifting grating of a given (spatial) frequency and orientation is measured. This is more or less the same thing as the analysis that we are conducting here on our model simple cells. The fact that the gratings move in time may be necessary in neurophysiology because movement greatly enhances the cell responses, and so this method allows faster and more accurate measurement of the optimal orientation and frequency. However, it complicates the analysis because we have an additional parameter, the temporal frequency of the grating, in the system. Fortunately, we do not need to use drifting gratings in our analysis.

When we have found the optimal frequency and orientation parameters, we can analyze the selectivities by changing one of the parameters in the grating, and computing again the total response to two gratings that have the new parameters and are in quadrature phase. Such analysis of selectivity (tuning curves) is routinely performed in visual neuroscience.

In the same way, we can analyze the selectivity to phase. Here, we must obviously take a slightly different approach since we cannot take two filters in quadrature phase since then the total response would not depend on the phase at all. In neurophysiology, this is analyzed by simply plotting the response as a function of time when the input is a drifting grating with the optimal frequency and orientation. We can simulate the response by simply taking the dot-product of W_i with gratings whose phase goes through all possible values, and still keeping the orientation and frequency at optimal values. (The real utility of the analysis of phase selectivity will be seen when the responses of linear features are compared with non-linear ones in Chap. 10.)

In Fig. 6.7 we show the results of the analysis for the first ten features in Fig. 6.6, i.e. the first ten receptive fields on the first row. What we see is that all the cells are tuned to a specific values of frequency, orientation, and phase: any deviation from the optimal value decreases the response.

It is also interesting to look at how the (optimal) orientations and frequencies are related to each other. This is shown in Fig. 6.8. One can see that the model tries to

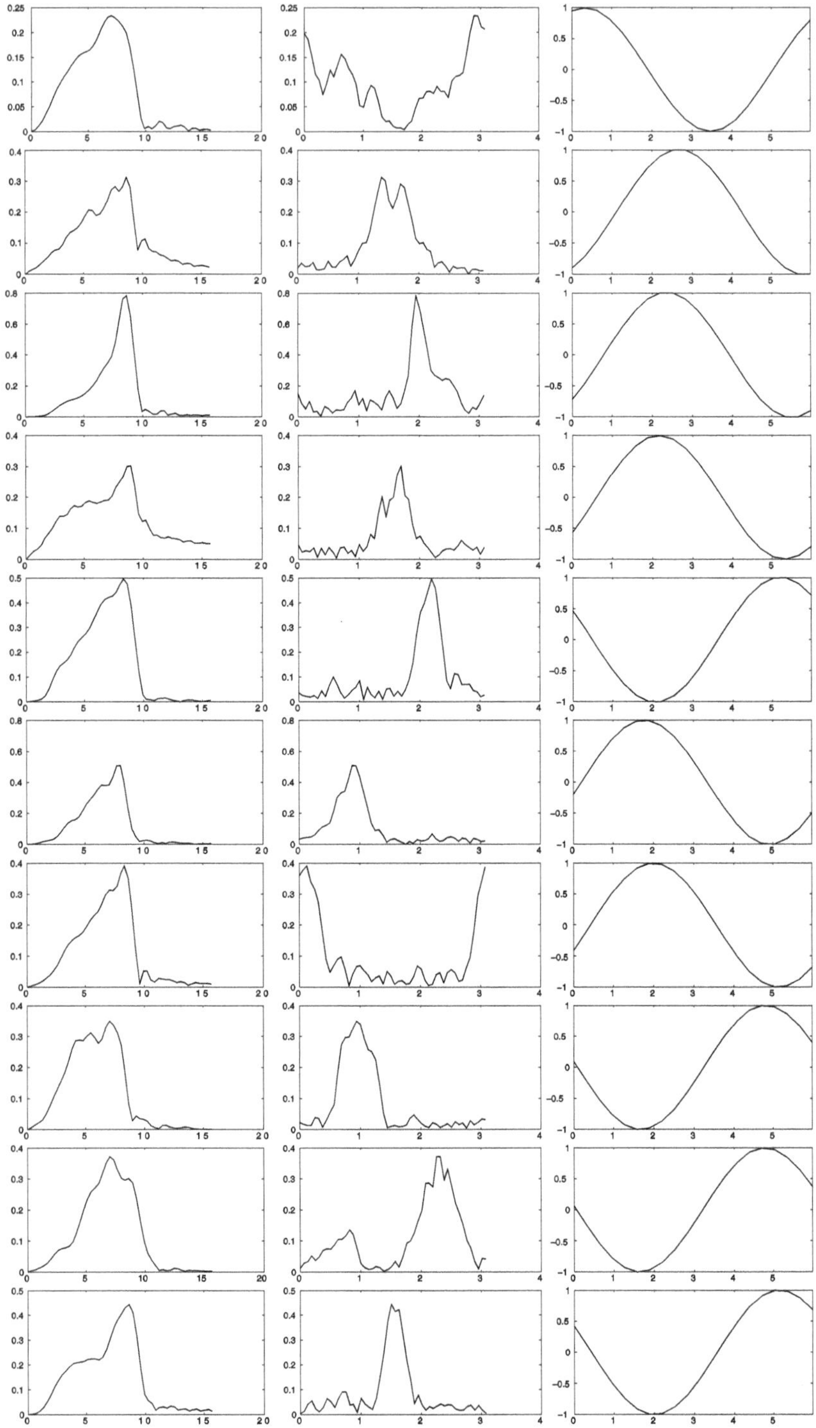

Fig. 6.7 Tuning curves of the ten first sparse coding features W_i in Fig. 6.6. *Left*: change in frequency (the unit is cycles in the window of 32×32 pixels, so that 16 means wavelength of 2 pixels). *Middle*: change in orientation. *Right*: change in phase

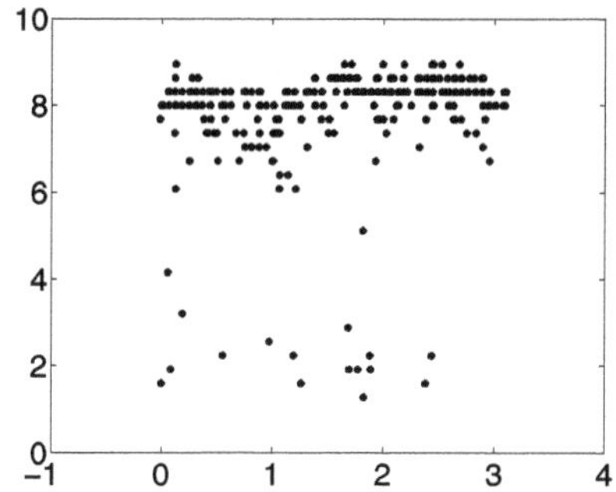

Fig. 6.8 Scatter plot of the frequencies and orientations of the sparse coding features. *Horizontal axis*: orientation, *vertical axis*: frequency

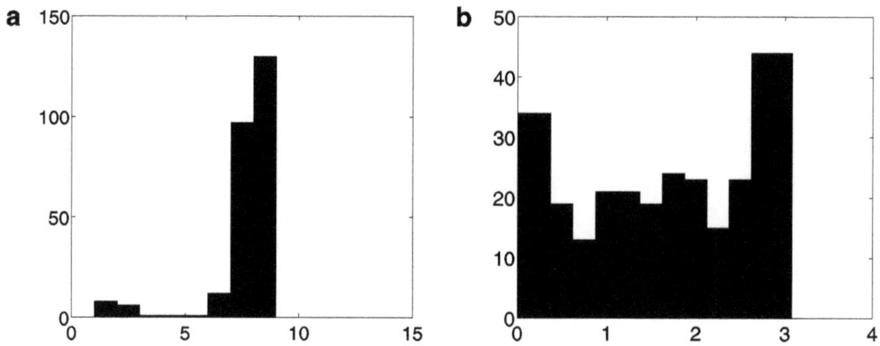

Fig. 6.9 Histograms of the optimal **a** frequencies and **b** orientations of the linear features obtained by sparse coding

cover all possible combinations of these variables. However, there is a strong emphasis on the highest frequencies that are present in the image. Note that preprocessing by PCA removed the very highest frequencies, so the highest frequencies present are much lower (approx. 9 cycles per patch) than the Nyquist frequency ($32/2 = 16$ cycles per patch).

Another way of looking at the distributions is to plot the histograms of the two parameters separately, as shown in Fig. 6.9. Here, we see again that most of the features have very high frequencies. The orientations are covered rather uniformly, but there are more features with horizontal orientation (0 or, equivalently, π). This is another expression of the anisotropy of natural images, already seen in the correlations in Sect. 5.7.

6.5 How Is Sparseness Useful?

6.5.1 Bayesian Modeling

The central idea in this book is that it is useful to find good statistical models for natural images because such models provide the prior probabilities needed in Bayesian inference or, in general, the prior information that the visual system needs on the environment. These tasks include de-noising and completion of missing data.

So, sparse coding models are useful for the visual system simply because they *provide a better statistical model of the input data*. The outputs of filter detectors are sparse, so this sparseness should be accounted for by the model. We did not really show that we get a better statistical model this way, but this point will be considered in the next chapter.

A related viewpoint is that of information theory: sparseness is assumed to lead to a more efficient code of the input. This viewpoint will be considered in Chap. 8.

6.5.2 Neural Modeling

Another viewpoint is to just consider the power of the statistical models to account for the properties of the visual system. From the viewpoint of computational neuroscience, *sparse coding leads to the emergence of receptive fields similar to simple cells*, so sparse coding is clearly a better model of the visual cortex in this respect than, say, PCA. Results in Chaps. 15 and 16 give even more support to this claim. This viewpoint does not consider *why* the visual system should use sparseness.

6.5.3 Metabolic Economy

However, there are other, additional, reasons as well why it would be advantageous for the visual system to use sparse coding, and these reasons have nothing to do with the statistics of the input stimuli. The point is that firing of cells consumes energy, and energy is one of the major constraints on the biological "design" of the brain. A sparse code means that most cells do not fire more than their spontaneous firing rate most of the time. Thus, sparse coding is *energy-efficient*.

So, we have a fortunate coincidence where those linear features that are optimal statistically are also optimal from the viewpoint of energy consumption. Possibly, future research will show some deep connections between these two optimality properties.

6.6 Concluding Remarks and References

In this chapter, we learned feature detectors which maximize the sparseness of their outputs when the input is natural images. Sparseness is a statistical property which is completely unrelated to variance, which was the criterion in PCA in the preceding chapter. Maximization of sparseness yield receptive fields which are quite similar to those of simple cells. This fundamental result is the basis of all the developments in the rest of this book.

Early work on finding maximally sparse projection can be found in Field (1987, 1994). Estimating a whole basis for image patches was first accomplished in the seminal paper (Olshausen and Field 1996) using a method considered in Sect. 13.1. A detailed comparison with simple cell receptive fields is in van Hateren and van der Schaaf (1998); see also van Hateren and Ruderman (1998). A discussion on sparseness of features vs. sparseness of representation is in Willmore and Tolhurst (2001).

The idea of increasing metabolic efficiency by sparse coding dates back to Barlow (1972); for more recent analysis, see e.g. Levy and Baxter (1996), Balasubramaniam et al. (2001), Attwell and Laughlin (2001), Lennie (2003).

Some researchers have actually measured the sparseness of real neuron outputs, typically concluding that they are sparse; see Baddeley et al. (1997), Gallant et al. (1998), Vinje and Gallant (2000, 2002), Weliky et al. (2003).

An approach that is popular in engineering is to take a fixed linear basis and then analyze the statistics of the coefficients in that basis. Typically, one takes a wavelet basis (see Sect. 17.3.2) which is not very much unlike the sparse coding basis. See Simoncelli (2005) for reviews based on such an approach.

Approaches for sparse coding using concepts related to spike trains instead of mean firing rates include Olshausen (2002), Smith and Lewicki (2005, 2006).

6.7 Exercises

Mathematical Exercises

1. Show that if $f(x)$ is a (strictly) convex function, i.e. fulfills (6.6), $f(x)+ax+b$ has the same property, for any constants a, b.
2. Show that the kurtosis of a Gaussian random variable is zero. (For simplicity, assume the variable is standardized to zero mean and unit variance. Hint: try partial integration to calculate the fourth moment.)
3. The Gram–Schmidt orthogonalization algorithm is defined as follows. Given n feature detector vectors $W_i(x, y)$ which have been normalized to unit norm, do
   a. Set $i \to 1$.
   b. Compute the new value of the vector w_i as
   $$W_i(x, y) \leftarrow W_i(x, y) - \sum_{j=1}^{i-1} \sum_{x', y'} W_j(x', y') W_i(x', y') W_j(x, y). \tag{6.31}$$
   c. Re-normalize W_i: $W_i(x, y) \leftarrow W_i(x, y)/\sqrt{\sum_{x',y'} W_i(x', y')^2}$.
   d. Increment i by one and go back to Step 1, if i is not yet larger than n.

   Show that the set of vectors is orthogonal after application of this algorithm.

Computer Assignments

1. Take some images. Take samples of 10×10 pixels. Construct a simple edge detector. Compute its output. Plot the histogram of the output, and compute its kurtosis.

Chapter 7
Independent Component Analysis

In this chapter, we discuss a statistical generative model called independent component analysis. It is basically a proper probabilistic formulation of the ideas underpinning sparse coding. It shows how sparse coding can be interpreted as providing a Bayesian prior, and answers some questions which were not properly answered in the sparse coding framework.

7.1 Limitations of the Sparse Coding Approach

In the preceding chapter, we showed that by finding linear feature detectors that maximize the sparseness of the outputs, we find features that are localized in space, frequency, and orientation, thus being similar to Gabor functions and simple cell receptive fields. While that approach had intuitive appeal, it was not completely satisfactory in the following respects:

1. The choice of the sparseness measure was rather ad hoc. It would be interesting to find a principled way of determining the optimal non-linear function h used in the measures.
2. The learning of many features was done by simply constraining the outputs of feature detectors to be uncorrelated. This is also quite ad hoc, and some justification for the decorrelation is needed.
3. The main motivation for this kind of statistical modeling of natural images is that the statistical model can be used as a prior distribution in Bayesian inference. However, just finding maximally sparse features does not give us a prior distribution.

A principled approach that also solves these problems is using generative models. A generative model describes how the observed data (natural images) is generated as transformations of some simple original variables. The original variables are called *latent* since they cannot usually be observed directly.

The generative model we propose here for modeling natural image patches is called independent component analysis. This model was originally developed to solve rather different kinds of problems, in particular, the so-called blind source separation problem, see the References section below for more information. However, it turns out that the same model can be interpreted as a form of sparse coding, and is more or less equivalent to finding linear features that are maximally sparse, as we will see in this chapter.

A. Hyvärinen, J. Hurri, P.O. Hoyer, *Natural Image Statistics*,
Computational Imaging and Vision 39,

7.2 Definition of ICA

7.2.1 Independence

The latent variables in independent component analysis (ICA) are called independent components. While the term "component" is mainly used for historical reasons (inspired by the expression "principal components"), the word "independence" tells what the basic starting point of ICA is: the latent variables are assumed to be statistically independent.

Let us consider two random variables, say s_1 and s_2. Basically, the variables s_1 and s_2 are statistically independent if information on the value of s_1 does not give any information on the value of s_2, and vice versa. In this book, whenever the word "independent" is used, it always refers to statistical independence, unless otherwise mentioned.

Section 4.5 gave a more extensive treatment of independence. Here, we recall the basic definition. Let us denote by $p(s_1, s_2)$ the joint probability density function of s_1 and s_2. Let us further denote by $p_1(s_1)$ the marginal pdf of s_1, i.e. the pdf of s_1 when it is considered alone. Then we define that s_1 and s_2 are independent if and only if the joint pdf is *factorizable*, i.e. the pdf can be expressed as a product of the individual marginal pdf's

$$p(s_1, s_2) = p_1(s_1)p_2(s_2). \tag{7.1}$$

This definition extends naturally for any number n of random variables, in which case the joint density must be a product of n terms. (Note that we use here a simplified notation in which s_i appears in two roles: it is the random variable, and the value taken by the random variable—often these are denoted by slightly different symbols.)

It is important to understand the difference between independence and uncorrelatedness. If the two random variables are independent, they are necessarily uncorrelated as well. However, it is quite possible to have random variables that are uncorrelated, yet strongly dependent. Thus, correlatedness is a special kind of dependence. In fact, if the two variables s_1 and s_2 were independent, any non-linear transformation of the outputs would be uncorrelated as well:

$$\operatorname{cov}\big(g_1(s_1), g_2(s_2)\big) = E\big\{g_1(s_1)g_2(s_2)\big\} - E\big\{g_1(s_1)\big\}E\big\{g_2(s_2)\big\} = 0 \tag{7.2}$$

for any two functions g_1 and g_2. When probing the dependence of s_i and s_j, a simple approach would thus be to consider the correlations of some non-linear functions. However, for statistical and computational reasons, we will develop a different approach below.

7.2.2 Generative Model

The generative model in ICA is defined by a linear transformation of the latent independent components. Let us again denote by $I(x, y)$ the pixel grey-scale values (point luminances) in an image, or in practice, a small image patch. In ICA, an image patch is generated as a linear superposition of some features A_i, as discussed in Sect. 2.3:

$$I(x, y) = \sum_{i=1}^{m} A_i(x, y)s_i \tag{7.3}$$

for all x and y. The s_i are coefficients that are different from patch to patch. They can thus be considered as random variables, since their values change randomly from patch to patch. In contrast, the features A_i are the same for all patches.

The definition of ICA is now based on three assumptions made regarding this linear generative model:

1. The fundamental assumption is that the s_i are *statistically independent* when considered as random variables.
2. In the next sections, we will also see that in order to be able to estimate the model, we will also have to assume that the distributions of the s_i are *non-Gaussian*. This assumption shows the connection to sparse coding since sparseness is a form of non-Gaussianity.
3. We will also usually assume that the linear system defined by the A_i is invertible but this is a technical assumption that is not always completely necessary. In fact, we will see below that we might prefer to assume that the linear system is invertible after canonical preprocessing, which is not quite the same thing.

These assumptions are enough to enable *estimation* of the model. Estimation means that given a large enough sample of image patches, I_t, $t = 1, \ldots, T$, we can recover some reasonable approximations of the values of A_i, without knowing the values of the latent components s_i in advance.

One thing that we cannot recover is the scaling and signs of the components. In fact, you could multiply a component s_i by any constant, say -2, and if you divide the corresponding A_i by the same constant, this does not show up in the data in any way. So, we can only recover the components up to a multiplicative constant. Usually, we simplify the situation by defining that the components have unit variance. This only leaves the signs of the components undetermined. So, for any component s_i, we could just as well consider the component $-s_i$.

As typical in linear models, estimation of the A_i is equivalent to determining the values of the W_i which give the s_i as outputs of linear feature detectors with some weights W_i:

$$s_i = \sum_{x,y} W_i(x, y)I(x, y) \tag{7.4}$$

for each image patch. The coefficients W_i are obtained by inverting the matrix of the A_i.

7.2.3 Model for Preprocessed Data

In practice, we will usually prefer to formulate statistical models for canonically preprocessed data (see Sect. 5.4). The data variables in that reduced representation are denoted by z_i. For a single patch, they can be collected to a vector $\mathbf{z} = (z_1, \ldots, z_n)$. Since a linear transformation of a linear transformation is still a linear transformation, the z_i are also linear transformations of the independent components s_i, although the coefficients are different from those in the original space. Thus, we have

$$z_i = \sum_{j=1}^{m} b_{ij} s_j \tag{7.5}$$

for some coefficients b_{ij} which can be obtained by transforming the features A_i using the same PCA transformation which is applied on the images.

We want to choose the number n of independent components so that the linear system can be inverted. Since we are working with preprocessed data, we will choose n so that it equals the number of variables after canonical preprocessing (instead of the number of original pixels). Then the system in (7.5) can be inverted in a unique way and we can compute the s_i as a linear function of the z_i:

$$s_i = \sum_{j=1}^{n} v_{ij} z_j = \mathbf{v}_i^T \mathbf{z}. \tag{7.6}$$

Here, the vector $\mathbf{v}_i = (v_{1i}, \ldots, v_{ni})$ allows a simple expression using vector products. The coefficients v_{ij} are obtained by inverting the matrix of the coefficients b_{ij}. The coefficients W_i in (7.4) are then obtained by concatenating the linear transformations given by v_{ij} and canonical preprocessing (i.e. multiplying the two matrices).

7.3 Insufficiency of Second-Order Information

When comparing the feature learning results by PCA and sparse coding, it is natural to conclude that the second-order information (i.e. covariances) used in PCA and other whitening methods is insufficient. In this section, we justify the same conclusion from another viewpoint: we show that second-order information is not sufficient for estimation of the ICA model, which also implies that the components should not be Gaussian.

7.3.1 Why Whitening Does Not Find Independent Components

It is tempting to think that if we just whiten the data, maybe the whitened components are equal to the independent components. The justification would be that

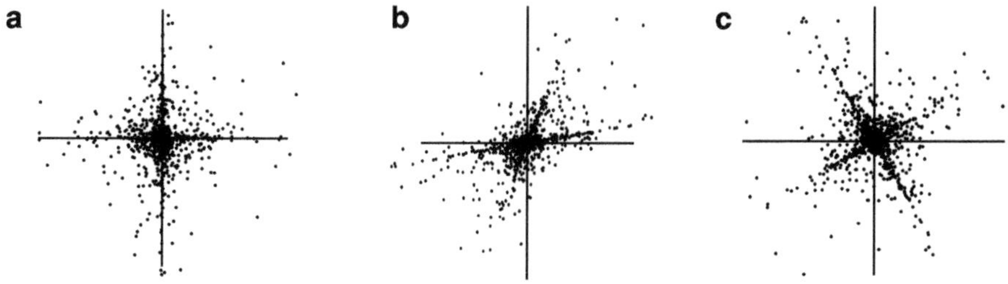

Fig. 7.1 **a** The joint distribution of the independent components s_1 and s_2 with sparse distributions. *Horizontal axis*: s_1, *vertical axis*: s_2. **b** The joint distribution of the observed data which are linear transformations of the s_1 and s_2. **c** The joint distribution of observed data after whitening by PCA

ICA is a whitening transformation because it gives components which are independent, and thus uncorrelated, and we defined their variances to be equal to one. The fundamental error with this logic is that there is an infinite number of whitening transformations because any orthogonal transformation of whitened data is still white, as pointed out in Sect. 5.3.2. So, if you whiten the data by, say, PCA, you get just one of those many whitening transformations, and there is absolutely no reason to assume that you would get the ICA transformation.

There is a reason why it is, in fact, not possible to estimate the ICA model using *any* method which is only based on covariances. This is due to the *symmetry* of the covariance matrix: $\text{cov}(z_1, z_2) = \text{cov}(z_2, z_1)$. Thus, the number of different covariances you can estimate from data is equal to $\frac{n(n+1)}{2}$, i.e. roughly one half of n^2. In contrast, the number of parameters b_{ij} we want to estimate (this refers to the model with preprocessed data in (7.5)) is equal to n^2. So, if we try to solve the b_{ij} by forcing the model to give just the right covariance structure, we have less equations (by one half!) than we have variables, so the solution is not uniquely defined! The same logic applies equally well to the original data before preprocessing.

This is illustrated in Fig. 7.1. We take two independent components s_1 and s_2 with very sparse distributions. Their joint distribution, in (a), has a "star-shape" because the data is rather much concentrated on the coordinate axes. Then we mix these variables linearly using randomly selected coefficients $b_{11} = 0.5$, $b_{12} = 1.5$, $b_{21} = 1$ and $b_{22} = 0.2$. The resulting distribution is shown in Fig. 7.1b. The star has now been "twisted". When we whiten the data with PCA, we get the distribution in (c). Clearly, the distribution is not the same as the original distribution in (a). So, whitening failed to recover the original components.

On the positive side, we see that the whitened distribution in Fig. 7.1c has the right "shape", because what remains to be determined is the right orthogonal transformation, since all whitening transformations are orthogonal transformations of each other. In two dimension, an orthogonal transformation is basically a rotation. So, we have solved part of the problem. After whitening, we know that we only need to look for the remaining orthogonal transformation, which reduces the space in which we need to search for the right solution.

Thus, we see why it was justified to constrain the different feature detectors to give uncorrelated outputs in the sparse coding framework in Sect. 6.3. Constraining the transformation to be orthogonal for whitened data is equivalent to constraining the features s_i to be uncorrelated (and to have unit variance). Even in the case of ICA estimation, the features are often constrained to be uncorrelated, because this simplifies the objective function, as discussed later in this chapter, and allows the development of very efficient algorithms (see Sect. 18.7). In contrast, in the ICA framework, it is not justified, for example, to constrain the original features A_i or the detector weights W_i to be orthogonal, since the mixing matrix (or rather, its inverse) is not necessarily orthogonal in the ICA model.

7.3.2 Why Components Have to Be Non-Gaussian

The insufficiency of second-order information also implies that the independent components must not be Gaussian because Gaussian data contains nothing else than second-order information. In this section, we explain a couple of different viewpoints which further elaborate this point.

7.3.2.1 Whitened Gaussian pdf is Spherically Symmetric

We saw above that after whitening, we have to find the right *rotation* (orthogonal transformation) which gives ICA. If the data is Gaussian, this is, in fact, not possible due to a symmetry property of Gaussian data.

To see why, let us consider the definition of the Gaussian pdf in (5.17) on page 109. Consider whitened variables, whose covariance matrix is the identity matrix by definition. The inverse of the identity matrix is the identity matrix, so $\mathbf{C}^{-1}$ is the identity matrix. Thus, we have $\sum_{ij} x_i x_j [\mathbf{C}^{-1}]_{ij} = \sum_i x_i^2$. Furthermore, the determinant of the identity matrix is equal to one. So, the pdf in (5.17) becomes

$$p(x_1,\dots,x_n) = \frac{1}{(2\pi)^{n/2}} \exp\left(-\frac{1}{2}\sum_i x_i^2\right) = \frac{1}{(2\pi)^{n/2}} \exp\left(-\frac{1}{2}\|\mathbf{x}\|^2\right) \tag{7.7}$$

This *pdf depends only on the norm* $\|\mathbf{x}\|$. Such a pdf is called spherically symmetric: It is the same in all directions. So, there is no information left in the data to determine the rotation corresponding to the independent components.

An illustration of this special property of the Gaussian distribution is in Fig. 7.2, which shows a scatter plot of two uncorrelated Gaussian variables of unit variance. The distribution is the same in all directions, except for random sampling effects. The circles show contours on which the pdf is constant. It is clear that if you rotate the data in any way, the distribution does not change, so there is no way to distinguish the right rotation from the wrong ones.

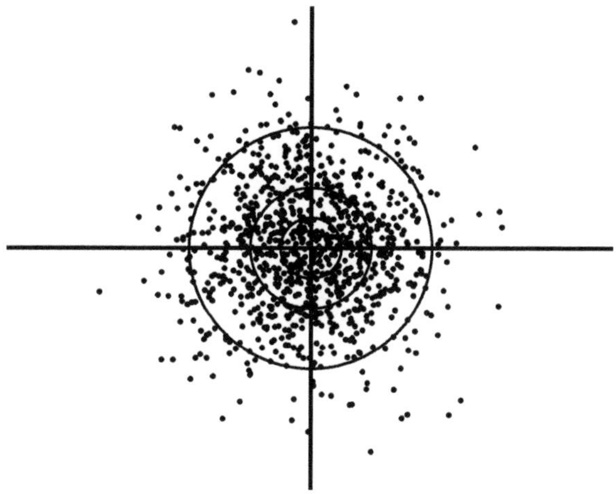

Fig. 7.2 A scatter plot of two uncorrelated Gaussian variables of unit variance. This is what any whitening method would give when applied on Gaussian data. The distribution is spherically symmetric, i.e. the same in all directions. This is also seen by looking at contours on which the pdf is constant: they are circles, as further plotted here

7.3.2.2 Uncorrelated Gaussian Variables Are Independent

A further justification why ICA is not possible for Gaussian variables is provided by a fundamental result in probability theory. It says that if random variables $s_1, \ldots, s_n$ have a Gaussian distribution and they are uncorrelated, then they are also independent. Thus, *for Gaussian variables, uncorrelatedness and independence are the same thing*, although in general uncorrelatedness does not imply independence. This further shows why ICA brings nothing new for Gaussian variables: The main interesting thing you can do to Gaussian variables is to decorrelate them, which is already accomplished by PCA and other whitening methods in Chap. 5.

It is easy to see from (7.7) why uncorrelated Gaussian variables are independent. Here, the variables are actually white, i.e. they have also been standardized to unit variance, but this makes no difference since such standardization obviously cannot change the dependencies between the variables. The point is that the pdf in (7.7) is something which can be factorized:

$$p(x_1, \ldots, x_n) = \prod_i \frac{1}{\sqrt{2\pi}} \exp\left(-\frac{1}{2}x_i^2\right) \tag{7.8}$$

where we have used the classic identity $\exp(a+b) = \exp(a)\exp(b)$. This form is factorized, i.e. it is a product of the one-dimensional standardized Gaussian pdf's. Such factorization is the essence of the definition of independence, as in (7.1). So, we have shown that Gaussian variables x_i are independent if they are uncorrelated.

Thus, the components in ICA have to be non-Gaussian in order for ICA to be meaningful. This also explains why models based on non-Gaussianity (such as ICA) are very new in the field of statistics: classic statistics is largely based on the assumption that continuous-valued variables have a Gaussian distribution—that is why it is called the "normal" distribution!

7.4 The Probability Density Defined by ICA

Now that we have a statistical generative model of the data, we can compute the probability of each observed image patch using basic results in probability theory. Then we can estimate the optimal features using classic estimation theory. This solves some of the problems we mentioned in the Introduction to this chapter: we will find the optimal measure of sparseness, and we will see why the constraint of uncorrelatedness of the features makes sense. And obviously, we can then use the model as a prior probability in Bayesian inference.

Let us assume for the moment that we know the probability density functions (pdf's) of the latent independent components s_i. These are denoted by p_i. Then by definition of independence, the multi-dimensional pdf of all the s_i is given by the product:

$$p(s_1, \ldots, s_n) = \prod_{i=1}^{n} p_i(s_i) \tag{7.9}$$

What we really want to find is the pdf of the observed preprocessed variables z_i, which is almost the same thing as having a pdf of the image patches $I(x, y)$. It is tempting to think that we could just plug the formula for the s_i given by (7.6) into (7.9). However, this is not possible. The next digression (which can be skipped by readers not interested in mathematical details) will show why not.

Short Digression to Probability Theory To see why we cannot just combine (7.9) and (7.6), let us consider what the pdf means in the one-dimensional case, where we have just one variable s with probability density p_s. By definition, the pdf at some point s_0 gives the probability that s belongs to a very small interval of length d as follows:

$$P\big(s \in [s_0, s_0 + d]\big) = p_s(s_0)d. \tag{7.10}$$

Now, let us consider a linearly transformed variable $x = as$ for $a > 0$. Here, s can be solved as $s = wx$ where $w = 1/a$ (note that we use a notation that is as close to the ICA case as possible). Let us just plug this in (7.10) and consider the probability at point $s_0 = wx_0$:

$$P\big(wx \in [wx_0, wx_0 + d]\big) = p_s(wx_0)d. \tag{7.11}$$

Obviously, $P(wx \in [x_1, x_2]) = P(x \in [x_1/w, x_2/w])$. So, we can express (7.11) as

$$p_s(wx_0)d = P\left(x \in \left[x_0, x_0 + \frac{d}{w}\right]\right) = p_x(x_0)\frac{d}{w} \tag{7.12}$$

Note that the length of the interval d changed to d/w, and so we changed the right-hand side of the equation to get the same term. Multiplying both sides of this equation by w/d, we get $p_s(wx_0)w = p_x(x_0)$. Thus, the actual pdf of x is $p_s(wx)w$, instead of simply $p_s(wx)$! This shows that in computing the pdf of a transformation, the *change in scale* caused by the transformation must be taken into account, by multiplying the probability density by a suitable constant that depends on the transformation.

In general, an important theorem in probability theory says that for any linear transformation, the probability density function should be multiplied by the absolute value of the *determinant* $\det \mathbf{V}$ of the matrix that gives the linear transformation. The determinant of a matrix is a measure of the associated change in scale (volume). The absolute value of the determinant is equal to the volume of the parallelepiped that is determined by its column vectors. (For more information on the determinant, see Sect. 19.4.)

Thus, the pdf of the preprocessed data $\mathbf{z}$ defined by ICA is actually given by

$$p(\mathbf{z}) = |\det(\mathbf{V})| \prod_{i=1}^{n} p_i(\mathbf{v}_i^T \mathbf{z}) = |\det(\mathbf{V})| \prod_{i=1}^{n} p_i\left(\sum_{j=1}^{n} v_{ij} z_j\right) \tag{7.13}$$

where $\mathbf{V}$ is a matrix whose elements are given by the coefficients v_{ij}; in other words, the rows of $\mathbf{V}$ are given by the vectors $\mathbf{v}_i$.

The pdf depends not only on the patch via $\mathbf{z}$, but also on the parameters of the model, i.e. the weights v_{ij}. Equivalently, we could consider the probability as a function of the features b_{ij}, but this does not make any difference, since the v_{ij} are uniquely determined by the b_{ij} and vice versa. The formula for the probability in (7.13) is more easily formulated as a function of the vectors $\mathbf{v}_i$.

7.5 Maximum Likelihood Estimation in ICA

Maximum likelihood estimation is a classic, indeed, *the* classic method for estimating parameters in a statistical model. It is based on a simple principle: Find those parameter values that would give the highest probability for the observed data. A brief description was provided in Sect. 4.8.

The *likelihood* is the probability of observing the data for given model parameters. For a given data set, it is thus a function of the parameters. Let us assume that we have observed T image patches $I_t(x, y)$, $t = 1, \ldots, T$ that are collected at random locations in some natural images. We consider here canonically preprocessed data, let us denote by $\mathbf{z}_t$ the vector obtained by canonically preprocessing the image patch I_t.

Because the patches are collected in random locations, we can assume that the patches are independent from each other. Thus, the probability of observing all these patches is the product of the probabilities of each patch. This gives the likelihood L of the observed data:

$$L(\mathbf{v}_1, \ldots, \mathbf{v}_n) = \prod_{t=1}^{T} p(\mathbf{z}_t) = \prod_{t=1}^{T} \left[|\det(\mathbf{V})| \prod_{i=1}^{n} p_i(\mathbf{v}_i^T \mathbf{z}_t) \right]. \tag{7.14}$$

It is much simpler to look at the logarithm of the likelihood, which is after some simple rearrangements:

$$\log L(\mathbf{v}_1, \ldots, \mathbf{v}_n) = T \log|\det(\mathbf{V})| + \sum_{i=1}^{n} \sum_{t=1}^{T} \log p_i(\mathbf{v}_i^T \mathbf{z}_t). \tag{7.15}$$

Since the logarithm is a increasing function, maximization of the likelihood is the same as maximization of this log-likelihood. Estimation by the maximum likelihood method now means that we maximize the log-likelihood in (7.15) with respect to the parameters, that is, the weights $\mathbf{v}_i$. (Choosing the functions $\log p_i$ will be discussed in Sect. 7.7.)

Maximization of the log-likelihood can be accomplished by numerical optimization methods. In addition to general-purpose methods, special tailor-made methods have been developed for this particular ICA maximization task. A thorough discussion of such optimization methods can be found in Chap. 18, and we will not go into detail here. Let us just note that the first term in (7.15) can be considered to be constant and omitted, as we will see in Sect. 7.7.1.

7.6 Results on Natural Images

7.6.1 Estimation of Features

Using maximum likelihood estimation on 50 000 image patches of size 32×32 pixels as in the preceding chapters, we obtain the results in Fig. 7.3. These features have the same qualitative properties as the feature detectors estimated by maximization of sparseness in Fig. 6.6 on page 144. That is, the features are spatially localized, oriented, and code for different scales (frequencies).

This is actually not surprising because as will be shown next, maximum likelihood estimation of ICA is mathematically almost equivalent to the sparse coding analysis we did in Sect. 6.4. The only difference is that we are here showing the (generating) features A_i instead of the feature detectors W_i. This difference is explained in detail in Sect. 7.10.

7.6.2 Image Synthesis Using ICA

Now that we have defined a generative model, we can generate image data from it. We generate the values of the s_i independently from each other, and multiply the estimated features A_i with them to get one generated image patch. One choice we have to make is what our model of the marginal (i.e. individual) distributions of the independent component is. We use here two distributions. In the first case, we simply take the histogram of the actual component in the natural images, i.e. the histogram of each $\sum_{x,y} W_i(x, y)I(x, y)$ when computed over the whole set of images. In the second case, we use a well-known sparse distribution, the Laplacian distribution (discussed in the next section), as the distribution of the independent components.

Figures 7.4 and 7.5 show the results in the two cases. The synthesis results are clearly different than those obtained by PCA on page 111: Here, we can see more oriented, edge-like structures. However, we are obviously far from reproducing all the properties of natural images.

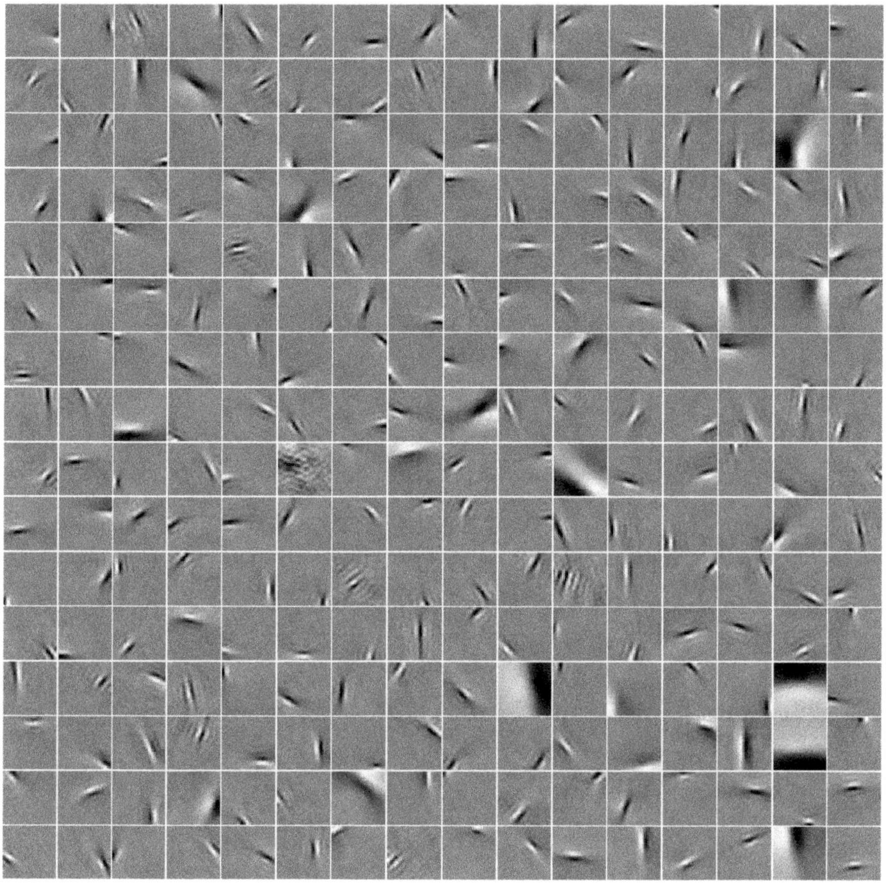

Fig. 7.3 The whole set of features A_i obtained by ICA. In this estimation, the functions $\log p_i$ were chosen as in (7.19) in Sect. 7.7

7.7 Connection to Maximization of Sparseness

In this section, we show how ICA estimation is related to sparseness, how we should model the $\log p_i$ in the log-likelihood in (7.15) and how this connection tells us how we should design the sparseness measure.

7.7.1 Likelihood as a Measure of Sparseness

Let us assume, as we typically do, that the linear features considered are constrained to be uncorrelated and to have unit variance. This is equivalent to assuming that the transformation given by $\mathbf{V}$ is orthogonal in the canonically preprocessed (whitened)

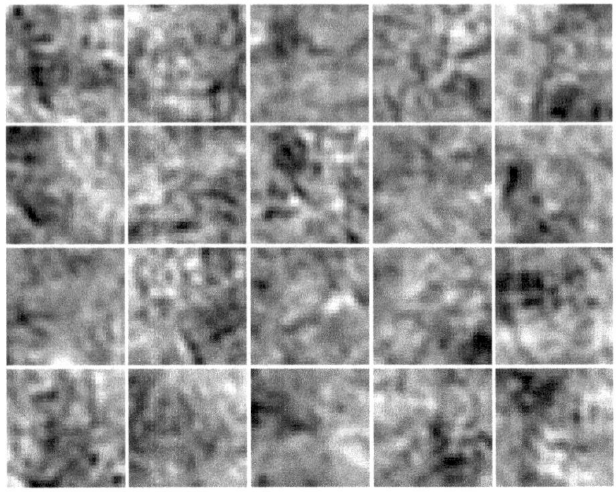

Fig. 7.4 Image synthesis using ICA. 20 patches were randomly generated using the ICA model whose parameters were estimated from natural images. In this figure, the marginal distributions of the components were those of the real independent components. Compare with real natural image patches in Fig. 5.2 on page 95, and the PCA synthesis results in Fig. 5.13 on page 111

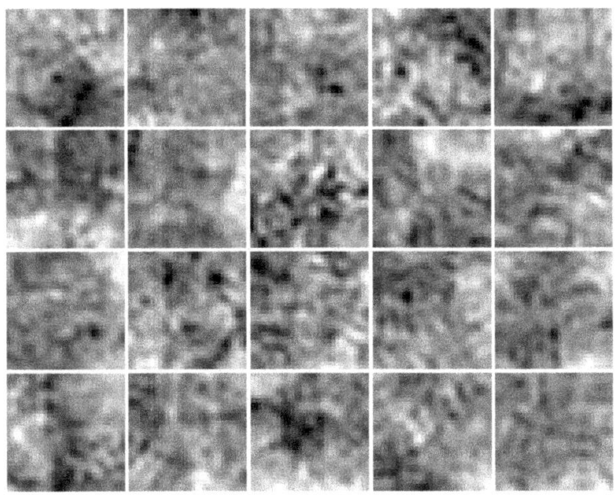

Fig. 7.5 Image synthesis using ICA, and a Laplacian approximation of the pdf of the independent components. Compare with Fig. 7.4, in which the real distributions were used for the independent components. The results are perhaps less realistic because the Laplacian distribution is less sparse than the real distributions

space. Thus, the matrix $\mathbf{V}$ is constrained orthogonal. It can be proven that the determinant of an orthogonal matrix is always equal to ± 1. This is because an orthogonal transformation does not change distances, and thus not volumes either; so the absolute value of the determinant which measures the change in volume must be equal to 1. Thus, the first term on the right-hand side of (7.15) is zero and can be omitted.

The second term on the right-hand side of (7.15) is the expectation (multiplied by T) of a non-linear function $\log p_i$ of the output s_i of the feature detector (more precisely, an estimate of that expectation, since this is computed over the sample). Thus, what the likelihood really boils down to is measuring the expectations of the form $E\{f(s_i)\}$ for some function f.

The connection to maximization of sparseness is now evident. If the feature outputs are constrained to have unit variance, maximization of the likelihood is equivalent to maximization of the sparsenesses of the outputs, *if* the functions $\log p_i$ are of the form required for sparseness measurements, i.e. if we can express them as

$$\log p_i(s) = h_i\left(s^2\right) \tag{7.16}$$

where the functions h_i are *convex*. In other words, the functions

$$h_i(u) = \log p_i\left(\sqrt{u}\right) \tag{7.17}$$

should be convex for $u \geq 0$. It turns out that this is usually the case in natural images, as will be seen in the next section.

Earlier, we considered using the negative of square root as h_i. In the probabilistic interpretation given by ICA, using the square root means that the pdf of a component s_i is of the form

$$p(s_i) = \frac{1}{\sqrt{2}} \exp\left(-\sqrt{2}|s_i|\right) \tag{7.18}$$

where the constants have been computed so that s_i has unit variance, and the integral of the pdf is equal to one, as is always required. This distribution is called Laplacian. It is also sometimes called the double exponential distribution, since the absolute value of s_i has the classic exponential distribution (see (4.71) on page 88). The Laplacian pdf was already illustrated in Fig. 6.2 on page 132.

As already pointed out in Chap. 6, using the Laplacian pdf maybe numerically problematic because of the discontinuity of its derivative. Thus, one might use a smoother version, where the absolute value function is replaced by the log cosh function. This also corresponds to assuming a particular pdf for the independent components, usually called the "logistic" pdf. When properly normalized and standardized to unit variance, the pdf has the form

$$\log p_i(s) = -2 \log\cosh\left(\frac{\pi}{2\sqrt{3}} s\right) - \frac{4\sqrt{3}}{\pi}. \tag{7.19}$$

In practice, the constants here are often ignored, and simply the plain log cosh function often is used to keep things simple.

7.7.2 Optimal Sparseness Measures

The maximum likelihood framework tells us what the non-linearities used in the sparseness measure really should be. They should be chosen according to (7.17). These non-linearities can be estimated from natural images. To really find the best non-linearities, we could first maximize the likelihood using some initial guesses of the h_i, then estimate the pdf's of the obtained independent components and recompute the h_i according to (7.17). In principle, we should then re-estimate the W_i using these new h_i, re-estimate the h_i using the latest W_i and so on until the process converges. This is because we are basically maximizing the likelihood with respect to two different groups of parameters (the W_i and the h_i) and the real maximum can only be found if we go on maximizing one of the parameters groups with the other fixed until no increase in likelihood can be obtained. However, in practice we do not need to bother to re-iterate this process because the h_i do not change that much after their initial estimation.

Figure 7.6 shows two h_i's estimated from natural images with the corresponding log-pdf's; most of them tend to be very similar. These are obtained by computing a histogram of distribution of two independent components estimated by a fixed h_i: the histogram gives an estimate of p_i from which h_i can be derived.

We see that the estimated h_i are convex, if we ignore the behavior in the tails, which are impossible to estimate exactly because they contain so few data points. The h_i estimated here are not very different from the square root function, although sometimes they tend to be more peaked.

If we want to use such optimal non-linearities in practice, we need to use parametric probability models for the (log-)pdf's. Using histogram estimates that we show here is not used in practice because such estimates can be very inexact and non-smooth. One well-known option for parameterized densities is the generalized Gaussian density (sometimes also called the generalized Laplacian density):

$$p(s) = \frac{1}{c} \exp\left(-\frac{|s_i|^\alpha}{b^\alpha}\right). \tag{7.20}$$

The parameters b and c are determined so that this is a pdf (i.e. its integral equals one) which has unit variance. The correct values are

$$b = \sqrt{\frac{\Gamma(\frac{1}{\alpha})}{\Gamma(\frac{3}{\alpha})}} \quad \text{and} \quad c = \frac{2b\sqrt{\pi}\Gamma(\frac{1}{\alpha})}{\alpha\Gamma(1/2)} \tag{7.21}$$

where Γ is the so-called "gamma" function which can be computed very fast in most software for scientific computation. The parameter $\alpha > 0$ controls the sparseness of the density. If $\alpha = 2$, we actually have the Gaussian density, and for $\alpha = 1$, the Laplacian density. What is most interesting when modeling images is that for $\alpha < 1$, we have densities that are sparser than the Laplacian density, and closer to the highly sparse densities sometimes found in image features.

Another choice is the following density:

$$p(s) = \frac{1}{2} \frac{(\alpha+2)[\alpha(\alpha+1)/2]^{(\alpha/2+1)}}{[\sqrt{\alpha(\alpha+1)/2} + |s|]^{(\alpha+3)}} \tag{7.22}$$

with a sparseness parameter α. When $\alpha \to \infty$, the Laplacian density is obtained as the limit. The strong sparsity of the densities given by this model can be seen, e.g. from the fact that the kurtosis of these densities is always larger than the kurtosis of the Laplacian density, and reaches infinity for $\alpha \leq 2$. Similarly, $p(0)$ reaches infinity as α goes to zero.

A problem with these highly peaked distributions is that they are not smooth, in particular their derivatives are discontinuous at zero. For the generalized Gaussian

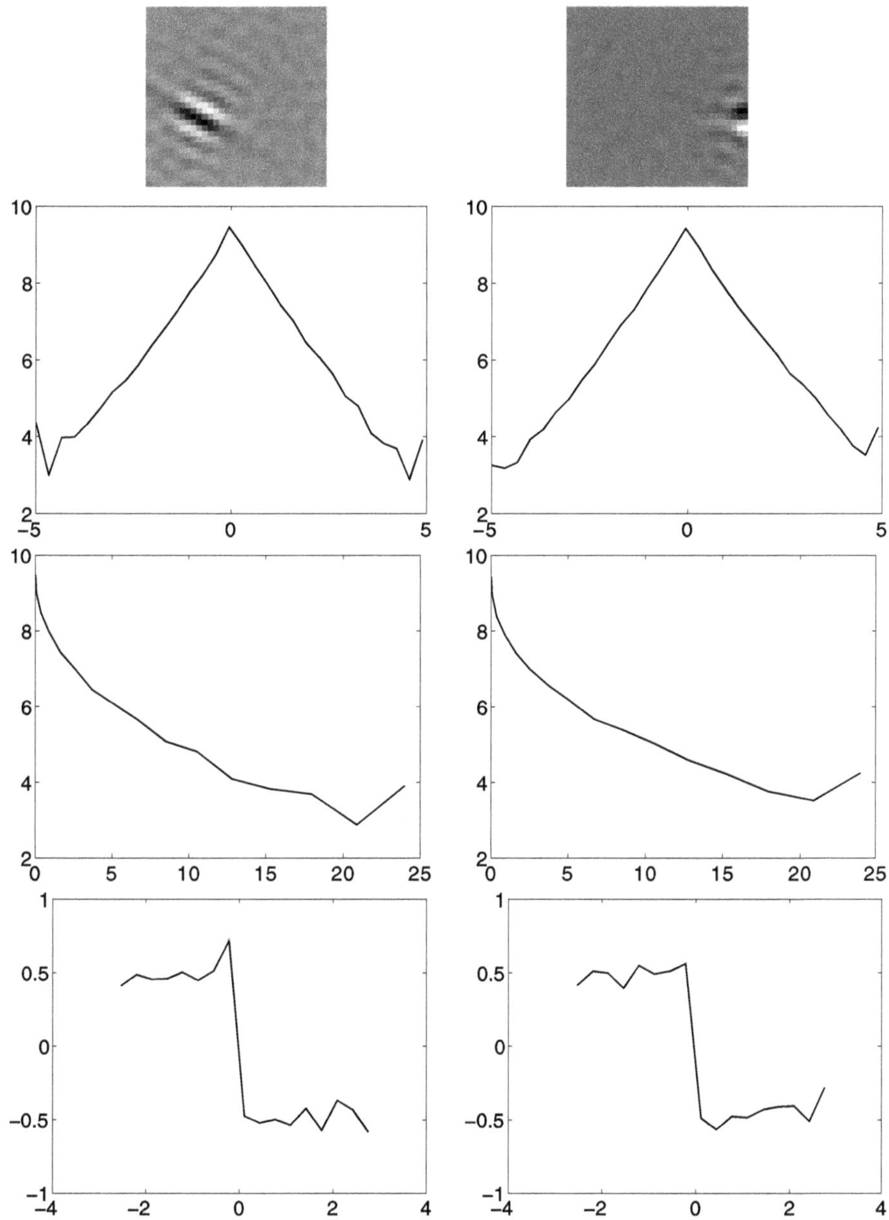

Fig. 7.6 Estimated optimal h_i from natural images. After doing ICA, the histograms of the component with the highest kurtosis and a component with kurtosis in the middle range were computed, and their logarithms taken. The feature corresponding to the highest kurtosis is on the *left*, and the one corresponding to the mid-range kurtosis is on the *right*. *Top row*: feature. *Second row*: logarithm of pdf. *Third row*: optimal h_i. *Bottom row*: the derivative of log-pdf for future reference

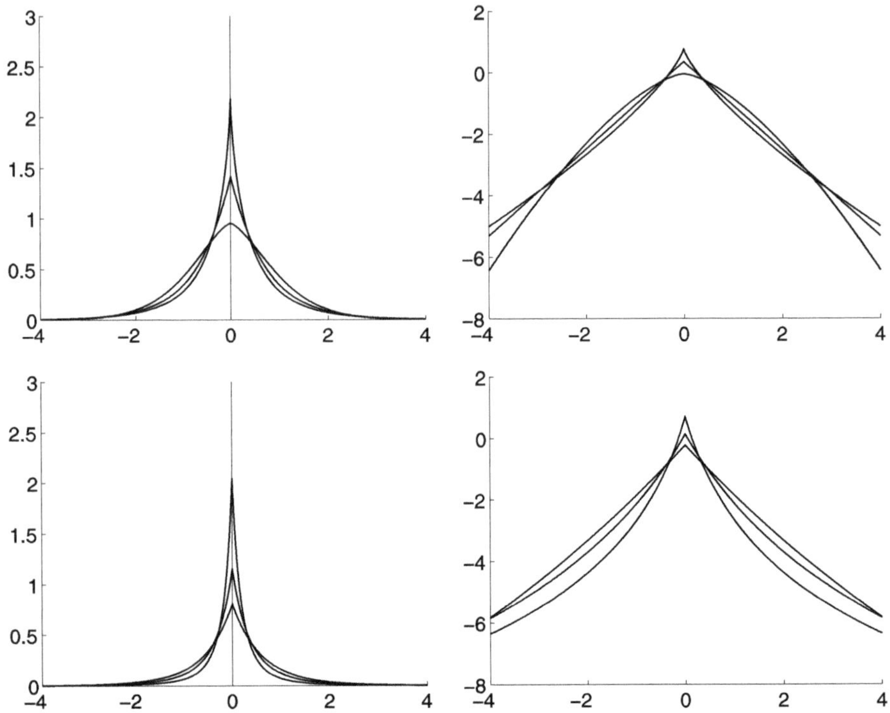

Fig. 7.7 *Top row*: Some plots of the density function (*left*) and its logarithm (*right*) given in (7.20), α is given values 0.75, 1 and 1.5. More peaked ones correspond to smaller α. *Bottom row*: Plots of the density function (7.22), α is given values 0.5, 2, 10. More peaked values correspond to smaller α

distribution, the derivative is actually infinite at zero for $\alpha < 1$. Thus, to avoid problems in the computational maximization of sparseness measures, it may not be a bad idea to use something more similar to a square root function in practical maximization of the sparseness measures. Actually, usually we use a smoothed version of the square root function as discussed in Sect. 7.7.1.

The two density families in (7.20) and (7.22) are illustrated in Fig. 7.7. While it does not seem necessary to use such more accurate density models in the estimation of the basis, they are likely to be quite useful in Bayesian inference where we really do need a good probabilistic model.

7.8 Why Are Independent Components Sparse?

There are many different ways in which random variables can be non-Gaussian. What forms do there exist, and why is it that independent components in images are always sparse—or are they? These are the questions that we address in this section.

7.8.1 Different Forms of Non-Gaussianity

While the forms of non-Gaussianity are infinite, most of the non-Gaussian aspects that are encountered in real data can be described as sub-Gaussianity, super-Gaussianity, or skewness.

Super-Gaussianity is basically the same as sparseness. Often, super-Gaussianity is defined as positive kurtosis (see (6.5) for a definition of kurtosis), but other definitions exist as well. The intuitive idea is that the probability density function has heavy tails and a peak at zero.

The opposite of super-Gaussianity is sub-Gaussianity, which is typically characterized by negative kurtosis. The density function is "flat" around zero. A good example is the uniform distribution (here standardized to unit variance and zero mean)

$$p(s) = \begin{cases} \frac{1}{2\sqrt{3}}, & \text{if } |s| \leq \sqrt{3}, \\ 0, & \text{otherwise.} \end{cases} \tag{7.23}$$

The kurtosis of this distribution equals $-6/5$, which is left as an exercise.

An unrelated form of non-Gaussianity is skewness which basically means the lack of symmetry of the probability density function. A typical example is the exponential distribution:

$$p(s) = \begin{cases} \exp(-s), & \text{if } s \geq 0, \\ 0, & \text{otherwise} \end{cases} \tag{7.24}$$

which is not symmetric with respect to any point on the horizontal (s) axis. Skewness is often measured by the third moment (assuming the mean is zero)

$$\text{skew}(s) = E\{s^3\}. \tag{7.25}$$

This is zero for a symmetrically-distributed random variable that has zero mean (this is left as an exercise). In fact, skewness is usually defined as exactly the third moment. However, any other non-linear odd function could be used instead of the third power, for example the function that gives the sign (± 1) of s.

7.8.2 Non-Gaussianity in Natural Images

Is it true that all the independent components in natural images are sparse, and no other forms of non-Gaussianity are encountered? This is almost true, but not quite.

The skewness of the components is usually very small. After all, natural images tend to be rather symmetric in the sense that black and white are equally probable. This may not be exactly so, since such symmetry depends on the measurement scale of the grey-scale values: a non-linear change of measurement scale will make symmetric data skewed. However, in practice, skewness seems to be so small that it can be ignored.

There are some sub-Gaussian components, though. In particular, the DC component, i.e. the mean luminance of the image patch is typically sub-Gaussian. In fact, its distribution is often not far from a uniform distribution. If we do not remove the DC component from the images (in contrast to what we usually do), and we use an ICA algorithm that is able to estimate sub-Gaussian components as well (not all of them are), the DC component actually tends to be estimated as one independent component. Depending on the window size and the preprocessing used, a couple of further very low-frequency components can also be sub-Gaussian.

7.8.3 Why Is Sparseness Dominant?

One reason why the independent components in images are mostly sparse is the variation of the local variance in different parts of an image. Some parts of the image have high variation whereas others have low variation. In fact, flat surfaces have no variation, which is sometimes called the blue-sky effect.

To model this change in local variance, let us model an independent component s_i as a product of an "original" independent component g_i of unit variance, and an independent, non-negative "variance" variable d_i:

$$s_i = g_i\, d_i. \tag{7.26}$$

We call d_i a variance variable because it changes the scale of each observation of g_i. Such a variance variables will be the central topic in Chap. 9.

Let us assume that the original component g_i is Gaussian with zero mean and unit variance. Then the distributions of the s_i is necessarily super-Gaussian, i.e. it has positive kurtosis. This can be shown using fact that for a Gaussian variable, kurtosis is zero, and thus $E\{g_i^4\} = 3$. So, we have

$$\begin{aligned} \operatorname{kurt} s_i &= E\{s_i^4\} - 3\big(E\{s_i^2\}\big)^2 = E\{d_i^4 g_i^4\} - 3\big(E\{d_i^2 g_i^2\}\big)^2 \\ &= E\{d_i^4\}E\{g_i^4\} - 3\big(E\{d_i^2\}\big)^2\big(E\{g_i^2\}\big)^2 = 3\big[E\{d_i^4\} - \big(E\{d_i^2\}\big)^2\big] \end{aligned} \tag{7.27}$$

which is always non-negative because it is the variance of d_i^2 multiplied by 3. It can be zero only if d_i is constant.

Thus, the changes in local variance are enough to transform the Gaussian distribution of g_i into a sparse distribution for s_i. The resulting distribution is called a Gaussian scale mixture.

7.9 General ICA as Maximization of Non-Gaussianity

Now, we can consider the problem of ICA estimation in more generality, in the case where the components are not necessarily sparse. In particular, we consider the following two questions: Does estimation of ICA for non-sparse components have

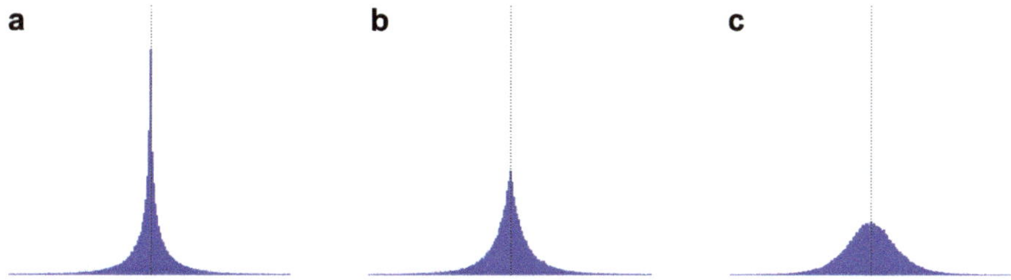

Fig. 7.8 **a** Histogram of a very sparse distribution. **b** Histogram of a sum of two independent random variables distributed as in **a**, normalized by dividing by $\sqrt{2}$. **c** Histogram of a normalized sum of ten variables with the same distribution as in **a**. The scale of the axes are the same in all plots. We see that the distribution goes toward Gaussianity

a simple intuitive interpretation, and is there a deeper reason why maximization of sparseness related to the estimation of the ICA model? These questions can be answered based on the central limit theorem, a most fundamental theorem in probability theory. Here, we explain this connection and show how it leads to a more general connection between independence and non-Gaussianity.

7.9.1 Central Limit Theorem

The Central Limit Theorem (CLT) basically says that when you take an average or sum of many independent random variables, it will have a distribution that is close to Gaussian. In the limit of an infinite number of random variables, the distribution actually tends to the Gaussian distribution, if properly normalized:

$$\lim_{N\to\infty} \frac{1}{\sqrt{N}} \sum_{n=1}^{N} s_n = \text{Gaussian} \tag{7.28}$$

where we assume that s_n have zero mean. We have to normalize the sum here because otherwise the variance of the sum would go to infinity. Note that if we normalized by $1/N$, the variance would go to zero.

Some technical restrictions are necessary for this results to hold exactly. The simplest choice is to assume that the s_n all have the same distribution, and that distribution has finite moments. The CLT is illustrated in Fig. 7.8.

7.9.2 "Non-Gaussian Is Independent"

What does the CLT mean in the context of ICA? Let us consider a linear combination of the observed variables, $\sum_i w_i z_i$. This is also a linear combination of the

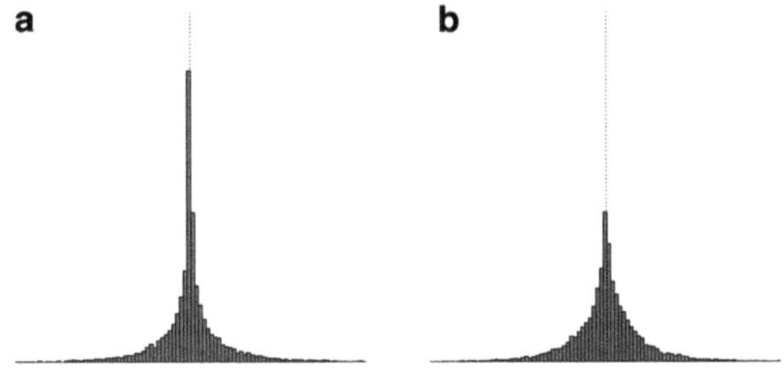

Fig. 7.9 **a** Histogram of one of the original components in Fig. 7.1. **b** Histogram of one of the whitened components in Fig. 7.1. The whitened component has a distribution which is less sparse, thus closer to Gaussian

original independent components:

$$\sum_i w_i z_i = \sum_i w_i \sum_j a_{ij} s_j = \sum_j \Big(\sum_i w_i a_{ij} \Big) s_j = \sum_j q_j s_j \tag{7.29}$$

where we have denoted $q_j = \sum_i w_i a_{ij}$. We do not know the coefficients q_j because they depend on the a_{ij}.

The CLT would suggest that this linear combination $\sum_j q_j s_j$ is closer to Gaussian than the original independent components s_j. This is not exactly true because the CLT is exactly true only in the limit of an infinite number of independent components, and there are restrictions on the distributions (for example, the variables $q_j s_j$ do not have identical distributions if the q_j are not equal). However, the basic idea is correct. This is illustrated in Fig. 7.9 which shows that the original independent components are more Gaussian than the observed data after whitening, shown in Fig. 7.1.

Thus, based on the central limit theorem, we can intuitively motivate a general principle for ICA estimation: find linear combinations $\sum_i w_i z_i$ of the observed variables that are maximally non-Gaussian.

Why would this work? The linear combination $\sum_i w_i z_i$ equals a linear combination of the independent components with some coefficients q_j. Now, if more than one of the q_j is non-zero, we have a sum of two independent random variables. Because of the CLT, we can expect that such a sum is closer to Gaussian that any of the original variables. (This is really only an intuitive justification and not exactly true.) Thus, the non-Gaussianity of such a linear combination is maximal when it equals one of the original independent components, and the maximally non-Gaussian linear combinations are the independent components.

Here, we have to emphasize that this connection between non-Gaussianity and independence only holds for linear transformations. In Chap. 9, we will see that for non-linear transformations, such a connection need not exist at all, and may in fact be reversed.

7.9.3 Sparse Coding as a Special Case of ICA

Estimation of ICA by maximization of sparseness can now be seen as a special case of maximization of non-Gaussianity. Sparseness is one form of non-Gaussianity, the one that is dominant in natural images. Thus, in natural images, maximization of non-Gaussianity is basically the same as maximization of sparseness. For other types of data, maximization of non-Gaussianity may be quite different from maximization of sparseness.

For example, in the theory of ICA, it has been proposed that the non-Gaussianity of the components could be measured by the sum of the squares of the kurtoses:

$$\sum_{i=1}^{n} \left[\text{kurt}\left(\mathbf{v}_i^T \mathbf{z}\right)\right]^2 \tag{7.30}$$

where, as usual, the data vector $\mathbf{z}$ is whitened and the feature vectors $\mathbf{v}_i$ are constrained to be orthogonal and to have unit norm. It can be shown that ICA estimation can really be accomplished by maximizing this objective function. This works for both sub-Gaussian and super-Gaussian independent components.

Now, if the components all have positive kurtoses, maximizing this sum is closely related to finding vectors $\mathbf{v}_i$ such that the $\mathbf{v}_i^T \mathbf{z}$ are maximally non-Gaussian. The square of kurtosis is, however, a more general measure of non-Gaussianity because there are cases where the kurtosis is negative as we saw above in Sect. 7.8.1. For such components, maximization of non-Gaussianity means *minimizing* kurtosis (and sparseness), because for negative values of kurtosis, maximization of the square means to minimize kurtosis.

In fact, maximization of sparseness may not always be the correct method for estimation ICA even on images. If we do not remove the DC component from the images, the DC component turns out to be one independent component, and it sometimes has negative kurtosis. For such data, simply maximizing sparseness of all the components will produce misleading results.

Thus, we see that there is a difference between basic linear sparse coding and ICA in the sense that ICA works for different kinds of non-Gaussianity and not just sparseness.

7.10 Receptive Fields vs. Feature Vectors

An important point to note is the relation between the feature vectors A_i and the feature detector weights W_i. The feature vectors are shown Fig. 7.3. However, it is often the W_i that are more interesting, since they are the weights that are applied to the image to actually compute the s_i, and in neurophysiological modeling; they are more closely connected to the receptive fields of neurons.

There is, in fact, a simple connection between the two: the A_i are basically low-pass filtered versions of the W_i. In fact, simple calculations show that the covariance $\text{cov}(I(x, y), I(x', y'))$ in images generated according to the ICA model equals

$$\sum_i A_i(x, y) A_i(x', y') \tag{7.31}$$

because the s_i are uncorrelated and have unit variance. Thus, we have

$$\sum_{x',y'} \operatorname{cov}\big(I(x,y), I(x',y')\big) W_i(x',y') = \sum_{x',y'} \sum_i A_i(x,y) A_i(x',y') W_i(x',y')$$
$$= \sum_i A_i(x,y) \sum_{x',y'} A_i(x',y') W_i(x',y')$$
$$= A_i(x,y) \tag{7.32}$$

by definition of the W_i as the inverse of the A_i. This means that the A_i can be obtained by multiplying the W_i by the covariance matrix of the data.

Such multiplication by the covariance matrix has a simple interpretation as a *low-pass filtering* operation. This is because the covariances are basically a decreasing function of the distance between (x, y) and (x', y'), as shown in Fig. 5.4. Thus, A_i and W_i have essentially the same orientation, location, and frequency tuning properties. On the other hand, the A_i are better to visualize because they actually correspond to parts of the image data; especially with data that is not purely spatial, as in Chap. 15, visualization of the detector weights would not be straightforward.

7.11 Problem of Inversion of Preprocessing

A technical point that we have to consider is computation of the A_i for original images based on ICA of canonically preprocessed data. There is actually a problem here: in order to get the A_i we have to invert the canonical preprocessing, because estimation of the model gives the vectors $\mathbf{v}_i$ in the reduced (preprocessed) space only. But canonical preprocessing is not invertible in the strict sense of the word because it reduces the dimension and therefore loses information!

Typically, a solution based on the idea of computing the best possible approximation of the inverse of the PCA/whitening transformation. Such best approximation can be obtained by the theory of multivariate regression, or, alternatively, by the theory of pseudo-inverses (see Sect. 19.8). Without going into details, the description of the solution is simple.

Denote by $\mathbf{U}$ the orthogonal matrix which contains the n vectors giving the directions of the principal components as its rows, i.e. the n dominant eigenvectors of the covariance matrix. Denote by λ_i the corresponding eigenvalues. Then, steps 3 and 4 of canonical preprocessing in Sect. 5.4 consist of multiplying the vectorized image patches by $\operatorname{diag}(1/\sqrt{\lambda_i})\mathbf{U}$.

We now define the inverse preprocessing as follows: After computing the feature vectors in the preprocessed space (the $\mathbf{v}_i$), the basis vectors are multiplied by $\mathbf{U}^T \operatorname{diag}(\sqrt{\lambda_i})$. These are the optimal approximations of the feature vectors in the original space. They can also be computed by taking the pseudo-inverse of the matrix of the features W_i, which is what we did in our simulations.

Note that we have no such problem with computation of the W_i for the original data because we just multiply the vectors $\mathbf{v}_i$ with the PCA/whitening matrix, and no inversion is needed.

7.12 Frequency Channels and ICA

A long-standing tradition in vision science is to talk about "frequency channels", and more rarely, about "orientation" channels. The idea is that in the early visual processing (something like V1), information of different frequencies is processed independently. The word "independence" as used here has nothing to do with statistical independence: it means that processing happens in different physiological systems that are more or less anatomically separate, and do not exchange information with each other.

Justification for talking about different channels is abundant in research on V1. In recordings from single cells, simple and complex cell receptive fields are band-pass, i.e. respond only to stimuli in a certain frequency range, and various optimal frequencies are found in the cells (see Chap. 3 and its references). In psychophysics, a number of experiments also point to such a division of early processing. For example, in Fig. 3.8 on page 59, the information on the high- and low-frequency parts are quite different, yet observes have no difficulty in processing (reading) them separately.

In the results of ICA on natural images, we see an interesting new interpretation of why the frequency channels might be independent in terms of physiological and anatomical separation in the brain. The reason is that the information in different frequency channels seems to be *statistically* independent, as measured by ICA. The feature vectors A_i given by ICA are band-pass, thus showing that a decomposition into statistically independent features automatically leads to frequency channels.

7.13 Concluding Remarks and References

Independent component analysis is a statistical generative model whose estimation boils down to sparse coding. It gives a proper probabilistic formulation of sparse coding and thereby solves a number of theoretical problems in sparse coding (in particular: optimal ways of measuring sparseness, optimality of decorrelation), and gives a proper pdf to be used in Bayesian inference. The expression "independent component analysis" also points out another important property of this model: the components are considered statistically independent. This independence assumption is challenged in many more recent models which are the topics of Chaps. 9–11.

The first application of ICA, as opposed to sparse coding, on image patches was in Bell and Sejnowski (1997) based on the earlier sparse coding framework in Olshausen and Field (1996) considered in Sect. 13.1.

For more information on the ICA model, see Hyvärinen and Oja (2000) or Hyvärinen et al. (2001b). Some of the earliest historical references on ICA include Hérault and Ans (1984), Mooijaart (1985), Cardoso (1989), Jutten and Hérault (1991). Classic references include Delfosse and Loubaton (1995), which showed explicitly how maximization of non-Gaussianity is related to ICA estimation; Comon (1994), which showed the uniqueness of the decomposition, and the validity of the sum-of-squares-of-kurtosis in (7.30); the maximum likelihood framework was introduced in Pham et al. (1992), Pham and Garrat (1997).

For more information on the central limit theorem, see any standard textbook on probability theory, for example Papoulis and Pillai (2001).

A related way of analyzing the statistics of natural images is to look at the cumulant tensors (Thomson 1999, 2001). Cumulants (strictly speaking, higher-order cumulants) are statistics which can be used as measures of non-Gaussianity; for example, kurtosis and skewness are one of the simplest cumulants. Cumulant tensors are generalizations of the covariance matrix to higher-order cumulants. Analysis of cumulant tensors is closely related to ICA, as discussed in detail in Chap. 11 of Hyvärinen et al. (2001b). Equivalently, one can analyze the poly-spectra (usually the trispectrum) which are obtained by Fourier transformations of the cumulant spectra, in a similar way as the ordinary Fourier spectrum function can be obtained as the Fourier transform of the autocovariance function; see, e.g. Nikias and Mendel (1993), Nikias and Petropulu (1993) for further information.

7.14 Exercises

Mathematical Exercises

1. Prove (7.2).
2. Based on (7.2), prove that two independent random variables are uncorrelated.
3. Calculate the kurtosis of the uniform distribution in (7.23).
4. Calculate the kurtosis of the Laplacian distribution in (7.18).
5. Show that the skewness of a random variable with a pdf which is even-symmetric (i.e. $p(-x) = p(x)$) is zero.
6. In this exercise, we consider a very simple case of Gaussian mixtures (see Sect. 7.8.3). Assume that a component follows $s = vz$ where z is Gaussian with zero mean and unit variance. Let us assume that in 50% of the natural images the variance coefficient v has value α. In the remaining 50% of natural images, v has value β.
   a. What is the distribution of the random variable s in the set of all natural images? (Give the density function $p(s)$).
   b. Show that $E\{s^2\} = \frac{1}{2}(\alpha^2 + \beta^2)$.
   c. Show that $E\{s^4\} = \frac{3}{2}(\alpha^4 + \beta^4)$.
   d. What is the kurtosis of this distribution?
   e. Show that the kurtosis is positive for almost any parameter values.
7. Prove (7.31).

Computer Assignments

1. Using numerical integration, compute the kurtoses of the Laplacian and uniform distributions.

2. Load the FastICA program from the web. This will be used in the following assignments. You will also need some images; try to find some that have not been compressed, since compression can induce very nasty effects.
3. Take patches of 16×16 of the image. 10 000 is a sufficient amount of patches. Input these to FastICA.
4. Plot histograms of the independent components. Compute their kurtoses. Plot (some of the) RFs. Look at the RFs and comment on whether they look like V1 RFs or not. If not, why not?

Chapter 8
Information-Theoretic Interpretations

So far, we have been operating within the theoretical framework of Bayesian inference: the goal of our models is to provide priors for Bayesian inference. An alternative framework is provided by information theory. In information theory, the goal is to find ways of coding the information as efficiently as possible. This turns out to be surprisingly closely connected to Bayesian inference. In many cases, both approaches start by estimation of parameters in a parameterized statistical model.

This chapter is different from the previous ones because we do not provide any new models for natural image statistics. Rather, we describe a new framework for interpreting some of the previous models.

8.1 Basic Motivation for Information Theory

In this section, we introduce the basic ideas of information theory using intuitive examples illustrating the two principal motivations: data compression and data transmission.

8.1.1 Compression

One of the principal motivations of information theory is data compression. Let us begin with a simple example. Consider the following string of characters:

BABABABADABACAABAACABDAAAAABAAAAAAAADBCA

We need to code this using a binary string, consisting of zeros and ones, because that's the form used in computer memory. Since we have four different characters, a basic approach would be to assign the four possible two-digit codewords for each of them:

$$\text{A} \rightarrow 00 \quad (8.1)$$

$$\text{B} \rightarrow 01 \quad (8.2)$$

$$\text{C} \rightarrow 10 \quad (8.3)$$

$$\text{D} \rightarrow 11 \quad (8.4)$$

A. Hyvärinen, J. Hurri, P.O. Hoyer, *Natural Image Statistics*,
Computational Imaging and Vision 39,

Replacing each of the characters by its two-digit codeword gives the code

01000100010001001100010010000000100001000
01110000000000001000000000000000011011000

However, a shorter code can be obtained by using the fundamental insight of information theory: *frequent characters should be given shorter codewords.* This makes intuitive sense: if frequent characters have short codewords, even at the expense of giving less frequent characters longer codewords, the average length could be shortened.

In this example, the character A is the most frequent one: approximately one-half of the letters are A's. The letter B is the next, with a proportion of approximately one quarter. So, let us consider the following kind of codeword assignment:

$$\mathrm{A} \rightarrow 0 \tag{8.5}$$

$$\mathrm{B} \rightarrow 10 \tag{8.6}$$

$$\mathrm{C} \rightarrow 110 \tag{8.7}$$

$$\mathrm{D} \rightarrow 111 \tag{8.8}$$

Now, the code becomes

1001001001001110100110001000110010111000
0010000000000111101100

This is more than 10% shorter than the basic code given above. (In real-world applications, the saving is often much larger, sometimes reaching more than 90% in image compression.) Note that the codeword assignment has been cleverly constructed so that the original string can be recovered from this code without any ambiguity.

Compression was possible because some of the characters were more common than others. In other words, it was due to the statistical regularities of the data (or redundancy, which will be defined later). Thus, it is not surprising that the methods developed in information theory have also been applied in the field of natural image statistics.

8.1.2 Transmission

A rather different application of information theory is in data transmission. In transmission, the central problem is *noise*. That is, the transmission will introduce random errors in the data. The goal here is to code the data so that the receiving end of the system can correct as many of the random errors as possible.

As a simple example, consider the following binary string which we want to transmit:

1111101100100011100110100101110110100101

To code this string, we use a very simple method: we simply repeat all the digits three times. Thus, we get the code

111111111111111100011111100000001110000000
001111111111000000111111000111000000011100
01111111110001111110001110000000111000111

This is transmitted through a channel which has relatively strong noise: 25% of the digits are randomly changed to the opposite. So, the receiving end in the channel receives the following string:

111110011111111100001111100000001110100001
1000110100100100111010101011100001010100
1111111111000110101000111010010111100111

Now, we can use the repetitions in the string in the following way: we look at all consecutive groups of three digits, and guess that the original string probably had the digit which has the most occurrences among each group. Thus, we obtain the following string

1111101100100101000110100101110110100101

This string has less than 10% wrong digits. Thus, our encoding scheme reduced the number of errors introduced by the channel from 25% to less than 10%. (Actually, we were a bit lucky with this string: on the average the errors would be of the order of 16%.)

So, the central idea in transmission is very different from compression. In order to combat noise, we want to code the data so that we introduce redundancy (to be defined later), which often means making the code longer than the original data.

8.2 Entropy as a Measure of Uncertainty

Now, we introduce the concept of entropy, which is the foundation of information theory. First, we give its definition and a couple of examples, and then show how it is related to data compression.

8.2.1 Definition of Entropy

Consider a random variable z which takes values in a discrete set $a_1, \dots, a_N$ with probabilities $P(z = a_i)$, $i = 1, \dots, N$. The most fundamental concept of information theory is *entropy*, denoted by $H(z)$, which is defined as

$$H(z) = -\sum_{i=1}^{N} P(z = a_i) \log_2 P(z = a_i). \tag{8.9}$$

If the logarithms to base 2 are used, the unit of entropy is called a *bit*. Entropy is a measure of the *average uncertainty* in a random variable. It is always non-negative. We will next present a couple of examples to illustrate the basic idea.

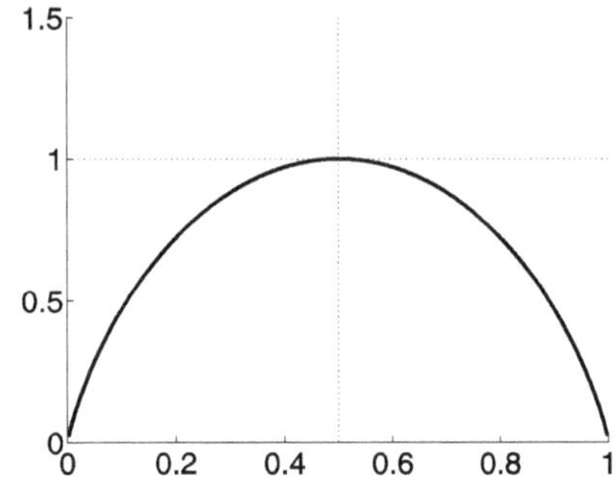

Fig. 8.1 Entropy of a random variable which can only take two values, plotted as a function of the probability p_0 of taking one of those values

Example 1 Consider a random variable which takes only two values, A and B. Denote the probability $P(z = A)$ by p_0. Then entropy of z equals

$$H(z) = -p_0 \log_2 p_0 - (1 - p_0)\log_2(1 - p_0). \tag{8.10}$$

We can plot this as a function of p_0, which is shown in Fig. 8.1 The plot shows that the maximum entropy is obtained when $p_0 = 0.5$, which means that both outcomes are equally probable. Then the entropy equals one bit. In contrast, if p_0 equals 0 or 1, there is no uncertainty at all in the random variable, and its entropy equals zero.

Example 2 Consider a random variable z which takes, for some given n, any of 2^n values with equal probability, which is obviously $1/2^n$. The entropy equals

$$H(z) = -\sum_{i=1}^{2^n} \frac{1}{2^n} \log_2 \frac{1}{2^n} = -2^n \frac{1}{2^n}(-n) = n. \tag{8.11}$$

Thus, the entropy equals n bits. This is also the number of binary digits (also called bits) that you would need to represent the random variable in a basic binary representation.

Example 3 Consider a random variable z which always takes the same value, i.e. it is not really random at all. Its entropy equals

$$H(z) = -1 \times \log_2 1 = 0. \tag{8.12}$$

Again, we see that the entropy is zero since there is no uncertainty at all.

8.2.2 Entropy as Minimum Coding Length

An important result in information theory shows that the numerical value of entropy directly gives the *average code length for the shortest possible code* (i.e. the one giving maximum compression). It is no coincidence that the unit of entropy is called a bit, since the length of the code is also given in bits, in the sense of the number of zero-one digits used in the code.

The proof of this property is very deep, so we will only try to illustrate it with an example.

Example 4 Now, we will return back to the compression example in the preceding section to show how entropy is connected to compression. Consider the characters in the string as independent realizations of a random variable which takes values in the set {A, B, C, D}. The probabilities used in generating this data are

$$P(\mathrm{A}) = 1/2, \tag{8.13}$$

$$P(\mathrm{B}) = 1/4, \tag{8.14}$$

$$P(\mathrm{C}) = 1/8, \tag{8.15}$$

$$P(\mathrm{D}) = 1/8. \tag{8.16}$$

We can compute the entropy, which equals 1.75 bits. Thus, the entropy is smaller than 2 bits, which would be (according to Example 2), the maximum entropy for a variable with four possible values. Using the definition of entropy as minimum coding length, we see that the saving in code length can be at most $(2 - 1.75)/2 = 12.5\%$ for data with these characteristics. (This holds in the case of infinite strings in which random effects are averaged out. Of course, for a finite-length string, the code would be a bit shorter or longer due to random effects.)

Note also it is essential here that the characters in the string are generated independently at each location; otherwise, the code length might be shorter. For example, if the characters would be generated in pairs, as in AACCBBAA..., this dependency could obviously be used to reduce the code length, possibly beyond the bound that entropy gives.

8.2.3 Redundancy

Redundancy is a word which is widely used in information theory, as well as in natural image statistics. It is generally used to refer to the statistical regularities, which make part of the information "redundant", or unnecessary. Unfortunately, when talking about natural images, different authors use the word in slightly different ways.

An information-theoretic definition of redundancy is based on entropy. Given a random variable which has 2^n different values, say $1, \ldots, 2^n$, we compare the length of the basic n-bit code and the length of the shortest code, given by entropy:

$$\text{redundancy} = n - H(z). \tag{8.17}$$

This is zero only if all the values $1, \ldots, 2^n$ are equally likely. Using this terminology, we can say that compression of the string in Sect. 8.1.1 was possible because the string had some redundancy (n was larger than $H(z)$). Compression is possible by removing, or at least reducing, redundancy in the data.

This definition is actually more general than it seems, because it can also consider dependencies *between* different variables (say, pixels). If we have two variables z_1 and z_2, we can simply define a new variable z whose possible values correspond to all the possible *combinations*[1] of the values of z_1 and z_2. Then we can define entropy and redundancy for this new variable. If the variables are highly dependent from each other, some combinations will have very small probabilities, so the entropy will be small and redundancy large.

It is important to understand that you may want to either reduce or increase redundancy, depending on your purpose. In compression, you want to reduce it, but in information transmission, you actually want to increase it. The situation is even more complicated than this because usually before transmission, you want to compress the data in order to reduce the time needed for data transmission. So, you first try to remove the redundancy to decrease the length of the data to be transmitted, and then introduce new redundancy to combat noise. This need not be contradictory because in introducing new redundancy, you do it is a controlled way using carefully designed codes which increase code length as little as possible for a required level of noise-resistance. A result called "source-channel coding theorem" lays out the conditions under which such a two-stage procedure is, in fact, optimal.

8.2.4 Differential Entropy

The extension of entropy to continuous-valued random variables or random vectors is algebraically straightforward. For a random variable with probability density function p_z, we define the *differential entropy*, denoted by H just like entropy, as follows:

$$H(z) = -\int p_z(z) \log p_z(z)\, dz. \tag{8.18}$$

So, basically we have just replaced the summation in the original definition in (8.9) by an integral. The same definition also applies in the case of a random vector.

It is not difficult to see what kind of random variables have small differential entropies. They are the ones whose probability densities take large values, since these give strong negative contributions to the integral in (8.18). This means that certain small intervals are quite probable. Thus, we again see that entropy is small when the variable is not very random, that is, it is typically contained in some limited intervals with high probabilities.

---

[1] More precisely: if z_1 can take values in the set $\{1, 2\}$ and z_2 can take values in the set $\{1, 2, 3\}$, we define z so that it takes values in the Cartesian product of those sets, i.e. $\{(1, 1), (1, 2), (1, 3), (2, 1), (2, 2), (2, 3)\}$, so that the probability $z = (a_1, a_2)$ simply equals the probability that $z_1 = a_1$ *and* $z_2 = a_2$.

Differential entropy is related to the shortest code length of a *discretized* version of the random variable z. Suppose we discretize z so that we divide the real line to bins (intervals) of length d, and define a new discrete-valued random variable $\tilde{z}$ which tells which of these bins the value of z belongs to. This is similar to using a limited number of decimals in the representation, for example using only one decimal place as in "1.4" to represent the values of the random variable. Then the entropy of $\tilde{z}$ is approximately equal to the differential entropy of z plus a constant which only depends on the size of the bins, d.

Example 5 Consider a random variable z which follows a uniform distribution in the interval $[0, a]$: Its density is given by

$$p_z(z) = \begin{cases} 1/a, & \text{for } 0 \leq z \leq a, \\ 0, & \text{otherwise.} \end{cases} \tag{8.19}$$

Differential entropy can be evaluated as

$$H(z) = -\int_0^a \frac{1}{a} \log \frac{1}{a} \, dz = \log a. \tag{8.20}$$

We see that the entropy is large if a is large, and small if a is small. This is natural because the smaller a is, the less randomness there is in z. Actually, in the limit where a goes to 0, differential entropy goes to $-\infty$, because in the limit, z is no longer random at all: it is always 0. This example also shows that differential entropy, in contrast to basic entropy, need not be non-negative.

8.2.5 Maximum Entropy

An interesting question to ask is: What kind of distributions have maximum entropy?

In the binary case in Example 1, we already saw that it was the distribution with 50%–50% probabilities which is clearly consistent with the intuitive idea of maximum uncertainty. In the general discrete-valued case, it can be shown that the uniform distribution (probabilities of all possible values are equal) has maximum entropy.

With continuous-valued variables, the situation is more complicated. Differential entropy can become infinite; consider, for example when $a \to \infty$ in Example 5 above. So, some kind of constraints are needed. The simplest constraint would perhaps be to constrain the random variable to take values only inside a finite interval $[a, b]$. In this case, the distribution of maximum entropy is again the uniform distribution, i.e. a pdf which equals $\frac{1}{b-a}$ in the whole interval and is zero outside of it. However, such a constraint may not be very relevant in most applications.

If we consider random variables whose *variance is constrained* to a given value (e.g. to 1), the distribution of maximum entropy is, interestingly, the Gaussian distribution. This is why the Gaussian distribution can be considered the least "informative", or the least "structured", of continuous-valued distributions.

This also means that differential entropy can be considered a measure of *non-Gaussianity*. The smaller the differential entropy, the further away the distribution is from the Gaussian distribution. However, it must be noted that differential entropy depends on the scale as well. Thus, if entropy is used as a measure of non-Gaussianity, the variables have to normalized to unit variance first, just like in the case of kurtosis. We will see in Sect. 8.4 how differential entropy is, in fact, closely related to the sparseness measures we used in Chap. 6.

8.3 Mutual Information

In information transmission, we need a measure of how much information the output of the channel contains about the input. This the purpose of the concept of mutual information.

Let's start with the concept of *conditional entropy*. It is simply the average entropy calculated for the conditional distribution of the variable z, where the conditioning is by the observation of another variable y:

$$H(z|y) = -\sum_y P(y) \sum_z P(z|y) \log P(z|y). \tag{8.21}$$

That is, it measures how much entropy there is left in z when we know (observe) the value of y; this is averaged over all values of y.

If z and y are independent, the conditional distribution of z given y is the same as the distribution of z alone, so conditional entropy is the same as the entropy of z. If z and y are not independent, conditional entropy is smaller than the entropy of z, because then knowledge of the value of y reduces the uncertainty on z. In the extreme case where $z = y$, the conditional distribution z given y is such that all the probability is concentrated on the observed value of y. The entropy of such a distribution is zero (see Example 3 above), so the conditional entropy is zero.

Let us assume that z is the message input to a transmission channel and y is the output, i.e. the received signal. Basically, if transmission is very good, knowledge of y will tell us very much about what z was. In other words, the conditional distribution of z given y is highly concentrated on some values. So, we could measure the transmitted information by the change in entropy which is due to measurement of y. It is called the mutual information, which we denote[2] by J:

$$J(z, y) = H(z) - H(z|y). \tag{8.22}$$

---

[2]We use a non-conventional notation J for mutual information because the conventional one, I, could be confused with the notation for an image.

Just as entropy gives the code length of the optimal code, mutual information is related to the amount of information which can be obtained about z, based on observation of the channel output y.

Note that in practice mutual information depends not only on the noise in the channel, but also on how we code the data as the variable z in the first place. Therefore, to characterize the properties of the channel itself, we need to consider the maximum of mutual information over all possible ways of coding z. This is called *channel capacity* and gives the maximum amount of information which can be transmitted through the channel.

A generalization of mutual information to many variables is often used in the theory of ICA. In that case, the interpretation as information transmitted over a channel is no longer directly applicable. The generalization is based on the fact that mutual information can be expressed in different forms:

$$J(z, y) = H(z) - H(z|y) = H(y) - H(y|z) = H(z) + H(y) - H(z, y) \quad (8.23)$$

where $H(z, y)$ is the *joint entropy*, which is simply obtained by defining a new random variable so that it can take all the possible combinations of values of z and y. Based on this formula, we define mutual information of n random variables as

$$J(z_1, z_2, \ldots, z_n) = \sum_{i=1}^{n} H(z_i) - H(z_1, z_2, \ldots, z_n). \quad (8.24)$$

The utility in this quantity is that it can be used a measure of independence: it is always non-negative and zero only if the variables z_i are independent.

Conditional entropy, joint entropy, and mutual information can all be defined for continuous-valued variables by using differential entropy in the definitions instead of ordinary entropy.

8.4 Minimum Entropy Coding of Natural Images

Now, we discuss the application of the information-theoretic concepts in the context of natural image statistics. This section deals with data compression models.

8.4.1 Image Compression and Sparse Coding

Consider first the engineering application of image compression. Such compression is routinely performed when images are transmitted over the Internet or stored on a disk. Most successful image compression methods begin with a linear transformation of image patches. The purpose of such a transformation is to reduce (differential) entropy. Grey-scale values of single pixels have a much larger entropy than, for example, the coefficients in a Fourier or discrete cosine transform (DCT) basis (at least when these are applied on small patches). Since the coefficients in those bases have smaller differential entropy, discretized (quantized) versions of the coefficients are easier to code: the quantization error, i.e. the error in quantizing the coefficients for a fixed code length, is reduced. This is why such a transformation is done as the first step.

This means that the optimal linear transformation of images as the first step of a compression method would be a transformation which minimizes the differential entropy of the obtained components. This turns out to be related to sparse coding, as we will now show.

Let us consider the differential entropy of a linear component s. How can we compute the value $H(s)$ using the definition in (8.18) in practice? The key is to understand that entropy is actually the expectation of a non-linear function of s. We have

$$H(s) = E\{G(s)\} \tag{8.25}$$

where the function G is the negative of the log-pdf: $G(s) = -\log p_s(s)$. In practice, we have a sample of s, denote it by $s(t)$, $t = 1, \dots, T$. Assume we also have reasonable approximation of G, that is, we know rather well what the log-pdf is like. Then differential entropy can be estimated as the sample average for a fixed G as

$$H(s) = \frac{1}{T} \sum_t G(s(t)). \tag{8.26}$$

Comparing this with (6.2) on page 133, we see that differential entropy is similar the sparseness measures we used in sparse coding. In fact, in Sect. 7.7.2, it was shown that the optimal sparseness measures are obtained when we use exactly the log-pdf of s as the non-linearity. Thus, *differential entropy is the optimal sparseness measure* in the sense of Sect. 7.7.2: it provides the maximum likelihood estimator of the ICA model.

However, there is one important point which needs to be taken into account. The sparseness measures assumed that the variance of s is constrained to be equal to one. This is consistent with the theory of ICA which tells that the transformation should be orthogonal in the whitened space. In contrast, in image compression, the transformation giving the components is usually constrained to be orthogonal (in the original image space). One reason is that then the quantization error in the image space is the same as the quantization error in the component space. This makes sense because then minimizing quantization error for the components is directly related to the quantization error of the original image. In contrast, any method which whitens the data amplifies high-frequency components which have low variance, thus emphasizing errors in their coding. So, the quantization error in the components is not the same as the error in the original image—which is what we usually want to minimize in engineering applications.

If we consider transformations which are orthogonal in the original space, the constraint of unit variance of a component s is not at all fulfilled. For example, PCA is an orthogonal transformation which finds components with maximally different variances. From the viewpoint of information theory, $\log p$ changes quite a lot as a function of variance, so using a fixed G may give a very bad approximation of entropy. So, while sparse coding, as presented in Chap. 6, is related to finding an optimal basis for image compression, it uses a rather unconventional constraint which means it does not optimize the compression in the usual sense.

8.4.2 Mutual Information and Sparse Coding

Information-theoretic concepts allow us to see the connection between sparse coding and ICA from yet another viewpoint. Consider a linear invertible transformation $\mathbf{y} = \mathbf{V}\mathbf{z}$ of a random vector $\mathbf{z}$ which is *white*. The mutual information between the components y_i is equal to

$$J(y_1, \ldots, y_n) = \sum_{i=1}^{n} H\left(\mathbf{v}_i^T \mathbf{z}\right) - H(\mathbf{V}\mathbf{z}). \tag{8.27}$$

Recall that mutual information can be interpreted as a measure of dependence. Now, let us constrain $\mathbf{V}$ to be orthogonal. Then we have $H(\mathbf{V}\mathbf{z}) = H(\mathbf{z})$ because the *shape* (in an intuitive sense) of the distribution is not changed at all: an orthogonal transformation simply rotates the pdf in the n-dimensional space, leaving its shape intact. This means that the values taken by $p_\mathbf{z}$ and $\log p_\mathbf{z}$ in the definition in (8.18) are not changed; they are just taken at new values of $\mathbf{z}$. That is why differential entropy is not changed by an orthogonal transformation of the data.[3]

So, to minimize the mutual information, we simply need to find an orthogonal transformation which minimizes the differential entropies of the components; this is the same as maximizing the non-Gaussianities of the components. And for sparse data, maximizing non-Gaussianity is usually the same as maximizing sparseness. Thus, we see that under the constraint of orthogonality, *sparse coding is equivalent to minimization of the dependence of the components* if the data is white. This provides another deep link between information theory, sparse coding, and independence.

8.4.3 Minimum Entropy Coding in the Cortex

A very straightforward application of the data compression principle is then to assume that V1 "wants" to obtain a minimum entropy code. This is very well in line

[3] A rigorous proof is as follows: denoting $\mathbf{y} = \mathbf{V}\mathbf{z}$, we simply have $p_\mathbf{y}(\mathbf{y}) = p_\mathbf{z}(\mathbf{V}^T\mathbf{y})$ for an orthogonal $\mathbf{V}$. The basic point is that the absolute value of the determinant of the transformation matrix needed in transforming pdf's, or variables in an integral formula, (see Sect. 7.4) is equal to one for an orthogonal transformation, so it can be omitted. Thus, we have $\int p_\mathbf{y}(\mathbf{y}) \log p_\mathbf{y}(\mathbf{y})\, d\mathbf{y} = \int p_\mathbf{z}(\mathbf{V}^T\mathbf{y}) \log p_\mathbf{z}(\mathbf{V}^T\mathbf{y})\, d\mathbf{y}$. In this integral, we make a change of variables $\tilde{\mathbf{z}} = \mathbf{V}^T\mathbf{y}$, and we get $\int p_\mathbf{z}(\tilde{\mathbf{z}}) \log p_\mathbf{z}(\tilde{\mathbf{z}})\, d\tilde{\mathbf{z}}$.

with the results on sparse coding and ICA in Chaps. 6 and 7 because we have just shown that the objective functions optimized there can be interpreted as differential entropies and code lengths. Basically, what information theory provides is a new interpretation of the objective functions used in learning simple cell receptive fields.

Yet, it is not quite clear whether such an entropy-based arguments are relevant to the computational tasks facing the visual cortex. A critical discussion on the analogy between compression and cortical coding is postponed to Sect. 8.6.

8.5 Information Transmission in the Nervous System

Following the fundamental division of information theory into compression and transmission, the second influential application of information theory to visual coding considers maximization of data transmission, usually called simply infomax.

8.5.1 Definition of Information Flow and Infomax

Assume that $\mathbf{x}$ is a continuous-valued random vector. It is the input to a neural system, which is modeled using a linear/non-linear model (see Sect. 3.4.1), with additive noise. Thus, outputs are of the form

$$y_i = \phi_i\left(\mathbf{b}_i^T \mathbf{x}\right) + \mathbf{n} \tag{8.28}$$

where the ϕ_i are some scalar functions, the $\mathbf{b}_i$ are the connection weight vectors of the neurons, and $\mathbf{n}$ is a vector of white Gaussian noise. That is, the neural network first computes a linear transformation of the input data, with the coefficients given by network connection weights $\mathbf{b}_i$; then it transforms the outputs using scalar functions ϕ_i, and there is noise in the system.

Let us consider information flow in such a neural network. Efficient information transmission requires that we maximize the mutual information between the inputs $\mathbf{x}$ and the outputs $\mathbf{y}$, hence the name "infomax". This problem is meaningful only if there is some information loss in the transmission. Therefore, we assume that there is some noise in the network; in practice, we have to assume that the noise is infinitely small to be able to derive clear results. We can then ask how the network parameters should be adapted (learned) so as to maximize information transmission.

8.5.2 Basic Infomax with Linear Neurons

To begin with, we shall consider the very basic case where there are actually no nonlinearities: we define $\phi(u) = u$, and the noise has constant variance. (It may seem odd to say that white noise has constant variance, because that seems obvious.

However, in Sect. 8.5.4, we will consider a model where the variance is not constant because that is the case in neural systems.)

By definition of mutual information, we have

$$J(\mathbf{x}, \mathbf{y}) = H(\mathbf{y}) - H(\mathbf{y}|\mathbf{x}). \tag{8.29}$$

In the present case, the conditional distribution of $\mathbf{y}$ given $\mathbf{x}$ is simply the distribution of the Gaussian white noise. So, the entropy $H(\mathbf{y}|\mathbf{x})$ does not depend on the weights $\mathbf{b}_i$ at all: it is just a function of the noise variance. This means that for the purpose of finding the $\mathbf{b}_i$ which maximizes information flow, we only need to consider the output entropy $H(\mathbf{y})$. This is true as long as the noise variance is constant.

To simplify the situation, let us assume just for the purposes of this section that the transformation matrix $\mathbf{B}$, with the $\mathbf{b}_i$ as its rows, is orthogonal. Then $\mathbf{y}$ is just an orthogonal transformation of $\mathbf{x}$, with some noise added.

Furthermore, in all infomax analysis, we consider the limit where the noise variance goes to zero. This is because simple analytical results can only be obtained in that limit.

So, combining these assumptions and results, infomax for linear neurons with constant noise variance boils down to the following: we maximize the entropy $H(\mathbf{y})$, where $\mathbf{y}$ is an orthogonal transformation of $\mathbf{x}$. Noise does not need to be taken into account because we consider the limit of zero noise. But, as shown in Sect. 8.4.2, an orthogonal transformation does not change differential entropy, so the information flow does not depend on $\mathbf{B}$ at all!

Thus, we reach the conclusion that for linear neurons with constant noise variance, the infomax principle does not really give anything interesting. Fortunately, more sophisticated variants of infomax are more interesting. In the next subsections, we will consider the two principal cases: noise of constant variance with non-linear neurons, and noise of non-constant variance with linear neurons.

8.5.3 Infomax with Non-linear Neurons

8.5.3.1 Definition of Model

First, we consider the case where

1. The functions ϕ_i are non-linear. One can build a more realistic neuron model by taking a non-linearity which is saturating, and has no negative outputs.
2. The vector $\mathbf{n}$ is additive Gaussian white noise.This is the simplest noise model to begin with.

Maximization of this mutual information $J(\mathbf{x}, \mathbf{y})$ is still equivalent to maximization of the output entropy, as in the previous subsection. Again, we take the limit where the noise has zero variance. We will not go into detail here, but it can be shown that the output entropy in this non-linear infomax model then equals

$$H(\mathbf{y}) = \sum_i E\left\{\log \phi_i'(\mathbf{b}_i^T \mathbf{x})\right\} + \log|\det \mathbf{B}|. \tag{8.30}$$

It turns out that this has a simple interpretation in terms of the ICA model. Now, we see that the output entropy is of the same form as the expectation of the likelihood as in (7.15). The pdf's of the independent components are here replaced by the functions ϕ_i'. Thus, if the non-linearities ϕ_i used in the neural network are chosen as the cumulative distribution functions corresponding to the densities p_i of the independent components, i.e. $\phi_i'(\cdot) = p_i(\cdot)$, the output entropy is actually equal to the likelihood. This means that infomax is equivalent to maximum likelihood estimation of the ICA model.

Usually, the logistic function

$$\phi_i(u) = \frac{1}{1+\exp(-u)} \tag{8.31}$$

is used in the non-linear infomax model (see Fig. 8.2a). This estimates the ICA model in the case of sparse independent components, because if we interpret ϕ_i' as a pdf, it is sparse. In fact, the log-pdf given by $\log \phi_i'$ is nothing else than the familiar $\log\cosh$ function (with negative sign and some unimportant constants), which we have used as a measure of sparseness in Chap. 6, and as a model of a smooth sparse log-pdf in (7.19).

8.5.4 Infomax with Non-constant Noise Variance

Here, we present some critique of the non-linear infomax model, and propose an alternative formulation.

8.5.4.1 Problems with Non-linear Neuron Model

Using a logistic function as in (8.31) is a correct way of estimating the ICA model for natural image data in which the components really are super-Gaussian. However, if the transfer function ϕ_i is changed to the Gaussian cumulative distribution function, the method does not estimate the ICA model anymore, since this would amount to assuming Gaussian independent components, which makes the estimation impossible. An even worse situation arises if we change the function ϕ_i so that ϕ_i' is the pdf of a sub-Gaussian (anti-sparse) distribution. This amounts to estimating the ICA model assuming sub-Gaussian independent components. Then the estimation fails completely because we have made a completely wrong assumption on the distribution of the components.

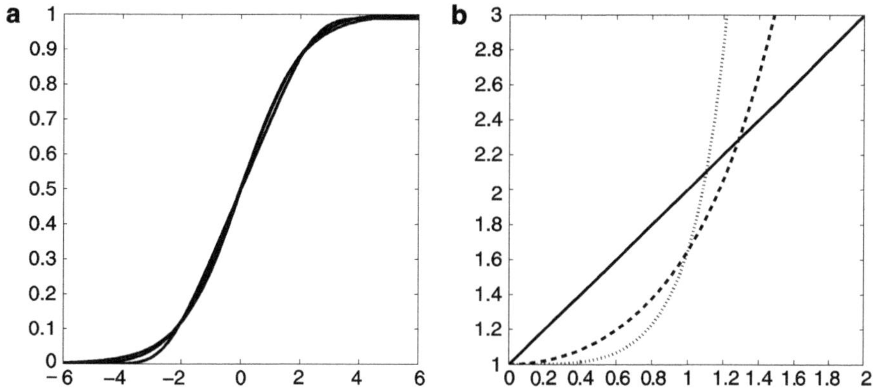

Fig. 8.2 **a** Three sigmoidal non-linearities corresponding to logistic, Gaussian, and sub-Gaussian (with log-pdf proportional to $-x^4$) prior cumulative distributions for the independent components. The non-linearities are practically indistinguishable. (Note that we can freely scale the functions along the x-axis since this has no influence on the behavior in ICA estimation. Here, we have chosen the scaling parameters so as to emphasize the similarity.) **b** Three functions h that give the dependencies of noise variance (functions h) which are equivalent to different distributions. *Solid line*: basic Poisson like variance as in (8.34), corresponding to a sparse distribution. *Dashed line*: the case of Gaussian distribution as in (8.37). *Dotted line*: the sub-Gaussian distribution used in **a**. Here, the function in the basic Poisson-like variance case is very different from the others, which indicates better robustness for the model with changing noise variance. From Hyvärinen (2002), Copyright ©2002 Elsevier, used with permission

Unfortunately, the three non-linear functions ϕ_i corresponding to (8.31), the Gaussian case, and one particular sub-Gaussian case look all very similar, however. This is illustrated in Fig. 8.2a. All the three functions have the same kind of qualitative behavior. In fact, all cumulative distribution functions look very similar after appropriate scaling along the x-axis.

It is not very likely that the neural transfer functions (which are only crude approximations anyway) would consistently be of the type in (8.31), and not closer to the two other transfer functions. Thus, *the model can be considered to be non-robust, that is, too sensitive to small fluctuations in its parameters.*[4]

8.5.4.2 Using Neurons with Non-constant Variance

A possible solution to the problems with non-linear infomax is to consider a more realistic noise model. Let us thus take the linear function as ϕ_i, and change the definition of the noise term instead.

What would be a sensible model for the noise? Many classic models of neurons assume that the output is coded as the mean firing rate in a spike train, which follows

[4]It could be argued that the non-linear transfer function can be estimated from the data and it need not be carefully chosen beforehand, but this only modifies this robustness problem because then that estimation must be very precise.

a *Poisson process*. Without going into details, we just note that in that case, the variance of mean firing rate has a variance that is equal to its mean. Thus, we have

$$\mathrm{var}(n_i|\mathbf{x}) \propto r + \left|\mathbf{b}_i^T\mathbf{x}\right| \tag{8.32}$$

where r is a constant that embodies the spontaneous firing rate which is not zero (and hence does not have zero noise). We take the absolute value of $\mathbf{b}_i^T\mathbf{x}$ because we consider the output of a signed neuron to be actually coded by two different neurons, one for the negative part and one for the positive part. The distribution of noise in mean firing rate is non-Gaussian in the Poisson process case. However, in the following, we approximate it as Gaussian noise: The fundamental property of this new type of noise is considered to be the variance behavior given in (8.32), and not its non-Gaussianity. Therefore, we call noise with this kind of variance behavior "noise with *Poisson-like variance*" instead of Poisson noise.

A more general form of the model can be obtained by defining the variance to be a non-linear function of the quantity in (8.32). To investigate the robustness of this model, we do in the following all the computations in the more general case where

$$\mathrm{var}(n_i|\mathbf{x}) = h\left(\mathbf{b}_i^T\mathbf{x}\right) \tag{8.33}$$

where h is some arbitrary function with non-negative values, for example

$$h(u) = r + |u| \tag{8.34}$$

in the case of (8.32).

It then can be shown that the mutual information in the limit of zero noise is equal to

$$J(\mathbf{x},\mathbf{y}) = \log|\det\mathbf{B}| - \sum_i E\left\{\log\sqrt{h\left(\mathbf{b}_i^T\mathbf{x}\right)}\right\} + \text{const.} \tag{8.35}$$

where terms that do not depend on $\mathbf{B}$ are grouped in the constant. A comparison of (8.35) with (8.30) reveals that in fact, mutual information is of the same algebraic form in the two cases. By taking $h(u) = 1/\phi_i'(u)^2$, we obtain an expression of the same form. Thus, we see that *considering noise with non-constant variance, we are able to reproduce the same results as with a non-linear transfer function.*

If we consider the basic case of Poisson-like variance, which means defining the function h so that we have (8.32); this is equivalent to the non-linear infomax with

$$\phi_i'(u) = \frac{1}{\sqrt{r+|u|}}. \tag{8.36}$$

In the non-linear infomax, ϕ_i' corresponds to the probability density function assumed for the independent component. The function in (8.36) is an improper probability density function, since it is not integrable. However, its qualitative behavior is typically super-Gaussian: very heavy tails and a peak at zero.

Thus, in the basic case of Poisson-like variance, the infomax principle is equivalent to estimation of the ICA model with this improper prior density for the components. Since the choice of non-linearity is usually critical only along the sub-Gaussian vs. super-Gaussian axis, this improper prior distribution can still be expected to properly estimate the ICA model for most super-Gaussian components.[5]

To investigate the robustness of this model, we can consider what the noise variance structure should be like to make the estimation of super-Gaussian components fail. As with the non-linear infomax, we can find a noise structure that corresponds to the estimation of Gaussian independent components. In the limit of $r = 0$, we have the relation $h(u) = 1/\phi'(u)^2$, and we see that the Gaussian case corresponds to

$$h(u) \propto \exp(u^2). \tag{8.37}$$

This is a fast-growing ("exploding") function which is clearly very different from the Poisson-like variance structure given by the essentially linear function in (8.34). In the space of possible functions h that define the noise structure in this model, the function in (8.37) can be considered as a borderline between those variance structures that enable the estimation of super-Gaussian independent components, and those that do not. These two different choices for h, together with one corresponding to sub-Gaussian independent components (see caption) are plotted in Fig. 8.2b. The Poisson-like variance is clearly very different from the other two cases.

Thus, we may conclude that the model with Poisson-like variance is quite robust against changes of parameters in the model, since the main parameter is the function h, and this can change qualitatively quite a lot before the behavior of the model with respect to ICA estimation changes. This is in contrast to the non-linear infomax principle where the non-linearity has to be very carefully chosen according to the distribution of the data.

8.6 Caveats in Application of Information Theory

We conclude this chapter with a discussion on some open problems encountered in the application of information theory to model cortical visual coding in the context of natural image statistics.

Classic information theory is fundamentally a theory of compression and transmission of binary strings. It is important to ask *Is this theory really useful in the study of cortical visual representations?* Often, the concepts of information theory are directly applied in neuroscience simply because it is assumed that the brain processes "information". However, the concept of information, or the way it is processed, may be rather different in the two cases.

---

[5]There is, however, the problem of scaling the components. Since the improper density has infinite variance, the estimates of the components (and the weight vectors) grow infinitely large. Such behavior can be prevented by adding a penalty term of the form $\alpha \sum_i \|\mathbf{w}_i\|^2$ in the objective function. An alternative approach would be to use a saturating non-linearity as ϕ_i, thus combining the two infomax models.

In the data compression scheme, we start with a binary string, i.e. a sequence of zeros and ones. We want to transform the vectors into a another string, so that the string is as short as possible. This is basically accomplished by coding the most frequent realizations by short substrings or codewords, and using longer codewords for rarely occurring realizations. Such an approach has been found immensely useful in storage of information in serial digital computers.

However, if the processing of information is massively parallel, as in the brain, it is not clear what would be the interpretation of such reduction in code *length*. Consider an image that is coded in the millions of neurons in V1. A straightforward application of information theory would suggest that for some images, we only use k_1 neurons, where each neuron codes for one digit in the string, whereas others need k_2 neurons where $k_2 > k_1$. Furthermore, an optimal image code would be one where the average number of neurons is minimized. Yet, the number of neurons that are located in, say, the primary visual cortex is just the same for different stimuli. It would be rather absurd to think that some region of V1 is not needed to represent the most probable images. Even if some cells are not activated above the spontaneous firing rate, this lack of activation is an important part of the code, and does not mean that the neuron is not "part of the code".

In fact, in sparse coding the active neurons are assumed to be different for different stimuli, and each neuron is more or less equally important for representing some of the stimuli. While sparseness, when interpreted in terms of entropy, has some superficial similarity to information theoretic arguments, reducing the length of a string is very different from sparse coding because sparse coding is fundamentally a parallel scheme where no sequential order is given to the neurons, and the outputs of all neurons are needed to reconstruct the stimulus. That is, there is no reduction of code "length" because the number of coding units needed for reconstructing the stimulus is always the same, i.e. the total number of neurons. The whole concept of "length" is not well defined in the case of massively parallel and distributed processing. Reducing the length of a string is fundamentally an objective in *serial* information processing.

Another motivation for application of information theory in learning optimal representations comes from transmission of data. Optimal transmission methods are important in systems where data have to be sent through a noisy channel of limited capacity. Again, the basic idea is to code different binary sequences using other binary strings, based on their probabilities of occurrence. This allows faster and more reliable transmission of a serial binary signal.

Such "limited-capacity channel" considerations may be quite relevant in the case of the retina and optic nerve as well as nerves coming from other peripheral sensory organs. Another important application for this theory is in understanding coding of signals using spike trains. However, in V1, a limited capacity channel may be difficult to find. A well-known observation is that the visual input coming from the lateral geniculate nucleus (LGN) is expanded in the V1 by using a representation consisting of many more neurons than there are in the LGN. So, the transmission of information from LGN to V1 may not be seriously affected by the limited capacity of the "channel". Yet, the limited capacity of the channel coming to V1 is the basic assumption in infomax models.[6]

Thus, we think application of information-theoretical arguments in the study of *cortical* visual coding has to be done with some caution. Borrowing of concepts originally developed in electrical engineering should be carefully justified. This is an important topic for future research.

8.7 Concluding Remarks and References

Information theory provides another viewpoint to the utility of statistical modeling of images. The success in tasks such as compression and transmission depends on finding a useful representation of the data, and information theory points out that the optimal representation is the one which provides the best probabilistic model. Some studies therefore apply information-theoretic concepts to the study of natural image statistics and vision modeling. The idea of minimum-entropy coding gives some justification for sparse coding, and information transmission leads to objective functions which are sometimes equivalent to those of ICA. Nevertheless, we take here a more cautious approach because we think it is not clear if information theoretical concepts can be directly applied in the context of neuroscience, which may be far removed from the original digital communications setting in which the theory was originally developed.

A basic introduction to information theory is Mackay (2003). A classic reference which can be read as an introduction as well is Cover and Thomas (2006).

Basic and historical references on the infomax principle are Laughlin (1981), van Hateren (1992), Linsker (1988), Fairhall et al. (2001). The non-linear infomax principle was introduced in Nadal and Parga (1994), Bell and Sejnowski (1995). Infomax based on noise models with non-constant variance were introduced by van Vreeswijk (2001), Hyvärinen (2002), using rather different motivations. Poisson models for spike trains are discussed in Dayan and Abbott (2001). Information content in spike trains in considered in, e.g. Rieke et al. (1997). Another critique of the application of infomax principles to cortical coding can be found in Ringach and Malone (2007).

8.8 Exercises

Mathematical Exercises

1. Consider the set of all possible probability distributions for a random variable which takes values in the set $\{1, 2, \dots, 100\}$. Which distribution has minimum entropy?

[6]Possibly, the channel *from* V1 to V2 and other extrastriate areas could have a very limited capacity, but that is not the usual assumption in current infomax models.

2. Prove (8.23).
3. Consider a general (not standardized) one-dimensional Gaussian distribution, with pdf given by

$$p(z) = \frac{1}{\sqrt{2\pi}\sigma} \exp\left(-\frac{1}{2\sigma^2}(z-\mu)^2\right). \tag{8.38}$$

Compute its differential entropy. When it is maximized? When minimized?
4. Consider a random variable z with pdf

$$\frac{1}{\sigma} p_0\left(\frac{z}{\sigma}\right) \tag{8.39}$$

where z takes values on the whole real line, and the function p_0 is fixed. Compute the differential entropy as a function of σ and p_0.
5. * Assume we have a random vector $\mathbf{z}$ with pdf $p_{\mathbf{z}}$, and differential entropy $H(\mathbf{z})$. Consider a linear transformation $\mathbf{y} = \mathbf{Mz}$. What is the differential entropy of $\mathbf{y}$? Hint: Don't forget to use the probability transformation formula involving the determinant of $\mathbf{M}$, as in (7.13); and note that the preceding exercise is a special case of this one.

Computer Assignments

1. Let's consider discrete probability distributions which take values in the set $\{1, 2, \ldots, 100\}$. Create random probabilities for each of those values taken (remember to normalize). Compute the entropy of the distribution. Repeat this 1000 times. Find the distribution which had the largest and smallest entropies. What do they look like? Compare with results in Examples 2 and 3.

Part III
Nonlinear Features and Dependency of Linear Features

Chapter 9
Energy Correlation of Linear Features and Normalization

It turns out that when we estimate ICA from natural images, the obtained components are not really independent. This may be surprising since after all, in the ICA model, the components are assumed to be independent. But it is important to understand that while the components in the theoretical model are independent, the *estimates* of the components of *real image data* are often not independent. What ICA does is that it finds the most independent components that are possible by a linear transformation, but a linear transformation has so few parameters that the estimated components are often quite far from being independent. In this chapter and the following ones, we shall consider some dependencies that can be observed between the estimated independent components. They turn out to be extremely interesting both from the viewpoint of computational neuroscience and image processing. Like in the case of ICA, the models proposed here are still very far from providing a complete description of natural image statistics, but each model does exhibit some very interesting new phenomena just like ICA.

9.1 Why Estimated Independent Components Are Not Independent

9.1.1 Estimates vs. Theoretical Components

A paradox with ICA is that in spite of the name of the method, the estimated components need not be independent. That is, when we have a sample of real image patches, and estimate the independent components by an ICA algorithm, we get components which are usually not independent. The key to this paradox is the distinction between the estimated components and theoretical components. The theoretical components, which do not really exist because they are just a mathematical abstraction, are assumed to be independent. However, what an ICA algorithm gives, for any real data, is estimates of those theoretical components, and the estimates do not have all the properties of the theoretical components. In particular, the estimates need not be independent.

Actually, it is not surprising that the components estimated by ICA are not independent. If they were, the statistical structure of natural images would be completely described by the simple ICA model. If we knew the linear features A_i and the distributions of the independent components, we would know everything there is to know about the statistical structure of natural images. This would be rather absurd

Computational Imaging and Vision 39,

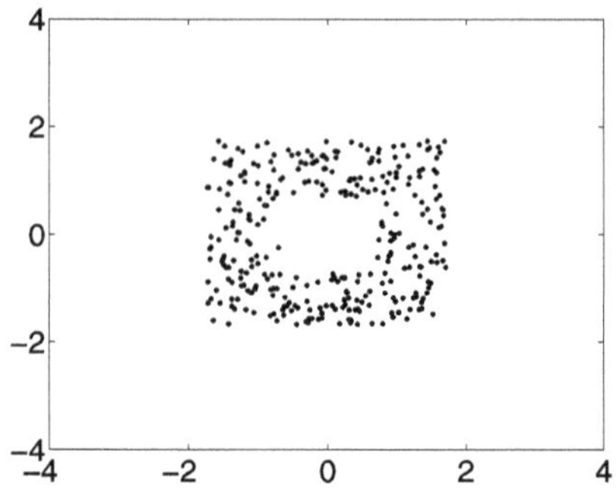

Fig. 9.1 Scatter plot of a distribution which cannot be linearly decomposed to independent components. Thus, the estimated components (given by the *horizontal* and *vertical axes*) are dependent

because natural images are obviously an extremely complex data set; one could say it is as complex as our world.

There are two different reasons why the estimates need not have the properties assumed for the theoretical components. First, the real data may not fulfill the assumptions of the model. This is very often the case, since models are basically abstractions or approximations of reality. (We will see below how the ICA model does not hold for natural image data.) The second reason is random fluctuation, called sampling effect in statistics: When we estimate the model for a finite number of image patches, we have only a limited amount of information about the underlying distribution, and some errors in the estimates are bound to occur because of this.

Consider, for example, the two-dimensional data in Fig. 9.1. This data is white (uncorrelated and unit variance), so we can constrain the ICA matrix to be orthogonal. If we input this data into an ICA algorithm, the algorithm says that the horizontal and vertical axis (say, s_1 and s_2) are the "independent components". However, it is easy to see that these components are not independent: If we know that s_1 is zero, we know that s_2 cannot be zero. Thus, information on one of the components gives information on the other component, so the components cannot be independent. This data does not follow the ICA model for *any* parameter values.

9.1.2 Counting the Number of Free Parameters

Another way of looking at this paradox is to think of the number of free parameters. A linear transformation of n variables to n new variables has n^2 free parameters. We can think of the problem of finding really independent components as a large system of equations which express the independence of the obtained components. How many equations are there? In Sect. 4.6, we saw that two independent components have the following non-linear uncorrelatedness property:

$$\operatorname{cov}\big(f_1(s_i), f_2(s_j)\big) = 0 \tag{9.1}$$

for any non-linear functions f_1 and f_2. Now, there are an infinite number of different non-linearities we could use. So, based on (4.42), we can form an infinite number of different equations (constraints) that need to be fulfilled, but we only have a finite number of free parameters, namely n^2. Thus, it is clear[1] that usually, no solution can be found!

Note that the situation with respect to independence is in stark contrast to the situation with whitening. As we saw in Chap. 5, we can *always* find a linear transformation which gives uncorrelated, and further whitened, components. This is because whitening only needs to consider the covariance matrix, which has a relatively small number of free parameters. In fact, the number of equations we get is $n(n+1)/2$ (because the covariance is a symmetric operation, this is the number of free parameters), which is smaller than n^2, so we can find a transformation which whitens the data.

9.2 Correlations of Squares of Components in Natural Images

Now, let us consider the dependencies of the components estimated in Chap. 7. The components are forced to be exactly uncorrelated by the ICA method we used. So, any dependencies left between the components must take the form of some kind of non-linear correlations. Let us compute correlations of the type in (9.1) for different non-linear functions f (we use the same function as both f_1 and f_2). Because the variances of the non-linear transformations are not necessarily equal to one, it is a good idea to normalize the covariance to yield the correlation coefficient:

$$\operatorname{corr}\big(f(s_i), f(s_j)\big) = \frac{E\{f(s_i)f(s_j)\} - E\{f(s_i)\}E\{f(s_j)\}}{\sqrt{\operatorname{var}(f(s_i))\operatorname{var}(f(s_j))}}. \tag{9.2}$$

In Fig. 9.2, we show the correlation coefficients for several different functions f. The figure shows the histograms of all the correlation coefficients between different pairs of independent components estimated in Chap. 7.

It turns out that we have a strong correlation for even-symmetric functions, i.e. functions for which

$$f(-s) = f(s). \tag{9.3}$$

Typical examples are the square function or the absolute value (b and a in the figure).

9.3 Modeling Using a Variance Variable

Intuitively, the dependency of two components that was described above is such that the components tend to be "active", i.e. have non-zero outputs at the same time.

[1] Strictly speaking, we should show that we can form an infinite number of equations which cannot be reduced to each other. This is too difficult to show, but it is likely to be true when we look at some arbitrary data distribution, such as the distribution of natural images. Of course, the situation is different when the data actually follows the ICA model: in that case, we know that there is a solution. A solution is then possible because in this very special case, the equations can be reduced, just as if we needed to solve a system of linear equations where the matrix is not full rank.

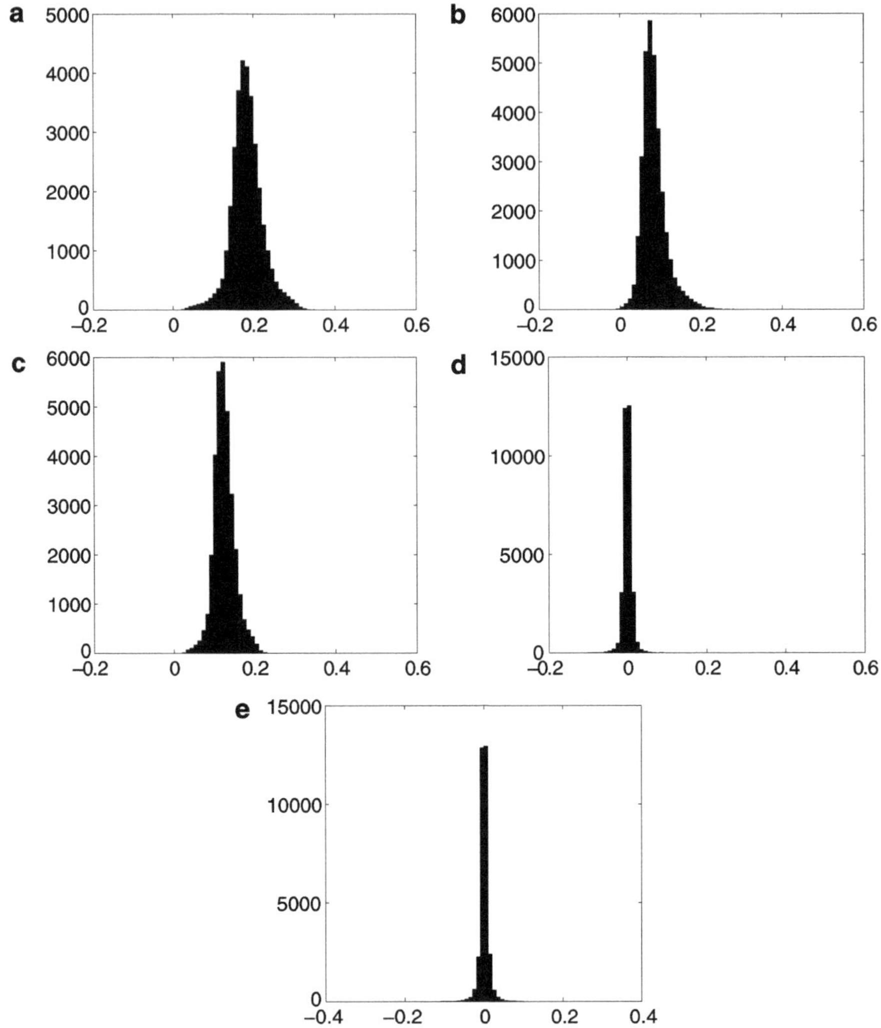

Fig. 9.2 Histograms of correlation coefficients of non-linear functions of independent components estimated from natural images. **a** $f(s) = |s|$, **b** $f(s) = s^2$, **c** $f(s)$ is a thresholding function that gives 0 between -1 and 1 and gives 1 elsewhere, **d** $f(s) = \text{sign}(s)$, **e** $f(s) = s^3$. Note that linear correlations, i.e. the case $f(s) = s$ are zero up to machine precision by definition

However, the actual values of the components are not easily predictable from each other. To understand this kind of dependency, consider a case where the components s_i are defined as products of "original" independent variables $\tilde{s}_i$ and a common "variance" variable d, which is independent of the $\tilde{s}_i$. For simplicity, let us define that the means of the $\tilde{s}_i$ are zeros and the variances are equal to one. Thus, we define the distribution of the components as follows:

$$
\begin{aligned}
s_1 &= \tilde{s}_1 d, \\
s_2 &= \tilde{s}_2 d, \\
&\vdots \\
s_n &= \tilde{s}_n d.
\end{aligned}
\tag{9.4}
$$

Now, s_i and s_j are uncorrelated for $i \neq j$, but they are not independent. The idea is that d controls the overall activity level of the two components: if d is very small, s_1 and s_2 are probably both very small, and if d is very large, both components tend to have large absolute values.

Such dependency can be measured by the correlation of their squares s_i^2, sometimes called the "energies". This means that

$$\operatorname{cov}(s_i^2, s_j^2) = E\{s_i^2 s_j^2\} - E\{s_i^2\}E\{s_j^2\} > 0. \tag{9.5}$$

In fact, assuming that $\tilde{s}_i$ and $\tilde{s}_j$ have zero mean and unit variance, the covariance of the squares equals

$$
\begin{aligned}
&E\{\tilde{s}_i^2 d^2 \tilde{s}_j^2 d^2\} - E\{\tilde{s}_i^2 d^2\}E\{\tilde{s}_j^2 d^2\} \\
&\quad = E\{\tilde{s}_i^2\}E\{\tilde{s}_j^2\}E\{d^2 d^2\} - E\{\tilde{s}_i^2\}E\{d^2\}E\{\tilde{s}_j^2\}E\{d^2\} \\
&\quad = E\{d^4\} - E\{d^2\}^2.
\end{aligned}
\tag{9.6}
$$

This covariance is positive because it equals the variance of d^2, and the variance of a random variable is always positive (unless the variable is constant).

Moreover, if the $\tilde{s}_i$ are Gaussian, the resulting components s_1 and s_2 can be shown to be sparse (leptokurtic). This is because the situation for each of the components is just the same as in Sect. 7.8.3: changing the variance of a Gaussian variable creates a sparse variable. However, we do not assume here that original components $\tilde{s}_i$ are Gaussian, so the effect of the variance variables is to *increase* their sparseness.

What is not changed from basic ICA is that the components s_i are uncorrelated in the ordinary sense. This is because we have

$$E\{s_i s_j\} = E\{\tilde{s}_i\}E\{\tilde{s}_j\}E\{d^2\} = 0 \tag{9.7}$$

due to the independence of the d from $\tilde{s}_j$. One can also define that the s_i have variance equal to one, which is just a scaling convention as in ICA. Thus, the vector $(s_1, s_2, \ldots, s_n)$ can be considered to be white.

9.4 Normalization of Variance and Contrast Gain Control

To reduce the effect of the variance dependencies, it is useful to normalize the local variance. Let us assume that the image patch is generated as a linear combination of independent features, as in ICA. However, now the variances of the components change from patch to patch as above. This can be expressed as

$$I(x, y) = \sum_{i=1}^{m} A_i(x, y)(d\tilde{s}_i) = d \sum_{i=1}^{m} A_i(x, y)\tilde{s}_i \tag{9.8}$$

where d is the variance variables that gives the standard deviation at each patch. It is a random variable because its value changes from patch to patch.

Here, we made the strong assumption that the variances of all the components are determined by a single variance variable d. This may be not a bad approximation when considering small image patches. It simplifies the situation considerably, since now we can simply estimate d and divide the image patch by d:

$$\bar{I}(x, y) \leftarrow \frac{I(x, y)}{\hat{d}}. \tag{9.9}$$

Assuming that we have a perfect estimator $\hat{d} = d$, the normalized images $\bar{I}$ then follow the basic ICA model

$$\bar{I}(x, y) = \frac{d}{d} \sum_{i=1}^{m} A_i(x, y)\tilde{s}_i = \sum_{i=1}^{m} A_i(x, y)\tilde{s}_i \tag{9.10}$$

with the original components $\tilde{s}_i$. In practice, we don't have a perfect estimator, so the data will follow the ICA model only approximatively. Also, it is preferable to do

$$\bar{I}(x, y) \leftarrow \frac{I(x, y)}{\hat{d} + \varepsilon} \tag{9.11}$$

where ε is a relatively small constant that prevents division by zero or a very small number.[2]

This kind of normalization of variances is called *contrast gain control*. It can be compared with the subtraction of the DC component: an unrelated (irrelevant) variable that has a strong effect on the statistics of the features we are modeling is removed so that we have a more direct access to the statistics.

For small image patches, one rather heuristic approach is to simply estimate d using the norm of the image patch

$$\hat{d} = c \sqrt{\sum_{x,y} I(x, y)^2} \tag{9.12}$$

where c is a constant that is needed to make the s_i have unit variance after normalization; it depends of the covariance matrix of the data. Usually, however, we do not need to compute c because it only changes the overall scaling of the data.

---

[2] A reasonable method for determining ε might be to take the 10% quantile of the values of d, which we did in our simulations.

When we normalize the contrast as described here, and compute the output of linear feature detector, the result is closely related to the neurophysiological model of *divisive normalization;* see (3.9) on page 64. The output of the linear feature detector is then computed as

$$\tilde{s} = \frac{\sum_{x,y} W(x,y)I(x,y)}{\sqrt{\sum_{x,y} I(x,y)^2 + \varepsilon}}. \tag{9.13}$$

In the divisive normalization model, the denominator was essentially the sum of the squares of the outputs of linear feature detectors. Here, we have the norm of the image patch instead. However, these two can be closely related to each other if the set of linear feature detectors form an orthogonal basis for a small image patch; then the sum of squares of pixel values and feature detectors are exactly equal.

The approach to estimating d in this section was rather ad hoc. A more principled method for contrast gain control could be obtained by properly defining a probability distribution for d together with the $\tilde{s}$, and estimating all those latent variables using maximum likelihood estimation or some other principled methods. Furthermore, if the model is to be used on whole images instead of small patches, a single variance variable is certainly insufficient. This is still an area of ongoing research; see the References section for more information.

Let us just mention here a slight modification of the divisive normalization in (9.11) and (9.12) which has been found useful in some contexts. The idea is that one could compute a weighted sum of the pixel values $I(x, y)$ to estimate the variance variable. In particular, low frequencies dominate (9.12) because they have the largest variances. This effect could be eliminated by computing a whitening matrix and using the norms of the patches in the whitened space as $\hat{d}$. Note that the divisive normalization destroys the whiteness of the data, so after such normalization, the whitening matrix has to be recomputed, and the data has to be whitened with this new whitening matrix.

9.5 Physical and Neurophysiological Interpretations

Why are the variances, or general activity levels, so strongly correlated, and what is the point in contrast gain control? A number of intuitive explanations can be put forward.

9.5.1 Canceling the Effect of Changing Lighting Conditions

The illumination (lighting) conditions can drastically change from one image to another. The same scene can be observed under very different lighting conditions, think for example of daylight, dusk, and indoor lighting. The light coming onto the retina is a function of the reflectances of the surfaces in the scene (R), and the light (illuminance) level (L). In fact, the reflectances are multiplied by the illuminance to give the luminance I arriving at the retina:

$$I(x, y) = L(x, y)R(x, y). \tag{9.14}$$

In bright daylight, the luminance levels are uniformly larger than in an indoor room, and so are the contrasts. The average luminance level is not visible in our images because we have removed the DC component, which is nothing else than the mean luminance in an image patch. But the general illuminance level still has a clear effect on the magnitude of the contrasts in the image, and these are seen in the values of the independent components. In a whole scene, the illuminance may be quite different in different parts of the image due to shadows, but in a small image patch illuminance is likely to be approximately the same for all pixels. Thus, the single variance variable d, which does not depend on x or y, could be interpreted as the general illuminance level in an image patch.

In this interpretation, the utility of divisive normalization is that it tries to estimate the reflectances R of the surfaces (objects). These are what we are usually interested in, because they are needed for object recognition. Illuminance L is usually not of much interest.

9.5.2 Uniform Surfaces

A second reason for the correlated changes in the variances of features outputs is what is called the "blue sky effect". Natural images contain large areas of almost zero contrast, such as the sky. In such areas, the variances of all the independent components should be set to almost zero. Thus, the variance variable d is related to whether the image patch is in a uniform surface or not. This would partly explain the observed changes in the variances of the components, but this does not seem to explain utility of contrast gain control.

9.5.3 Saturation of Cell Responses

Mechanisms related to gain control have been observed in many parts of the visual system, from the retina to the visual cortex (see the References section below). Before the advent of statistical modeling, their existence was usually justified by the limited response range of neurons. As discussed in Sect. 3.4.1, the neurons cannot fire above a certain firing rate. The range of contrasts that are present in the stimuli coming to the retina is huge because of the changes in illuminance condition: the incoming signal can differ in several orders of magnitude. Contrast gain control is assumed to solve this problem by dividing the contrasts (to be coded by cell responses) by a measure of the general contrast level. This leads to a very similar computation to what our statistical modeling proposes, although our model does not consider limited response ranges of neurons—linear RFs have no such limitation.

9.6 Effect of Normalization on ICA

Although we considered the dependencies after estimating ICA, it makes sense to do the variance normalization before ICA. This would be theoretically optimal because then the ICA estimation would be performed on data whose distribution is closer to the distribution given by the ICA model. In fact, the method given above in Sect. 9.4 can actually normalize the patches without computing independent components.

A valid question then is: does variance normalization affect the independent components? Let us now estimate ICA after variance normalization to see what effect there may be. The obtained W_i and A_i are shown in Figs. 9.3 and 9.4. The similar-

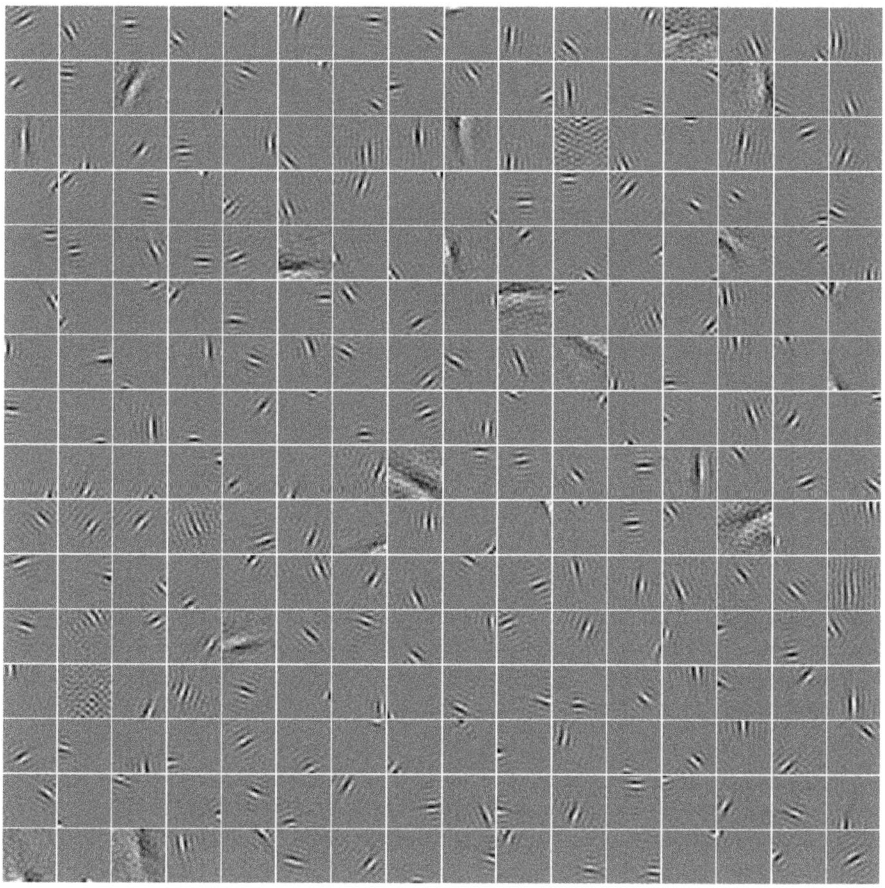

Fig. 9.3 The whole set of detector weights W_i obtained by ICA after the variances have been normalized as in (9.11) and (9.12)

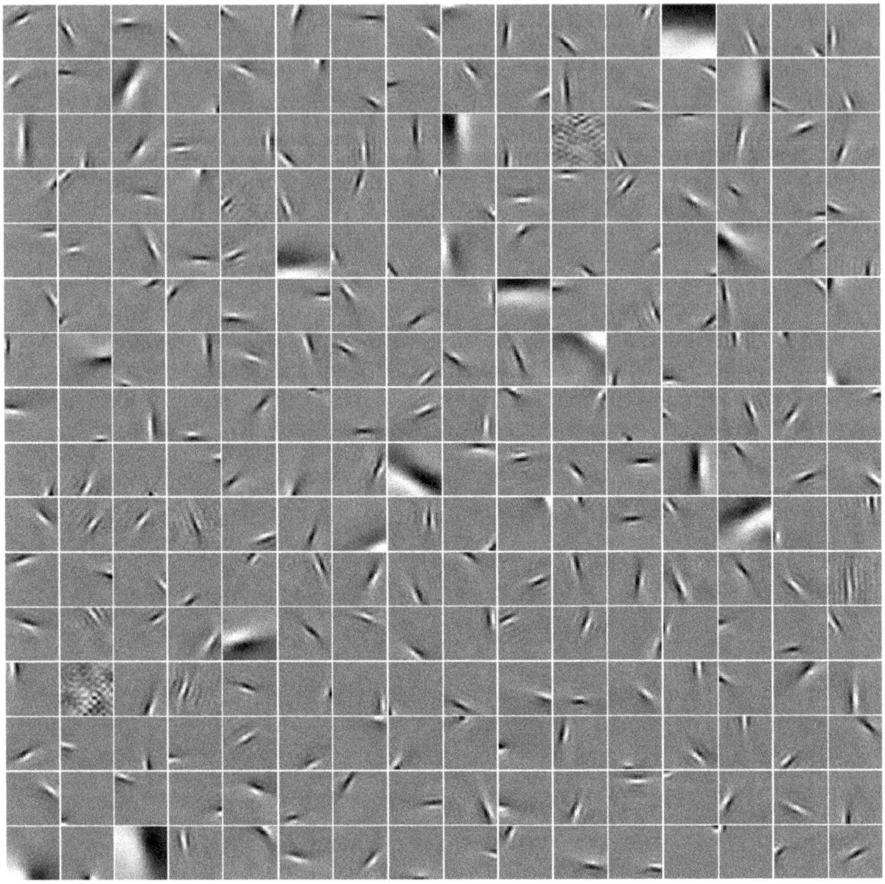

Fig. 9.4 The whole set of features A_i obtained by ICA after the variances have been normalized

ity to the results obtained without variance normalization (Fig. 6.6 on page 144 and Fig. 7.3 on page 161) is striking.[3]

There is one important difference, however. Variance normalization makes the components less sparse. In Fig. 9.5, we have plotted the histogram of the kurtoses of the independent components, estimated either with and without variance normalization. Variance normalization clearly reduces the average kurtosis of the components. The components after variance normalization correspond to estimates $\tilde{s}_i$.

This reduction in kurtosis is not very surprising if we recall the results in Sect. 7.8.3. There, it was shown that changing variance of Gaussian variables is one mechanism for creating sparse variables. The variance variable in (9.8) does exactly that. Here, the variance variable is the same for all components, but that does

[3] However, there is some difference as well: the vectors A_i now have some spurious oscillations. The reason for this phenomenon remains to be investigated.

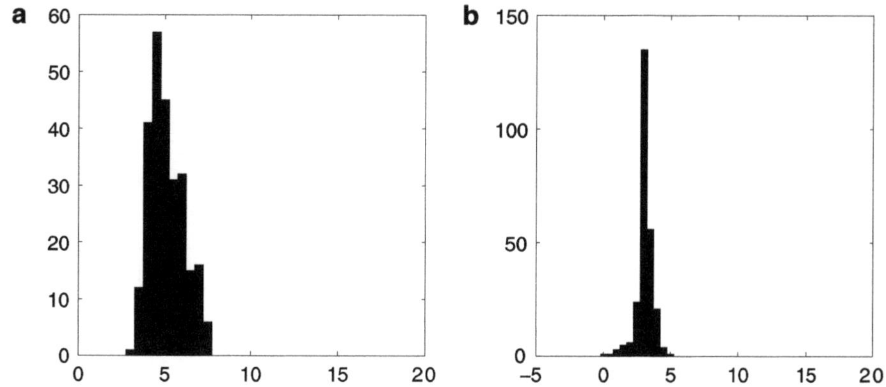

Fig. 9.5 Histograms of kurtoses of independent components. **a** Estimated *without* variance normalization, **b** estimated *with* variance normalization

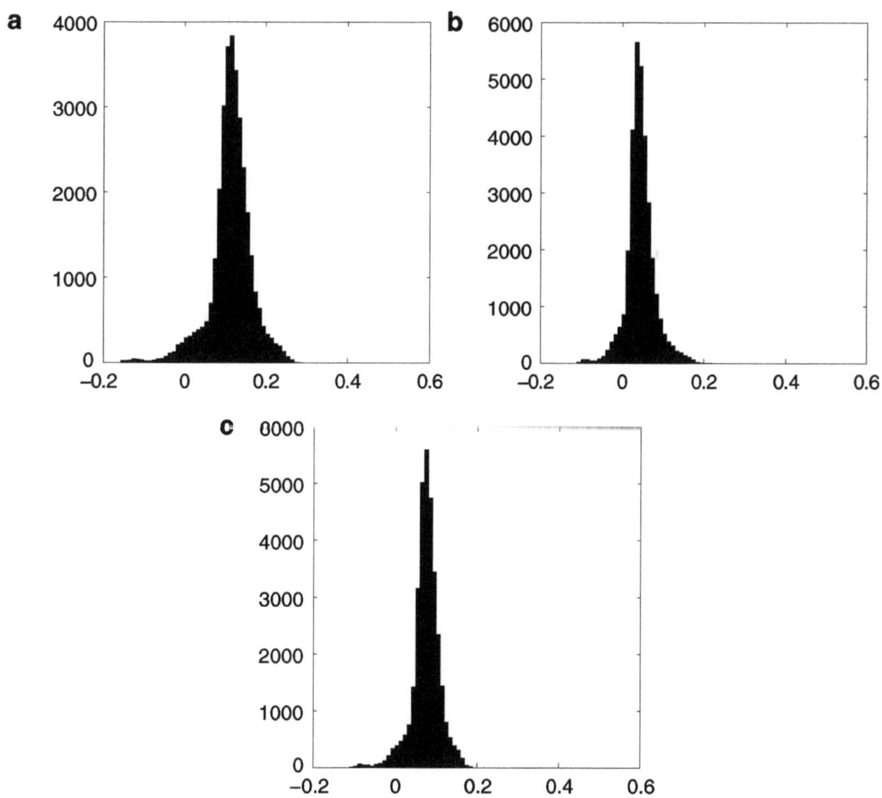

Fig. 9.6 Histograms of correlation coefficients of non-linear functions of independent components estimated *with* variance normalization. **a** $f(s) = |s|$, **b** $f(s) = s^2$, **c** $f(s)$ is a thresholding function that gives 0 between -1 and 1 and gives 1 elsewhere. Compare with **a–c** in Fig. 9.2

not change the situation regarding the marginal distributions of the single components: multiplication by the variance variable makes them more sparse. Thus, if we cancel the effect of the variance variable, it is natural that the components become less sparse.

In practice, normalization of the image patches only reduces the variance dependencies but does not eliminate them. The process described above for modeling the variances of the components was only a very rough approximation. Let us do the same measurements on the variance dependencies that we did above before normalization. The results are shown in Fig. 9.6. We see that energy correlations still remain, although they are now smaller.

9.7 Concluding Remarks and References

This chapter focused on the simple empirical fact that the "independent components" estimated from natural images are not independent. This seemingly paradoxical statement is due to the slightly misleading way the expression "independent component" is used in the context of ICA. While ICA finds the most independent components possible by a linear transformation, there is no guarantee that they would be completely independent. The dependencies observed in the case of natural images can be partly explained by the concept of a global variance variable which changes from patch to patch. Attempts to cancel the dependencies generated by such a changing variance lead to divisive normalization or gain control models. However, this is merely the beginning of a new research direction—modeling dependencies of "independent" components—which will be continued in the following chapters.

The results in this chapter also point out an interesting, and very important, change in the relationship between sparseness and independence. With linear models, maximization of sparseness is equivalent to maximization of independence, if the linear projections are sparse (super-Gaussian). But in Sect. 9.6, we saw that divisive normalization *increases independence*, as measured by correlations of squares, while *decreasing sparseness* as measured by kurtosis. Thus, sparseness and independence have a simple relation in linear models only; with non-linear processing, we cannot hope to maximize both simultaneously. This point is elaborated in Lyu and Simoncelli (2008); we will also return to this point in Sect. 17.2.1.

A seminal work on normalizing images to get more Gaussian distributions for the components is Ruderman and Bialek (1994b). Simoncelli and co-workers have proposed sophisticated methods for modeling variance dependencies in large images. They start with a fixed wavelet transform (which is similar to the ICA decomposition; see Sect. 17.3.2). Since the linear basis is fixed, it is much easier to build further models of the wavelet coefficients, which can then be used in contrast gain control (Schwartz and Simoncelli 2001a; Schwartz et al. 2005). More complex models include hidden Markov models (Wainwright et al. 2001; Romberg et al. 2001) as well as Markov Random Fields (Gehler and Welling 2005; Lyu and Simoncelli 2007). A recent experimental work which considers different alternatives for the functional form of divisive normalization is Bonin et al. (2006);

the authors conclude that something like the division by the norm is a good model of contrast gain control in the cat's LGN.

Finally, let us mention that some work considers the statistical properties of illumination itself, before its interaction with objects (Dror et al. 2004; Mury et al. 2007).

Gain control phenomena can be found in many different parts of the visual system. Some of the earliest quantitative work on the effect of natural stimuli statistics considered gain control in the retina (Laughlin 1981; van Hateren 1992). Recent work on gain control in the LGN (Bonin et al. 2006) confirms the idea that it is based on normalization by the Euclidean norm (i.e. root-mean-square contrast).

9.8 Exercises

Mathematical Exercises

1. Show that if two components s_1 and s_2 are created using a variance variable as in (9.4), then also their absolute values have a positive correlation, i.e. $\operatorname{cov}(|s_1|, |s_2|) > 0$ unless d is constant.
2. Consider a variance variable d which only takes two values: α with probability $1/2$ and β with probability $1/2$. Assume s_1 and s_2 follow (9.4) with Gaussian $\tilde{s}_1$ and $\tilde{s}_2$.
   a. Show that for $\alpha, \beta > 0$, the resulting joint pdf of (s_1, s_2) is a sum of two Gaussian pdf's.
   b. Take $\alpha = 0$ and $\beta > 0$. What is the distribution now like? Can you write the pdf?

Computer Assignments

1. Take some images and sample patches from them. Then build two edge detectors in orthogonal orientations. Compute the outputs of both edge detectors for all the patches. Normalize the outputs to unit variance. What are the
   a. The ordinary covariances and correlation coefficients of edge detector outputs.
   b. The covariances and correlation coefficients of the squares of the edge detector outputs?

Chapter 10
Energy Detectors and Complex Cells

In preceding chapters, we considered only linear features. In this chapter, we introduce non-linear features. There is an infinite variety of different kinds of non-linearities that one might use for computing such features. We will consider a very basic form inspired by models previously used in computer vision and neuroscience. This approach is based on the concepts of subspaces and energy detectors. The resulting model called "independent subspace analysis" gives non-linear features which turn out to be very similar to complex cells when the parameters are learned from natural images.

10.1 Subspace Model of Invariant Features

10.1.1 Why Linear Features Are Insufficient

In the previous chapters, we assumed that each feature is basically a linear entity. Linearity worked in two directions: The coefficient (strength) s_i of the feature in the image was computed by a linear feature detector as in (7.4) on page 153, and the image was composed of a linear superposition of the features A_i, as in (7.3) on page 153.

A problem with linear features is that they cannot represent *invariances*. For example, an ideal complex cell gives the same response for a grating irrespective of its phase. A linear feature detector cannot have such behavior because the response of a linear system to a grating depends on the match between the phase of the input and the phase of the detector, as discussed in Sect. 2.2.3. In higher levels of the visual system, there are cells which respond to complex object parts irrespective of their spatial location. This cannot be very well described by a single linear template (feature detector) either.

10.1.2 Subspaces or Groups of Linear Features

Linear combinations are a flexible tool that is capable of computing and representing invariant features. Let us consider a feature that consist of several vectors and all their linear combinations. Thus, one such feature corresponds to a group of simple linear features which are mixed together with some coefficients:

$$\text{invariant feature} = \text{set of} \sum_{i=1}^{q} A_i(x, y)s_i \quad \text{for all values of } s_i \tag{10.1}$$

A. Hyvärinen, J. Hurri, P.O. Hoyer, *Natural Image Statistics*,
Computational Imaging and Vision 39,

where q is the number of vectors in a single group. This grouping of components is closely related to the concept of *subspace* in linear algebra. The subspace spanned by the vectors A_i for $i = 1, \ldots, q$ is defined as the set of all the possible linear combinations of the vectors. Thus, the range of values that the invariant feature represents is a subspace.

The point is that each such subspace is representing an invariant feature by taking different linear combinations of the vectors. Let us consider, for example, the problem of constructing a detector for some shape, so that the output of the detector does not depend on location of that shape, i.e. it detects whether the shape occurs anywhere in the image. You could approach the problem by linear vectors (templates) A_i that represent the shape in many different locations. Now, if these vectors are dense enough in the image space, so that all locations are covered, the occurrence of the shape in any location can be represented as a trivial linear combination of the templates: take the coefficient at the right location to be equal to one, and all the other coefficients s_i to be zero. Thus, the subspace can represent the shape in a way that is invariant with respect to the location, i.e. it does not depend on the location. In fact, subspaces are even more powerful because to represent a shape that is not exactly in the locations given by the basic vectors, a satisfactory representation can often be obtained by taking the average of two or more templates in nearby locations (say, just to the left and just to the right of the actual location). It is this capacity of *interpolation* that makes the linear subspace representation so useful.

The whole image (patch) can be expressed as a linear superposition of these non-linear features. Let us denote by $S(k)$ the set of indices i of those A_i that belong to the kth group or subspace; for example, if all the subspaces have dimension two, we would have $S(1) = \{1, 2\}$, $S(2) = \{3, 4\}$, etc. Thus, we obtain the following model:

$$I(x, y) = \sum_k \sum_{i \in S(k)} A_i(x, y) s_i. \tag{10.2}$$

The image is still a linear superposition of the vectors $A_i(x, y)$, but the point is that these vectors are grouped together. The grouping is useful in defining non-linear feature detectors, as shown next.

10.1.3 Energy Model of Feature Detection

In (10.2), there was no quantity that would directly say what the strength (value, output) of a subspace feature is. It is, of course, important to define such a measure, which is the counterpart of the s_i in the case of simple linear features. We now define the value of the feature, i.e. the output of a feature detector as a particular *non-linear* function of the input image patch. We shall denote it by e_k.

First of all, since the model in (10.2) is still a linear superposition model, we can invert the system given by the vectors A_i as in the linear case. To be able to do this easily, we assume as in ICA, that the total number of vectors A_i is equal to the number of pixels (alternatively, equal to the number of dimensions after canonical preprocessing, as discussed below). Then each linear coefficient s_i can be computed by inverting the system, just as in (7.6):

$$s_i = \sum_{x,y} W_i(x,y)I(x,y) \tag{10.3}$$

for some detector weights W_i which are obtained just as with any linear image model (e.g. the ICA model).

Now, how do we compute the strength of the subspace feature as a function of the coefficients s_i that belong to that subspace? There are several reasons for choosing the square root of the sum of the squares:

$$e_k = \sqrt{\sum_{i \in S(k)} s_i^2}. \tag{10.4}$$

The first reason for using this definition is that the sum of squares is related to the norm of the vector $\sum_{i \in S(k)} A_i s_i$. In fact, if the A_i form an orthogonal matrix, i.e. they are orthogonal and have norms equal to one, the square root of the sum of squares is equal to the norm of that vector. The norm of the vector is an obvious measure of its strength; here it can be interpreted as the "total" response of all the s_i in the subspace.

Second, a meaningful measure of the strength e_k of a subspace feature in an image I could be formed by computing the distance of I from the best approximation (projection) that the subspace feature is able to provide:

$$\min_{s_i, i \in S(k)} \sum_{x,y} \left[I(x,y) - \sum_{i \in S(k)} s_i A_i(x,y) \right]^2. \tag{10.5}$$

Again, if the A_i form an orthogonal matrix, this can be shown to be equal to

$$\sum_{x,y} I(x,y)^2 - \sum_{i \in S(k)} s_i^2 \tag{10.6}$$

which is closely related to the sum of squares, because the first term does not depend on the coefficients s_i at all.

Third, a sum of squares is often used in Fourier analysis: the "energy" in a given frequency band is usually computed as the sum of squares of the Fourier coefficients in the band. This is why a feature detector using sum of squares is often called an energy detector. Note, however, that the connection to energy in the sense of physics is quite remote.

Fourth, the sum of squares seems to be a good model of the way complex cells in the visual cortex compute their outputs from outputs of simple cells; see Sect. 3.4.2. In the physiological model, a square root is not necessarily taken, but the basic idea of summing the squares is the same. In that context, the summation is often called "pooling".

Note that we could equally well talk about linear combinations of linear feature detectors W_i instead of linear combinations of the A_i. For an orthogonal basis, these

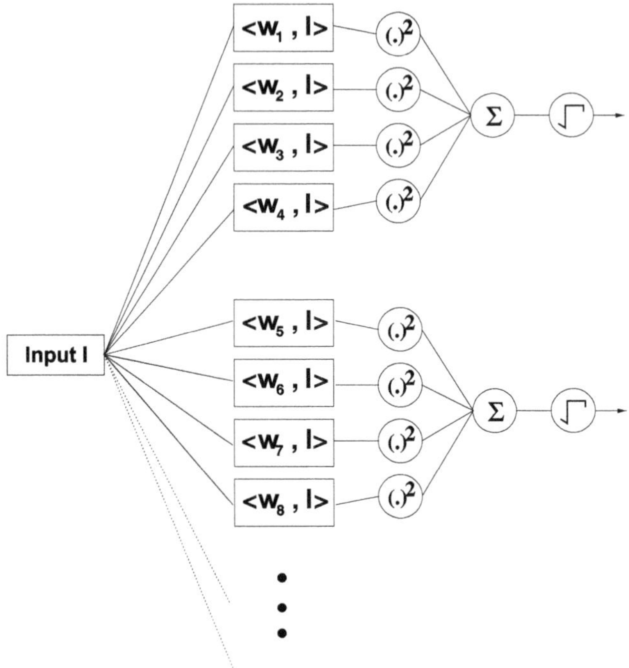

Fig. 10.1 Illustration of computation of complex cell outputs by pooling squares of linear feature detectors. From Hyvärinen and Hoyer (2000), Copyright ©2000 MIT Press, used with permission

are essentially the same thing, as shown in Sect. 19.6. Then linear combinations of the W_i in the same subspace give the forms of all possible linear feature detectors associated with that subspace.

Figure 10.1 illustrates such an energy pooling (summation) model.

Canonically Preprocessed Data Invariant features can be directly applied to data whose dimension has been reduced. Just as in the case of a basic linear decomposition, we can simply formulate the linear model as in (10.2) where the data on the left-hand side is the preprocessed data z_i, and the linear feature vectors are in the reduced space. Nothing is changed in the concept of subspaces. Likewise, the energy detector in (10.4) takes the same form.

10.2 Maximizing Sparseness in the Energy Model

10.2.1 Definition of Sparseness of Output

What we are now going to show is that we can learn invariant features from natural images by maximization of sparseness of the energy detectors e_k given by the subspace model. Sparseness can be measured in the same way as in the linear case. That is, we consider the expectation of a convex function of the square of the detector output.

First of all, we take a number of linear features that span a feature subspace. To keep things simple, let us take just two in the following. Let us denote the detector weight vectors, which work in the reduced space after canonical preprocessing, by $\mathbf{v}_1$ and $\mathbf{v}_2$. Since we are considering a single subspace, we can drop the index i of the subspace. So, what we want to maximize is a measure of sparseness of the form

$$E\{h(e^2)\} = E\{h((\mathbf{v}_1^T\mathbf{z})^2 + (\mathbf{v}_2^T\mathbf{z})^2)\} = E\left\{h\left(\left(\sum_{j=1}^{n} v_{1j}z_j\right)^2 + \left(\sum_{j=1}^{n} v_{2j}z_j\right)^2\right)\right\} \tag{10.7}$$

where h is a convex function just as in the linear feature case.

An important point that must be considered is how the relation between $\mathbf{v}_1$ and $\mathbf{v}_2$ should be constrained. If they are not constrained at all, it may easily happen that these two linear detectors end up being the equal to each other. Then we lose the capability of representing a subspace. Mathematically speaking, such a situation is violating our assumption that the linear system given by the vectors A_i is invertible, because this assumption implies that the W_i (or the $\mathbf{v}_i$) are not linear combinations of each other.

We constrain here $\mathbf{v}_1$ and $\mathbf{v}_2$ in the same way as in the linear feature case: the outputs of the linear feature detectors must be uncorrelated:

$$E\{(\mathbf{v}_1^T\mathbf{z})(\mathbf{v}_2^T\mathbf{z})\} = 0 \tag{10.8}$$

and, as before, we also constrain the output variances to be equal to one:

$$E\{(\mathbf{v}_i^T\mathbf{z})^2\} = 1, \quad \text{for } i = 1, 2. \tag{10.9}$$

Now, we can maximize the function in (10.7) under the constraints in (10.8) and (10.9).

10.2.2 One Feature Learned from Natural Images

To give some preview of what this kind of analysis means in practice, we show the results of estimation of a single four-dimensional subspace from natural images. The four vectors $\mathbf{v}_i$, converted back to the original image space (inverting the preprocessing), are shown in Fig. 10.2.

What is the invariance represented by the subspace like? A simple way to analyze this is to plot a lot of linear combinations of the weight vectors W_i belonging to the same subspace. Thus, we see many instances of the different features that together define the invariant feature. This is shown in Fig. 10.3 for the weight vectors in Fig. 10.2, using random coefficients inside the subspace.

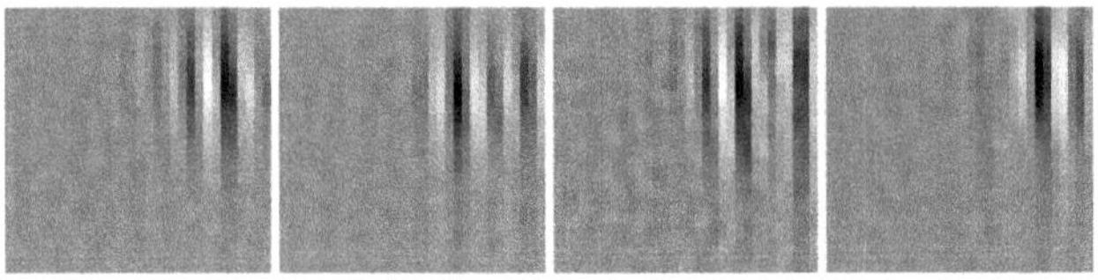

Fig. 10.2 A group of weight vectors W_i found by maximization of the non-linear energy detector in natural images

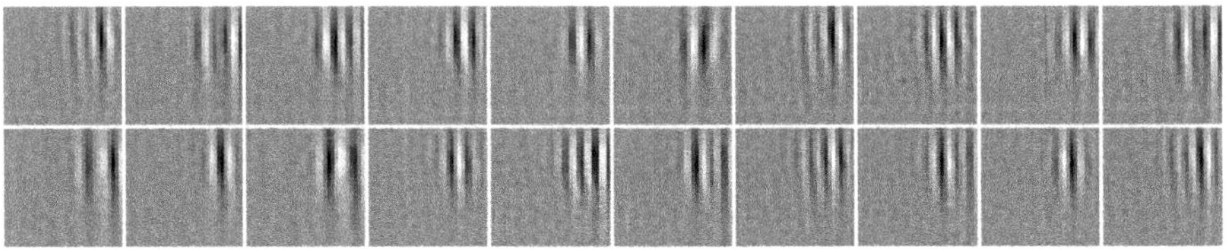

Fig. 10.3 Random combinations of the weight vectors W_i in the subspace shown in Fig. 10.2. These combinations are all particular instances of the feature set represented by the invariant feature

The resulting invariance has a simple interpretation: The invariant feature obtained by the algorithm is maximally invariant with respect to the *phase* of the input. This is because all the four linear features W_i are similar to Gabor functions which have quite similar parameters otherwise, but with the major difference that the phases of the underlying oscillations are quite different. In the theory of space-frequency analysis (Sect. 2.4), and in complex cell models (Sect. 3.4.2), invariance to phase is achieved by using two different linear feature detectors which are in quadrature-phase (as sine and cosine functions). Here, we have four linear feature detectors, but the basic principle seems to be the same.

Such phase-invariance does, in fact, practically always emerge for the feature subspaces estimated from natural image data; see Sect. 10.7 for a more detailed analysis. The invariant features are thus similar to complex cells in the visual cortex. This invariance appears because the linear features in the same subspace have similar orientations, and frequencies, whereas they have quite different phases, and slightly different positions. Note that it is not easy to distinguish the effects of different phases and slightly different positions, since they result in very much the same transformations in the overall shape of the features (something that looks like a small displacement of the feature).

These results indicate that from a statistical viewpoint, the invariance to phase is a more important feature of natural images that, say, invariance to orientation. Such invariance to phase has been considered very important in visual neuroscience because it is the function usually attributed to complex cells: phase-invariance is the hallmark property that distinguished simple and complex cells.

To see that this "emergence" of phase-invariant features is not self-evident, we can consider some alternatives. A well-known alternative would be a feature subspace invariant to orientation, called "steerable filters" in computer vision. Actually, by taking a subspace of Gabor-like vectors that are similar in all other parameters than orientation, one can obtain exactly orientation-invariant features (see References and Exercises sections below). What our results show is that in representing natural images, invariance with respect to phase is more important in the sense that it gives a better statistical model of natural images. This claim will be justified in the next section, where we build a proper probabilistic model based on sparse, independent subspaces.

10.3 Model of Independent Subspace Analysis

Maximization of sparseness can be interpreted as estimation of a statistical model just as in the case of linear features. Assume that the pdf of the s_k is of the following form:

$$\log p(s_1,\ldots,s_n) = \sum_k h(e_k^2) = \sum_k h\Big(\sum_{i\in S(k)} s_i^2\Big). \tag{10.10}$$

(A constant needs to be added to make this a proper pdf if the function h is not properly normalized, but it has little significance in practice.) Denote by $\mathbf{z}_t$, $t = 1,\ldots,T$ a set of observed image patches after preprocessing. Then the likelihood of the model can be obtained in very much the same way as in the case of ICA in (7.15) on page 159. The log-likelihood is given by

$$\log L(\mathbf{v}_1,\ldots,\mathbf{v}_n) = T\log\left|\det(\mathbf{V})\right| + \sum_k \sum_{t=1}^{T} h\Big(\sum_{i\in S(k)} (\mathbf{v}_i^T\mathbf{z}_t)^2\Big). \tag{10.11}$$

Again, if we constrain the s_i to be uncorrelated and of unit variance, which is equivalent to orthogonality of the matrix $\mathbf{V}$, the term $\log|\det(\mathbf{V})|$ is constant. The remaining term is just the sum of the sparseness measures of all the energy detectors. Thus, we see that maximization of the sparsenesses is equivalent to estimation of the statistical generative model by maximization of likelihood.

As a concrete example, let us consider the case of two-dimensional subspaces, and choose $h(y) = -\sqrt{y}$. This defines a distribution inside each subspace for which $\log p(s_i,s_j) = -\sqrt{s_i^2+s_j^2}$. If we further normalize this pdf so that its integral is equal to one, and so that s_i and s_j have unit variance, we get the following pdf for s_i and s_j in the same subspace:

$$p(s_i,s_j) = \frac{2}{3\pi}\exp\Big(-\sqrt{3}\sqrt{s_i^2+s_j^2}\Big). \tag{10.12}$$

This could be considered as a two-dimensional generalization of the Laplacian distribution. If you assume s_j is given as $s_j = 0$, the conditional pdf of s_i is proportional to $\exp(-\sqrt{3}\sqrt{s_i^2})$, which is as in the Laplacian pdf in (7.18) up to some scaling constants.

What is the main difference between this statistical model and ICA? In ICA, the pdf was derived using the assumption of independence of the components s_i. Since we have here a rather different model, it must mean that some statistical dependencies exist among the components. In fact, the pdf above corresponds to a model where the *non-linear features e_k are independent, but the components (i.e. linear features) in the same subspace are not*. The independence of the non-linear features can be seen from the fact that the log-density in (10.10) is a sum of functions of the non-linear features. By definition, the non-linear features are then independent. This also implies that two components in two different subspaces are independent. Since the subspaces are independent in these two ways, this model is called independent subspace analysis (ISA).

The more difficult question is: What kind of dependencies exist between the components in a single subspace? This will be considered next.

10.4 Dependency as Energy Correlation

The basic result is that in the ISA model, the dependencies of the linear components in the same subspace take the form of energy correlations already introduced in Chap. 9. This result will be approached from different angles in the following.

10.4.1 Why Energy Correlations Are Related to Sparseness

To start this investigation on the statistical dependencies of components in ISA, we consider a simple intuitive explanation of why the sparseness of energy detectors is related to the correlations of energies of the underlying linear features.

Let us consider the following two cases. First, consider just two linear feature detectors which have the same output distributions, and whose output energies are summed (pooled) in a non-linear energy detector. If the outputs are statistically independent, the pooling reduces sparseness. This is because of the fundamental result given by the Central Limit Theorem (see Sect. 7.9.1). It says, roughly speaking, that the sum of independent random variables is closer to Gaussian (and, therefore, less sparse) than the original random variables themselves.

Second, consider the contrasting extreme case where the linear detector outputs are perfectly dependent, that is, equal. This means that the distribution of the pooled energies is equal to the distribution of the original energies (up to a scaling constant), and therefore there is no reduction in sparseness.

So, we see that maximization of the sparseness of the energy is related to maximization of the energy correlations (dependencies) of the underlying linear features.

10.4.2 Spherical Symmetry and Changing Variance

Next, we show how the ISA pdf can be interpreted in terms of a variance variable, already used in Chap. 9.

The distribution inside each subspace, as defined by (10.10), has the distinguishing property of being *spherically symmetric*. This simply means that the pdf depends on the norm $\sqrt{\sum_{i \in S(k)} s_i^2}$ only. Then any rotation (orthogonal transformation) of the variables in the subspace has exactly the same distribution.

Spherically symmetric distributions constitute one of the simplest models of non-linear correlations. If h is non-linear, the variables in the same subspace are dependent. In contrast, an important special case of spherically symmetric distributions is obtained when $h(u) = u$, in which case the distribution is just the ordinary Gaussian distribution with no dependencies or correlations.

Spherical symmetry is closely related to the model in which a separate variance variable multiplies two (or more) independent variables as in (9.4) on page 203. If the independent variables $\tilde{s}_1$ and $\tilde{s}_2$ are *Gaussian*, the distribution of the vector (s_1, s_2) is spherically symmetric. To show this, we use the basic principle that the marginal pdf of the vector (s_1, s_2) can be computed by integrating the joint pdf of (s_1, s_2, d) over d. First, note that we have $\tilde{s}_i^2 = s_i^2/d^2$. Since the $\tilde{s}_i$ are Gaussian and independent (let us say they have unit variance), and independent of d, the pdf can be computed as:

$$p(s_1, s_2) = \int p(s_1, s_2, d)\, dd = \int \frac{1}{2\pi d^2} \exp\left(-\frac{s_1^2 + s_2^2}{2d^2}\right) p(d)\, dd. \tag{10.13}$$

Even without actually computing the integral ("integrating d out") in this formula, we see that the pdf only depends on the (square of the) norm $s_1^2 + s_2^2$. Thus, the distribution is spherically symmetric. This is because the distribution of $(\tilde{s}_1, \tilde{s}_2)$ was spherically symmetric to begin with. The distribution of d, given by $p(d)$ in the equation above, determines what the distribution of the norm is like.

In the model estimation interpretation, h is obtained as the logarithm of the pdf, when it is expressed as a function of the square of the norm. Thus, based in (10.13), we have

$$h(e^2) = \log \int \frac{1}{2\pi d^2} \exp\left(-\frac{e^2}{2d^2}\right) p(d)\, dd. \tag{10.14}$$

Note that we obtain a spherically symmetric distribution only if the $\tilde{s}_i$ are Gaussian, because only Gaussian variables can be both spherically symmetrically distributed and independent. In Chap. 9, we did not assume that the $\tilde{s}_i$ are Gaussian; in fact, when we normalized the data we saw the estimated $\tilde{s}_i$ are still quite super-Gaussian. This apparent contradiction arises because in the ISA model, we have a different variance variable d_k for each subspace, whereas in Chap. 9, there was only one d for the whole image patch. If we estimated the $\tilde{s}_i$ in the ISA model, their distributions would presumably be much closer to Gaussian than in Chap. 9.

10.4.3 Correlation of Squares and Convexity of Non-linearity

Next, we consider the role of the non-linearity h in (10.11). In the model developed in this chapter, we don't have just any h, but h is assumed to be convex because we are considering measures of sparseness. Actually, it turns out that the h that can be derived from the model with a variance variable as in the preceding section, are necessarily convex. A detailed mathematical analysis of this connection is given in Sect. 10.8.

Conversely, if we define the pdf inside a subspace by taking a convex function h of the square of the norm, we usually get a positive covariance between the squares of the components. Again, a detailed mathematical analysis of this connection is given in Sect. 10.8, but we will discuss this connection here with an example.

As an illustrative example, consider two-dimensional subspaces with pdf defined as in (10.12). The covariance of the squares of s_i and s_j can be calculated, it is equal to 2/3. The kurtosis of either s_i or s_j is equal to 2, and the variables are uncorrelated. (This density has been standardized to that its mean is zero and variance equal to one.) Using this pdf we can investigate the conditional distribution of s_j for a given s_i:

$$p(s_j|s_i) = \frac{p(s_i, s_j)}{p(s_i)} = \frac{p(s_i, s_j)}{\int p(s_i, s_j)\, ds_j}. \tag{10.15}$$

This can be easily computed for our pdf, and is plotted in Fig. 10.4a. We see a shape that has been compared to a *bow-tie*: when going away from zero on the horizontal

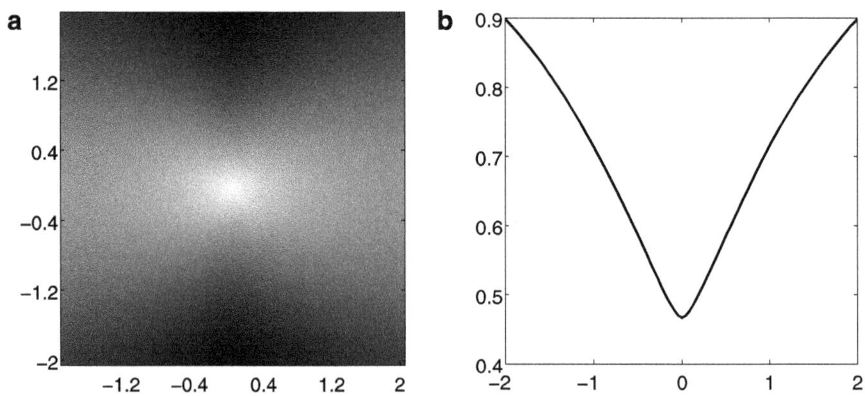

Fig. 10.4 Illustration of the correlation of squares in the probability density in (10.12). **a** The two-dimensional *conditional* density of s_j (*vertical axis*) given s_i (*horizontal axis*). The conditional density is obtained by taking vertical slices of the density function, and then normalizing each slice so that it integrates to one, and thus defines a proper probability density function. Black means low probability density and white means high probability density. We see that the conditional distribution get broader as s_i goes further from zero in either direction. This leads to correlation of energies since the expectation of the square is nothing but the variance. **b** The conditional variance of s_j (*vertical axis*) for a given s_i (*horizontal axis*). Here, we see that the conditional variance grows with the square (or absolute value) of s_i

axis (s_i), the distribution on the vertical axis (s_j) becomes wider and wider, i.e. its variance grows. This can be quantified by the conditional variance

$$\operatorname{var}(s_j|s_i) = \int s_j^2 p(s_j|s_i)\, ds_j. \tag{10.16}$$

The actual conditional variance of s_i, given s_j is shown in Fig. 10.4b. We see that the conditional variance grows with the absolute value of s_i.

What is the connection to energy correlations? Both increasing conditional variance and energy correlations try to formalize the same intuitive idea: when one of the variables has a large absolute values, the other(s) is likely to have a large absolute values as well. Correlations of squares or energies is something we can easily compute, whereas conditional variance is a more faithful formalization of the same idea.

Thus, we see that taking a convex h, or assuming the data to come from the variance variable model (with Gaussian "original" variables $\tilde{s}_i$) are closely related.

10.5 Connection to Contrast Gain Control

Both the ISA model and the model we used to motivate divisive normalization in Chap. 9 leads to a similar kind of dependency. This may give the impression that the two are actually modeling the same thing. This is not so because in ISA, the variance variables d are different for each subspace, whereas in the contrast gain control there was a single d for the whole patch.

In the ISA model, the variance variables d_i are actually closely related to the outputs of non-linear feature detectors. The sum of squares of the s_i inside one subspace (or rather, that sum divided by the dimension of the subspace) can be considered a very crude estimator of the d_i^2 of that subspace because, in general, the average of squares is an estimator of variance. No such interpretation of d can be made in the contrast gain control context, where the single d is considered an uninteresting "nuisance parameter", something whose influence we want to cancel.

Although the contrast gain control models could be generalized to the case where the patch is modeled using several variance variables, which possibly control the variances in different parts of the patch due to different illumination conditions, the basic idea is still that in ISA, there are many more energy detectors than there are variance variables due to illumination conditions in the contrast gain model in Chap. 9.

Because the dependencies in the two models are so similar, one could envision a single model that encompasses both models. Steps toward such a model are discussed in Sect. 11.8. On the other hand, we can use ISA to model the energy correlations that remain in the images *after* divisive normalization. In the image experiments below, we first reduce energy correlation by divisive normalization using (9.11), and then model the data by ISA. This has two different motivations:

1. We want to model energy correlations, or in general the statistical structure of images, as precisely as possible. So, it makes sense to first reduce the overall value of energy correlations to be able better to see fine details. This can be compared with removal of the DC component, which makes the details in second-order correlations more prominent. Just like in ICA, one finds that the dependencies between the subspaces are reduced by divisive normalization, so the ISA model is then simply a better model of image data.
2. On a more intuitive level, one goal in image modeling is to find some kind of "original" independent features. Reducing the dependencies of linear features by divisive normalization seems a reasonable step toward such a goal.

10.6 ISA as a Non-linear Version of ICA

It is also possible to interpret the ISA model of independent subspace analysis as a non-linear invertible transformation of the data. Obviously, the transformation is non-linear, but how can we say that it is invertible? The point is to consider not just the norms of the coefficients in the subspaces, but also the angles inside the subspaces. That is, we look at what is called the *polar* coordinates inside each subspace. For simplicity, let us consider just two-dimensional subspaces, although this discussion also applies in higher-dimensional subspaces.

The point is that if we express the coordinates s_1, s_2 in a two-dimensional subspace as a function of the norm $r = \sqrt{s_1^2 + s_2^2}$ and the angle $\theta = \arctan s_2/s_1$ with respect to one of the axes. This is an invertible transformation; the inverse is given by $s_1 = r\cos\theta$ and $s_2 = r\sin\theta$.

The fundamental point is that the two variables r and θ are *independent* under the ISA model. This is precisely because of the assumption that the pdf is spherically symmetric, i.e. it depends on the norm only. Intuitively, this is easy to see: since the pdf only depends on the norm, it can be factorized, as required by the definition of statistical independence, to two factors. The first depends on the norm only, and the second, completely trivial factor is equal to 1. The constant factor can be interpreted to be a (constant) function of θ, corresponding to a uniform distribution of the angles. So, we see that the pdf can be factorized into a product of a function of r and a function θ, which proves the independence. (Note that this proof is not quite correct because we have to take into account the determinant of the Jacobian, as always when we transform pdf's. The rigorous proof is left as an exercise for mathematically very sophisticated readers.)

Thus, we see that we can think of the generative model of ISA as a non-linear, invertible transformation, which is, in the case of two-dimensional subspaces, as follows:

$$\begin{pmatrix} r_1 \\ \theta_1 \\ \vdots \\ r_{n/2} \\ \theta_{n/2} \end{pmatrix} \rightarrow \begin{pmatrix} z_1 \\ z_2 \\ \vdots \\ z_{n-1} \\ z_n \end{pmatrix} \tag{10.17}$$

where the components on the left-hand side are all independent from each other. The same idea holds for subspaces of any dimensions; we just need to parameterize arbitrary rotations in those subspaces (which is rather complicated).

10.7 Results on Natural Images

10.7.1 Emergence of Invariance to Phase

10.7.1.1 Data and Preprocessing

We took the same 50 000 natural image patches of size 32×32 as in the ICA case. We performed contrast gain control by divisive normalization as in (9.11), as motivated in Sect. 10.5. Then we preprocessed the normalized patches in the same ("canonical") way as with ICA, reducing the dimension to 256.

The non-linearity h used in the likelihood or sparseness measure was chosen to be a smoothed version of the square root as in (6.14) on page 136. We then estimated the whole set of feature subspaces using subspace size 4 for natural images, which means $64 = 256/4$ subspaces.

10.7.1.2 Features Obtained

The results are shown in Figs. 10.5 and 10.6 for the W_i and the A_i, respectively. Again, the feature detectors are plotted so that the grey-scale value of a pixel means the value of the coefficient at that pixel. Grey pixels mean zero coefficients. As with linear independent components, the order of the subspaces is not defined by the model. For further analysis, the subspaces are ordered according to the sparsenesses of the subspaces as measure by the term $\sum_{t=1}^{T} h(\sum_{i \in S(k)} (\mathbf{v}_i^T \mathbf{z}_t)^2)$ in the likelihood.

Visually, one can see that these feature detectors have interesting localization properties. First, they are localized in space: most of the coefficients are practically zero outside of a small receptive field. This is true of the individual feature detectors in the same way as in the case of linear feature detectors estimated by sparse coding or ICA. What is important here is that it is also true with respect to the whole subspace, because the non-zero coefficients are more or less in the same spatial location for all feature detectors corresponding to the same subspace. The linear feature detectors and the invariant features are also oriented and multiscale in exactly the same way: the optimal orientations and frequencies seem to be the same for all the linear features in the same subspace.

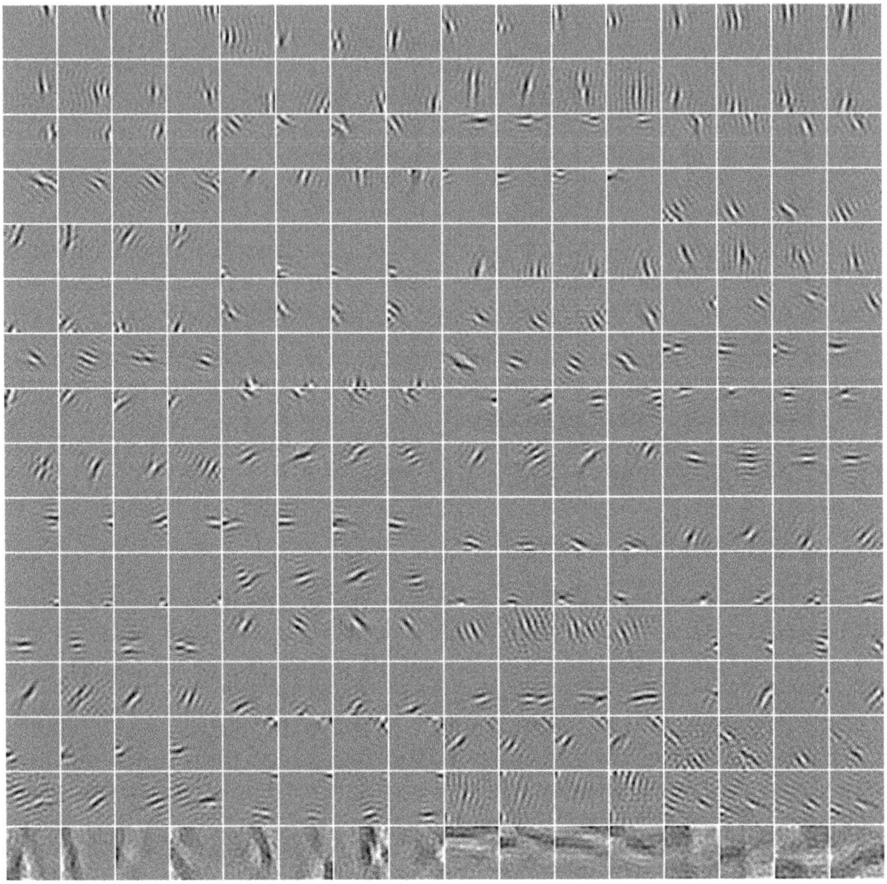

Fig. 10.5 The whole set of vectors W_i obtained by independent subspace analysis. The four vectors in the same subspace are shown consecutively on the same row. The subspaces have been ordered so that the sparsest ones are first (*top rows*)

10.7.1.3 Analysis of Tuning and Invariance

We can analyze these features further by fitting Fourier gratings, just as in Sect. 6.4. In determining the optimal orientation and frequency for a subspace, we find the grating that has maximum energy response, i.e. maximum sum of squares of linear dot-products inside the subspace. The analysis is made a bit more complicated by the fact that for these non-linear features; we cannot find the maximum response over all phases by using two filters in quadrature-phase and taking the square of the responses as we did in Sect. 6.4. We have to compute the responses over different values of orientation, frequency, *and* phase. Thus, we take many different values of α, β and θ in

$$f(x,y) = \sin\big(2\pi\alpha\big(\sin(\theta)x + \cos(\theta)y\big) + \beta\big). \tag{10.18}$$

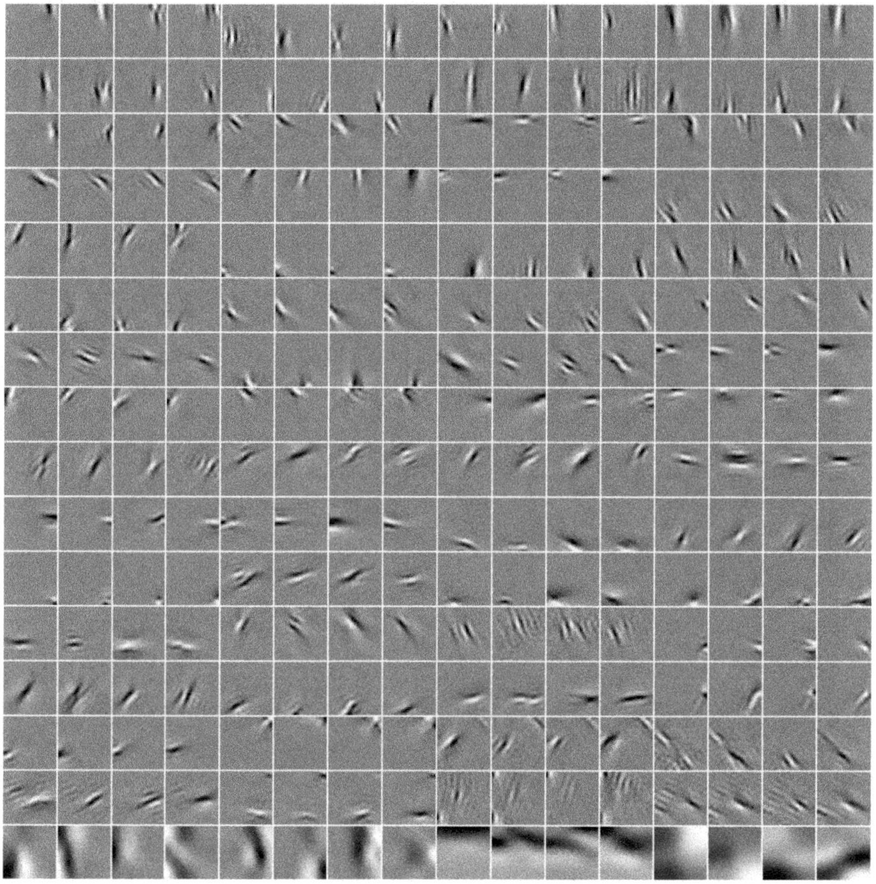

Fig. 10.6 The whole set of vectors A_i obtained by independent subspace analysis

Then we compute the responses of the energy detectors and find the α, θ that maximize the sum of responses over the different β for each subspace.

We can then investigate the selectivities of the features by changing one of the parameters, while the others are fixed to the optimal values. This gives the tuning curves for each of the parameters. Note that when computing the responses for varying orientation or frequency, we again take the sum over all possible phases to simulate the total response to a drifting grating. On the other hand, when we compute the tuning curve for phase, we do not take a sum over different phases.

In Fig. 10.7, we have show the results of the analysis for the first ten (i.e. the ten sparsest) subspaces in Fig. 10.5. We can clearly see that estimated energy detectors are still selective to orientation and frequency. However, they are less selective to phase. Some of the features are rather completely insensitive to phase, whereas in other, some selectivity is present. This shows that the model successfully produces the hallmark property of complex cells: invariance to phase—at least in some of the

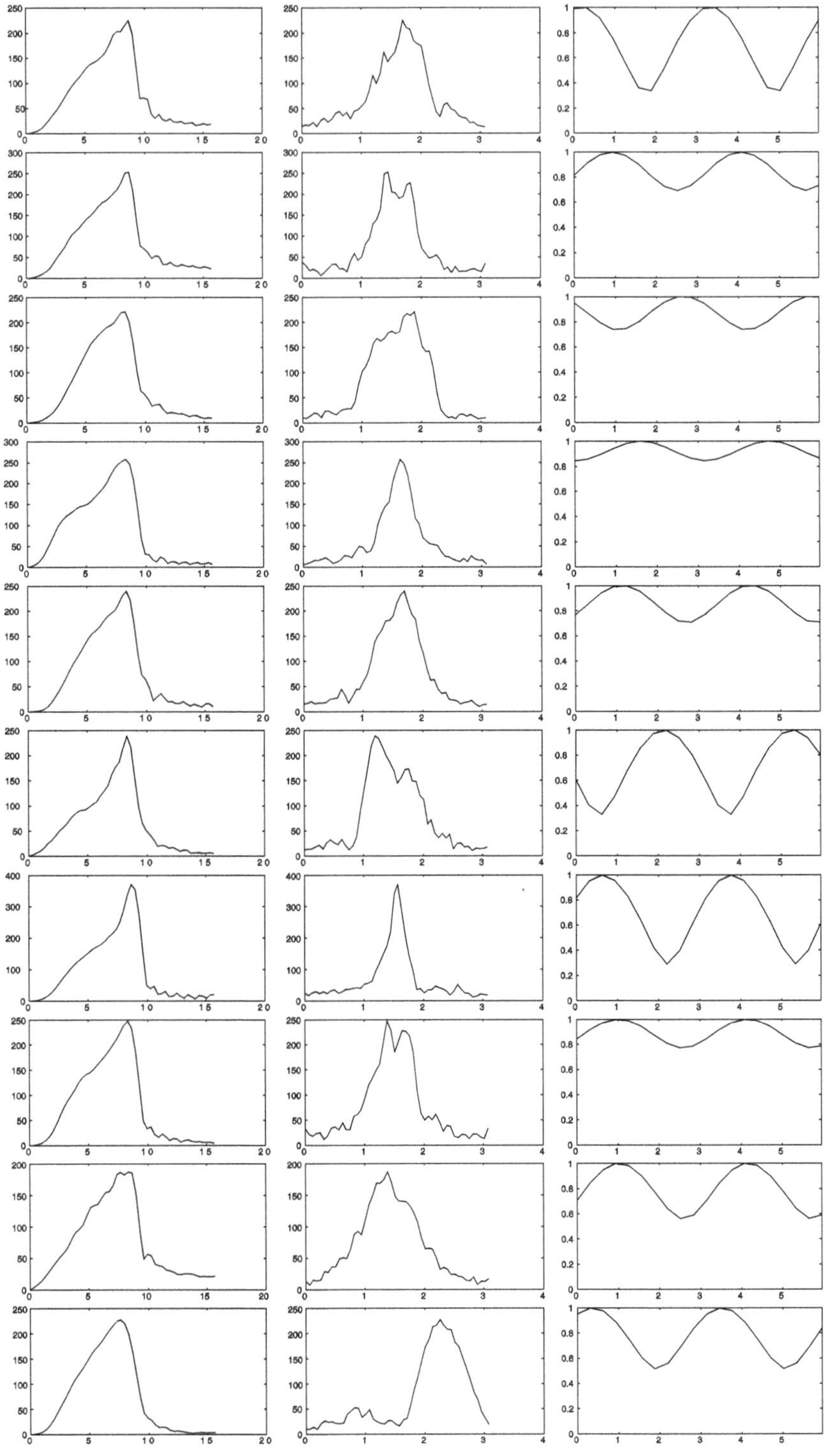

Fig. 10.7 Tuning curves of the ISA features W_i. *Left*: change in frequency (the unit is relative to the window size of 32 pixels, so that 16 means wavelength of 2 pixels). *Middle*: change in orientation. *Right*: change in phase

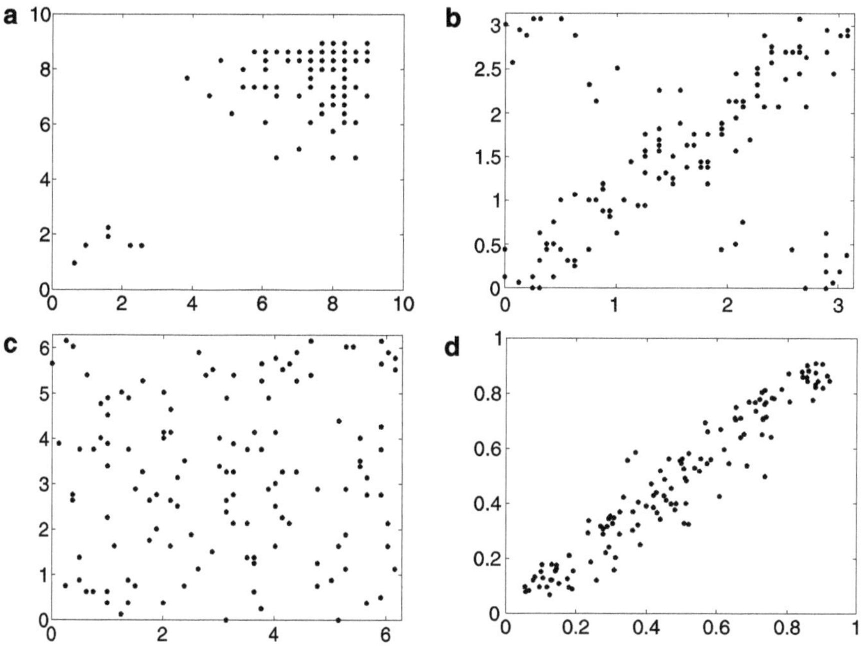

Fig. 10.8 Correlation of parameters characterizing the linear features in the same independent subspace. In each plot, we have divided the subspaces into two pairs, and plotted the optimal parameter values for the two linear features in a scatter plot. **a** Scatter plot of frequencies, **b** scatter plot of orientations, **c** scatter plot of phases, **d** scatter plot of locations (x-coordinate of centerpoints)

cells.[1] Thus, the invariance and selectivities that emerge from natural images by ISA is just the same kind that characterize complex cells.

The selectivity to orientation and frequency is a simple consequence of the fact that the orientation and frequency selectivities of the underlying linear feature detectors are similar in a given subspace. This can be analyzed in more detail by visualizing the correlations of the optimal parameters for two linear features in the same subspace. In Fig. 10.8, we see that the orientations (b) are strongly correlated. In the case of frequencies the correlation is more difficult to see because of the overall concentration to high frequencies. As for phases, no correlation (or any kind of statistical dependency) can be seen.

In this analysis, it is important to reduce the dimension by PCA quite a lot. This is because as explained in Sect. 5.3.3.2, the phase is not a properly defined quantity

[1]It should be noted that the invariance to phase of the sum of squares of linear filter responses is not an interesting property in itself. Even taking receptive fields with random coefficients gives similar phase-response curves as in Fig. 10.7 for the sum of squares. This is because the phase-responses are always sinusoids, and so are their squares, so if the phases of different filters are different enough, their sum often ends up being relatively constant. What is remarkable, and needs sophisticated learning, is the *combination* of selectivity to orientation and frequency with phase-invariance.

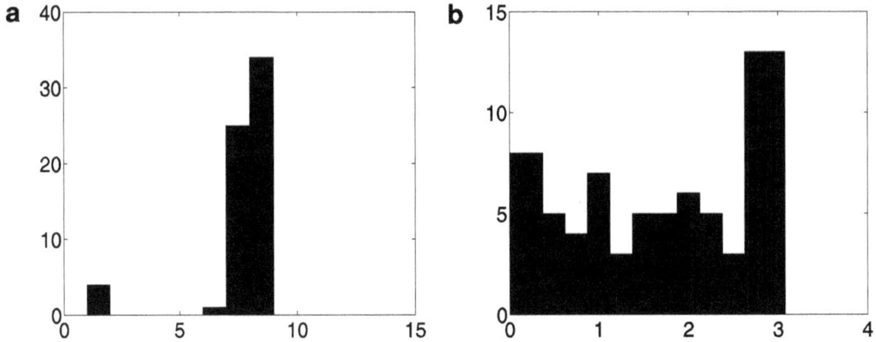

Fig. 10.9 Histograms of the optimal **a** frequencies and **b** orientations of the independent subspaces

at the very highest frequencies (the Nyquist frequency) that can be represented by the sampling lattice, i.e. the pixel resolution.

Finally, we can analyze the distribution of the frequencies and orientations of the subspace features. The plot in Fig. 10.9 shows that while all orientations are rather equally present (except for the anisotropy seen even in ICA results), the frequency distribution is strongly skewed: most invariant features are tuned to high frequencies.

10.7.1.4 Image Synthesis Results

Results of synthesizing image with the ISA model are shown in Fig. 10.10. This is based on the interpretation as a non-linear ICA in Sect. 10.6, but the model with variance dependencies in Sect. 10.4 would give the same results.

Here, the norms r_i, i.e. the values of the invariance features, were chosen to be equal to those actually observed in the data. The angles inside the subspace were then randomly generated.

The synthesized images are quite similar to those obtained by ICA. The invariance is not really reflected in the visual quality of these synthesized images.

10.7.2 The Importance of Being Invariant

What is the point in features that are invariant to phase? In general, the variability of how objects are expressed in the retinal image is one of the greatest challenges, perhaps the very greatest, to the visual system. Objects can be seen in different locations in the visual space (= retinal space). They can appear at different distances to the observer, which changes their size in the retinal image. Objects can rotate, turn around, and transform in a myriad of ways. And that's not all: the environment can change, moving light sources and changing their strength, casting shadows, and even occluding parts of the object.

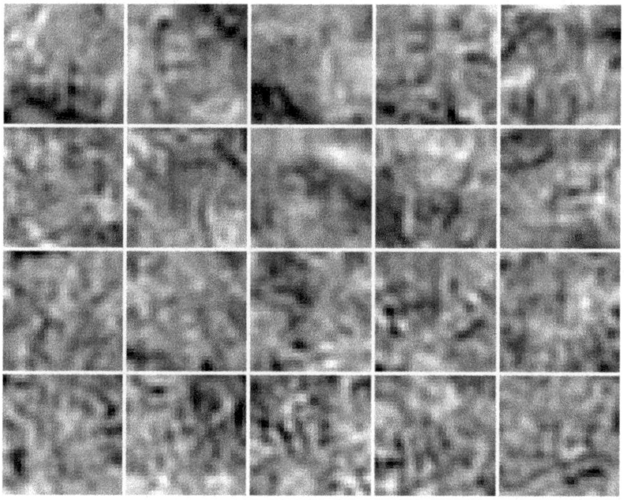

Fig. 10.10 Image synthesis using ISA. Compare with the ICA results in Fig. 7.4 on page 162

The visual system has learned to recognize objects despite these difficulties. One approach used in the early parts of the system is to compute features that are invariant to some of such changes. Actually, in Chap. 9, we already saw one such operation: contrast gain control attempts to cancel some effects of changes in lighting, and removal of the DC component is doing something similar.

With energy detectors, we find phase-invariant features, similar to those in complex cells. It is usually assumed that the point in such an invariance is to make recognition of objects less dependent on the exact position where they appear. The point is that a change in phase is very closely related to a change in position. In fact, it is rather difficult to distinguish between phase-invariance and position-invariance (which is often called translation- or shift-invariance). If you look at the different feature vectors A_i inside the same subspace in Fig. 10.6, you might say that they are the same feature in slightly different positions.

Changing the phase of a grating, and in particular of a Gabor function is indeed very similar to moving the stimulus a bit. However, it is not movement in an arbitrary direction: It is always movement in the direction of oscillations. Thus, phase-invariance is rather a special case of position-invariance. And, of course, the position-invariance exhibited by these energy detectors is very limited. If the stimulus is spatially localized (say, a Gabor function as always!), only a small change in the position is allowed, otherwise the stimulus goes out of the receptive field and the response goes to zero. Even this very limited position-invariance can be useful as a first step, especially if combined with further invariant computations in the next processing layers.

Figure 10.11 shows a number of Gabor stimuli that have all other parameters fixed at the same values but the phase is changed systematically. An ideal phase-invariant feature detector would give the same response to all these stimuli.

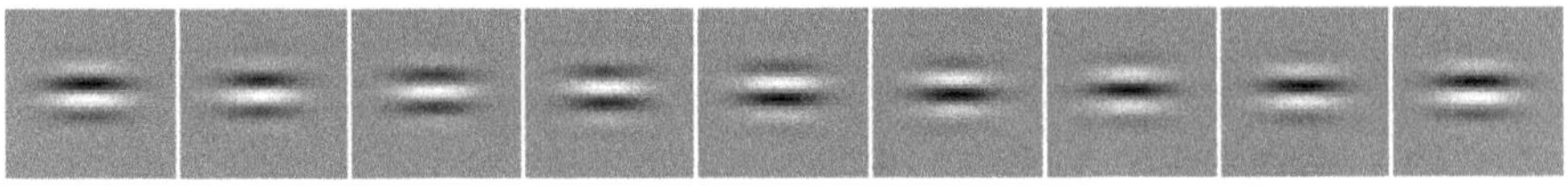

Fig. 10.11 A Gabor stimulus whose phase is changed

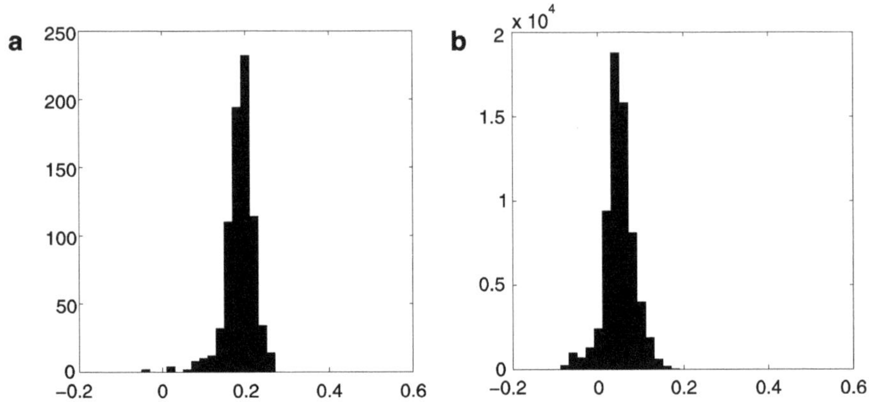

Fig. 10.12 Correlation coefficients of the squares of all possible pairs of components estimated by ISA. **a** Components in the same subspace, **b** components in different subspaces

10.7.3 Grouping of Dependencies

Next, we analyze how grouping of dependencies can be seen in the ISA results on natural images. A simple approach is to compute the correlation coefficients of the squares of components. This is done separately for components which belong to the same subspace, and for components which belong to different subspaces. When this is computed for all possible component pairs, we can plot the histogram of the correlation coefficients in the two cases. This is shown in Fig. 10.12. We see that square correlations are much stronger for components in the same subspace. According to the model definition, square correlations should be zero for components in different subspaces, but again we see that the real data does not exactly respect the independence assumptions.

Another way of analyzing the results is to visualize the square correlations. This is done in Fig. 10.13 for the first 80 components, i.e. 20 first subspaces. Visually, we can see a clear grouping of dependencies.

10.7.4 Superiority of the Model over ICA

How do we know if the ISA model really is better for natural images when compared to the ICA model? The first issue to settle is what it means to have a better model.

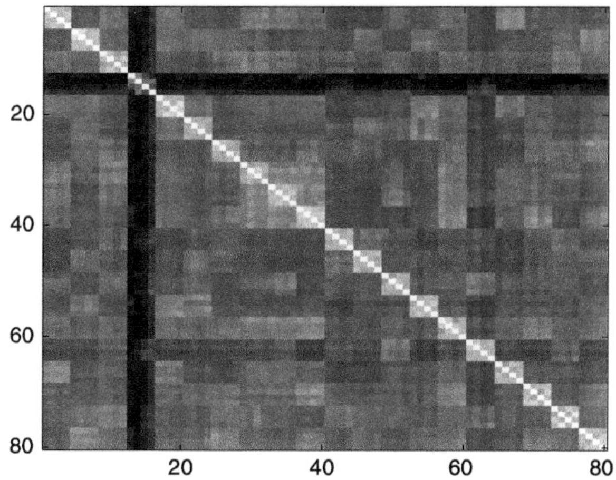

Fig. 10.13 Correlation coefficients of the squares of the 80 first components in Fig. 10.5. The correlation coefficients are shown in grey-scale. To improve the visualization, values larger than 0.3 have been set to 0.3

Of course, ISA is better than ICA in the sense that it shows emergence of new kinds of phenomena. However, since we are building statistical models, it is important to ask if the new model we have introduced, in this case ISA, really is better than ICA in a purely statistical sense. One useful way of approaching this is to compute the maximum likelihood. In a Bayesian interpretation, the likelihood is the probability of the parameters given the data, if the prior is flat. This helps us in comparing ICA and ISA models because we can consider the subspace size as another parameter. The ICA model is obtained in the case where the subspace size equals one. So, we can plot the maximum likelihood as a function of subspace size, always recomputing the W_i so as to maximize the likelihood for each subspace size. If the maximum is obtained for a subspace size larger than one, we can say that ISA is a better model than ICA.[2]

It is important to note that we need to use a measure which is in line with the theory of statistics. One might think that comparison of, say, sparsenesses of the ICA and ISA features could be used to compare the models, but such a comparison would be more problematic. First, ISA has fewer features, so how to compare the total sparseness of the representations? Second, we would also encounter the more fundamental question: Which sparseness measure to use? If we use likelihood, statistical theory automatically shows how to compute the quantities used in the comparison.

---

[2]Comparison of models in this way is actually a bit more complicated. One problem is that if the models may have a different number of parameters, a direct comparison of the likelihoods is not possible because having more parameters can lead to overfitting. Here, this problem is not serious because the number of parameters in the two models is essentially the same (it may be a bit different if the non-linearities h_i are parameterized as well). Furthermore, Bayesian theory proposes a number of more sophisticated methods for comparing models; they consider the likelihood with many different parameter values and not only at the maximal point. Such methods are, however, computationally quite complicated, so we don't use them here.

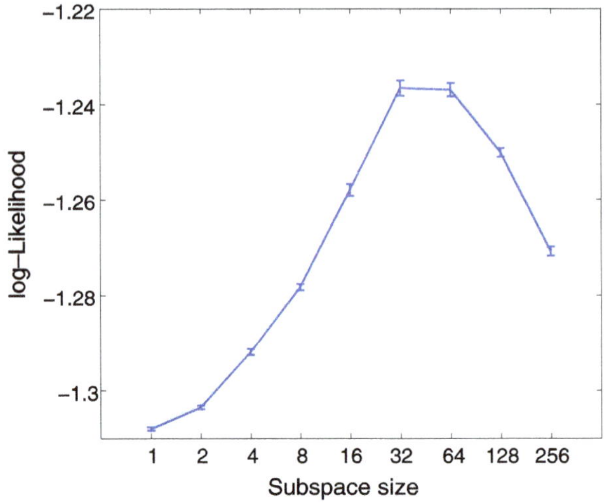

Fig. 10.14 Maximum likelihood of natural image data as a function of subspace dimensionality in ISA. Subspace size equal to 1 corresponds to the ICA model. The *error bars* are computed by doing the estimation many times for different samples of patches. Adapted from Hyvärinen and Köster (2007)

In Fig. 10.14, likelihood is given as a function of subspace size for the ISA model, for image patches is 24 × 24. What we see here is that the likelihood grows when the subspace size is made larger than one—a subspace size of one is the same as the ICA model. Thus, ISA gives a higher likelihood. In addition, the graph shows that the likelihood is maximized when the subspace size is 32, which is quite large. However, this maximum depends quite a lot on how contrast gain control is performed. Here, it was performed by dividing the image patches by their norms, but as noted in Chap. 9, this may not a very good normalization method. Thus, the results in Fig. 10.14 should not be taken too seriously. Combining a proper contrast gain control method with the ICA and ISA model is an important topic for future research.

10.8 Analysis of Convexity and Energy Correlations*

In this section, we show more detailed mathematical analysis on the connection of the correlation of squares and convexity of h discussed in Sect. 10.4.3. It can be omitted by readers not interested in mathematical details.

10.8.1 Variance Variable Model Gives Convex h

First, we show that the dependency implied by the model with a convex h typically takes the form of energy correlations. To prove that h in (10.14) is always convex, it is enough to show that the second derivative of h is always positive. We can ignore the factor $1/2\pi$. Using simple derivation under the integral sign, we obtain

$$h''(u) = \frac{\int \frac{1}{d^6}\exp(-\frac{u}{2d^2})p(d)\,dd \int \frac{1}{d^2}\exp(-\frac{u}{2d^2})p(d)\,dd - [\int \frac{1}{d^4}\exp(-\frac{u}{2d^2})p(d)\,dd]^2}{[\int \frac{1}{d^2}\exp(-\frac{u}{2d^2})p(d)\,dd]^2}. \tag{10.19}$$

Since the denominator is always positive, it is enough to show that the numerator is always positive. Let is consider $\exp(-\frac{u}{2d^2})p(d)$ as a new pdf of d, for any fixed u, after it has been normalized to have unit integral. Then the numerator takes the form $(E\{1/d^6\}E\{1/d^2\} - E\{1/d^4\}^2)$. Thus, it is enough that we prove the following general result: for any random variable $z \geq 0$, we have

$$\left(E\{z^2\}\right)^2 \leq E\{z^3\}E\{z\}. \tag{10.20}$$

When we apply this on $z = 1/d^2$, we have shown that the numerator is positive. The proof of (10.20) is possible by the classic Cauchy–Schwarz inequality, which says that for any $x, y \geq 0$, we have

$$E\{xy\} \leq E\{x^2\}^{1/2}E\{y^2\}^{1/2}. \tag{10.21}$$

Now, choose $x = z^{3/2}$ and $y = z^{1/2}$. Then taking squares on both sides, (10.21) gives (10.20).

10.8.2 Convex h Typically Implies Positive Energy Correlations

Next, we show why convexity of h implies energy correlations in the general case. We cannot show this exactly. We have to use a first-order approximation. Let us consider two variables, and look at the conditional pdf of s_2 near the point $s_2 = 0$. This gives

$$h(s_1^2 + s_2^2) = h(s_1^2) + h'(s_1^2)s_2^2 + \text{smaller terms}. \tag{10.22}$$

Let us interpret this as the logarithm of the pdf of s_2, given a fixed s_1. Some normalization term should then be added, corresponding to the denominator in (10.15), but it is a function of s_1 alone. This first-order approximation of the conditional pdf is Gaussian, because only the Gaussian distribution has a log-pdf that is a quadratic function. The variance of the distribution is equal to $2/|h'(s_1^2)|$. Because of convexity, h' is increasing. Usually, h' is also negative, because the pdf must go to zero (and its log to $-\infty$) when s_2 goes infinite. Thus, $|h'(s_1^2)|$ is a decreasing function, and $2/|h'(s_1^2)|$ is increasing. This shows that the conditional variance of s_2 increases with s_1^2, if h is convex. Of course, this was only an approximation, but it justifies the intuitive idea that a convex h leads to positive energy correlations.

Thus, we see that using a convex h in the ISA model is closely related to assuming that the s_i inside the same subspace have positive energy correlations.

10.9 Concluding Remarks and References

Independent subspace analysis is for complex cells what ICA was for simple cells. When estimated from natural image data, it learns an energy detector model which is similar to what is computed by complex cells in V1. The resulting features have a relatively strong phase-invariance, while they retain the simple-cell selectivities for frequency, orientation, and to a lesser degree, location. A restriction in the model is that the pooling in the second layer is fixed; relaxing this restriction is an important topic of current research and will be briefly considered in Sect. 11.8. Another question is whether the squaring non-linearity in computation of the features is better than, say, the absolute value; experiments in Hyvärinen and Köster (2007) indicate that it is.

Steerable filters (orientation-invariant features) are discussed in the exercises and computer assignments below. The earliest references include Koenderink and van Doorn (1987), Freeman and Adelson (1991), Simoncelli et al. (1992). An alternative viewpoint on using quadratic models on natural images is in Lindgren and Hyvärinen (2007), which uses a different approach and finds very different features.

Early and more recent work on energy detectors can be found in Pollen and Ronner (1983), Mel et al. (1998), Emerson et al. (1992), Gray et al. (1998). It is also possible to directly incorporate energy detectors in wavelets using complex-valued wavelets (Romberg et al. 2000). The idea of transforming the data into polar coordinates can be found in Zetzsche et al. (1999). Using position-invariant features in pattern recognition goes back to at least Fukushima (1980); see, e.g. Fukushima et al. (1994), Riesenhuber and Poggio (1999) for more recent developments.

Only recently, reverse correlation methods have been extended to estimation energy models (Touryan et al. 2005; Rust et al. 2005; Chen et al. 2007). These provide RFs for linear subunits in an energy model. The obtained results are quite similar to those we learned in this chapter. However, such reverse-correlation studies are quite scarce at the moment, so a detailed comparison is hardly possible. Alternative approaches to characterizing complex cells is presented in Felsen et al. (2005), Touryan et al. (2002).

10.10 Exercises

Mathematical Exercises

1. Show (10.6).
2. This exercise considers the simplest case of steerable filters. Consider the Gaussian function

$$\varphi(x, y) = \exp\left(-\frac{1}{2}(x^2 + y^2)\right). \tag{10.23}$$

a. Compute the partial derivatives of φ with respect to x and y. Denote them by φ_x and φ_y.
b. Show that φ_x and φ_y are orthogonal:

$$\int \varphi_x(x, y)\varphi_y(x, y)\,dx\,dy = 0. \tag{10.24}$$

c. The two functions φ_x and φ_y define a pair of steerable filters. The subspace they span has an invariance property which will be shown next. Define an orientation angle parameter α. Consider a linear combination

$$\varphi_\alpha = \varphi_x \cos\alpha + \varphi_y \sin\alpha. \tag{10.25}$$

The point is to show that φ_α has just the same shape as φ_x or φ_y, the only difference being that they are all rotated versions of each other. Thus, φ_x and φ_y form an orthogonal basis for a subspace which consists of simple edge detectors with all possible orientations. The proof can be obtained as follows. Define a rotated version of the variables as

$$\begin{pmatrix} x' \\ y' \end{pmatrix} = \begin{pmatrix} \sin\beta & \cos\beta \\ -\cos\beta & \sin\beta \end{pmatrix} \begin{pmatrix} x \\ y \end{pmatrix}. \tag{10.26}$$

Express φ as a function of x' and y'. Show that this is equivalent to φ_α for a suitably chosen β.

Computer Assignments

1. Create two two-dimensional Gabor filters in quadrature-phase and plot random linear combinations of them.
2. Next, we consider steerable filters.
   a. Plot the partial derivatives φ_x and φ_y defined in the Mathematical Exercise 2 above.
   b. For a couple of different values of alpha, plot their linear combinations φ_α. Compare visually the shapes of the functions plotted.

Chapter 11
Energy Correlations and Topographic Organization

The energy detection model in the preceding chapter can easily be modified to incorporate topography, i.e. an arrangement of the features on a two-dimensional grid. This is very interesting because such organization is one of the most prominent phenomena found in the primary visual cortex. In this chapter, we shall investigate such a topographic version of the ICA model. It is, mathematically, a rather simple modification of the independent subspace analysis model.

11.1 Topography in the Cortex

Topography means that the cells in the visual cortex are not in any random order; instead, they have a very specific spatial organization. When moving on the cortical surface, the response properties of the neurons change in systematic ways. The phenomenon can also be called topological organization, and sometimes the term "columnar organization" is used in almost the same sense.

Fundamentally the cortex is, of course, three-dimensional. In addition to the surface coordinates, which we denote by x_c and y_c, there is the depth dimension z_c. The depth "axis" goes from the very surface of the cortex through different layers of the grey matter to the white matter.

However, the depth dimension is usually assumed to be different from the other two dimensions. In the most simplistic interpretations, the cells that are on the same surface location (x_c, y_c) are similar irrespective of how deep they are on the cortex. This is most clearly expressed in the classic "ice cube" model of V1. Such a simplistic view has been challenged, and it is now well known that at least some properties of the cells are clearly different in different (depth) layers. In particular, input to V1 is received in some of the layers and others are specialized in outputting the results. Still, it seems that the response properties which we consider in this book, such as location, frequency, and orientation selectivities, depend mainly on the coordinates (x_c, y_c) of the cell with respect to the surface.

Looking at the spatial organization of response properties as a function of the surface coordinates x_c and y_c, the most striking aspect of topographic organization is *retinotopy*, which means that the location of the receptive field in the retinal space is closely correlated with the x_c and y_c coordinates. The global correspondence of the retinal coordinates and the cortical coordinates is somewhat complicated due to such phenomena as the magnification factor (the area in the center of the visual field has a relatively larger representation on the cortex), the division into two hemispheres, some unexpected discontinuities, and so on. The correlation is, therefore, more of a local nature.

A. Hyvärinen, J. Hurri, P.O. Hoyer, *Natural Image Statistics*,
Computational Imaging and Vision 39,

The second important topographic property is the gradual change of orientation tuning. The preferred orientation of simple and complex cells mostly changes smoothly. This phenomenon is often referred to as *orientation columns*. They can be seen most clearly in optical imaging experiments where one takes a "photograph" of the cortex that shows which regions are active when the input consists of a grating of a given orientation. Such activity patterns take the form of stripes (columns).

The third important property of spatial organization is that frequency selectivity seems to be arranged topographically into low-frequency blobs so that the blobs (or at least their centers) contain predominantly cells that prefer low-frequency cells and the inter-blob cells prefer higher frequencies. These low-frequency blobs seem to coincide with the well-known cytochrome oxidase blobs.

A final point to note is that phase is not arranged topographically. In fact, phase seems to be completely random: there is no correlation between the phase parameters in two neighboring cells.

11.2 Modeling Topography by Statistical Dependence

Now, we show how to extend the models of natural image statistics to include topography. The key is to consider the dependencies of the components. The model is thus closely related to the model of independent subspace analysis in Chap. 10. In fact, ISA can be seen as a special case of this model.

11.2.1 Topographic Grid

To model topographic organization, we have to first define which features are "close to each other" on the cortical surface. This is done by arranging the features s_i on a two-dimensional grid or lattice. The restriction to 2D is motivated by cortical anatomy, but higher dimensions are equally possible. The spatial organization on the grid models the organization on the cortical surface. The arrangement on the lattice is illustrated in Fig. 11.1.

The topography is formally expressed by a neighborhood function $\pi(i, j)$ that gives the proximity of the features (components) with indices i and j. Typically, one defines that $\pi(i, j)$ is 1 if the features are sufficiently close to each other (they are "neighbors"), and 0 otherwise. Typically, the neighborhood function is chosen by defining the neighborhood of a feature to be square. For example, $\pi(i, j)$ is 1 if the feature j is in a 5×5 square centered on feature i; otherwise, $\pi(i, j)$ is zero.

11.2.2 Defining Topography by Statistical Dependencies

Consider a number of features s_i, $i = 1, \ldots, n$. How can we order the features on the topographic grid in a meaningful way? The starting point is to define a measure of

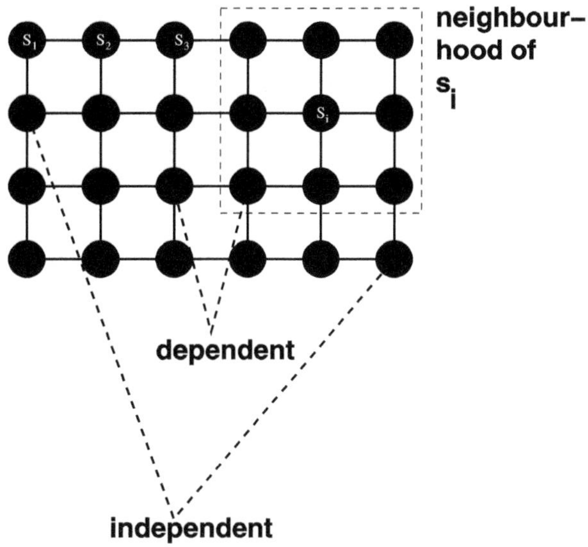

Fig. 11.1 Illustration of topography and its statistical interpretation. The neurons (feature detectors) are arranged on a two-dimensional grid that defines which neurons are near to each other and which are far from each other. It also defines the neighborhood of a cell as the set of cells which are closer than a certain radius. In the statistical model, neurons that are near to each other have statistically dependent outputs, neurons that are far from each other have independent outputs

similarity between two features, and then to order the features so that features that are similar are close to each other on the grid. This is a general principle that seems fair enough. But then, what is a meaningful way of defining similarity between two features? There are actually a couple of different possibilities.

In many models, the similarity of features is defined by similarity of the features weights or receptive fields W_i. Typically, this means the dot-product (also called, somewhat confusingly, the correlation of the receptive fields). This is the case in Kohonen's self-organizing map and related models. However, this seems rather inadequate in the case of the visual cortex. For example, two features of the same frequency need not exhibit large dot-products of weight vectors; in fact, the dot-product can be zero if the features are of orthogonal orientations with otherwise similar parameters. Yet, since the V1 exhibits low-frequency blobs, low-frequency features should be considered similar to each other even if they are quite different with respect to other parameters. What's even worse is that since the phases change randomly when moving a bit on the cortical surface, the dot-products between neighboring components also change rather randomly since the phase has a large influence on the shape of the receptive fields and on the dot-products.

Another candidate for a similarity measure would be correlation of the feature detector outputs s_i when the input consists of natural images. However, this is no good either, since the outputs (components) are typically constrained to be exactly uncorrelated in ICA and related models. Thus, they would all be maximally dissimilar if similarity is based on correlations.

Yet, using correlations seems to be a step in the right direction. The central hypothesis used in this book—visual processing in the cortex is strongly influenced by the statistical structure of the natural input—would suggest that we have to look at the statistics of feature detector outputs in order to find a meaningful measure of similarity to be used in a model of topography. We just need more information than the ordinary linear correlations.

Our statistical approach to topography thus concentrates on the pattern of statistical dependencies between the s_i, assuming that the joint distribution of the s_i is dictated by the natural image input. The basic idea is that *similarity is defined by the statistical dependency of the outputs*. Thus, features that have strong statistical dependencies are defined to be similar, and features that are independent or weakly dependent are defined to be dissimilar.

The application of this principle is illustrated in Fig. 11.1. The linear feature detectors (simple cells) have been arranged on a grid (cortex) so that any two feature detectors that are close to each other have dependent outputs, whereas feature detectors that are far from each other have independent outputs.

Actually, from Chaps. 9 and 10, we know what are the most prominent statistical dependencies that remains after ordinary ICA: the correlations of squares (or absolute values, which seems to be closely related). Thus, we do not need to model the whole dependency structure of the s_i, which would be most complicated. We can just concentrate on the dependencies of the squares s_i^2.

11.3 Definition of Topographic ICA

As in the ICA and ISA models, we model the image as a linear superposition of features A_i with random coefficients s_i:

$$I(x, y) = \sum_{i=1}^{m} A_i(x, y) s_i. \tag{11.1}$$

As in ICA and ISA, the s_i are obtained as the outputs of linear feature detectors as

$$s_i = \sum_{x,y} W_i(x, y) I(x, y) = \sum_{j=1}^{n} v_{ij} z_j = \mathbf{v}_i^T \mathbf{z} \tag{11.2}$$

where the z_j denotes the jth variable obtained from the image patch by canonical preprocessing.

The point is now to define the joint pdf of the s_i so that it expresses the topographic ordering. First, we define the "local energies" as

$$c_i = \sum_{j=1}^{n} \pi(i, j) s_j^2. \tag{11.3}$$

This is basically the general activity level in the neighborhood of the linear feature s_i. The weighting by $\pi(i, j)$ means that we only sum over s_j which are close to s_i in the topography.

Next, we define the likelihood of the topographic ICA model by a simple modification of the log-likelihood in the ISA model, given in (10.11) on page 219. We replace the subspace energies e_k by these local energies. (The connection between the two models is discussed in more detail later.) Thus, define the pdf of the s_i as

$$\log p(s_1, \ldots, s_n) = \sum_{i=1}^{n} h\left(\sum_{j=1}^{n} \pi(i, j) s_j^2\right) \tag{11.4}$$

where h is a convex function as in the preceding chapters, e.g. Sect. 6.2.1. Assuming we have observed a set of image patches, represented by z_t, $t = 1, \ldots, T$ after canonical preprocessing, we obtain the likelihood

$$\log L(\mathbf{v}_1, \ldots, \mathbf{v}_n) = T \log\left|\det(\mathbf{V})\right| + \sum_{i=1}^{n} \sum_{t=1}^{T} h\left(\sum_{j=1}^{n} \pi(i, j)\left(\mathbf{v}_j^T \mathbf{z}_t\right)^2\right). \tag{11.5}$$

The topography given by $\pi(i, j)$ is considered fixed, and only the linear feature weights $\mathbf{v}_j$ are estimated, so this likelihood is a function of the $\mathbf{v}_j$ only. As in earlier models, the vectors $\mathbf{v}_j$ are constrained to form an orthogonal matrix, so the determinant is constant (one) and the term $T \log|\det(\mathbf{V})|$ can be ignored.

The central feature of this model is that the responses s_i of near-by simple cells are *not* statistically independent in this model. The responses are still linearly uncorrelated, but they have non-linear dependencies. In fact, the energies s_i^2 are strongly positively correlated for neighboring cells. This property is directly inherited from the ISA model; that connection will be discussed next.

11.4 Connection to Independent Subspaces and Invariant Features

Topographic ICA can be considered a generalization of the model of independent subspace analysis. The likelihood of ISA (see (10.11)) can be expressed as a special case of the likelihood in (11.5) with a neighborhood function which is one if the components are in the same subspace and zero otherwise, or more formally:

$$\pi(i, j) = \begin{cases} 1, & \text{if there is some subspace with index } q \text{ so that } i, j \in S(q), \\ 0, & \text{otherwise.} \end{cases}$$

This shows that topographic ICA is closely connected to the principle of invariant feature subspaces in Chap. 10. In topographic ICA, every component has its own neighborhood, which corresponds to a subspace in ISA. Each of the local energies c_i could be considered as the counterpart of the energies e_k in ISA. Thus, the local energies, possibly after a non-linear transform, can be interpreted as the values of invariant features. The pooling process is controlled by the neighborhood function

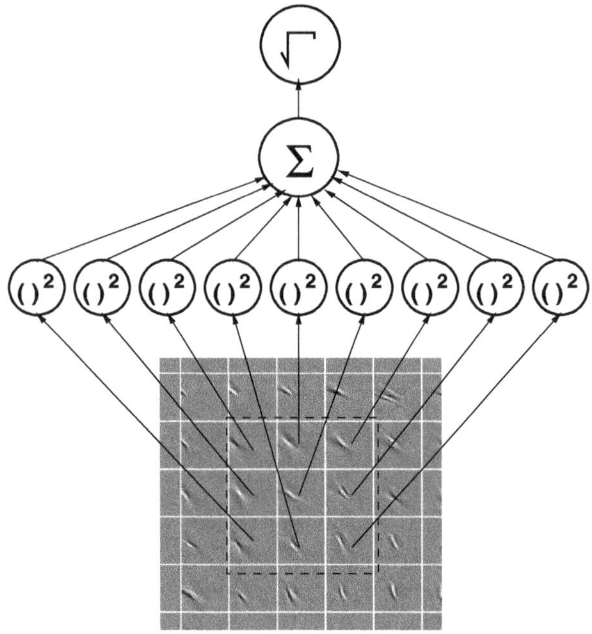

Fig. 11.2 Computation of invariant features in the topographic ICA model. Invariant features (complex cell outputs) are obtained by summing the squares of linear features (simple cell outputs) in a neighborhood of the topographic grid. From Hyvärinen et al. (2001a), Copyright ©2001 MIT Press, used with permission

$\pi(i, j)$. This function directly gives the pooling weights, i.e. the connections between the linear features with index i and the invariant feature cell with index j. Note that the number of invariant features is here equal to the number of underlying linear features.

The dependencies of the components can also be deduced from this analogy with ISA. In ISA, components which are in the same subspace have correlations of energies. In topographic ICA, components which are close to each other in the topographic grid have correlations of squares. Thus, all the features in the same neighborhood tend to be active (non-zero) at the same time.

In a biological interpretation, our definition of the pooling weights from simple cells to complex cells in topographic ICA is equivalent to the assumption that complex cells only pool outputs of simple cells that are near-by on the topographic grid. Neuroanatomic measurements indicate that the wiring of complex cells may indeed be so constrained; see References below. Such a two-layer network is illustrated in Fig. 11.2.

11.5 Utility of Topography

What is the computational utility of a topographic arrangement? A widely used argument is that such a spatial arrangement is useful to *minimize wiring length*. Wiring length means here the length of the physical connections (axons) needed to send signals from one neuron to another. Consider, for example, the problem of designing the connections from simple cells to complex cells so that the "wires" are as short as possible. It is rather obvious that topographic ICA is related to minimizing that wiring length because in topographic ICA all such connections are very local in the sense that they are not longer that the radius of the neighborhoods. A more general task may be to pool of responses to reduce noise: if a cell in a higher area wants to "read", say, the orientation of the stimulus, it could reduce noise in V1 cell responses by looking at the average of the responses of many cells which have the same orientation selectivity.

In general, if we assume that two cells need to communicate with each other if (and only if) their outputs are statistically dependent, topographic ICA provides optimal wiring. The same applies if the responses of two cells are combined by a third cell only if the outputs of the two cells are statistically dependent. Such assumptions are reasonable because if the cells represent pieces of information which are related (in some intuitive sense), it is likely that their outputs are statistically dependent, and vice versa; so, statistical dependence tells which cells contain related information which has to be combined in higher levels.

Minimization of wiring length may be important for keeping the total brain volume minimal: a considerable proportion of the brain volume is used up in interconnecting axons. It would also speed up processing because the signal travels along the axons with limited speed.

11.6 Estimation of Topographic ICA

A fundamental similarity to ISA is that we do *not* specify what parameters should be considered as defining the topographic order. That is, the model does not specify, for example, that near-by neurons should have receptive fields that have similar locations, or similar orientations. Rather, we let the natural images decide what the topography should be like, based on their statistical structure.

Since we have already defined the likelihood in (11.5), estimation needs hardly any further comment. We use whitened (canonically preprocessed) data, so we constrain $\mathbf{V}$ to be orthogonal just like in ICA and ISA. We maximize the likelihood under this constraint. The computational implementation of such maximization is discussed in detail in Chap. 18, in particular, Sect. 18.5.

The intuitive interpretation of such estimation is that we are maximizing the sparsenesses of the local energies. This is completely analogue to ISA, where we maximize sparsenesses of complex cell outputs. The learning process is illustrated in Fig. 11.3.

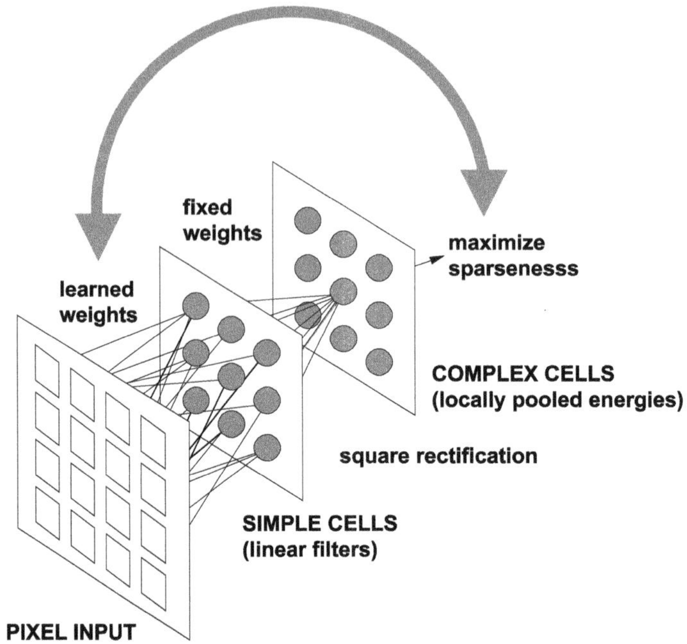

Fig. 11.3 Illustration of learning in the topographic ICA model. From Hyvärinen and Hoyer (2001), Copyright ©2001 Elsevier, used with permission

11.7 Topographic ICA of Natural Images

11.7.1 Emergence of V1-like Topography

11.7.1.1 Data and Preprocessing

We performed topographic ICA on the same data as in previous chapters. We took the same 50 000 natural image patches of size 32×32 as in the preceding chapters. We preprocessed the data in the same way as in the ISA case: This means divisive normalization using (9.11), and reducing the dimension to 256 by PCA. The nonlinearity h was chosen to be a smoothed version of the square root as in (6.14), just like in the ISA experiments.

The topography was chosen so that $\pi(i, j)$ is 1 if the cell j is in a 5×5 square centered on cell i; otherwise $\pi(i, j)$ is zero. Moreover, it was chosen to be cyclic (toroidal) so that the left edge of the grid is connected to the right edge, and the upper edge is connected to the lower edge. This was done to reduce border artifacts due to the limited size to the topographic grid.

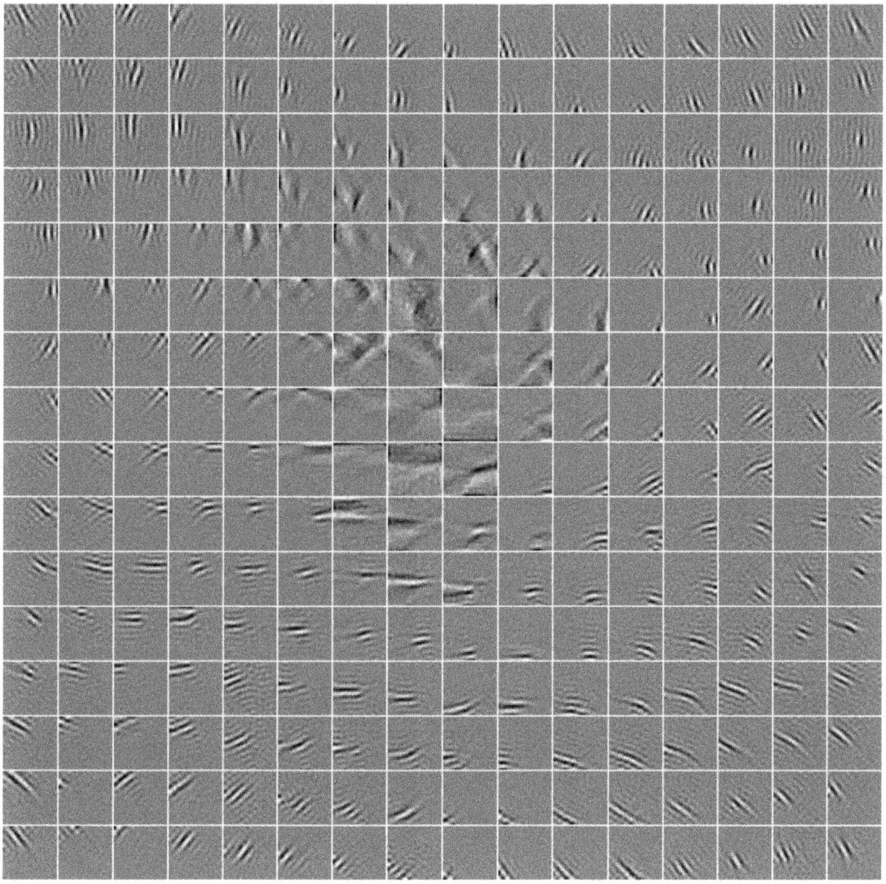

Fig. 11.4 The whole set of vectors W_i obtained by topographic independent component analysis, in the topographic order

11.7.1.2 Results and Analysis

The linear detector weights W_i obtained by topographic ICA from natural images are shown in Fig. 11.4, and the corresponding feature vectors A_i are in Fig. 11.5. The topographic ordering is visually obvious. The underlying linear features are tuned for the three principal parameters: orientation, frequency, and location. Visual inspection of the map shows that orientation and location mostly change smoothly as a function of position on the topographic grid. A striking feature of the map is a "blob" grouping low-frequency features. Thus, the topography is determined by the same set of parameters for which the features are selectively tuned; these are just the same as in ICA and ISA. These are also the three parameters with respect to which a clear spatial organization has been observed in V1.

The topography can be analyzed in more detail by either a global or a local analysis. A local analysis is done by visualizing the correlations of the optimal Gabor

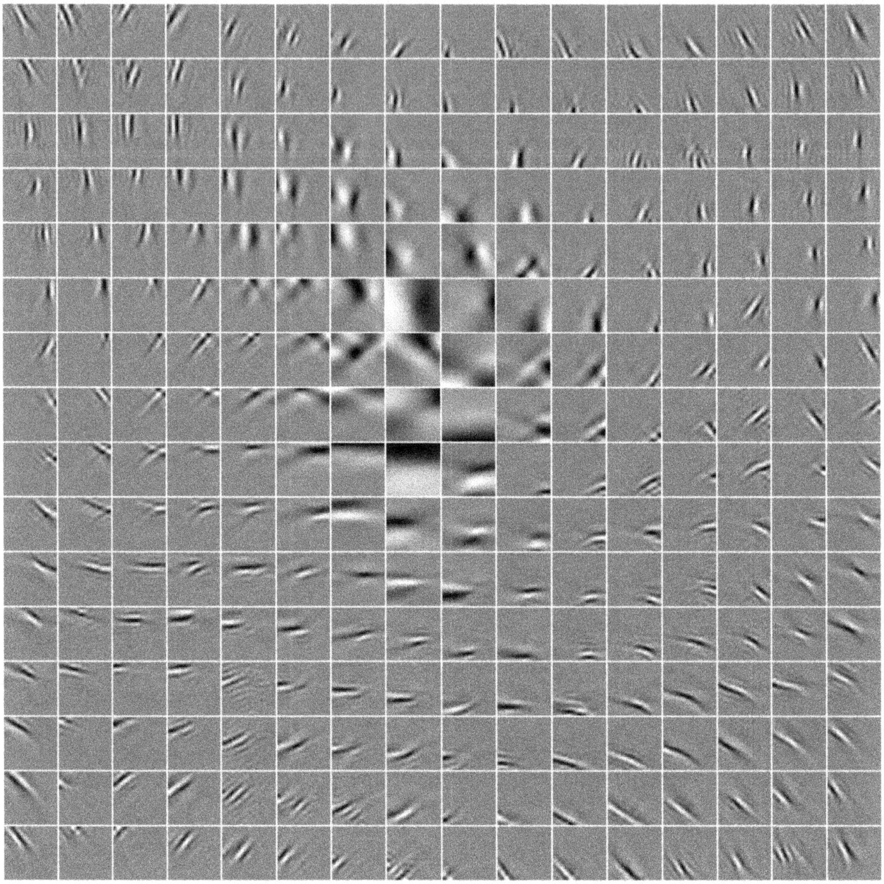

Fig. 11.5 The whole set of vectors A_i obtained by topographic independent component analysis

parameters for two linear features that are immediate neighbors. In Fig. 11.6, we see that the locations (a, b) and orientations (c) are strongly correlated. In the case of frequencies (d), the correlation is more difficult to see because of the overall concentration to high frequencies. As for phases (e), no correlation (or any kind of statistical dependency) can be seen, which is again similar to what has been observed in V1. Furthermore, all these correlations are similar to the correlations inside independent subspaces in Fig. 10.8 on page 229. This is not surprising because of the intimate connection between the two models, explained above in Sect. 11.4.

A global analysis is possible by color-coding the Gabor parameters of linear features. This gives "maps" whose smoothness shows the smoothness of the underlying parameter. The maps are shown in Fig. 11.7. The locations (a and b) can be seen to change smoothly, which is not obvious from just looking at the features in Fig. 11.4. The orientation and frequency maps (c and d) mainly change smoothly, which was rather obvious from Fig. 11.4 anyway. In some points, the orientation

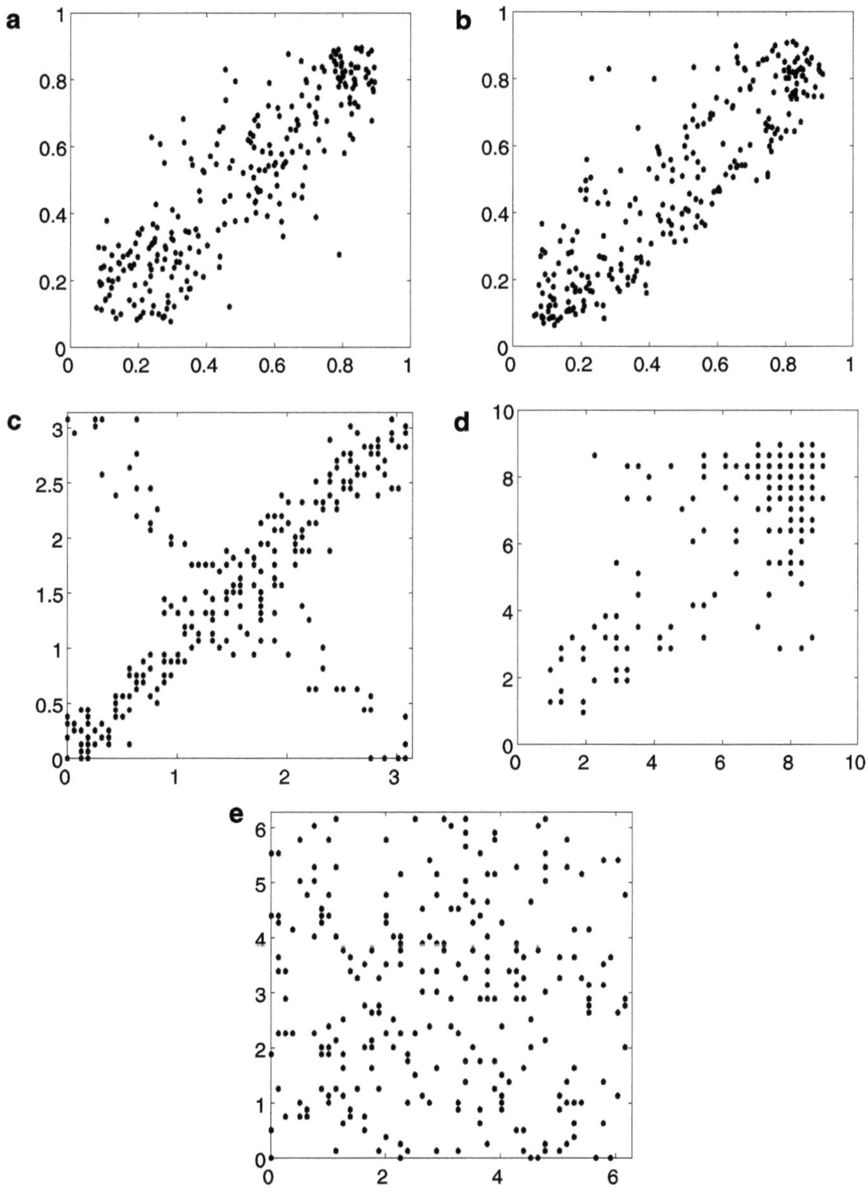

Fig. 11.6 Correlation of parameters characterizing the linear features of two neighboring features in Fig. 11.4. An immediate neighbor for each cell chosen as the one immediately to the right. Each point in the scatter plots is based on one such couple. **a** scatter plot of locations along x-axis, **b** locations along y-axis, **c** orientations, **d** frequencies, and **e** phases. The plots are very similar to corresponding plots for ISA in Fig. 10.8 on page 229; the main visual difference is simply due to the fact that here we have twice the number of dots in each plot

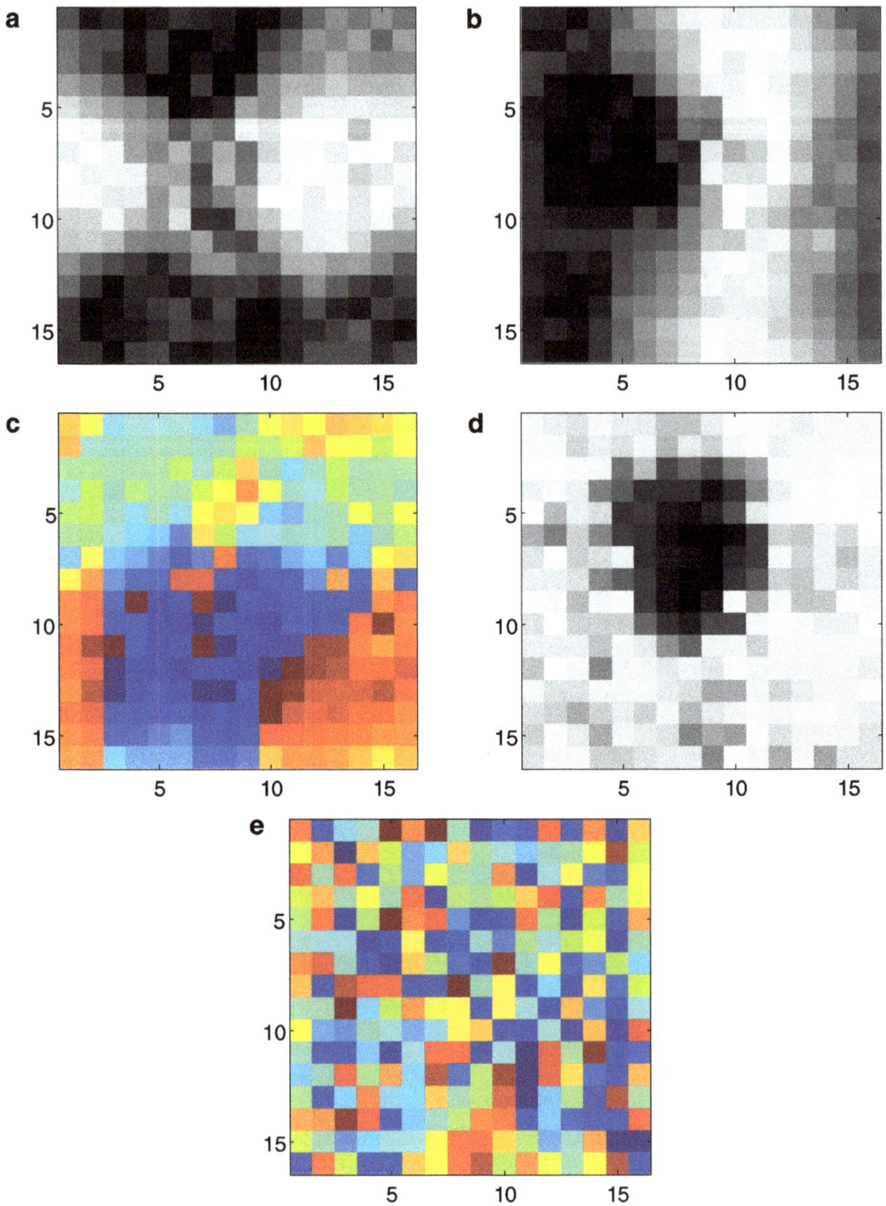

Fig. 11.7 Global structure of the topography estimated from natural images in Fig. 11.4. Each parameter of the Gabor functions describing the features is plotted grey-scale or color-coded. Color-coding is used for parameters which are cyclic: orientation and phase, since the color spectrum is also cyclic. The actual values of the parameters are not given because they have little importance. **a** Locations along x-axis, **b** locations along y-axis, **c** orientations, **d** frequencies, and **e** phases

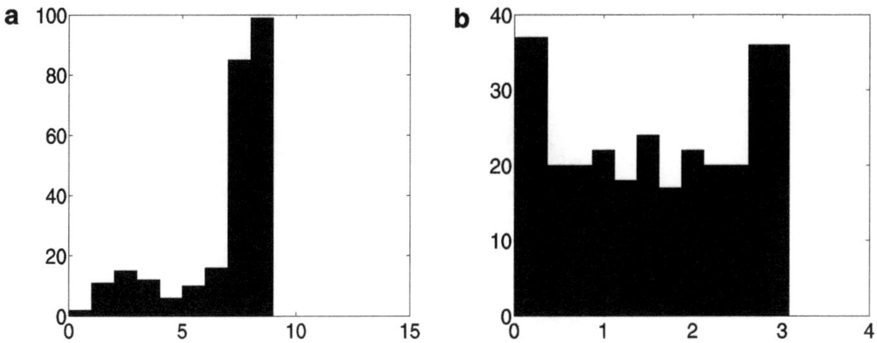

Fig. 11.8 Histograms of the optimal **a** frequencies and **b** orientations of the linear features in topographic ICA

seems to change abruptly, which may correspond to so-called "pinwheels", which are points in which many different orientations can be found next to each other, and have been observed on the cortex. As for phases, the map in (e) shows that they really change randomly.

We can also analyze the distribution of the frequencies and orientations of the features. The plot in Fig. 11.8 shows the histograms preferred orientations and frequencies for the linear features. We see that all orientations are almost equally present, but the horizontal orientation is slightly overrepresented. This is the same anisotropy we have seen all preceding models. In contrast, the frequency distribution is very strongly skewed: most linear features are tuned to high frequencies. However, the distribution of frequencies is a bit closer to uniform than in the cases of ICA (Fig. 6.9) or ISA (Fig. 10.9).

The connection of the model to ISA suggests that the local energies can be interpreted as invariant features. What kind of invariances do we see emerging from natural images? Not surprisingly, the invariances are similar to what we obtained with ISA, because the neighborhoods have the same kinds of parameters correlations (Fig. 11.6) as in ICA; we will not analyze them in more detail here. The main point is that *local energies are like complex cells*. That is, the topographic ICA model automatically incorporates a complex cell model.

Basically, the conclusion to draw from these results is that the topographic ICA model produces a spatial topographic organization of linear features that is quite similar to the one observed in V1.

11.7.1.3 Image Synthesis Results and Sketch of Generative Model

Next, we will synthesize images from the topographic ICA model. This is a bit tricky because, in fact, we did not yet introduce a proper generative model for topographic ICA. Such a model can be obtained as a special case of the framework introduced later in Sect. 11.8.2. We will here briefly describe how such a generative model can be obtained.

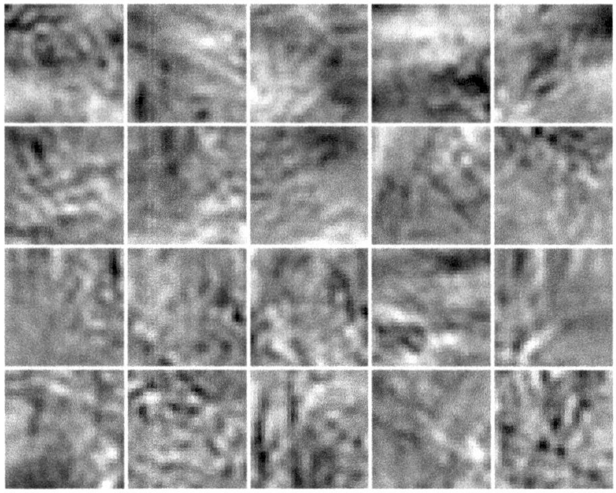

Fig. 11.9 Image synthesis using topographic ICA. Compare with the ICA results in Fig. 7.4 on page 162 and ISA results in Fig. 10.10 on page 231

Basically, the idea is a simple generalization of the framework using variance variables as in Sects. 10.4 and 9.3. Here, we have a separate variance variable d_i for each component s_i:

$$s_i = \tilde{s}_i d_i \tag{11.6}$$

where the $\tilde{s}_i$ are Gaussian and independent from each other (and from the d_i). The point is to generate the d_i so that their dependencies incorporate the topography. This can be accomplished by generating them using a higher-order ICA model, where the mixing matrix is given by the neighborhood function. Denoting the higher-order components by u_i, we simply define

$$d_i = \sum_j \pi(i, j) u_i. \tag{11.7}$$

This produces approximately the same distribution as the pdf which we used to define the topographic ICA model earlier in this chapter. See Sect. 11.8.2 for details. A problem we encounter here is that it is not obvious how to estimate the distributions of the u_i. So, we have to fix them rather arbitrarily, which means the results are not quite directly comparable with those obtained by ISA and ICA where we could use the observed histograms of the features.

Results of synthesizing images with this generative model are shown in Fig. 11.9. The u_i were generated as the fourth powers of Gaussian variables. The synthesized images seem to have more global structure than those obtained by ICA or ISA, but as we just pointed out, this may be related to the way we fixed the distributions of the u_i.

11.7.2 Comparison with Other Models

When compared with other models on V1 topography, we see three important properties in the topographic ICA model:

1. The topographic ICA model shows emergence of a topographic organization using the above-mentioned three principal parameters: location, frequency, and orientation. The use of these particular three parameters is not predetermined by the model, but determined by the statistics of the input. This is in contrast to most models that only model topography with respect to one or two parameters (usually orientation possibly combined with binocularity) that are chosen in advance.
2. No other model has shown the emergence of a low-frequency blob.
3. Topographic ICA may be the first one to explicitly show a connection between topography and complex cells. The topographic, columnar organization of the simple cells is such that complex cell properties are automatically created when considering local activations. This is related to the randomness of phases, which means that in each neighborhood, there are linear features with very different phases, like in the subspaces in ISA.

It is likely that the two latter properties (blobs and complex cells) can only emerge in a model that is based on simultaneous activation (energy correlation) instead of similarity of receptive fields as measured by Euclidean distances or receptive field correlations. This is because Euclidean distances or correlations between feature vectors of different frequencies, or of different phases, are quite arbitrary: they can obtain either large or small values depending on the other parameters. Thus, they do not offer enough information to qualitatively distinguish the effects of phase vs. frequency, so that phase can be random and frequency can produce a blob.

11.8 Learning Both Layers in a Two-Layer Model *

In this section, we discuss estimation of a two-layer model which is a generalization of the topographic ICA. The section is quite sophisticated mathematically, and presents ongoing work with a lot of open problems, so it can be skipped by readers not interested in mathematical details.

11.8.1 Generative vs. Energy-Based Approach

Many of the results in the preceding chapters are related to a two-layer generative model. In the model, the observed variables $\mathbf{z}$ are generated as a linear transformation of components $\mathbf{s}$, just as in the basic ICA model: $\mathbf{z} = \mathbf{A}\mathbf{s}$. The point is to define the joint density of $\mathbf{s}$ so that it expresses the correlations of squares that seem to be dominant in image data.

There are two approaches we can use. These parallel very much the sparse coding and ICA approaches in Chaps. 6 and 7. In the first approach, typically called "energy-based" for historical reasons,[1] we just define an objective function which expresses sparseness or some related statistical criterion, and maximize it. In the second approach, we formulate a generative model which describes how the data is generated starting from some elementary components. We shall consider here first the generative-model approach; the energy-based model is considered in Sect. 11.8.5.

11.8.2 Definition of the Generative Model

In the generative-model approach, we define the joint density of $\mathbf{s}$ as follows. The variances d_i^2 of the s_i are not constant, instead they are assumed to be random variables. These random variables d_i are, in their turn, generated according to a model to be specified. After generating the variances d_i^2, the variables s_i are generated independently from each other, using some conditional distributions to be specified. In other words, the s_i are *independent given their variances*. Dependence among the s_i is implied by the dependence of their variances.

This is a generalization of the idea of a common variance variable presented in Sect. 7.8.3. Here, there is no single common variance variable, since there is a separate variance variable d_i^2 corresponding to each s_i. However, these variance variables are correlated, which implies that the squares of the s_i are correlated. Consider the extreme case where the d_i are completely correlated. Then the d_i^2 are actually the same variable, possibly multiplied by some constants. Thus, in this extreme case, we actually have just a single variance variable as in the divisive normalization model in Chap. 9.

Many different models for the variances d_i^2 could be used. We prefer here to use an ICA model followed by a non-linearity:

$$d_i = r\left(\sum_{k=1}^{n} \pi(i,k) u_k\right). \tag{11.8}$$

Here, the u_k are the "higher-order" independent components used to generate the variances, and r is some scalar non-linearity (possibly just the identity $r(z) = z$). The coefficients $\pi(i,k)$ are the entries of a higher-order feature matrix. It is closely related to the matrix defining the topography in topographic ICA, which is why we use the same notation.

This particular model can be motivated by two facts. First, taking sparse u_i, we can model a two-layer generalization of sparse coding, where the activations (i.e. the variances) of the components s_i are sparse, and constrained to some groups of "related" components. Related components means here components whose variances are strongly influenced by the same higher-order components u_i.

[1] Note that the word "energy" here has nothing to do with Fourier energy; it comes from a completely different physical analogy.

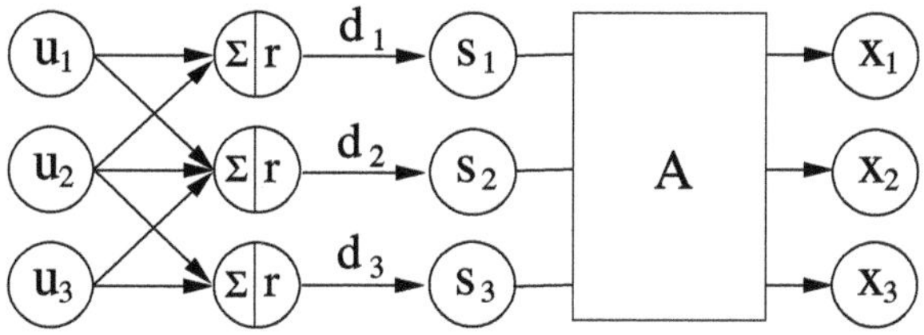

Fig. 11.10 An illustration of the two-layer generative model. First, the "variance-generating" variables u_i are generated randomly. They are then mixed linearly. The resulting variables are then transformed using a non-linearity r, thus giving the local variances d_i^2. Components s_i are then generated with variances d_i^2. Finally, the components s_i are mixed linearly to give the observed variables x_i (which are subsequently whitened to give the z_i)

In the model, the distributions of the u_i and the actual form of r are additional parameters; some suggestions will be given below. It seems natural to constrain the u_k to be non-negative. The function r can then be constrained to be a monotonic transformation in the set of non-negative real numbers. This ensures that the d_i's are non-negative, so is a natural constraint since they give the standard deviation of the components.

The resulting two-layer model is summarized in Fig. 11.10. Note that the two stages of the generative model can be expressed as a single equation, analogously to (9.4), as follows:

$$s_i = r\left(\sum_k \pi(i,k)u_k\right)\tilde{s}_i \tag{11.9}$$

where $\tilde{s}_i$ is a random variable that has the same distribution as s_i given that d_i is fixed to unity. The u_k and the $\tilde{s}_i$ are all mutually independent.

11.8.3 Basic Properties of the Generative Model

Here, we discuss some basic properties of the generative model just defined.

11.8.3.1 The Components s_i Are Uncorrelated

This is because according to (11.9) we have

$$E\{s_i s_j\} = E\{\tilde{s}_i\}E\{\tilde{s}_j\}E\left\{r\left(\sum_k \pi(i,k)u_k\right)r\left(\sum_k \pi(j,k)u_k\right)\right\} = 0 \tag{11.10}$$

due to the independence of the u_k from $\tilde{s}_i$ and $\tilde{s}_j$. (Recall that $\tilde{s}_i$ and $\tilde{s}_j$ are zero-mean.) To simplify things, one can define that the marginal variances (i.e. integrated over the distribution of d_i) of the s_i are equal to unity, as in ordinary ICA. In fact, we have

$$E\{s_i^2\} = E\{\tilde{s}_i^2\}E\left\{r\left(\sum_k \pi(i,k)u_k\right)^2\right\}, \tag{11.11}$$

so we only need to rescale $\pi(i, j)$ (the variance of $\tilde{s}_i$ is equal to unity by definition).

11.8.3.2 The Components s_i Are Sparse

This is true in the case where component s_i is assumed to have a Gaussian distribution when the variance is given. This follows from the proof given in Sect. 7.8.3: the logic developed there still applies in this two-layer model, when the marginal distribution of each component s_i is consider separately. Then the marginal, unconditional distributions of the components s_i are called Gaussian scale mixtures.

11.8.3.3 Topographic Organization Can Be Modeled

This is possible simply by constraining the higher-order matrix $\pi(i, j)$ to equal a topographic neighborhood matrix as in Sect. 11. We can easily prove that components which are far from each other on the topography are then independent. Assume that s_i and s_j are such that their neighborhoods have no overlap, i.e. there is no index k such that both $\pi(i, k)$ and $\pi(j, k)$ are non-zero. Then their variances d_i and d_j are independent because no higher-order component influences both of these variances. Thus, the components s_i and s_j are independent as well.

11.8.3.4 Independent Subspaces Are a Special Case

This is more or less implied by the discussion in Sect. 11.4 where independent subspace analysis was shown to be a special case of topographic ICA. A more direct connection is seen by noting that each variance variable could determine the variance inside a single subspace, with no interactions between the variance variables. Then we get the ISA model as explained in Sect. 10.4.

11.8.4 Estimation of the Generative Model

11.8.4.1 Integrating Out

In this section, we discuss the estimation of the two-layer model introduced in the previous section. In principle, this can be done by "integrating out" the latent variables. Integrating out is an intuitive appealing method: since the likelihood depends on the values of the variance variables u_i which we don't know, why not just compute the likelihood averaged over all possible values of u_i? Basically, if we have the joint density of the s_i and the u_i, we could just compute the integral over the u_i to get the density over s_i alone:

$$p(\mathbf{s}) = \int p(\mathbf{s}, \mathbf{u})\, d\mathbf{u} \tag{11.12}$$

The problem is, as always with integration, that we may not be able to express this integral with a simple formula, and numerical integration may be computationally impossible.

In our case, the joint density of **s**, i.e. the topographic components, and **u**, i.e. the higher-order independent components generating the variances, can be expressed as

$$p(\mathbf{s}, \mathbf{u}) = p(\mathbf{s}|\mathbf{u})p(\mathbf{u}) = \prod_i p_i^s\left(\frac{s_i}{r(\sum_k \pi(i,k)u_k)}\right)\frac{1}{r(\sum_k \pi(i,k)u_k)}\prod_j p_j^u(u_j) \tag{11.13}$$

where the p_i^u are the marginal densities of the u_i and the p_i^s are the densities of p_i^s for variance fixed to unity. The marginal density of **s** could be obtained by integration:

$$p(\mathbf{s}) = \int \prod_i p_i^s\left(\frac{s_i}{r(\sum_k \pi(i,k)u_k)}\right)\frac{\prod_j p_j^u(u_j)}{r(\sum_k \pi(i,k)u_k)}\, d\mathbf{u} \tag{11.14}$$

Possibly, for some choices of the non-linearity r and the distributions p_i^u, this integral could be computed easily, but no such choices are known to us.

11.8.4.2 Approximating the Likelihood

One thing which we can do is to *approximate* the likelihood by an analytical expression. This approximation actually turns out to be rather useless for the purpose of estimating the two-layer model, but it shows an interesting connection to the likelihood of the topographic ICA model.

To simplify the notation, we assume in the following that the densities p_i^u are equal for all i, and likewise for p_i^s. To obtain the approximation, we first fix the density $p_i^s = p_s$ to be Gaussian, as discussed in Sect. 11.8.3, and we define the non-linearity r as

$$r\left(\sum_k \pi(i,k)u_k\right) = \left(\sum_k \pi(i,k)u_k\right)^{-1/2}. \tag{11.15}$$

The main motivation for these choices is algebraic simplicity that makes a simple approximation possible. Moreover, the assumption of conditionally Gaussian s_i, which implies that the unconditional distribution of s_i super-Gaussian, is compatible with the preponderance of super-Gaussian variables in ICA applications.

With these definitions, the marginal density of **s** equals:

$$p(\mathbf{s}) = \int \frac{1}{\sqrt{2\pi}^n} \exp\left(-\frac{1}{2}\sum_i s_i^2 \left[\sum_k \pi(i,k)u_k\right]\right) \prod_i p_u(u_i) \sqrt{\sum_k \pi(i,k)u_k}\, d\mathbf{u} \tag{11.16}$$

which can be manipulated to give

$$p(\mathbf{s}) = \int \frac{1}{\sqrt{2\pi}^n} \exp\left(-\frac{1}{2}\sum_k u_k \left[\sum_i \pi(i,k)s_i^2\right]\right) \prod_i p_u(u_i) \sqrt{\sum_k \pi(i,k)u_k}\, d\mathbf{u}. \tag{11.17}$$

The interesting point in this form of the density is that it is a function of the "local energies" $\sum_i \pi(i,k)s_i^2$ only. The integral is still intractable, though. Therefore, we use the simple approximation:

$$\sqrt{\sum_k \pi(i,k)u_k} \approx \sqrt{\pi(i,i)u_i}. \tag{11.18}$$

This is actually a lower bound, and thus our approximation will be a lower bound of the likelihood as well. This gives us the following approximation $\tilde{p}(\mathbf{s})$:

$$\tilde{p}(\mathbf{s}) = \prod_k \exp\left(G\left(\sum_i \pi(i,k)s_i^2\right)\right) \tag{11.19}$$

where the scalar function G is obtained from the p_u by

$$G(y) = \log \int \frac{1}{\sqrt{2\pi}} \exp\left(-\frac{1}{2}uy\right) p_u(u)\sqrt{\pi(i,i)u}\, du. \tag{11.20}$$

Recall that we assumed $\pi(i,i)$ to be constant.

Next, using the same derivation as in ICA, we obtain the likelihood of the data as

$$\log \tilde{L}(\mathbf{V}) = \sum_{t=1}^{\mathrm{T}} \sum_{j=1}^{n} G\left(\sum_{i=1}^{n} \pi(i,j)\left(\mathbf{v}_i^{\mathrm{T}}\mathbf{z}(t)\right)^2\right) + T\log|\det \mathbf{V}|. \tag{11.21}$$

where $\mathbf{V} = (\mathbf{v}_1, \ldots, \mathbf{v}_n)^{\mathrm{T}} = \mathbf{A}^{-1}$, and the $\mathbf{z}(t)$, $t = 1, \ldots, T$ are the observations of $\mathbf{z}$. It is here assumed that the neighborhood function and the non-linearity r as well as the densities p_i^u and p_i^s are known. This approximation is a function of local energies. Every term $\sum_{i=1}^n \pi(i,j)(\mathbf{v}_i^{\mathrm{T}}\mathbf{z}(t))^2$ could be considered as the energy of a neighborhood, related to the output of a higher-order neuron as in complex cell models. The function G has a similar role as the log-density of the independent components in ICA; the corresponding function h is basically obtained as $h(u) = G(\sqrt{|u|})$.

The formula for G in (11.20) can be analytically evaluated only in special cases. One such case is obtained if the u_k are obtained as squares of standardized Gaussian variables. Straightforward calculation then gives the following function

$$G_0(y) = -\log(1+y) + \text{const.} \tag{11.22}$$

However, in ICA, it is well known that the exact form of the log-density does not affect the consistency of the estimators, as long as the overall shape of the function is correct. This is probably true in topographic ICA as well.

11.8.4.3 Difficulty of Estimating the Model

What we have really shown in deriving the approximation of likelihood in (11.21) is that the heuristically justified objective function in (11.5) can be obtained from the two-layer generative model as an approximation. But we have not really got any closer to the goal of estimating both layers of weights. This is because the approximation used here approximates the dependence of the likelihood from π quite badly. To see why, consider maximization of the approximative likelihood in (11.21) with respect to the $\pi(i, j)$. Take G as in (11.22). Now, $\sum_{i=1}^{n} \pi(i, j)(\mathbf{v}_i^T \mathbf{z}(t))^2$ is always non-negative. On the other hand, G attains its maximum at zero. So, if we simply take $\pi(i, j) = 0$ for all i, j, G is actually always evaluated at zero and the approximative likelihood is maximized. So, taking all zeros in π is the maximum, which is absurd!

One approach would be to find the values of the latent variables u_i which maximize the likelihood, treating the u_i like the parameters. Thus, we would not try to integrate out the u_i, but rather just formulate the joint likelihood of $\mathbf{V}$, $\pi(i, j)$, $u_i(t)$ for all i, j and all $t = 1, \ldots, T$. This is computationally very difficult because the latent variables d_i are different for each image patch, so there is a very large number of them. The situation could be simplified by first estimating the first layer by ordinary ICA, and then fixing $\mathbf{V}$ once and for all (Karklin and Lewicki 2005). However, this does not reduce the number of dimensions.

So, we see that the estimation of both layers in a generative two-layer model is quite difficult. However, abandoning the generative-model approach simplifies the situation, and provides a promising approach, which will be treated next.

11.8.5 Energy-Based Two-Layer Models

A computationally simpler alternative to estimation of the two layers is provided by an "energy-based" approach. The idea is to take the likelihood in (11.5) as the starting point. As pointed out above, it does not make sense to try to maximize this with respect to the π, because the maximum is obtained by taking all zeros as the second layer weights.

There is a deep mathematical reason why we cannot maximize the likelihood in (11.5) with respect to the π. The reason is that the likelihood is *not normalized*. That is, when we interpret the likelihood as a pdf, its integral over the data variables is not equal to one: the integral depends on the values of the π. This means it is not a properly defined pdf because a pdf must always integrate to one, so the likelihood is not a properly defined likelihood either. To alleviate this, we have to introduce what is called a *normalization constant* or a *partition function* in the likelihood. The normalization constant, which is actually not a constant but a function of the model parameters, is chosen so that it makes the integral equal to one. Denoting the normalization constant by $Z(\pi)$, we write

$$\log L(\mathbf{v}_1, \ldots, \mathbf{v}_n) = \sum_{t=1}^{T} \sum_{i} h\left(\sum_{j=1}^{n} \pi(i,j)\left(\mathbf{v}_i^T \mathbf{z}_t\right)^2 \right) - \log|\det \mathbf{V}| - \log Z(\pi). \tag{11.23}$$

See Sect. 13.1.5 and Chap. 21 for more discussion on the normalization constant.

In principle, the normalization constant can be computed by computing the integral of the underlying pdf over the space of the $\mathbf{v}$, but this is extremely complicated numerically. Fortunately, there is a way around this problem, which is to use special estimation methods which do not require the normalization constant. Thus, we abandon maximization of likelihood because it requires that we compute the normalization constant. See Chap. 21 for information on such methods.

Attempts to estimate both layers in a two-layer model, using an energy-based approach, and estimation methods which circumvent the need for a normalization constant, can be found in Osindero et al. (2006), Köster and Hyvärinen (2007, 2008). This is a very active area of research (Karklin and Lewicki 2008). Some more remotely related work is in Köster et al. (2009a).

11.9 Concluding Remarks and References

A simple modification of the model of independent subspace analysis leads to emergence of topography, i.e. the spatial arrangement of the features. This is in contrast to ICA and ISA, in which the features are in random order. (In ISA, it is the subspaces which are in random order, but the linear features have some organization because of their partition to subspaces.) The basic idea in modeling topography is to consider subspaces which are overlapping, so that the neighborhood of each cell is one subspace. It is also possible to formulate a proper generative model which incorporates the same kind of statistical dependencies using variance variables which are generated by a higher-order ICA model, but that approach is mathematically difficult and still under construction.

Basic though old papers on topography are Hubel and Wiesel (1968), DeValois et al. (1982). Optical imaging results are shown in Blasdel (1992), and a recent high-resolution imaging study is in Ohki et al. (2005). Topography with respect to spatial frequency is investigated in Tootell et al. (1988), Silverman et al. (1989), Edwards et al. (1996). Seminal papers on pinwheels are Bonhoeffer and Grinvald (1991), Maldonado et al. (1997). A most interesting recent paper is DeAngelis et al. (1999) that also shows that the phases are not correlated in neighboring cells. The relationships of the topographic representation for different parameters are considered in Hübener et al. (1997). An important point is made in Yen et al. (2007), who show that the topography of responses is not so clear when the stimuli are complex, presumable due to non-linear interactions. The connection between topography and complex cell pooling is discussed in Blasdel (1992), DeAngelis et al. (1999).

The idea of minimum wiring length, or wiring economy, goes back to Ramón y Cajal, cited in Chen et al. (2006). The metabolic advantages of topography are further considered in Durbin and Mitchison (1990), Mitchison (1992), Koulakov and Chklovskii (2001), Attwell and Laughlin (2001). Comparisons between white and grey matter volume also point out how brain (skull) size limits the connectivity (Zhang and Sejnowski 2000).

Original papers describing the topographic ICA models are Hyvärinen and Hoyer (2001), Hyvärinen et al. (2001a). Kohonen's famous self-organizing map is also closely related (Kohonen 1982, 2001), but it has not been shown to produce a realistic V1-like topography; reasons for this were discussed in Sect. 11.7.2. A model which produces more a realistic topography (but still no low-frequency blobs) is Kohonen's ASSOM model (Kohonen 1996; Kohonen et al. 1997). However, in that model the nature of the topography is strongly influenced by an artificial manipulation of the input (a sampling window that moves smoothly in time), and it does not really emerge from the structure of images alone.

A related idea on minimization of wiring length has been proposed in Vincent and Baddeley (2003), Vincent et al. (2005), in which it is proposed that the retinal coding minimizes wiring, whereas cortical coding maximizes sparseness of activities.

Chapter 12
Dependencies of Energy Detectors: Beyond V1

All the models in this book so far have dealt with the primary visual cortex (V1). In this chapter, we show how statistical models of natural images can be extended to deal with properties in the extrastriate cortex, i.e. those areas which are close to V1 (also called the striate cortex) and to which the visual information is transmitted from V1.

12.1 Predictive Modeling of Extrastriate Cortex

Most of the experimental results in early cortical visual processing have considered V1. The function of most extrastriate areas is still rather much a mystery. Likewise, most research in modeling natural image statistics has been on low-level features, presumably corresponding to V1.

However, the methodology that we used in this book could possibly be extended to such extrastriate areas as V2, V3(A), V4, and V5. Actually, since the function of most extrastriate areas is not well understood, it would be most useful if we could use this modeling endeavor in a *predictive* manner, so that we would be able to predict properties of cells in the visual cortex, in cases where the properties have not yet been demonstrated experimentally. This would give testable, quantitative hypotheses that might lead to great advances in visual neuroscience.

In the next sections, we attempt to accomplish such predictive modeling in order to predict properties of a third processing step, following the simple and complex cell layers. The predictions should be based on the statistical properties of modeled complex-cell outputs. Our method is to apply ordinary independent component analysis to modeled outputs of complex cells whose input consists of natural images.[1]

12.2 Simulation of V1 by a Fixed Two-Layer Model

The basic idea in this chapter is to fix a model of complex cells and then learn a representation for complex cell outputs using a statistical model. The resulting three-layer network is depicted in Fig. 12.1.

This approach is rather different from the one used in previous chapters, in which we learned first the simple cells and then the complex cells from the data. Here, to simplify the model and the computations, we do not attempt to learn everything

[1] This chapter is based on the article (Hyvärinen et al. 2005a), originally published in BMC Neuroscience. The experiments were done by Michael Gutmann. Copyright retained by the authors.

A. Hyvärinen, J. Hurri, P.O. Hoyer, *Natural Image Statistics*,
Computational Imaging and Vision 39,

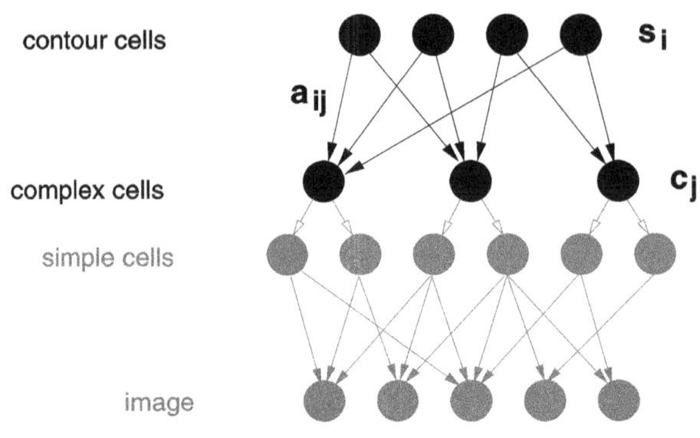

Fig. 12.1 The simplified hierarchical model investigated in this chapter. Modeled complex-cell responses are calculated in a feedforward manner, and these responses are subsequently analyzed by a higher-order feature layer in the network ("contour" layer). To emphasize that the lower layers are fixed and not learned, these layers have been greyed out in the figure. The direction of the arrows is from higher features to lower ones which is in line with the interpretation of our analysis as a generative model

at the same time. Instead, we fix the first two layers (simple and complex cells) according to well-known models, and learn only the third layer.

The classic complex-cell model is based on Gabor functions. As explained in Sect. 3.4.2, complex cells can be modeled as the sum of squares of two Gabor functions which are in quadrature phase. Quadrature phase means simply that if one of them is even-symmetric, the other one is odd-symmetric. This is related to computation of the Fourier energy locally, as explained in Sect. 2.4.

Complex-cell responses c_k to natural images were thus modeled with a Gabor energy model of the following form:

$$c_k = \left(\sum_{x,y} W_k^{\mathrm{o}}(x, y) I(x, y) \right)^2 + \left(\sum_{x,y} W_k^{\mathrm{e}}(x, y) I(x, y) \right)^2 \tag{12.1}$$

where W_k^{e} and W_k^{o} are even- and odd-symmetric Gabor receptive fields; the equation shows that their squares (energies) are pooled together in the complex cell. The complex cells were arranged on a 6×6 spatial grid. They had $6 \times 6 = 36$ different spatial locations, and at each location, four different preferred orientations and three different frequency selectivities ("bands"). The aspect ratio (ratio of spatial length to width) was fixed to 1.5. The frequency selectivities of the Gabor filters are shown in Fig. 12.2, in which all the filters W were normalized to unit norm for visualization purposes. The actual normalization we used in the experiments consisted of standardizing the variances of the complex cell outputs so that they were equal to unity for natural image input. The number of complex cells totaled $36 \times 4 \times 3 = 432$.

As the basic data, we used 1008 grey-scale natural images of size 1024×1536 pixels from van Hateren's database.[2] We manually chose natural images in the nar-

[2]Available at http://hlab.phys.rug.nl/imlib/index.html, category "deblurred" (van Hateren and van der Schaaf 1998).

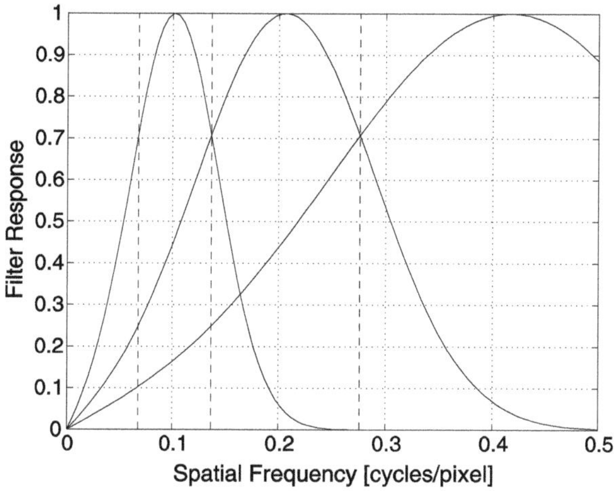

Fig. 12.2 We used fixed complex cells with three different frequency selectivities. The amplitudes of the Fourier Transforms of the odd-symmetric Gabor filters are shown here. The selectivities are such that each cell is sensitive to a certain frequency "band". The underlying Gabor filters had logarithmically spaced frequency peaks. Peak spatial frequencies were chosen as follows: $f_1 = 0.1$ cycles/pixel, $f_2 = 0.21$ cycles/pixel and $f_3 = 0.42$ cycles/pixel

rower sense, i.e. only wildlife scenes. From the source images, 50 000 image patches of size 24 × 24 pixels were randomly extracted. The mean grey value of each image patch was subtracted and the pixel values were rescaled to unit variance. The resulting image patch will be denoted by $I(x, y)$.

12.3 Learning the Third Layer by Another ICA Model

After fixing the first two layers, we learned the feature weights in the third layer by doing a simple ICA of the complex cell (second-layer) outputs denoted by c_k. No PCA dimension reduction was done here, so the number of independent components equals the number of complex cells, K. Thus, ICA was performed on the vector $\mathbf{c} = (c_1, \ldots, c_K)$ using the FastICA algorithm (see Sect. 18.7). In ICA, the orthogonalization approach was symmetric. Different non-linearities g were used, see Table 12.1. (The non-linearities are related to the non-Gaussianity measures used; see Sect. 18.7.)

Thus, we learned (estimated) a linear decomposition of the form

$$c_k = \sum_{i=1}^{K} a_{ki} s_i \quad \text{for all } k = 1, \ldots, K \tag{12.2}$$

or in vector form

$$\mathbf{c} = \sum_{i=1}^{K} \mathbf{a}_i s_i = \mathbf{A}\mathbf{s} \tag{12.3}$$

Table 12.1 The measures of non-Gaussianity used, i.e. the different functions $G = \log p_s$ used in the likelihood of the ICA model. These correspond to different non-linearities g in the FastICA algorithm, and to different sparseness measures h. The measures probe the non-Gaussianity of the estimated components in different ways

Non-Gaussianity measure	FastICA non-linearity	Motivation
$G_1(y) = \log\cosh y$	$g_1(y) = \tanh(y)$	Basic sparseness measure
$G_2(y) = -\exp(-y^2/2)$	$g_2(y) = y\exp(-y^2/2)$	More robust variant of g_1
$G_3(y) = \frac{1}{3}y^3$	$g_3(y) = y^2$	Skewness (asymmetry)
$G_4(y) =$ Gaussian cum. distr. function	$g_4(y) = \exp(-y^2/2)$	Robust variant of g_3

where each vector $\mathbf{a}_i = (a_{1i}, \ldots, a_{Ki})$ gives a higher-order feature vector. The s_i define the values of the higher-order features in the third cortical processing stage. Recall that the input to the system was natural images, so the statistics of $\mathbf{c}$ reflect natural image statistics.

Note that the signs of the feature vectors are not defined by the ICA model, i.e. the model does not distinguish between $\mathbf{a}_i$ and $-\mathbf{a}_i$ because any change in sign of the feature vector can be canceled by changing the sign of s_i accordingly. Here, unlike in the original natural images, the features will not be symmetric with respect to such a change of sign, so it makes sense to define the signs of the $\mathbf{a}_i$ based on that asymmetry. We defined the sign for each vector $\mathbf{a}_i$ so that the sign of the element with the maximal absolute value was positive.

This model can be interpreted as a generative model of image patches, following the interpretation of ISA as a non-linear ICA in Sect. 10.6. The higher-order independent component (here denoted by s_i) are generated according to (12.2). Then the activity of the complex cell is expressed as activities of simple cells with random division of the activity to the simple cells, using a random angle variable as in (10.17) on page 225. Finally, the simple cell activities are linearly transformed to image patches as in ICA or ISA models. This provides a complete generative model from the higher-order features to image pixel values.

12.4 Methods for Analyzing Higher-Order Components

We need to introduce some special methods to analyze the "higher-order" components obtained by this method, because the resulting higher-order feature vectors $\mathbf{a}_i$ cannot be simply plotted in the form of image patches.

We visualize the vectors $\mathbf{a}_i$ by plotting an ellipse at the centerpoint of each complex cell. The orientation of the ellipse is the orientation of the complex cell with index k, and the brightness of the ellipse with index i is proportional to the coefficient a_{ki} of the feature vector $\mathbf{a}_i$, using a grey-scale coding of coefficient values. We plotted complex cells in each frequency band (i.e. with the same frequency selectivity) separately.

We are also interested in the frequency pooling of complex cells in different higher-order features. We quantified the pooling over frequencies using a simple measure defined as follows. Let us denote by $a_i(x, y, \theta, f_n)$ the coefficient in the higher-order feature vector $\mathbf{a}_i$ that corresponds to the complex cell with spatial location (x, y), orientation θ and preferred frequency f_n. We computed a quantity which is similar to the sums of correlations of the coefficients over the three frequency bands, but normalized in a slightly different way. This measure P_i was defined as follows:

$$P_i = \sum_{m<n} \frac{|\sum_{x,y,\theta} a_i(x, y, \theta, f_m) a_i(x, y, \theta, f_n)|}{C_m C_n} \tag{12.4}$$

where the normalization constant C_m is defined as

$$C_m = \sqrt{\frac{1}{K} \sum_{j,x,y,\theta} a_j(x, y, \theta, f_m)^2} \tag{12.5}$$

and likewise for C_n.

For further analysis of the estimated feature vectors, we defined the preferred orientation of a higher-order feature. First, let us define for a higher-order feature of index i the hot-spot $(x_i, y_i)^*$ as the center location (x, y) of complex cells where the higher-order component s_i generates the maximum amount of activity. That is, we sum the elements of $\mathbf{a}_i$ that correspond to a single spatial location, and choose the largest sum. This allows us to define the tuning to a given orientation of a higher-order feature i by summing over the elements of $\mathbf{a}_i$ that correspond to the spatial hotspot and a given orientation; the preferred orientation is the orientation for which this sum is maximized. We also computed the length of a higher-order feature by least-squares fitting a Gaussian kernel to the patterns $\mathbf{a}_i$ (Hoyer and Hyvärinen 2002).

It is also possible to perform an image synthesis from a higher-order feature vector. However, the mapping from image to complex-cell outputs is not one-to-one. This means that the generation of the image is not uniquely defined given the activities of higher-order features alone. A unique definition can be achieved by constraining the phases of the complex cells. For the purposes of image synthesis, we assume that only odd-symmetric Gabor filters are active. Furthermore, we make the simplifying assumptions that the receptive fields W in simple cells are equal to the corresponding feature vectors, and that all the elements in the higher-order feature vector are non-negative (or small enough to be ignored). Then the synthesized image I^i_{synth} for higher-order feature vector $\mathbf{a}_i$ is given by

$$I^i_{\text{synth}}(x, y) = \sum_{k \in H} W^{\text{o}}_k(x, y) \sqrt{a_{ki}} \tag{12.6}$$

where the square root cancels the squaring operation in the computation of complex-cell responses, and H denotes the set of indices that correspond to complex cells of the preferred orientation at the hotspot. Negative values of a_{ki} were set to zero in this formula.

Since we are applying ICA on data which has been heavily processed (by the complex cell model), we have to make sure that the model is not only analyzing the artifacts produced by that processing. To obtain a baseline with which to compare our results, and to show which part of the results is due to the statistical properties of natural images instead of some intrinsic properties of our filterbank and analysis methods, we did exactly the same kind of analysis for 24×24 image patches that consisted of white Gaussian noise, i.e. the grey-scale value in each pixel was randomly and independently drawn from a Gaussian distribution of zero mean and unit variance. The white Gaussian noise input provides a "chance level" for any quantities computed from the ICA results. In a control experiment, such white noise patches were thus fed to the complex cell model, and the same kind of ICA was applied on the outputs.

12.5 Results on Natural Images

12.5.1 Emergence of Collinear Contour Units

In the first experiment, we used only the output from complex cells in a single frequency band, f_2 in Fig. 12.2.

The higher-order features are represented by their feature vectors $\mathbf{a}_i$ which show the contribution of the third-stage feature of index i on the activities of complex cells. A collection of the obtained feature vectors is shown in Fig. 12.3 for the non-linearity g_1 (see Table 12.1), visualized as described above. We can see emergence of collinear features. That is, the higher-order features code for the simultaneous activation of complex cells that together form something similar to a straight line segment.

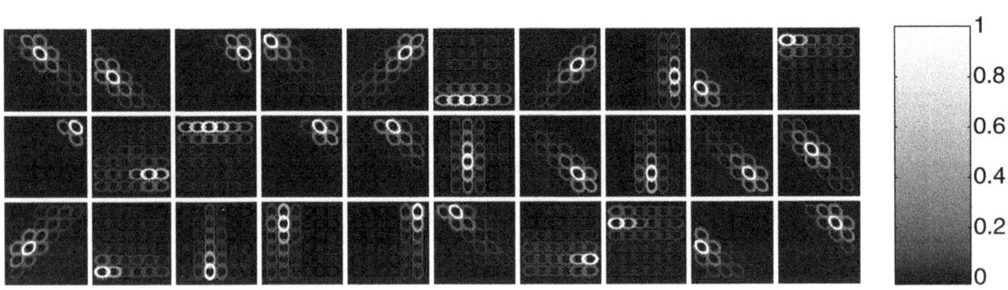

Fig. 12.3 Random selection of learned feature vectors $\mathbf{a}_i$ when the complex cells are all in a single frequency band. ICA non-linearity g was the tanh non-linearity g_1. Each patch gives the coefficients of one higher-order feature. Each ellipse means that the complex cell in the corresponding location and orientation is present in the higher-order feature, the brightness of the ellipse is proportional to the coefficient a_{ki}

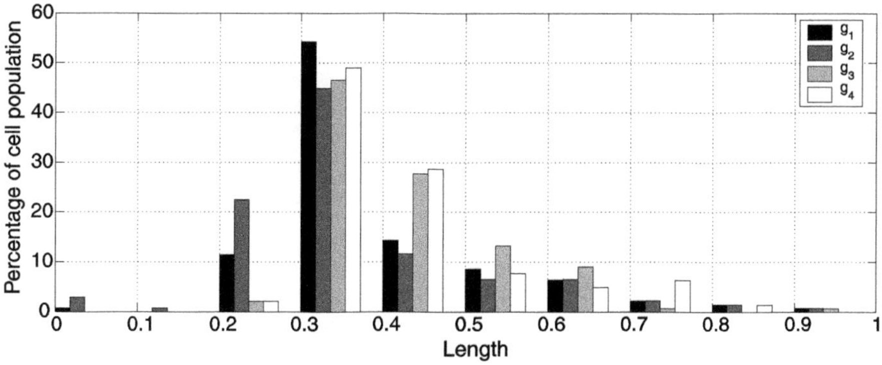

Fig. 12.4 Comparison of different measures of non-Gaussianity (FastICA non-linearities) in the first experiment. The histogram gives the lengths of the contour patterns for the four different non-linearities $g_1, \ldots, g_4$ in Table 12.1

Those coefficients that are clearly different from zero have almost always the same sign in a single feature vector. Defining the sign as explained above, this means that the coefficients are essentially non-negative.[3]

Other measures of non-Gaussianity (FastICA non-linearities) led to similar feature vectors. However, some led to a larger number of longer contours. Figure 12.4 shows the distribution of lengths for different non-linearities. The non-linearity g_4 (robust skewness) seems to lead to the largest number of long contours. The outputs of complex cells are skewed (non-symmetric), so it makes sense to use a skewness-based measure of non-Gaussianity, as discussed in Sect. 7.9. In this experiment, the results were very similar to those obtained by sparseness, however.

12.5.2 Emergence of Pooling over Frequencies

In the second experiment, the complex cell set was expanded to include cells of three different preferred frequencies. In total, there were now 432 complex cells. We performed ICA on the complex cell outputs when their input consisted of natural images. Thus, we obtained 432 higher-order feature vectors (features) $\mathbf{a}_i$ with corresponding activities s_i.

We visualized a random selection of higher-order features learned from natural images in Fig. 12.5. The visualization shows that the features tend to be spatially localized and oriented, and show collinearity as in the single-channel experiment

[3]In earlier work (Hoyer and Hyvärinen 2002), we actually imposed a non-negativity constraint on the coefficients; see Sect. 13.2. The results reported here show that those results can be replicated using ordinary ICA methods. The constraint of non-negativity of the feature vectors has little impact on the results: even without this constraint, the system learns feature vectors which are mainly non-negative.

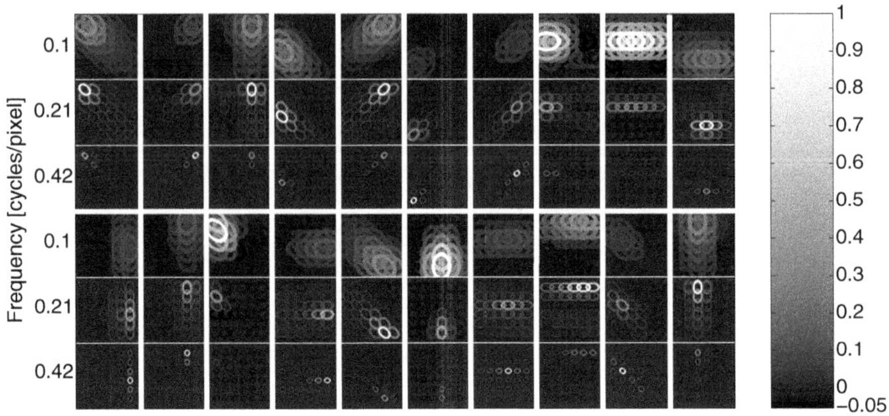

Fig. 12.5 A random selection of higher-order feature vectors $\mathbf{a}_i$ estimated from natural images using complex cells of multiple frequencies in the second experiment. ICA non-linearity g was the tanh non-linearity g_1. Each display of three patches gives the coefficients of one higher-order feature. Each patch gives the coefficients of one higher-order feature in one frequency band. Each ellipse means that the complex cell in the corresponding location, and of the corresponding orientation and frequency is present in the higher-order feature, brightness of ellipse is proportional to coefficient a_{ki}

above. What is remarkable in these results is that many cells pool responses over different frequencies. The pooling is coherent in the sense that the complex cells that are pooled together have similar locations and orientations. A smaller number of cells is shown in more detail in Fig. 12.6, where the coefficients in all orientations are shown separately.

We computed the frequency pooling measure P_i in (12.4) for the learned feature vectors. The distribution of this measure for natural image input and white Gaussian noise input is shown in Fig. 12.7. The figure shows that frequency pooling according to this measure was essentially non-existent for white Gaussian noise input, but relatively strong for many feature vectors when the input consisted of natural images. To express this more quantitatively, we computed the 99% quantile for the white Gaussian noise input. Then 59% of the basis vectors for natural image input had a pooling index P_i that was larger than this quantile. (For the 95% quantile, the proportion was 63%.) Thus, we can say that more than half of the higher-order basis vectors, when learned from natural images, have a pooling over frequencies that is significantly above chance level.

To show that the pooling measure is valid, and to further visualize the frequency pooling in the higher-order features, we chose randomly feature vectors learned from natural images that have pooling significantly over chance level (P_i above its 99% quantile for white Gaussian noise). These are plotted in Fig. 12.8. Visual inspection shows that in this subset, all basis vectors exhibit pooling over frequencies that respects the orientation tuning and collinearity properties.

The corresponding results when the input is white Gaussian noise are shown in Fig. 12.9, for a smaller number of higher-order cells. (To make the comparison fair,

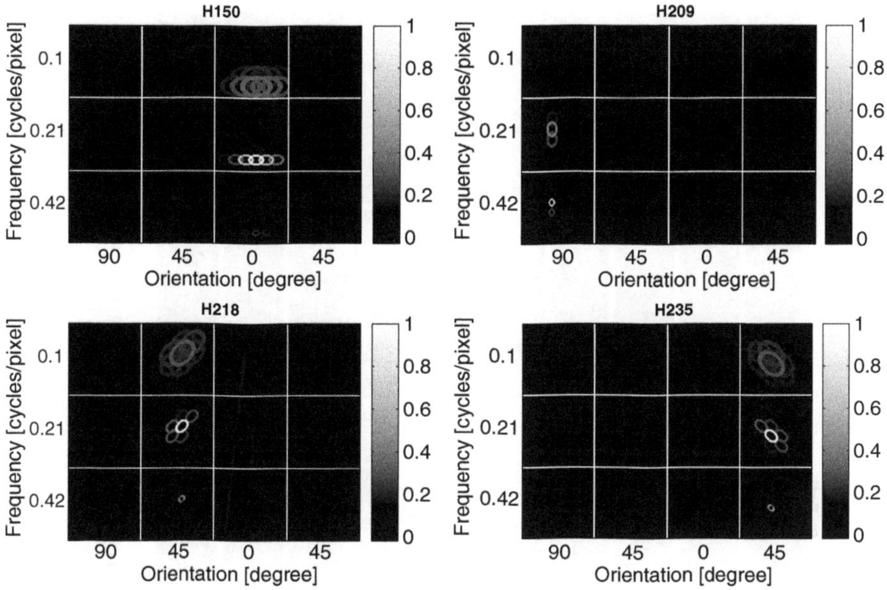

Fig. 12.6 Higher-order feature vectors of four selected higher-order features in the second experiment, shown in detail. The coefficients in each orientation and frequency band are plotted separately

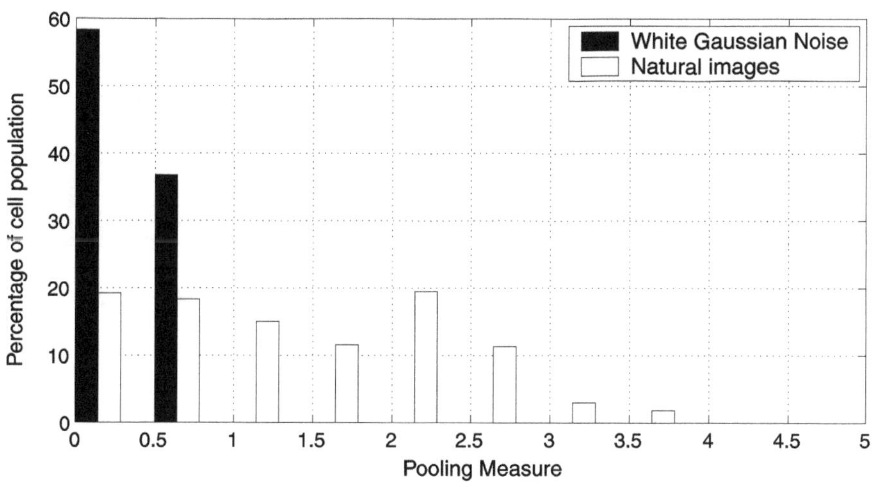

Fig. 12.7 The distributions of the frequency pooling measure in (12.4) for natural images and white Gaussian noise

these were randomly chosen among the 59% that had higher pooling measures, the same percentage as in Fig. 12.8.) Pooling over frequencies as well as collinearity are minimal. Some weak reflections of these properties can be seen, presumably

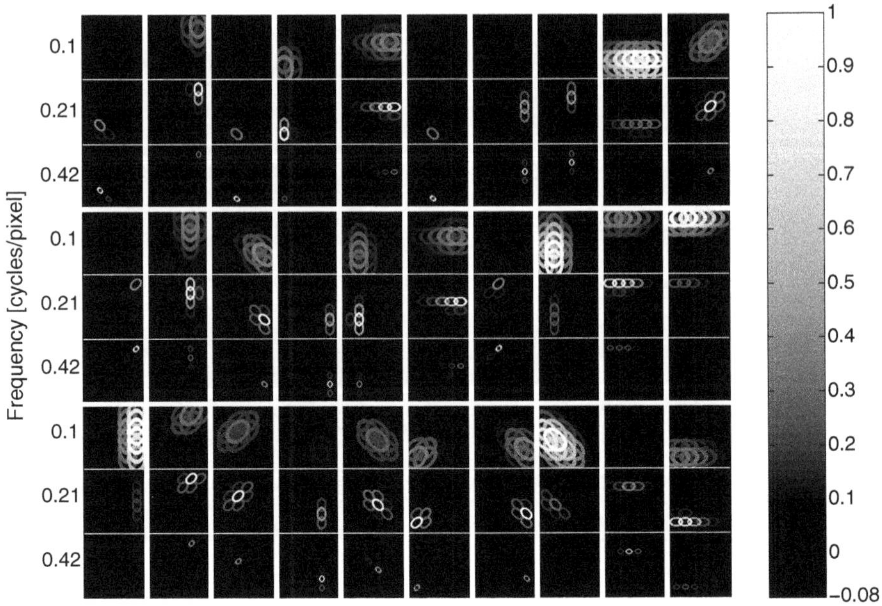

Fig. 12.8 A selection of higher-order feature vectors $\mathbf{a}_i$ estimated from natural images in the second experiment. These basis vectors were chosen randomly among those that have frequency pooling significantly above chance level

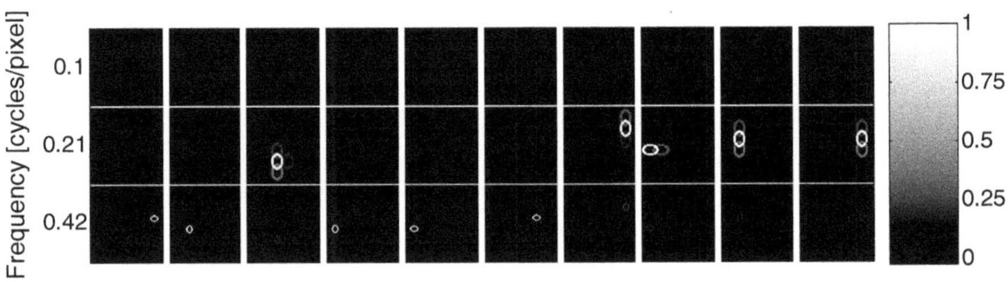

Fig. 12.9 For comparison, higher-order feature vectors estimated from white Gaussian noise, with each frequency band shown separately

due to the small overlap of the filters in space and frequency, which leads to weak statistical correlations between complex cells that are spatially close to each other or in neighboring frequency bands.

We also examined quantitatively whether the higher-order features are tuned to orientation. We investigated which complex cell has the maximum weight in $\mathbf{a}_i$ for each i in each frequency band. When the data used in learning consisted of natural images, in 86% of the cells the maximally weighted complex cells were found to be located at the hotspot $(x_i, y_i)^*$ (i.e., point of maximum activity, see above) and tuned to the preferred orientation of the higher-order feature for *every* frequency f. This shows how the higher-order features are largely selective to a single orientation.

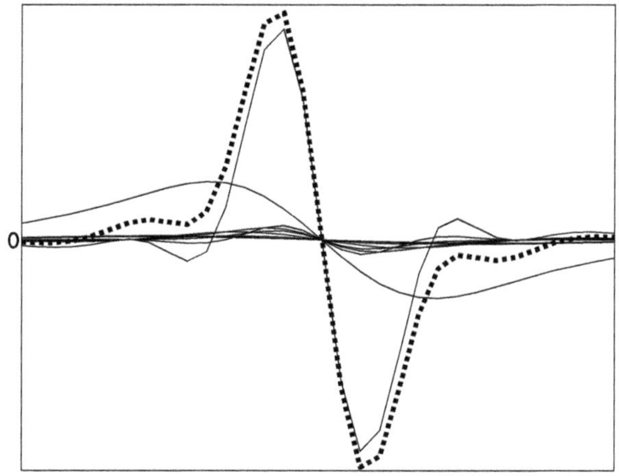

Fig. 12.10 Local image synthesis from the three odd-symmetric Gabor elements that have preferred orientation at the hotspot of a higher-order feature vector (H209 in Fig. 12.6). The *thick dotted curve* shows the synthesis using coefficients from natural images, and the *solid curves* show various synthesis results using coefficients learned from white Gaussian noise input

When the data used in learning consisted of Gaussian white noise, only 34% of the cells were found to be orientation-selective according to this criterion.

Finally, we synthesized images from higher-order feature activities. Figure 12.10 shows a slice orthogonal to the preferred orientation of one higher-order feature vector (H209 in Fig. 12.6). The intensity of the synthesized image shows no side-lobes (unnecessary oscillations), while representing a sharp, localized edge. In contrast, synthesis in the white Gaussian noise case (also shown in Fig. 12.10) gives curves that have either side-lobes like the underlying Gabor filters, or do not give a sharp localized edge. Thus, the curve obtained from synthesis of the features learned from natural images corresponds better to the notion of an edge.

12.6 Discussion of Results

12.6.1 Why Coding of Contours?

The result of the first experiment, using a single frequency channel (Sect. 12.5.1), is that simple ICA of simulated complex cell outputs leads to emergence of units coding for collinear contours (Fig. 12.3). First, we have to note that this result is not logically necessary: It is not obvious that the higher-order representation should necessarily code for contours. Multi-layer mechanisms similar to the one used here have been proposed in the context of texture segregation as well (Sperling 1989; Malik and Perona 1990). A priori, one could have expected such texture boundary detectors to emerge from this model. Our results seem to indicate that contour coding is, at least in this sparse coding sense, more fundamental than texture segregation.

The higher-order neurons which represent long contours bear many similarities to 'collator' (or 'collector') units, proposed in the psychophysical literature (Mussap and Levi 1996; Moulden 1994). Such units are thought to integrate the responses of smaller, collinear filters, to give a more robust estimate of global orientation than could be achieved with elongated linear mechanisms.[4]

12.6.2 Frequency Channels and Edges

In the second experiment using multiple frequency channels (Sect. 12.5.2), we saw emergence of pooling of contour information across multiple frequencies (Figs. 12.5, 12.6, 12.8). What is the functional meaning of this frequency pooling? One possibility is that this spatially coherent pooling of multiple frequencies leads to a representation of an edge that is more realistic than the edges given by typical Gabor functions. Presumably, this is largely due to the fact that natural images contain many sharp, step-like edges that are not contained in a single frequency band. Thus, representation of such "broad-band" edges is difficult unless information from different frequency bands is combined.

In terms of frequency channels, the model predicts that frequency channels should be pooled together after complex cell processing. Models based on frequency channels and related concepts have been most prominent in image coding literature in recent years, both in biological and computer vision circles. The utility of frequency channels in the initial processing stages is widely acknowledged, and it is not put into question by these results—in fact, the results in Chaps. 6–10 show that using frequency-selective simple and complex cells is statistically optimal. However, the question of when the frequency channels should be pooled or otherwise combined has received little attention. The results in this chapter (second experiment) indicate that a statistically optimal way is to pool them together right after the complex cell "stage", and this pooling should be done among cells of a given orientation which form a local, collinear configuration.

12.6.3 Toward Predictive Modeling

As we explained in the beginning of the chapter, the present results are an instance of predictive modeling, where we attempt to predict properties of cells and cell assemblies that have not yet been observed in experiments. To be precise, the prediction is that in V2 (or some related area) there should be cells whose optimal stimulus is a

[4]In principle, long contours could be represented by long feature vectors on the level of simple cells as well. However, the representation by these higher-order contour coding cells has the advantage of being less sensitive to small curvature and other departures from strict collinearity. Even very small curvature can completely change the response of an elongated linear filter (simple cell), but it does not change the representation on this higher level, assuming that the curvature is so small that the line stays inside the receptive fields of the same complex cells. Thus, higher-order contour cells give a more robust representation of the contours. Of course, the intermediate complex cell layer also confers some phase-invariance to the contour detectors.

broad-band edge that has no side lobes while being relatively sharp, i.e. the optimal stimulus is closer to a step-edge than the Gabor functions that tend to be optimal for V1 simple and complex cells. The optimal stimulus should also be more elongated (Polat and Tyler 1999; Gilbert and Wiesel 1985) than what is usually observed in V1, while being highly selective for orientation.

Statistical models of natural images offer a framework that lends itself to predictive modeling of the visual cortex. First, they offer a framework where we often see emergence of new kinds of feature detectors—sometimes very different from what was expected when the model was formulated. Second, the framework is highly constrained and data-driven. The rigorous theory of statistical estimation makes it rather difficult to insert the theorist's subjective expectations in the model and, therefore, the results are strongly determined by the data. Third, the framework is very constructive. From just a couple of simple theoretical specifications, e.g. non-Gaussianity, natural images lead to the emergence of complex phenomena.

We hope that the present work as well as future results in the same direction will serve as a basis for a new kind of synergy between theoretical and experimental neuroscience.

12.6.4 References and Related Work

Several investigators have looked at the connection between natural image statistics, Gestalt grouping rules, and local interactions in the visual cortex (Geisler et al. 2001; Sigman and Gilbert 2000; Elder and Goldberg 2002; Krüger 1998). However, few have considered the statistical relations between features of different frequencies. It should be noted that some related work on interactions of different frequencies does exist in the models of contrast gain control; see Chap. 9 or Schwartz and Simoncelli (2001a).

Recent measurements from cat area 18 (somewhat analogous to V2) emphasize responses to "second-order" or "non-Fourier" stimuli, typically sine-wave gratings whose amplitudes are modulated (Mareschal and Baker 1998a, 1998b). These results and the proposed models are related to our results and predictions, yet fundamentally different. In the model in Mareschal and Baker (1998b), a higher-order cell pools outputs of complex cells in the same frequency band to find contours that are defined by texture-like cues instead of luminance. The same cell also receives direct input from simple cells of a different frequency, which enables the cell to combine luminance and second-order cues. This is in stark contrast to higher-order cells in the model we used in this chapter, which pool outputs of complex cells of different frequencies. They can hardly find contours defined by second-order cues; instead they seem to be good for coding broad-band contours. Furthermore, in Mareschal and Baker (1998a, 1998b), any collinearity of pooling seems to be absent. This naturally leads to the question: Why are our predictions so different from these results from area 18? We suspect this is because it is customary to think of visual processing in terms of division into frequency channels—"second-order" stimuli are just an extension of this conceptualization. Therefore, not much attempt has been made to find cells that break the division into frequency channels according to our prediction. On the other hand, one can presume that the cells found in area 18 in Mareschal and Baker (1998a, 1998b) are different from our predictions because they use a learning strategy which is different from sparse coding used in our model, perhaps related to the temporal aspects of natural image sequences, see Chap. 16.

Another closely related line of work is by Zetzsche and co-workers (Zetzsche and Krieger 1999; Zetzsche and Röhrbein 2001) who emphasize the importance of decomposing the image information to local phase and amplitude information. The local amplitude is basically given by complex cell outputs, whereas the physiological coding of the local phases is not known. An important question for future work is how to incorporate phase information in the higher-order units. Some models by Zetzsche et al. actually predict some kind of pooling over frequencies, but rather directly after the simple cell stage; see Fig. 16 in Zetzsche and Röhrbein (2001).

Related models in which edge detection uses phase information pooled over different frequencies are in Morrone and Burr (1988), Kovesi (1999). An interesting investigation into the relation of edges and space-frequency analysis filter outputs in natural images is in Griffin et al. (2004). A psychophysical study on the integration of information over different frequencies is Olzak and Wickens (1997).

The model in this chapter opens the way to highly non-linear multilayer models of natural image statistics. While this seems like a most interesting direction of research, not much work has been done so far. Related attempts to construct very general, non-linear models of natural image statistics include Pedersen and Lee (2002), Lee et al. (2003), Malo and Gutiérrez (2006), Chandler and Field (2007), Griffin (2007).

12.7 Conclusion

Experiments in this chapter show that two different kinds of pooling over complex cells emerge when we model the statistical properties of natural images. First, the higher-order features group collinear complex cells which form a longer contour. Second, they group complex cells of different frequency preferences. This is accomplished by applying ordinary ICA on a set of modeled complex cells with multiple frequencies, and inputting natural images to the complex cells. Thus, statistical modeling of natural stimuli leads to an interesting hypothesis on the existence of a new kind of cells in the visual cortex.

Chapter 13
Overcomplete and Non-negative Models

In this chapter, we discuss two generalizations of the basic ICA and sparse coding models. These do not reject the assumption of independence of the components but change some of the other assumptions in the model. Although the generative models are linear, the computation of the features is non-linear. In the overcomplete basis model, the number of independent components is larger than the number of pixels. In the non-negative model, the components, as well as the feature vectors, are constrained to be non-negative.

13.1 Overcomplete Bases

13.1.1 Motivation

An important restriction of most of the models treated so far is that the number of features cannot be larger than the dimension of the data. The dimension of the data is at most equal to the number of pixels, and it is actually smaller after canonical preprocessing including PCA. This was for two reasons:

1. In the sparse coding models the feature detector weights were constrained to be orthogonal. In a space with n dimensions, we can have at most n orthogonal vectors, so this constrains the number of features.
2. In the generative models such as ICA, we had to assume that the matrix $\mathbf{A}$, which has the features as its columns, is invertible. Again, a matrix can be invertible only if it is square: thus, the number of features cannot be larger than the number of pixels.

However, it can be argued that the number of features should be larger than the dimension of the image data. The computational justification for such a claim goes as follows:

1. The processing of an image part, corresponding perhaps to an object, should not depend on which location of the image it happens to occupy. That is, if a face is in on the left side of the visual field, it should be processed in the same way as if it were on the right side; and if the object is moved one pixel to the left, its processing should not change either.[1]

[1]The resolution of the retinal image changes as a function of eccentricity (the distance from the centerpoint), so talking about moving "one pixel to the left" is an oversimplification. However, this does not change the underlying logic very much, if one simply thinks of photo-receptors or ganglion cells instead of pixels.

A. Hyvärinen, J. Hurri, P.O. Hoyer, *Natural Image Statistics*,
Computational Imaging and Vision 39,

2. Thus, any feature the system computes should be computed at each possible location—at the minimum at the location corresponding to each pixel. For example, if we have an edge detector, the output of that edge detector should be computed at each possible (x, y) location possible. Denote their number by N.
3. So, any feature should basically have N replicates in the system, one for each location. Possibly it could be a bit less because we may not want to replicate the feature very close to borders where they could not be replicated completely, but this does not change the basic argument.
4. What all this implies is that if we just take one feature, say a vertical odd-symmetric Gabor of a given frequency and envelope, copy it in all different locations, we already have N different features, supposedly the maximum number!
5. Of course, we would actually like to have many different Gabors with different orientations, different phases, different frequencies, and maybe something else as well. Actually, the argument in point 1 can be applied equally well to different orientations and frequencies, which should be processed equally well. So, in the end, the number of features must be many times greater than the number of pixels.

A neuroanatomical justification for the same phenomenon is the following calculation: the number of simple cells in V1 seems to be much larger than the number of retinal ganglion cells which send out the information on the retina, perhaps by a factor of 25 (Olshausen 2003). So, if we consider the number of ganglion cells as the "dimension" of input to V1, the number of features seems to be much larger than the number of dimensions.[2]

13.1.2 Definition of Generative Model

Now, we define a generative model which has more features than the data has dimensions. In this context, to avoid any confusion, we call the feature vectors A_i *basis vectors*. A set of basis vectors which contains more vectors than the space has dimensions is called an *overcomplete basis* (Simoncelli et al. 1992; Olshausen and Field 1997).

The definition of a generative model with an overcomplete basis is rather straightforward. We just need to express the image as a linear superposition

$$I(x, y) = \sum_{i=1}^{m} A_i(x, y) s_i \tag{13.1}$$

[2]This point is a bit complicated by the fact that the number of photo-receptors in the retina is approximately 100 times larger than the number of ganglion cells. Thus, ganglion cells reduce the dimension of the data, and V1 seems to increase it again. Nevertheless, if we consider the computational problem faced by V1, it does seem justified to say that it uses an overcomplete basis because it can only receive the outputs of ganglion cells.

where the only difference to previous models is that the number of features m is arbitrarily large. We also need to specify the statistical properties of the components s_i. In the basic case, we assume that they are sparse and statistically independent.

For technical reasons, another modification is also usually introduced at this point: we assume that the image is not exactly a linear sum of the features, but there is noise as well. That is, Gaussian noise $N(x, y)$ is added to each pixel:

$$I(x, y) = \sum_{i=1}^{m} A_i(x, y)s_i + N(x, y). \tag{13.2}$$

This does not change the behavior of the model very much, especially if the noise level is small, but it simplifies the computations in this case. (In the case of basic ICA, introduction of noise in the model just complicates things, so it is usually neglected.)

Note that the meaning of overcompleteness changes when the dimension is reduced by PCA. From the viewpoint of statistical modeling, the dimension of the data is then the dimension given by PCA. So, even a basis which has the same number of vectors as there are pixels can be called overcomplete because the number of pixels is larger than the PCA-reduced dimension.

Despite the simplicity of the definition of the model, the overcomplete basis model is much more complicated to estimate. What is interesting is that it has a richer behavior than the basic sparse coding and ICA models because it leads to some non-linearities in the computation of the features. We will treat this point first.

13.1.3 Nonlinear Computation of the Basis Coefficients

Consider first the case where the basis vectors A_i are given, and we want to compute the coefficients s_i for an input image I. The fundamental problem is that the linear system given by the basis vectors A_i is not invertible: If one tries to solve for the s_i given an I, there are more unknowns s_i than there are equations. So, computation of the s_i seems impossible. Indeed, it is impossible in the sense that even if the image were created as a linear sum of the A_i for some coefficient values s_i, we cannot recover those original coefficients from the input image alone, without some further information.

As an illustration, consider an image with two pixels with values $(1, 1)$. Assume we use a basis with three vectors: $(0, 1)$, $(1, 0)$, and $(1, 1)$. Thus, we have

$$(1, 1) = (0, 1)s_1 + (1, 0)s_2 + (1, 1)s_3. \tag{13.3}$$

Obviously, we could represent the image by setting $s_1 = 0$, $s_2 = 0$, and $s_3 = 1$. But equally well, we could set $s_1 = 1$, $s_2 = 1$, and $s_3 = 0$. Even if the image was exactly generated using one of these choices for s_i, we cannot tell which one it was by using information in the image alone. Actually, there is an infinite number of different solutions: you could take any weighted average with of the two solution just given, and it would be a solution as well.

However, there is a partial solution to this problem. The key is to use sparseness. Since we know that the s_i are sparse, we can try decide to find the *sparsest solution*. In the illustration above, we would choose the solution $s_1 = 0$, $s_2 = 0$, and $s_3 = 1$ because it is the sparsest possible in the sense that only one coefficient is different from zero.[3]

There is a clear probabilistic justification for such a procedure. Basically, we can find the most probable values for the coefficients s_i, under the assumption that the s_i have sparse distributions. This is possible by using conditional probabilities in a manner similar to Bayes' rule (see Sect. 4.7). Now we will derive the procedure based on probabilistic reasoning. By the definition of conditional pdf's, we have

$$p(\mathbf{s}|I) = \frac{p(\mathbf{s}, I)}{p(I)} = \frac{p(I|\mathbf{s})p(\mathbf{s})}{p(I)} \tag{13.4}$$

which is the basis for Bayes' rule. The formula can be simplified because $p(I)$ does not depend on $\mathbf{s}$. Since our goal is to find the $\mathbf{s}$ which maximizes $p(\mathbf{s}|I)$, we can just ignore this constant. We can also maximize its logarithm instead because it is often simpler, and equivalent because logarithm is a strictly increasing function. This gives us the following objective function to maximize:

$$\log p(I|\mathbf{s}) + \log p(\mathbf{s}). \tag{13.5}$$

Such estimation of the $\mathbf{s}$ is called maximum a posteriori (MAP) estimation, as discussed in Sect. 4.8.2.

Now, we have to compute the probabilities $\log p(I|\mathbf{s})$ and $\log p(\mathbf{s})$ needed. The first thing we consider is the *prior* distribution $p(\mathbf{s})$ of the s_i. In Bayesian inference, the prior distribution (or prior for short) incorporates the knowledge we have before making any observations. What prior knowledge do we have here? First, we know that the components are sparse. Second, we assume that they are independent, which is a simple approximation although it is not terribly precise. Thus, $\log p(\mathbf{s})$ is similar to what was used in ordinary ICA estimation and linear sparse coding. It can be expressed as

$$\log p(\mathbf{s}) = \sum_{i=1}^{m} G(s_i) \tag{13.6}$$

where the function G is the same kind of function we used in ICA estimation; see e.g. (7.19) on page 163.

To compute $p(I|\mathbf{s})$, we will use the noisy version of the model in (13.2). Assume that we know the variance of the Gaussian noise, and denote it by σ^2. Then the conditional probability of $I(x, y)$ given all the s_i is the Gaussian pdf of $N(x, y) =$

---

[3]Another solution would be to use the Moore–Penrose pseudo-inverse; see Sect. 19.8. However, that method is less justified by statistical principles, and less useful in practice.

$\sum_{i=1}^{m} A_i(x, y)s_i - I(x, y)$. By definition of the Gaussian pdf, the pdf of a single noise variable is thus

$$p\big(N(x, y)\big) = \frac{1}{\sqrt{2\pi}} \exp\left(-\frac{1}{2\sigma^2} N(x, y)^2\right). \tag{13.7}$$

So, the conditional log-pdf for one pixel is

$$\begin{aligned}\log p\big(I(x, y)|\mathbf{s}\big) &= -\frac{1}{2\sigma^2} N(x, y)^2 - \frac{1}{2}\log 2\pi \\ &= -\frac{1}{2\sigma^2}\left[I(x, y) - \sum_{i=1}^{m} A_i(x, y)s_i\right]^2 - \frac{1}{2}\log 2\pi. \end{aligned} \tag{13.8}$$

We assume that the noise is independent in all pixels, so the conditional pdf of the whole image I is the sum of these log-pdf's:

$$\log p(I|\mathbf{s}) = -\frac{1}{2\sigma^2}\sum_{x,y}\left[I(x, y) - \sum_{i=1}^{m} A_i(x, y)s_i\right]^2 - \frac{n}{2}\log 2\pi. \tag{13.9}$$

The constant $\frac{n}{2}\log 2\pi$ can be omitted for simplicity.

Putting all this together: To find the most probable $s_1, \dots, s_m$ that generated the image, we maximize

$$\begin{aligned}\log p(\mathbf{s}|I) &= \log p(I|\mathbf{s}) + \log p(\mathbf{s}) + \text{const.} \\ &= -\frac{1}{2\sigma^2}\sum_{x,y}\left[I(x, y) - \sum_{i=1}^{m} A_i(x, y)s_i\right]^2 + \sum_{i=1}^{m} G(s_i) + \text{const.}\end{aligned} \tag{13.10}$$

where the "const" means terms which do not depend on $\mathbf{s}$. Maximization of this objective function is usually not possible in closed form, and numerical optimization methods have to be used. We have here assumed that the A_i are known; their estimation will be considered below.

Maximization of such an objective function leads to a *non-linear* computation of the cell activities s_i. This is in stark contrast to ordinary (non-overcomplete) models, in which the s_i are a linear function of the $I(x, y)$. The implications of such a non-linearity will be considered in more detail in Chap. 14.

13.1.4 Estimation of the Basis

Estimation of the basis vectors A_i can be performed using the same principle as estimation of the s_i. Basically, the solution is hidden in (13.10). First, note that the pdf in (13.9) depends on the A_i as well. So, that equation actually describes $p(I|\mathbf{s}, A_1, \dots, A_m)$ instead of just $p(I|\mathbf{s})$. Further, if we backtrack in the logic that lead us to (13.10), we see that the conditional probability in (13.10), when considered as a function of both $\mathbf{s}$ and the A_i, is equal to $p(\mathbf{s}, A_1, \ldots, A_m | I)$, if we assume a flat (constant) prior for the A_i. This is the conditional probability of *both* $\mathbf{s}$ and the A_i, given the image I. Thus, the conditional log-pdf can be interpreted as essentially the likelihood of the A_i.

Estimation of the A_i can now be performed by maximizing the conditional pdf in (13.10) for a *sample* of images $I_1, I_2, \ldots, I_T$. (Obviously, we cannot estimate a basis from a single image.) As usual, we assume that the images in the sample have been collected independently from each other, in which case the log-pdf for the sample is simply the sum of the log-pdf. So, we obtain the final objective function

$$\sum_{t=1}^{T} \log p\big(\mathbf{s}(t), A_1, \ldots, A_m | I_t\big) = -\frac{1}{2\sigma^2} \sum_{t=1}^{T} \sum_{x,y} \left[I_t(x, y) - \sum_{i=1}^{m} A_i(x, y) s_i(t) \right]^2 + \sum_{t=1}^{T} \sum_{i=1}^{m} G\big(s_i(t)\big) + \text{const.} \tag{13.11}$$

When we maximize this objective function with respect to all the basis vectors A_i and cell outputs $s_i(t)$ (the latter are different for each image), we obtain at the same time, the estimates of the components and the basis vectors.[4] In other words, we compute both the non-linear cell outputs and the features A_i.

Note that it is not straightforward to define the receptive fields of the cell anymore. This is because computation of the cell outputs is non-linear, and receptive fields are simple to define for linear cells only. Actually, if we collect the basis vectors A_i into a matrix $\mathbf{A}$ as we did earlier in the ordinary ICA case, that matrix is simply not invertible, so we cannot define the receptive fields as the rows of its inverse, as we did earlier.

13.1.5 Approach Using Energy-Based Models

An alternative approach for estimating an overcomplete representation is the following: We give up a generative model and concentrate on generalizing the sparseness criteria. Basically, we take the log-likelihood of the basic ICA model, and relax the constraint that there cannot be too many linear feature detectors. This approach is computationally more efficient because we do not need to compute the non-linear estimates of the components s_i which requires another optimization.

[4] One technical problem with this procedure is that the scales of the independent components are not fixed, which leads to serious problems. This problem can be solved simply by normalizing the variances of the independent components to be equal to unity at every optimization step. Alternatively, one can normalize the basis vector A_i to unit norm at every step.

Consider the log-likelihood of the basic ICA model in (7.15), which we reproduce here for convenience:

$$\log L(\mathbf{v}_1, \ldots, \mathbf{v}_n; \mathbf{z}_1, \ldots, \mathbf{z}_T) = T \log\left|\det(\mathbf{V})\right| + \sum_{i=1}^{m} \sum_{t=1}^{T} G_i\left(\mathbf{v}_i^T \mathbf{z}_t\right) \tag{13.12}$$

where $\mathbf{z}_t$ is the canonically preprocessed data sample, and the $\mathbf{v}_i$ are the feature detector vectors in the preprocessed space. We have changed the number of feature detectors to m in line with the notation in this section. Moreover, we use here general functions G_i, which in the case of basic ICA is equal to $\log p_i$, the log-pdf of the independent component. (In this section, we revert to using canonically preprocessed data, but this does not really change anything in the mathematical developments. Overcompleteness then means that the number of features is larger than the PCA-reduced dimension.)

Now, could we just use the formula in (13.12) with more features than dimensions? Let us denote the dimension of the data by n. Then this means that we just take $m > n$ to achieve an overcomplete representation.

Unfortunately, this is not possible. The problem is the term $\log|\det(\mathbf{V})|$. The simple reason is that if $m > n$, the matrix $\mathbf{V}$, which collects the $\mathbf{v}_i$ as its rows, would not be square, and the determinant is only defined for a square matrix.

On the other hand, the second term on the right-hand side in (13.12) is just a sum of measures of sparseness of the features, so this term need not be changed if we want to have an overcomplete representation.

So, we have to understand the real meaning of the term $\log|\det(\mathbf{V})|$ to obtain a model with an overcomplete representation. This term is actually the logarithm of what is called the *normalization constant* or a *partition function*. It is a function of the model parameters which makes the pdf of the data fulfill the fundamental constraint that the integral of the pdf is equal to one—a constraint that every pdf must fulfill. A likelihood is nothing else than a pdf interpreted as a function of the parameters, and computed for the whole sample instead of one observation. So, the likelihood must fulfill this constraint as well.

The normalization constant is, in theory, obtained in a straightforward manner. Let us define the pdf (for one observation) by replacing the first term in (13.12) by the proper normalization constant, which we denote by Z:

$$\log L(\mathbf{z}; \mathbf{v}_1, \ldots, \mathbf{v}_n) = -\log Z(\mathbf{V}) + \sum_{i=1}^{n} G_i\left(\mathbf{v}_i^T \mathbf{z}\right). \tag{13.13}$$

Normalization of the pdf means that we should have

$$\int L(\mathbf{z}; \mathbf{v}_1, \ldots, \mathbf{v}_n)\, d\mathbf{z} = 1. \tag{13.14}$$

In the present case, this means

$$\int L(\mathbf{z}; \mathbf{v}_1, \ldots, \mathbf{v}_n)\, d\mathbf{z} = \frac{1}{Z(\mathbf{V})} \int \prod_{i=1}^{n} \exp\left(G_i\left(\mathbf{v}_i^T \mathbf{z}\right)\right) d\mathbf{z} = 1. \tag{13.15}$$

So, in principle, we just need to take

$$Z(\mathbf{V}) = \int \prod_{i=1}^{n} \exp\left(G_i\left(\mathbf{v}_i^T \mathbf{z}\right)\right) d\mathbf{z} \tag{13.16}$$

because this makes the integral in (13.15) equal to one.

However, in practice, evaluation of the integral in (13.16) is extremely difficult even with the best numerical integration methods. So, the real problem when we take more feature detector vectors than there are dimensions in the data, is the computation of the normalization constant.

Estimation of the model by maximization of likelihood requires that we know Z. If we omit Z and maximize only the first term in (13.13), the estimation goes completely wrong: If the G_i have a single peak at zero (like the negative log cosh function), as we have assumed in earlier chapters, the maximum of such a truncated likelihood is obtained when the $W_i(x, y)$ are all zero, which is quite absurd!

So, the model becomes much more complicated to estimate since we don't know how to normalize the pdf as a function of the vectors $\mathbf{v}_i$. This is in stark contrast to the basic case where the number of feature detector vectors equals the number of input variables: the function Z is simply obtained from the determinant of the matrix collecting all the vectors $\mathbf{v}_i$, as seen in (7.15).

Fortunately, there are methods for estimating models in the case where Z cannot be easily computed. First of all, there is a number of methods for computing Z approximately, so that the maximum likelihood estimation is computationally possible. However, in our case, it is probably more useful to look at methods which estimate the model directly, avoiding the computation of the normalization constant. Score matching and contrastive divergence are two methods for estimating such "non-normalized" models. The mathematical details of score matching are described in Chap. 21.

One point to note is that we are really estimating linear receptive fields W_i using this method. Thus, the result is not really an overcomplete *basis* but rather an overcomplete representation using an overcomplete set of receptive fields.

This approach is sometimes called "energy-based" due to complicated historical reasons. The model in (13.13) has also been called a "Products of Experts" model (Hinton 2002). Further related methods are considered in Hyvärinen and Inki (2002). See Utsugi (2001) for an overcomplete version of the ISA model.

13.1.6 Results on Natural Images

We estimated an overcomplete representation from natural images using the method in Sect. 13.1.5. Thus, we defined the model using the non-normalized log-likelihood in (13.13). We basically used the classic (negative) log cosh function as G, but we allowed a bit more flexibility by allowing rescaling of the G_i by defining $G_i(u) = -\alpha_i \log\cosh(u)$, where α_i are parameters that are estimated at the same time as the $\mathbf{v}_i$. We also constrained the norms of the $\mathbf{v}_i$ to be equal to one. We used the score matching approach (see above or Chap. 21) to estimate the parameters without computation of the normalization constant.

To reduce the computational load, we took patches of 16×16 pixels. We preprocessed the data just like with ICA in Chap. 7, but the dimension reduction was less strong: we retained 128 principal components, i.e. one half of the dimensions. Then we estimated a representation with 512 receptive fields. The representation is thus 4 times overcomplete when compared to the PCA dimension, and two times overcomplete when compared with the number of pixels.

The resulting receptive fields are shown in Fig. 13.1. To save space, only a random sample of 192 receptive fields is shown. The receptive fields are quite similar to those estimated by basic ICA or sparse coding. Some are more oscillatory, though.

13.1.7 Markov Random Field Models *

The approach of energy-based overcomplete representations can be readily extended to models which cover the whole image using the principle of Markov random fields. Here, we provide a very brief description of this extension for readers with some background in MRFs.

A very important question for any image-processing application is how the models for image patches can be used for whole images which have tens of thousands, or even millions of pixels. One approach for this is to use Markov random fields (MRF). What this means is that we define what is called in that theory a *neighborhood* for each pixel, and define the probability density for the image as a function of each pixel value and the values of the pixels in the neighborhood. The central idea is that we compute the same function in all possible locations of the image.

In our context, the neighborhood of a pixel can be defined to be an image patch taken so that the pixel in question is in the very middle. To extend our models to a MRF, we can also use the outputs of linear feature detectors to define the pdf.

This leads to a pdf of the following form:

$$\log p(I; W_1, \ldots, W_n) = \sum_{x,y} \sum_{i=1}^{n} G\Big(\sum_{\xi,\eta} W_i(\xi, \eta) I(x+\xi, y+\eta)\Big) - \log Z(W_1, \ldots, W_n). \tag{13.17}$$

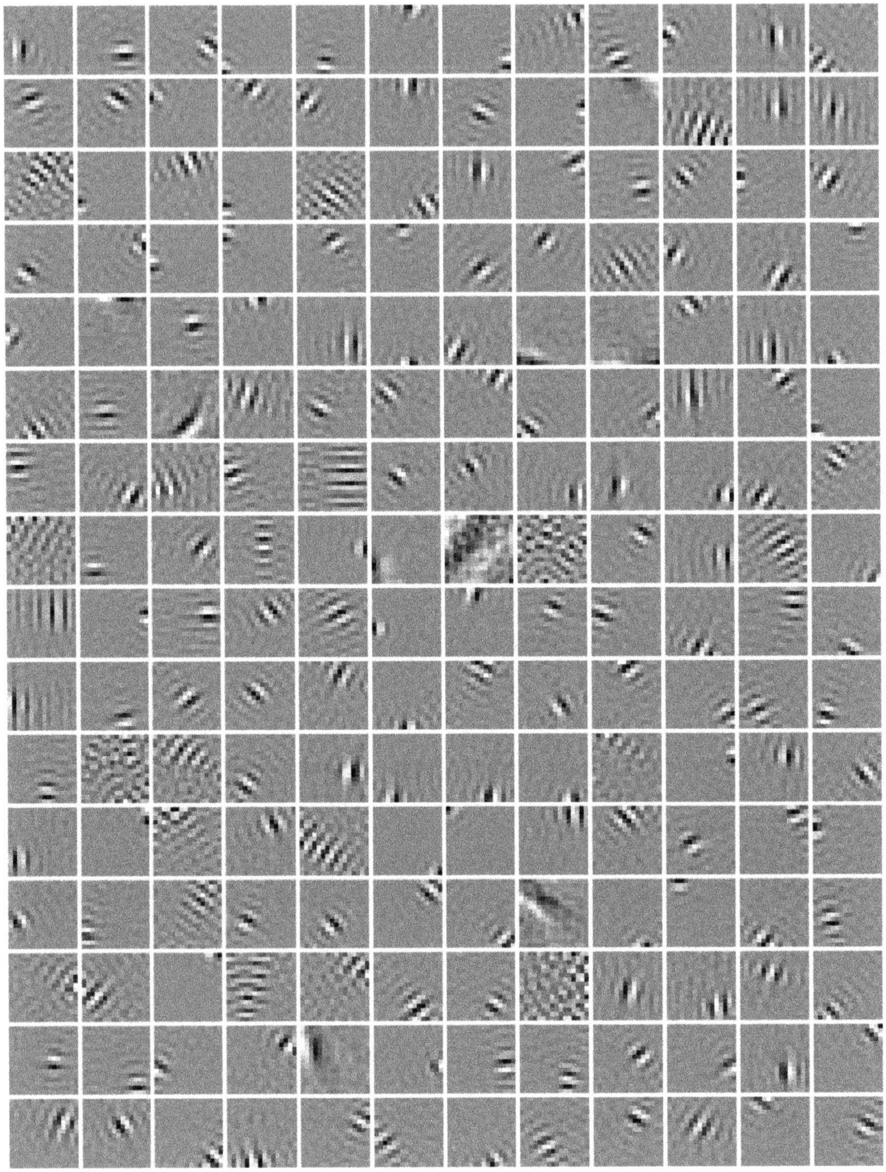

Fig. 13.1 Receptive fields W_i in a four times overcomplete basis for canonically preprocessed data, estimated using the model in (13.13) and score matching estimation. Only a random sample of the W_i is shown to save space

Here, the first sum over x, y goes over all possible image locations and neighborhoods. For each location, we compute the outputs of n linear feature detectors so that they are always centered around the location x, y. The function G is the same kind of function, for example log cosh, as used in sparse coding.

An important point is that the indices ξ, η only take values inside a small range, which is the neighborhood size. For example, we could define that they belong to the range $-5, \ldots, 5$, in which case the patch size would be 11×11 pixels.

One interpretation of this pdf is that we are sliding a window over the whole image and computing the outputs of the feature detectors in those windows. In other words, we compute the convolution of each of the W_i with the image, and then apply the non-linear function G on the results of the convolution. Summation over x, y and over i then simply means that the log-pdf is the sum over the whole convolved, non-linearly processed image, and all the filters.

As in the case of the model in Sect. 13.1.5, the log-pdf includes a normalization constant Z, which is a function of the feature detector weights W_i. Again, the computation of the normalization constant is most difficult, and the model is probably best estimated using methods which avoid computation of the normalization constant (see, e.g. Chap. 21).

In fact, we can see a direct connection with the overcomplete basis framework as follows. Define the translated feature detector $W^{(a,b)}$ as a feature detector whose weights have been translated by the amount given by a and b, so that $W^{(a,b)}(x, y) = W(x-a, y-b)$. Also, redefine indices as $x+\xi = x'$, $y+\eta = y'$. Then we can write the log-pdf as

$$\log p(I; W_1, \ldots, W_n) = \sum_{i=1}^{n} \sum_{x,y} G\Big(\sum_{x',y'} W_i^{(x,y)}(x', y') I(x', y')\Big) - \log Z. \quad (13.18)$$

This model is just like the overcomplete model in Sect. 13.1.5, but the feature weights are *constrained* so that they are copies of a small number of feature weights W_i in all the different locations, obtained by the translation operation $W_i(x, y)$. Due to the summation over the translation parameters x, y, each weight vector is copied to all different locations. (We are here neglecting any border effects which appear because for those weight in the W_i which go over the edges of the image.) Furthermore, the normalization constant is computed in a slightly different way because the integration is over the whole image.

Learning feature detector weights of MRFs was proposed in Roth and Black (2005). A related approach was proposed in Zhu et al. (1997). At the time of this writing, the first successful attempt to estimate MRFs in the sense that we obtain Gabor-like features was obtained in Köster et al. (2009b). A review of classic MRF models, i.e. models in which the features are not learned but manually tuned, is in Li (2001); a more mathematical treatise is Winkler (2003).

Let us finally mention some completely different approaches to modeling whole images or scenes. One is to extract some global statistics, i.e. feature histograms, which can then be further analyzed by various statistical models, as in, e.g. Liu and Cheng (2003), Lindgren and Hyvärinen (2004). Yet another alternative is to compute a low-dimensional holistic representation by techniques related to PCA, as in, e.g. Torralba and Oliva (2003).

13.2 Non-negative Models

13.2.1 Motivation

Neural firing rates are never negative. Even if we consider the spontaneous firing rate as the baseline and define it to be zero in our scale, the firing in cortical cells cannot go much below zero because the spontaneous firing rates are so low; so, it may be useful to consider them non-negative anyway. It has been argued that this non-negativity of firing rates should be taken into account in statistical models. Non-negative matrix factorization (NMF) (Lee and Seung 1999) is a recent method for finding such a representation. It was originally introduced in a different context and called *positive* matrix factorization (Paatero and Tapper 1994), but the acronym NMF is now more widely used.[5]

13.2.2 Definition

Let us assume that our data consists of T of n-dimensional vectors, denoted by $\mathbf{x}(t)$ $(t = 1, \ldots, T)$. These are collected to a non-negative data matrix $\mathbf{X}$ which has $\mathbf{x}(t)$ as its columns. NMF finds an approximate factorization of $\mathbf{X}$ into non-negative factors $\mathbf{A}$ and $\mathbf{S}$. Thus, non-negative matrix factorization is a linear, non-negative approximate data representation, given by

$$\mathbf{x}(t) \approx \sum_{i=1}^{m} \mathbf{a}_i s_i(t) = \mathbf{A}\mathbf{s}(t) \quad \text{or} \quad \mathbf{X} \approx \mathbf{A}\mathbf{S}$$

where $\mathbf{A}$ is an $n \times m$ matrix containing the *basis vectors* $\mathbf{a}_i$ as its columns. This representation is, of course, similar in many respects to PCA and ICA. In particular, the dimension of the representation m can be smaller than the dimension of the data, in which the dimension is reduced as in PCA.

Whereas PCA and ICA do not in any way restrict the signs of the entries of $\mathbf{A}$ and $\mathbf{S}$, NMF requires all entries of both matrices to be non-negative. What this means is that the data is described by using additive components only. This constraint has been motivated in a couple of ways: First, in many applications one knows (for example by the rules of physics) that the quantities involved cannot be negative—firing rates are one example. In such cases, it can

[5]This section is based on the article (Hoyer 2004), originally published in Journal of Machine Learning Research. Copyright retained by the author.

be difficult to interpret the results of PCA and ICA (Paatero and Tapper 1994; Parra et al. 2000). Second, non-negativity has been argued for based on the intuition that parts are generally combined additively (and not subtracted) to *form a whole*; hence, these constraints might be useful for learning parts-based representations (Lee and Seung 1999).

Given a data matrix **X**, the optimal choice of matrices **A** and **S** are defined to be those non-negative matrices that minimize the reconstruction error between **X** and **AS**. Various error functions have been proposed (Paatero and Tapper 1994; Lee and Seung 2001), perhaps the most widely used is the squared error (Euclidean distance) function

$$D(\mathbf{A},\mathbf{S}) = \|\mathbf{X}-\mathbf{AS}\|^2 = \sum_{i,j}(x_{ij} - [\mathbf{AS}]_{ij})^2.$$

A gradient algorithm for this optimization was proposed by Paatero and Tapper (1994), whereas in Lee and Seung (2001) a multiplicative algorithm was devised that is somewhat simpler to implement and also showed good performance.

Although some theoretical work on the properties of the NMF representation exists (Donoho and Stodden 2004), much of the appeal of NMF comes from its empirical success in learning meaningful features from a diverse collection of real-life datasets. It was shown in Lee and Seung (1999) that, when the dataset consisted of a collection of face images, the representation consisted of basis vectors encoding for the mouth, nose, eyes, etc.; the intuitive features of face images. In Fig. 13.2a,we have reproduced that basic result using the same dataset. Additionally, they showed that meaningful topics can be learned when text documents are used as data. Subsequently, NMF has been successfully applied to a variety of datasets (Buchsbaum and Bloch 2002; Brunet et al. 2004; Jung and Kim 2004; Kim and Tidor 2003).

Despite this success, there also exist datasets for which NMF does not give an intuitive decomposition into parts that would correspond to our idea of the 'building blocks' of the data. It was shown by Li et al. (2001) that when NMF was applied to a different facial image database, the representation was global rather than local, qualitatively different from that reported by Lee and Seung (1999). Again, we have rerun that experiment and confirm those results; see Fig. 13.2b. The difference was mainly attributed to how well the images were hand-aligned (Li et al. 2001).

Another case where the decomposition found by NMF does not match the underlying elements of the data is shown in Fig. 13.2c. In this experiment, natural image patches were whitened, and subsequently split into positive ('ON') and negative ('OFF') contrast channels, simply by separating positive and negative values into separate channels (variables). This is somewhat similar to how visual information is processed by the retina. Each image patch of 12×12 pixels was thus represented by a $2 \times 12 \times 12 = 288$-dimensional vector, each element of which mimics the activity of an ON- or OFF-center neuron to the input patch. These vectors made up the columns of **X**. When NMF is applied to such a dataset, the resulting decomposition does not consist of the oriented filters which form the cornerstone of most of visual models and modern image processing. Rather, NMF represents these images using simple, dull, circular 'blobs'.

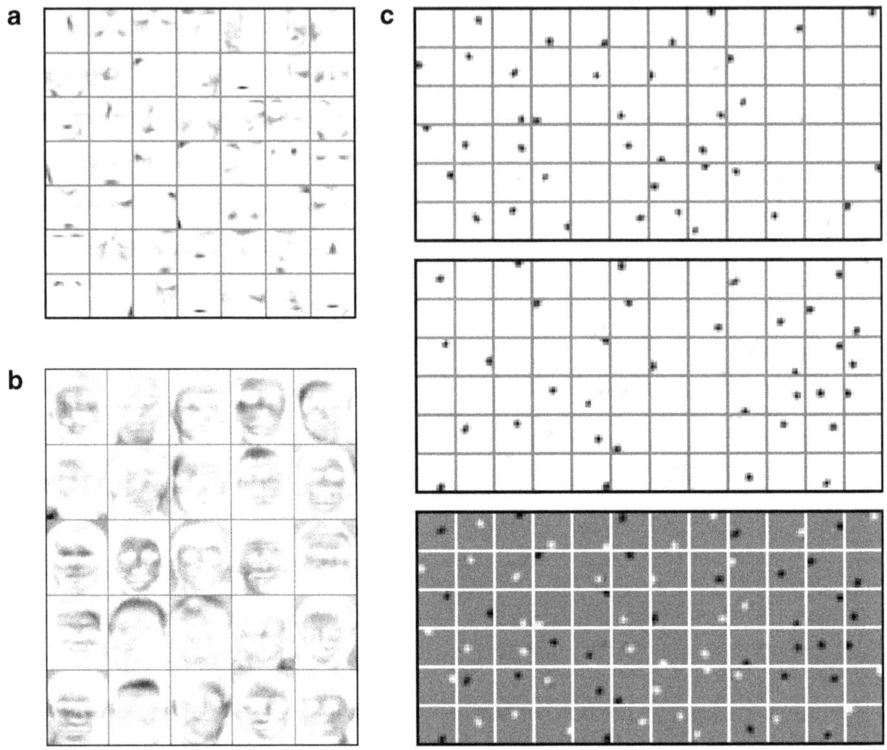

Fig. 13.2 NMF applied to various image dataset. **a** Basis images given by NMF applied to face image data from the CBCL database (http://cbcl.mit.edu/cbcl/software-datasets/FaceData2.html), following Lee and Seung (1999). In this case, NMF produces a parts-based representation of the data. **b** Basis images derived from the ORL face image database (http://www.uk.research.att.com/facedatabase.html), following Li et al. (2001). Here, the NMF representation is global rather than parts-based. **c** Basis vectors from NMF applied to ON/OFF-contrast filtered natural image data. *Top*: Weights for the ON-channel. Each patch represents the part of one basis vector $\mathbf{a}_i$ corresponding to the ON-channel. (*White pixels* denote zero weight, *darker pixels* are positive weights.) *Middle*: Corresponding weights for the OFF-channel. *Bottom*: Weights for ON minus weights for OFF. (Here, *grey pixels* denote zero.) NMF represents this natural image data using simple blobs

13.2.3 Adding Sparseness Constraints

Now we show, following Hoyer (2004), how explicitly controlling the sparseness of the representation leads to representations that are parts-based and match the intuitive features of the data. Here, we use a sparseness measure based on the relationship between the sum of absolute values and the sum of squares (Euclidean norm):

$$\text{sparseness}(\mathbf{s}) = \frac{\sqrt{m} - (\sum |s_i|)/\sqrt{\sum s_i^2}}{\sqrt{m} - 1},$$

where m is the dimensionality of $\mathbf{s}$. This function evaluates to unity if and only if $\mathbf{s}$ contains only a single non-zero component, and takes a value of zero if and only if all components are equal (up to signs), interpolating smoothly between the two extremes.

Our aim is to constrain NMF to find solutions with desired degrees of sparseness. The first question to answer is then: what exactly should be sparse? The basis vectors $\mathbf{A}$ or the coefficients $\mathbf{S}$? This is a question that cannot be given a general answer; it all depends on the specific application in question. Further, just transposing the data matrix switches the role of the two, so it is easy to see that the choice of which to constrain (or both, or none) must be made by the experimenter.

When trying to learn useful features from images, it might make sense to require both $\mathbf{A}$ and $\mathbf{S}$ to be sparse, signifying that any given object is *present* in few images and *affects* only a small part of the image. Or we could take the approach in Chap. 6 and only require the s_i to be sparse.

These considerations lead us to defining NMF with sparseness constraints as follows: Given a non-negative data matrix $\mathbf{X}$ of size $n \times T$, find the non-negative matrices $\mathbf{A}$ and $\mathbf{S}$ of sizes $n \times m$ and $m \times T$ (respectively) such that

$$D(\mathbf{A}, \mathbf{S}) = \|\mathbf{X} - \mathbf{A}\mathbf{S}\|^2 \tag{13.19}$$

is minimized, under *optional constraints*

$$\begin{aligned} \text{sparseness}(\mathbf{a}_i) &= S_a, \quad \forall i, \\ \text{sparseness}(\mathbf{s}_i) &= S_s, \quad \forall i, \end{aligned}$$

where $\mathbf{a}_i$ is the ith *column* of $\mathbf{A}$ and $\mathbf{s}_i$ is the ith *row* of $\mathbf{S}$. Here, m denotes the number of components, and S_a and S_s are the desired sparsenesses of $\mathbf{A}$ and $\mathbf{S}$ (respectively). These three parameters are set by the user.

Note that we did not constrain the scales of $\mathbf{a}_i$ or $\mathbf{s}_i$ yet. However, since $\mathbf{a}_i s_i = (\mathbf{a}_i \lambda)(s_i/\lambda)$ for any λ, we are free to arbitrarily fix any norm of either one. In our algorithm, we thus choose to fix the Euclidean norm (sum of squares) of $\mathbf{s}$ to unity, as a matter of convenience.

An algorithm for learning NMF with sparseness constraints is described in Hoyer (2004). In Fig. 13.2c, we showed that standard NMF applied to natural image data produces only circular features, not oriented features as have been observed in the cortex. Now, let us see the result of using additional sparseness constraints. Figure 13.3 shows the basis vectors obtained by putting a sparseness constraint on the coefficients ($S_s = 0.85$) but leaving the sparseness of the basis vectors unconstrained. In this case, NMF learns oriented Gabor-like features that represent edges and lines. This example illustrates how it is often useful to combine sparseness and non-negativity constraints to obtain a method which combines the biologically plausible results of low-level features with the purely additive learning of NMF. Such combinations may be useful in future models which attempt to go beyond the primary visual cortex because non-negativity may be an important property of complex cell outputs and other higher-order features, as was already pointed out in Chap. 12.

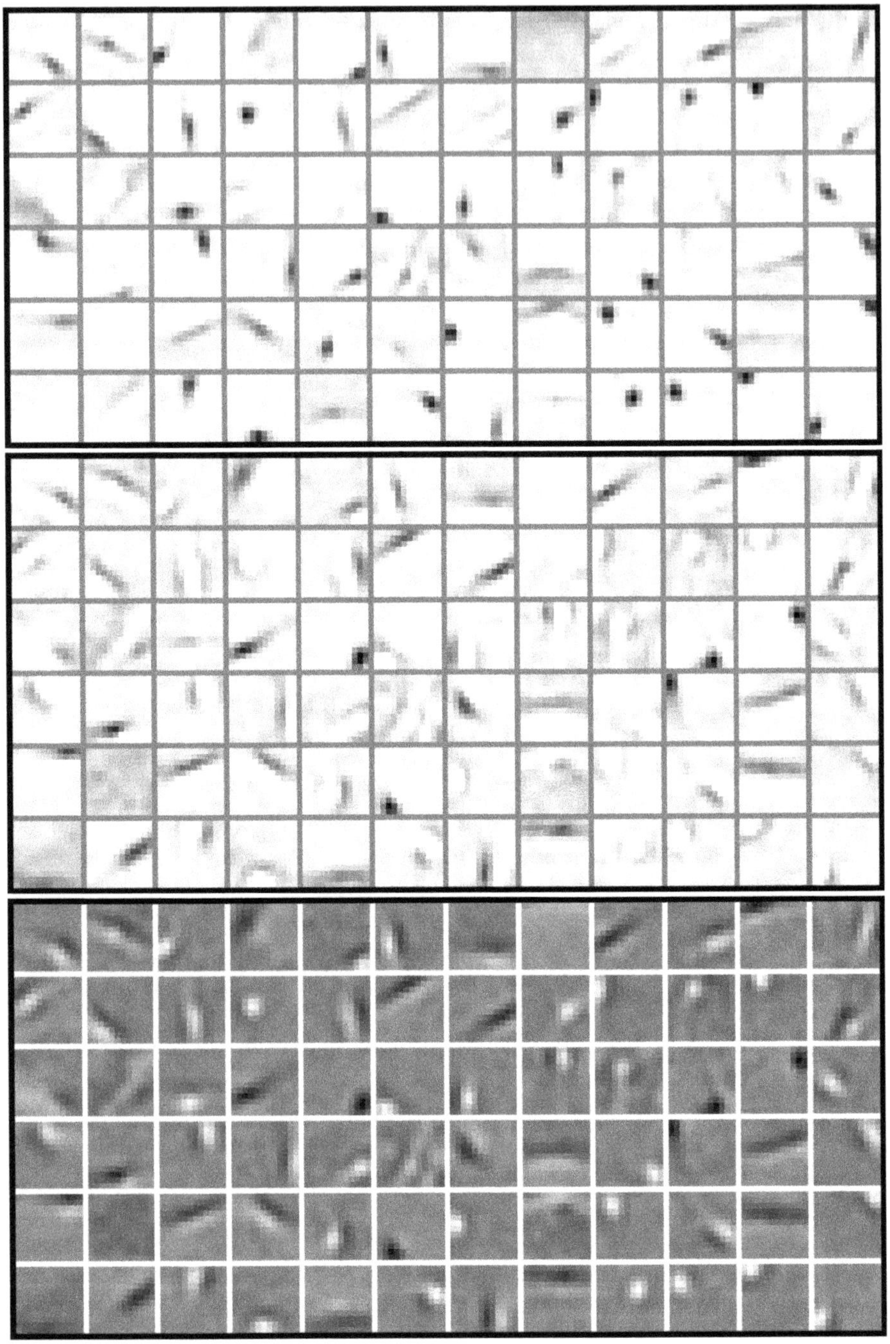

Fig. 13.3 Basis vectors from ON/OFF-filtered natural images obtained using NMF with sparseness constraints. The sparseness of the coefficients was fixed at 0.85, and the sparseness of the basis images was unconstrained. *Top*: weights in ON channel. *Middle*: weights in OFF channel. *Bottom*: weights in ON channel minus weights in OFF channel. As opposed to standard NMF (cf. Fig. 13.2c), the representation is based on oriented, Gabor-like, features

13.3 Conclusion

In this chapter, we saw two quite different extensions of the basic linear ICA model. The model with overcomplete basis is well motivated as a model of simple cells, and the next chapter will show some more implications of the principle.

In contrast, the utility of non-negative models for feature extraction is still to be explored. Possibly, non-negative models can be useful in learning higher-order features, which can be considered either to be "there" (positive values) or "not there" (zero value), negative values being less meaningful. On the other hand, negative values can often be interpreted as meaning that the feature is there "less strongly" or "less likely", possibly related to some baseline. In fact, after our initial work (Hoyer and Hyvärinen 2002) learning the third layer as in Chap. 12 using non-negativity constraints, we found out that the non-negativity constraints had little effect on the results, and the results in Chap. 12 do not use any such constraint.

Moreover, it is not clear if both the basis vectors and their coefficients should be constrained non-negative: A partly non-negative model in which either the basis vectors or the components are constrained non-negative may also be more meaningful. Non-negativity may, in the end, find its utility as one of the many properties of (some of the) parameters in a statistical model, instead of being very useful in itself.

Chapter 14
Lateral Interactions and Feedback

So far, we have almost exclusively considered a "bottom-up" or feedforward framework, in which the incoming image is processed in a number of successive stages, the information flowing in one direction only. However, it is widely appreciated in visual neuroscience that the brain is doing something much more complicated than just feedforward processing. There is a lot of evidence for

1. feedback from "higher" areas to lower areas, e.g., "top-down" connections from V2 back to V1, as well as
2. lateral (horizontal) interactions, by which we mean here connections between features in the same stage, e.g., connections between simple cells.

In this chapter, we will see how such phenomena are rather natural consequences of Bayesian inference in the models we have introduced. First, we will introduce a model of feedback based on thresholding, or shrinkage, of coefficients in the higher stage. Second, we will consider a lateral interaction phenomenon: end-stopping in simple cells. Finally, we will discuss the relationship of the principle of predictive coding to these phenomena.

14.1 Feedback as Bayesian Inference

A central question in visual neuroscience concerns the computational role of feedback connections. It has been suggested that the purpose of feedback is that of using information from higher-order units to modulate lower-level outputs, so as to selectively enhance responses which are consistent with the broader visual context (Lamme 1995; Hupé et al. 1998). In hierarchical generative models, this is naturally understood as part of the inference process: finding the most likely configuration of the network requires integrating incoming (bottom-up) sensory information with priors stored in higher areas (top-down) at each layer of the network (Hinton and Ghahramani 1997).

Why would this kind of feedback inference be useful? In many cases, there can be multiple conflicting interpretations of the stimulus even on the lowest level, and top-down feedback is needed to resolve such conflicts. In essence, feedback inference computes the most likely interpretation of the scene (Knill and Richards 1996; Lee and Mumford 2003; Yuille and Kersten 2006), combining bottom-up sensory information with top-down priors.

A. Hyvärinen, J. Hurri, P.O. Hoyer, *Natural Image Statistics*,
Computational Imaging and Vision 39,

14.1.1 Example: Contour Integrator Units

An example of Bayesian feedback inference can be constructed based on the model of higher-order units that integrate outputs of complex cells, introduced in Chap. 12. Basically, the idea is as follows: if enough collinear complex cells are active, they will activate a higher-order contour-coding unit. The activation of such a unit is then evidence for a contour at that location, and this evidence will strengthen responses of all complex cells lying on the contour, especially those whose bottom-up input is relatively weak.

The structure of the network was depicted in Fig. 12.1 in Chap. 12. In that chapter, we interpreted this network as performing feedforward computations only: first, the energy model for complex cells, and then a linear transformation. How can we then simulate the full network inference process to model feedback?

One approach is reduction of noise (Hupé et al. 1998). "Noise" in this context refers to any activity that is not consistent with the learned statistical model and is thus not only neural or photo-receptor noise. Such noise reduction essentially suppresses responses which are not typical of the training data, while retaining responses that do fit the learned statistical model. Denoting the complex cell responses by c_k, we model them by a linear generative model which includes a noise term:

$$c_k = \sum_{i=1}^{K} a_{ki} s_i + n_k \quad \text{for all } k \tag{14.1}$$

where n_k is Gaussian noise of zero mean and variance σ^2. The outputs of higher-order contour-coding units are still denoted by s_i.

We postulate that the outputs s_i of higher-order cells are computed by Bayesian inference in this generative model. Given an image, the complex-cell outputs are first computed in a feedforward manner; these initial values are denoted by c_k. (It is here assumed that the feature weights a_{ik} have already been learned.) Next, the outputs of higher-order cells are computed by finding the s_i which have the highest posterior probability—we use the Bayesian terminology "posterior probability (distribution)", which simply means the conditional probability given the observations. Let us denote the computed outputs as $\hat{\mathbf{s}}$:

$$\hat{\mathbf{s}} = \arg\max_{\mathbf{s}} \log p(\mathbf{s}|\mathbf{c}). \tag{14.2}$$

As is typical in Bayesian inference (see Sect. 4.7), we can formulate the posterior log-probability as the sum of two terms:

$$\log p(\mathbf{s}|\mathbf{c}) = \log p(\mathbf{c}|\mathbf{s}) + \log p(\mathbf{s}) - \text{const.} \tag{14.3}$$

where $p(\mathbf{s})$ is the *prior* pdf of $\mathbf{s}$. It incorporates our knowledge of the structure of the world, e.g. that the cell outputs are sparse. The term $\log p(\mathbf{c}|\mathbf{s})$ incorporates our knowledge of the image generation process; an example will be given below.

The important point here is that the outputs $\hat{s}_i$ of higher-order units are *non-linear* functions of the complex cell outputs. We will discuss below why this is so. This opens up the possibility of *reducing noise* in the complex cell outputs by reconstructing them using the linear generative model in (14.1), ignoring the noise. The obtained reconstructions, i.e. the outputs of the complex cells after they have received feedback, are denoted by $\hat{c}_k$, and computed as

$$\hat{c}_k = \sum_{i=1}^{K} a_{ki}\hat{s}_i \quad \text{for all } k. \tag{14.4}$$

Non-linearity is essential in these models. If the outputs $\hat{s}_i$ were simply linear transformations of the complex cell outputs, little would be gained by such feedback. This is because the reconstructed values $\hat{c}_k$ would still be linear transformations of the original feed-forward c_k. Thus, one could wonder why any feedback would really be needed to compute the $\hat{c}_k$ because a linear transformation could certainly be easily incorporated in the feedforward process which computes the c_k in the first place. However, the non-linear computations that emerge from the Bayesian inference process do need more complicated computing circuitry, so it is natural that feedback is needed.

The effect of this inference is that the top–down connections from the contour-coding units to the complex cells seek to adjust the complex cell responses toward that predicted by the contour units. To be more precise, such an effect can be obtained, for example, by sending a dynamic feedback signal of the form

$$u_{ki} = \left[\sum_{i=1}^{K} a_{ki}\hat{s}_i\right] - c_k \tag{14.5}$$

from the ith higher-order cell to the kth complex cell. When c_k is equal to its de-noised estimate, this signal is zero and equilibrium is achieved. Of course, this feedback signal is just one possibility and it is not known how this computation is actually achieved in the visual system. What is important here is that Bayesian inference gives an exact proposal on what the *purpose* of such feedback signals should be, thus providing a normative model.

In Fig. 14.1, we show a very basic example of how feedback noise reduction in this model results in the emphasis of smooth contours. We generated image patches by placing Gabor functions at random locations and orientations (for simplicity, we consider only a single frequency band here). In one case, there was a collinear alignment of three consecutive Gabors; in the other these same Gabors had random orientations. These image patches are shown in Fig. 14.1a. Next, we processed these by our model complex cells, as we had processed the natural image patches in our experiments in Chap. 12. The resulting c_k are shown in Fig. 14.1b. Finally, we calculated the contour-coding unit activities s_i (the actual method is discussed in the next subsection), and plotted the noise-reduced complex cell activity in Fig. 14.1c.

Note how the noise-reduction step suppresses responses to "spurious" edges, while emphasizing the responses that are part of the collinear arrangement. Such

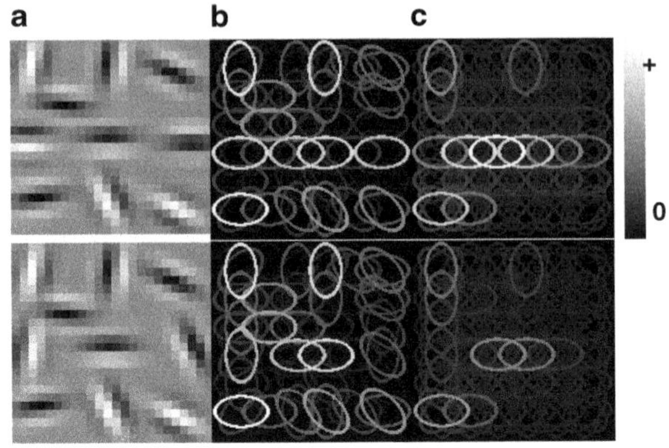

Fig. 14.1 Noise reduction and contour integration. **a** Two image patches containing Gabors at random locations and orientations. In the top patch there is a collinear set of three Gabors, whereas in the bottom patch these same Gabors had random orientations. **b** The response of the model complex cells to the images in **a**. **c** The response of the complex cells after feedback noise reduction using the learned network model. Note that the reduction of noise has left the activations of the collinear stimuli but suppressed activity that did not fit the learned sparse coding model well. From Hoyer and Hyvärinen (2002), Copyright ©2002 Elsevier, used with permission

response enhancement to contours is the defining characteristic of many proposed computational models of contour integration; see, for example Grossberg and Mingolla (1985), Li (1999), Neumann and Sepp (1999). Comparing the denoised responses (Fig. 12c) with each other one can also observe collinear contextual interactions in the model. The response to the central Gabor is stronger when it is flanked by collinear Gabors (*upper row*) than when the flankers have random orientations (*bottom row*), even though the flankers fall well outside the receptive field of the central neuron. This type of contextual interaction has been the subject of much study recently (Polat and Sagi 1993; Polat et al. 1998; Polat and Tyler 1999, Kapadia 1995, 2000, Kurki et al. 2006); see Fitzpatrick (2000) for a review. It is hypothesized to be related to contour integration, although such a relation is not certain (Williams and Hess 1998).

14.1.2 Thresholding (Shrinkage) of a Sparse Code

What is the non-linearity in the inference of the s_i like in (14.2)? Because the code is sparse, it turns out to be something like a thresholding of individual cell activities, as we will show next.

14.1.2.1 Decoupling of Estimates

Inference in an ICA model which contains Gaussian noise, as in (14.1), is a special case of the principle in Sect. 13.1.3, in which the coefficients in an overcomplete basis were estimated. We will see that the noise alone leads to non-linear computations even if the basis is not overcomplete as it was in Sect. 13.1.3. We can directly use the posterior pdf we calculated there, in (13.10) on page 281; instead of the original image I the observed data is the vector of complex cell outputs $\mathbf{c}$. Thus, we have

$$\log p(\mathbf{s}|\mathbf{c}) = -\frac{1}{2\sigma^2}\sum_k \left[c_k - \sum_{i=1}^m a_{ki}s_i \right]^2 + \sum_{i=1}^m G(s_i) + \text{const.} \tag{14.6}$$

where a_{ki} is the matrix of higher-order features weights, and the constant does not depend on $\mathbf{s}$. Now, let us assume that the number of complex cells equals the number of higher-order features. This is just the classic assumption that we usually make with ICA (with the exception of the overcomplete basis model in Sect. 13.1). Then the matrix $\mathbf{A}$, which has the a_{ki} as its entries, is invertible. Second, let us make the assumption that the matrix $\mathbf{A}$ is orthogonal. This assumption is a bit more intricate. It can be interpreted as saying that the noise is added on the whitened data because $\mathbf{A}$ is orthogonal after whitening. Since the noise is, in our case, an abstract kind of noise whose structure is not very well known in any case, this may not be an unreasonable assumption.

After these simplifying assumptions, the inference defined in (14.6) becomes quite simple. First, note that the sum of squares is, in matrix notation, equal to $\|\mathbf{c} - \mathbf{As}\|^2$. Because an orthogonal transformation does not change the norm, we can multiply the vector $\mathbf{c} - \mathbf{As}$ by $\mathbf{A}^T$ without changing the norm. Thus, we can replace the sum of squares in (14.6) by $\|\mathbf{A}^T\mathbf{c} - \mathbf{s}\|^2$, obtaining

$$\log p(\mathbf{s}|\mathbf{c}) = -\frac{1}{2\sigma^2}\sum_{i=1}^m \left[\sum_{k=1}^K a_{ki}c_k - s_i \right]^2 + \sum_{i=1}^m G(s_i) + \text{const.} \tag{14.7}$$

Now, we see the remarkable fact that this posterior log-pdf is a sum of functions of the form

$$\log p(s_i|\mathbf{c}) = -\frac{1}{2\sigma^2}\left[\sum_{k=1}^K a_{ki}c_k - s_i \right]^2 + G(s_i) + \text{const.} \tag{14.8}$$

which are functions of single s_i (higher-order features) only. Thus, we can maximize this posterior pdf separately for each s_i: we only need to do one-dimensional optimization. Each such one-dimensional maximum depends only on $\sum_{k=1}^K a_{ki}c_k$. This means that the estimates of the s_i which maximize this pdf are obtained by applying some one-dimensional non-linear function f on linear transformations of the complex cell outputs:

$$\hat{s}_i = f\left(\sum_{k=1}^K a_{ki}c_k \right) \tag{14.9}$$

where the non-linear function f depends on G, the log-pdf of the s_i.

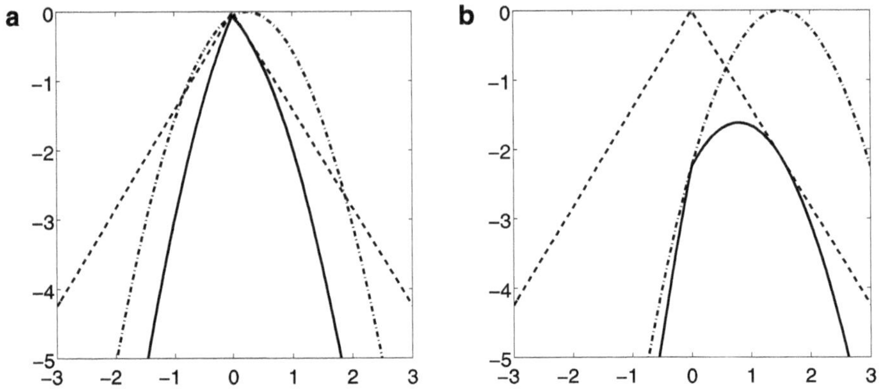

Fig. 14.2 Illustration of why noise reduction with a sparse prior for s_i leads to shrinkage. In both plots, the *dashed line* gives the Laplacian prior pdf $G(s) = -\sqrt{2}|s|$. The *dash-dotted line* gives the squared error term in (14.8). The *solid line* gives the sum of these two terms, i.e. the posterior log-probability $\log(s_i|\mathbf{c})$. The variance is fixed to $\sigma^2 = 0.5$. **a** The case where the feedforward signal is weak: $\sum_{k=1}^{K} a_{ki}c_k = 0.25$. We can see that the peak at zero of the Laplacian pdf dominates, and the maximum of the posterior is obtained at zero. This leads to a kind of thresholding. **b** The case where the feedforward signal is strong: $\sum_{k=1}^{K} a_{ki}c_k = 1.5$. Now, the sparse prior does not dominate anymore. The maximum of the posterior is obtained at a value which is clearly different from zero, but a bit smaller than the value given by the feedforward signal

14.1.2.2 Sparseness Leads to Shrinkage

What kind of nonlinearity f does noise reduction lead to? Intuitively, there are two forces at play in the posterior log-density $\log p(s_i|\mathbf{c})$. The first term, squared error, says that s_i should be close to $\sum_{k=1}^{K} a_{ki}c_k$, which can be thought of as the feed-forward linear estimate for s_i. The really interesting part is the prior given by the function G in (14.8). Now, for a sparse density, the log-density G is a peaked function. For example, it equals $G(s) = -\sqrt{2}|s|$ (plus some constant) for the Laplacian density which we have used previously (see (7.18)). This peakedness makes the inference non-linear so that if the linear estimate $\sum_{k=1}^{K} a_{ki}c_k$ is sufficiently close to zero, the maximum of $p(s_i|\mathbf{c})$ is obtained at zero. This is illustrated in Fig. 14.2.

Actually, the form of the function f can be obtained analytically for some choices of G. In particular, assume that G is the log-pdf of the Laplacian distribution. Then the function f becomes what is called a "shrinkage" function:

$$f(y) = \text{sign}(y)\max\big(|y| - \sqrt{2}\sigma^2, 0\big). \tag{14.10}$$

What this function means is that the linear transformation of complex cell outputs $\sum_{k=1}^{K} a_{ki}c_k$ is "shrunk" toward zero by an amount which depends on noise level σ^2. Such a function can be considered a "soft" form of thresholding. In fact, for some other choices of G, such as the one in (7.22), which correspond to much sparser pdf's, the non-linearity becomes very close to thresholding. See Fig. 14.3 for plots of such functions. For more details in shrinkage functions, see Hyvärinen (1999b),

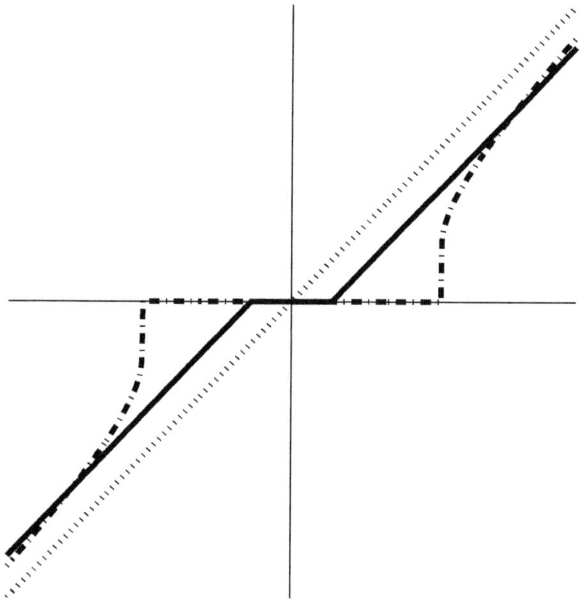

Fig. 14.3 Plots of the shrinkage functions f which modify the outputs of the higher-order contour cells. The effect of the functions is to reduce the absolute value of its argument by a certain amount which depends on the noise level. Small arguments are set to zero. This reduces Gaussian noise for sparse random variables. *Solid line*: shrinkage corresponding to Laplace density. *Dash-dotted line*: a thresholding function corresponding to the highly sparse density in (7.22). The line $x = y$ is given as the *dotted line*. The linear estimate to which this non-linearity is applied was normalized to unit variance, and noise variance was fixed to 0.3

Simoncelli and Adelson (1996), Johnstone and Silverman (2005) in a Bayesian context, and Donoho (1995) for a related method.

The non-linear behavior obtained by a sparse prior is in stark contrast to the case where the distribution of s_i is *Gaussian*: then G is quadratic, and so is $\log p(s_i|\mathbf{c})$. Minimization of a quadratic function leads to a linear function of the parameters. In fact, we can take the derivative of $\log p(s_i|\mathbf{c}) = -\frac{1}{2\sigma^2}[\sum_{k=1}^{K} a_{ki}c_k - s_i]^2 - s_i^2/2$ with respect to s_i and set it to zero, which gives as the solution $\hat{s}_i = \frac{1}{1+\sigma^2}\sum_{k=1}^{K} a_{ki}c_k$. This is a simple linear function of the feed-forward estimate. So, we see that it is sparseness, or non-Gaussianity, which leads to interesting non-linear phenomena.

Thus, we see that the function of Bayesian inference in this kind of a model is to *reduce small cell activities in the higher processing area to zero*. If there is not enough evidence that the feature encoded by a higher-order cell is there, the cell activity is considered pure noise, and set to zero. Feedback from higher areas modulates activity in lower areas by suppressing cells which are not consistent with the cell activity which is left after noise reduction on the higher level. In other words, activities of some cells in the lower-level area are suppressed because they are considered purely noise. At the same time, activities of some cells may even be enhanced so as to make them consistent with higher-level activities. Such a mechanism can work on many different levels of hierarchy. However, the mathematical difficulties constrain most analysis to a network where the feedback is only between two levels.

14.1.3 Categorization and Top-Down Feedback

The shrinkage feedback treated above is only one example of Bayesian inference in a noisy linear generative model. Different variants can be obtained depending on the assumptions of the marginal distributions of the latent variables, and their dependencies. Actually, even the generative model in (14.1) is applicable to any two groups of cells on two levels of hierarchy; the c_k need not be complex cells and s_i need not be contour-coding cells.

For example, consider the latent variables s_i as indicating *category membership*. Each of them is zero or one depending on whether the object in the visual field belongs to a certain category. For example, assume one of them, s_1, signals the category "face".

Thus, Bayesian inference based on (14.2) can again be used to infer the most probable values for the c_k variables for s_1. What is interesting here is that the binary nature of s_1 means that when the visual input is sufficiently close to the prototype of a face, the most likely value of s_1 will be exactly 1; at a certain threshold, it will jump from 0 to 1. This will affect the denoised estimates of the complex cell outputs c_k. The top-down feedback will now say that they should be similar to the first basis vector $(a_{11}, \ldots, a_{1n})$. Thus, a combination of excitatory and inhibitory feedback will be sent down to complex cells to drive the complex cell outputs in this direction.

For example, if the input is a face in which some of the contours have very low contrast, due to lighting conditions, this feedback will try to enhance them (Lee and Mumford 2003). Such feedback will be triggered if the evidence for s_1 being 1 is above the threshold needed. Otherwise, possibly the feedback from another category unit is activated.

14.2 Overcomplete Basis and End-stopping

A second kind of phenomenon which emerges from Bayesian inference and goes beyond a basic feedforward model is *competitive interactions*. This happens especially in the model with overcomplete basis; see Sect. 13.1, and can explain the phenomenon of end-stopping.

End-stopping refers to a phenomenon, already described by Hubel and Wiesel in the 1960s, in which the cell output is reduced when the optimal stimulus is made longer. That is, you first find a Gabor stimulus which gives maximum response in a simple cell. Then you simply make that Gabor longer, that is, more elongated, without changing anything else. You would expect that the response of the cell does not

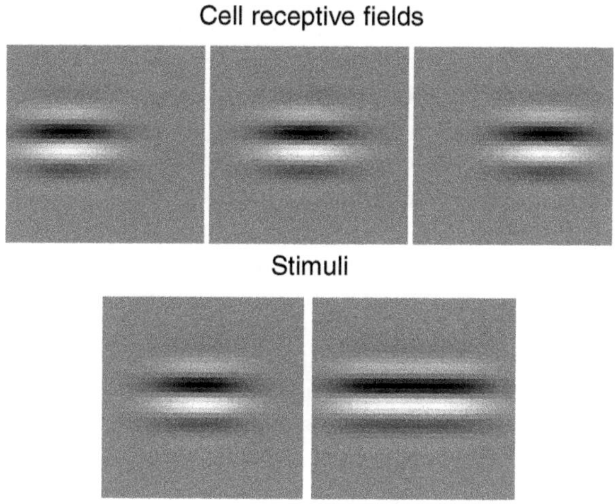

Fig. 14.4 The receptive fields and stimuli used in the end-stopping illustration. When the stimulus on the *left* is input to the system, the sparsest, i.e. the most probable pattern of coefficients is such that only the cell in the *middle* is activated $s_i > 0$. In contrast, when the stimulus is made longer, i.e. the stimulus on the *right* is input to the system, the inference leads to a representation in which only the cells on the *left* and the *right* are used $s_1, s_3 > 0$ whereas the cell in the *middle* has zero activity $s_2 = 0$

change because the new stuff that appears in the stimulus is outside of the receptive field. However, some cells (both simple and complex) actually reduce their firing rate when the stimulus is made more elongated. This is what is called end-stopping.

As discussed in Sect. 13.1, in an overcomplete basis there are often many different combinations of coefficients s_i which can give rise to the same image in a linear generative model $I(x, y) = \sum_i A_i(x, y)s_i$. Using Bayesian inference, the most likely coefficients s_i can be found, and this may provide a more accurate model for how simple cells in V1 compute their responses. This is related to end-stopping because such Bayesian inference in an overcomplete basis leads to dynamics which can be conceptualized as *competition*.

Here is an example of such competition. Consider only three Gabor-shaped basis vectors which are of the same shape but in slightly different locations, so that together they form an elongated Gabor. It is important that the Gabors are *overlapping*; this is necessary for the competition to arise. The three Gabors are depicted in Fig. 14.4.

First assume that the stimulus is a Gabor which is exactly the same as the feature coded by the cell in the middle. Then obviously, the sparsest possible representation of the stimulus is to set the coefficients of the left- and right-most features to zero ($s_1 = s_3 = 0$), and use only the feature in the middle. Next, assume that the stimulus is a more elongated Gabor, which is actually exactly the sum of the two Gabors on the left and the right sides. Now, the sparsest representation is such that the middle feature has zero activity ($s_2 = 0$), and the other two are used with equal coefficients.

Thus, the cell in the middle is first highly activated, but when the stimulus becomes more elongated, its activity is reduced, and eventually becomes zero. We can interpret this in terms of competition: The three cells are competing for the "right" to represent the stimulus, and with the first stimulus, the cell in the middle wins, whereas when the stimulus is elongated, the other two win. This competition provides a perfect example of end-stopping.

This kind of experiments also show that the classical concept of receptive field may need to be redefined, as already discussed in Sect. 13.1.4. After all, the concept of receptive field is based on the idea that the response of the cell only depends on the light pattern in a particular part of the retinal space. Now, end-stopping, and other phenomena such as contrast gain control, show that the cell response depends on stimulation outside of what is classically called the receptive field. Hence, the expression *classical* receptive field is used for the part which roughly corresponds to non-zero weights in $W(x, y)$, and the area from which signals of contrast gain control and end-stopping come is called the non-classical receptive field. See Angelucci et al. (2002) for an investigation of different kinds of receptive fields.

14.3 Predictive Coding

A closely related idea on the relation between feedback and feedforward processing is predictive coding. There are actually rather different ideas grouped under this title.

Firstly, one can consider prediction in time *or* in space, where "space" means different parts of a static image. Some of the earliest work in predictive coding considered prediction in both (Srivanivasan et al. 1982). Secondly, prediction can be performed between different processing stages (Mumford 1992; Rao and Ballard 1999) or inside a single stage (Srivanivasan et al. 1982; Hosoya et al. 2005). There is also a large body of engineering methods in which time signals, such as speech, are predicted in time in order to compress the signal (Spanias 1994); we shall not consider such methods here.

We constrain ourselves here to the case where *prediction happens between different levels of neural processing and for static stimuli*. The key idea here is that each neural level tries to predict the activity in the *lower* processing level. This is usually coupled with the idea that the lower level sends to the higher level the error in that prediction.

Prediction of the activities in lower levels is, in fact, implicit in the noisy generative model we have been using. As we saw in Sect. 13.1, estimation of the model in (14.1) can be accomplished by maximization of the objective function (the posterior probability) in (14.6) with respect to both a_{ki} and s_i. We can interpret $\sum_i a_{ki} s_i$ as the prediction that the higher level makes of lower-level activities. (In Sect. 14.1, we interpreted it as a denoised estimate which is closely related.) Then the first term in (14.6) can be interpreted as the prediction that the higher level makes of the lower level activities c_k. Thus, estimation of the model is, indeed, based on minimization of a prediction error as in predictive coding.

The idea that the lower level sends only the prediction error to the higher level needs some reinterpretation of the model. In Sect. 14.1, we showed how inference of s_i can, under some assumptions, be interpreted as shrinkage. Let us approach the maximization of the posterior probability in (14.8) by a basic gradient method. The partial derivative of the objective function with respect to s_i equals:

$$\frac{\partial \log p(\mathbf{s}|\mathbf{c})}{\partial s_i} = \frac{1}{\sigma^2} \sum_k a_{ki} \left[c_k - \sum_{i=1}^{m} a_{ki} s_i \right] + G'(s_i). \tag{14.11}$$

This derivative actually contains the prediction errors $c_k - \sum_{i=1}^{m} a_{ki} s_i$ of the lower-level activities c_k, and no other information on the c_k. Thus, if the higher level implements a gradient descent method to infer the most likely s_i, the information which it needs from the lower level can be conveyed by sending these prediction errors (multiplied by the weights a_{ki} which can be considered as feedforward connection weights).

The main difference between predictive coding and the generative modeling framework may thus be small from the viewpoint of statistical inference. The essential difference is in the interpretation of how the abstract quantities are computed and coded in the cortex. In the predictive modeling framework, it is assumed that the prediction errors $c_k - \sum_{i=1}^{m} a_{ki} s_i$ are actually the activities (firing rates) of the neurons on the lower level. This is a strong departure from the framework used in this book, where the c_k are considered as the activities of the neurons. Which one of these interpretations is closer to the neural reality is an open question which has inspired some experimental work; see Murray et al. (2002). Something like a synthesis of these views is to posit that there are two different kinds of neurons, each sending one of these signals (Roelfsema et al. 2002).

14.4 Conclusion

In this chapter, we have shown that in contrast to the impression one might get from preceding chapters, current models of natural images are not at all bound to a strict feed-forward thinking which neglects top-down influence. Quite on the contrary, Bayesian inference in these models leads to different kind of lateral interactions and feedback from higher cortical areas.

We have barely scratched the surface here. In many cases where connections between latent variables and images are not completely deterministic and one-to-one in both directions, such phenomena emerge. For example, the two-layer generative model in Sect. 11.8 would also give rise to such phenomena: If the latent variables are properly inferred from the input stimuli, some interesting dynamics might emerge.

Another very important case is contour integration by lateral (horizontal) connections between simple or complex cells. Basic dependencies between cells signaling contours which are typically part of a longer contour were pointed out in Krüger

(1998), Geisler et al. (2001), Sigman et al. (2001), Elder and Goldberg (2002). Probabilistic models incorporating horizontal connections can be found in Garrigues and Olshausen (2008), Osindero and Hinton (2008).

A very deep question related to feedback concerns the very definition of natural images. Any sufficiently sophisticated organism has an active mechanism, related to attention, which selects what kind of information it receives by its sensory organs. This introduces a complicated feedback loop between perception and action. It has been pointed out that the statistics in those image parts to which people attend, or direct their gaze, are different from the overall statistics (Reinagel and Zador 1999; Krieger et al. 2000); see Henderson (2003) for a review. The implications of this difference can be quite deep. A related line of work considers contours labeled by human subjects in natural images (Martin et al. 2004).

Part IV
Time, Color, and Stereo

Chapter 15
Color and Stereo Images

In this chapter, we show how we can model some other visual modalities, color, and stereopsis using ICA. We will see that ICA still finds features that are quite similar to those computed in the visual cortex.

15.1 Color Image Experiments

In this section, we extend the ICA image model from grey-scale (achromatic) to color (chromatic) images. Thus, for each pixel we have three values (red, green and blue), instead of one (grey-scale). The corresponding ICA model is illustrated in Fig. 15.1. First, we discuss the selection of data, then we analyze its second-order statistics and finally show the features found using ICA.

15.1.1 Choice of Data

Obviously, we should select as input data as "natural" images as possible if we wish to make any connection between our results and properties of neurons in the visual cortex. When analyzing colors, the spectral composition of the images becomes important in addition to the spatial structure.

It is clear that the color content of images varies widely with the environment in which the images are taken. Thus, we do not pretend to find some universally optimal features in which to code all natural color images. Rather, we seek the general qualitative properties of an ICA model of such images. In other words, we hope to find answers to questions such as: "How are colors coded in using such features; separate from, or mixed with achromatic channels?" and "What kind of spatial configuration do color-coding feature vectors have?"

We hope that as with grey-scale images, the ICA features are not too sensitive to the particular choice of color images, and that our data is realistic enough.

Neurons, of course, receive their information ultimately from the outputs of the photoreceptors in the retina. Color vision is made possible by the existence of photoreceptors called "cones" which come in three types, each sensitive to light of different wavelengths. Thus, our data should consist of the hypothetical outputs of the

---

A. Hyvärinen, J. Hurri, P.O. Hoyer, *Natural Image Statistics*, Computational Imaging and Vision 39,

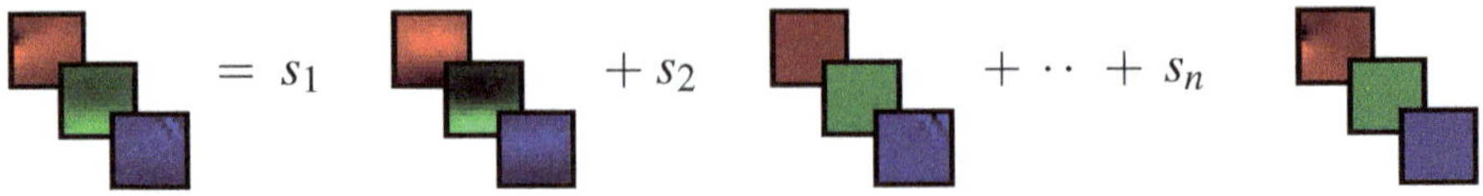

Fig. 15.1 The color image ICA model. As with grey-scale patches, we model the data as a linear combination of feature vectors A_i. Here, each feature vector consists of the three color planes (*red*, *green*, and *blue*), shown separately to clearly illustrate the linear model

three types of cones in response to our images. However, any three linear combinations of these outputs is just as good an input data, since we are applying ICA: Linearly transforming the data transforms the feature matrix $\mathbf{A}$, but does not alter the independent components.

We choose to use standard red/green/blue (RGB) values as inputs, assuming the transformation to cone outputs to be roughly linear. This has the advantage that the features found are directly comparable to features currently in use in image processing operations such as compression or denoising, and could straightforwardly be applied in such tasks. The drawback of using RGB values as inputs is of course that any non-linearities inherent in the conversion from RGB to cone responses will affect the ICA result and a comparison to properties of neurons may not be warranted. To test the effect of non-linearities, we have experimented with transforming the RGB values using the well-known gamma non-linearity[1] of cathode ray tubes used in computer screens. This did not qualitatively change the results and, therefore, we are confident that our results would be similar if we had used estimated cone outputs as inputs.

Our main data consists of color versions of natural scenes (depicting forest, wildlife, rocks, etc.) which we have used in previous work as well. The data is in the form of 20 RGB images (of size 384 × 256-pixels) in standard TIFF format.

15.1.2 Preprocessing and PCA

From the images, a total of 50 000 12-by-12 pixel image patches were sampled randomly. Since each channel yields 144 pixels, the dimensionality was now $3 \times 144 = 432$. Next, the mean value of each variable (pixel/color pair) was subtracted from that component, centering the dataset on the origin. Note that the DC component was not subtracted.

Then we calculated the covariance matrix and its eigenvectors, which gave us the principal components. These are shown in Fig. 15.2. The eigenvectors consist of global features, resembling 2D Fourier features. The variance decreases with increasing spatial frequency, and when going from grey-scale to blue/yellow to

[1]The gamma non-linearity is the most significant nonlinearity of the CRT monitor. After gamma-correction the transform from RGB to cone responses is roughly linear; see the Appendix in Wandell (1995).

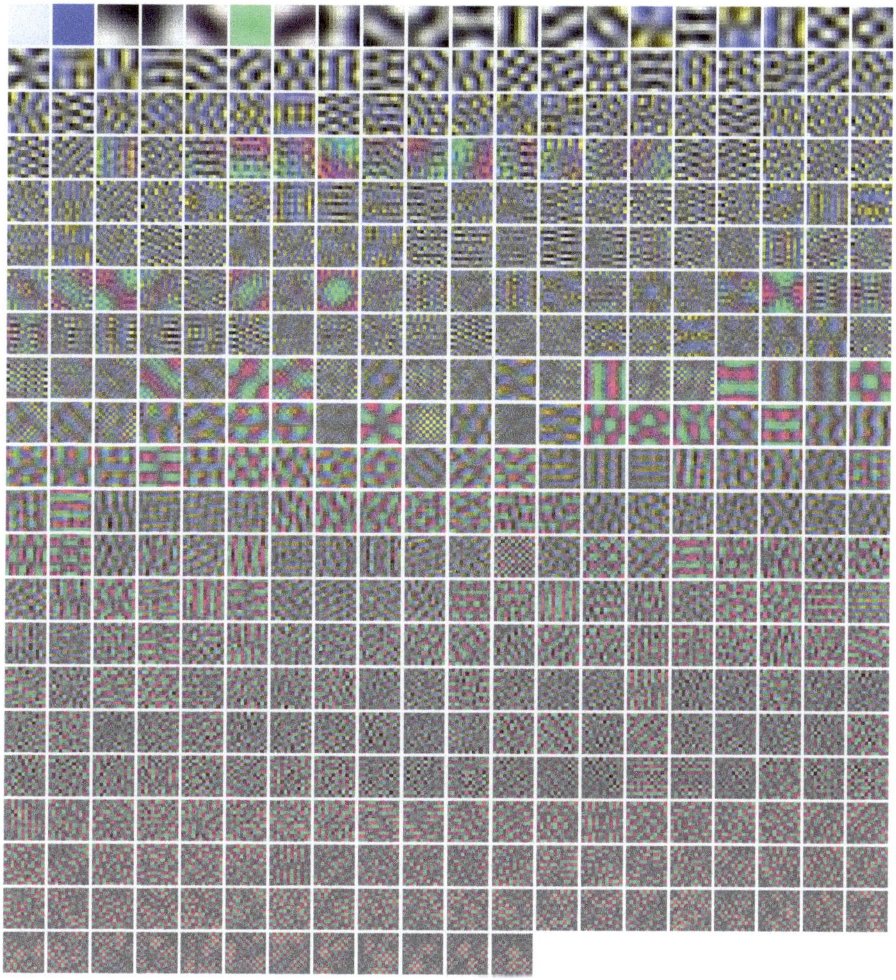

Fig. 15.2 PCA features of color images. These are the eigenvectors of the covariance matrix of the data, from left-to-right and top-to-bottom in order of decreasing corresponding eigenvalues. As explained in the main text, we projected the data on the first 160 principal components (top 8 rows) before performing ICA

red/green features.[2] These results were established by Ruderman et al. (1998) who used hyperspectral images (i.e. data with many more than the three spectral component in RGB data) as their original input data.

To analyze the color content of the PCA filters in more detail, we will show the pixels of a few filters plotted in a colored hexagon. In particular, each pixel (RGB-

[2]It should be noted that chromatic aberration in the eye might have an effect of additionally reducing signal energy at high spatial frequencies.

Fig. 15.3 The color hexagon used for analyzing the color content of the PCA and ICA features. The hexagon is the projection of the RGB cube onto a plane orthogonal to the luminance $(R+G+B)$ vector. Thus, achromatic RGB triplets map to the center of the hexagon while highly saturated ones are projected close to the edges

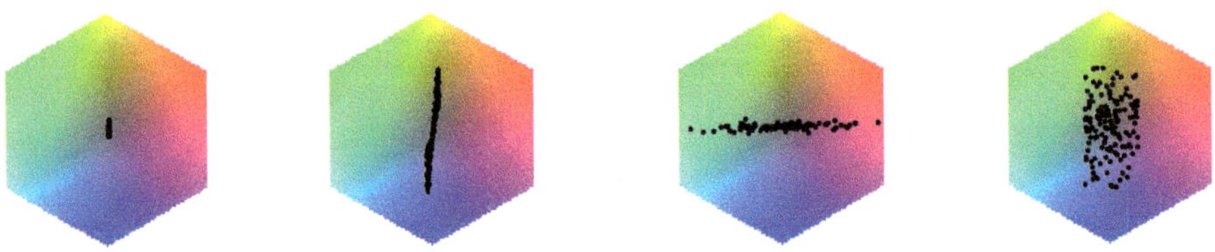

Fig. 15.4 Color content of four PCA filters. From left to right: Component nos. 3, 15, 432, and 67. All pixels of each filter have been projected onto the color hexagon shown in Fig. 15.3. See main text for a discussion of the results

triplet) is projected onto a plane given by

$$R + G + B = \text{constant}. \tag{15.1}$$

In other words, the luminance is ignored, and only the color content is used in the display. Figure 15.3 shows the colors in this hexagon. Note that this is a very simple 2D projection of the RGB color cube and should not directly be compared to any neural or psychophysical color representations.

Figure 15.4 shows a bright/dark filter (no. 3), a blue/yellow filter (no. 15), a red/green filter (no. 432, the last one), and a mixture (no. 67). Most filters are indeed exclusively opponent colors, as was found in Ruderman et al. (1998). However, there are also some mixtures of these in the transition zones of main opponent colors.

As described earlier, we project the data onto the n first principal components before whitening (we have experimented with $n = 100$, 160, 200, and 250). As can be seen from Fig. 15.2, dropping the dimension mostly discards blue/yellow features of high spatial frequency and red/green features of medium to high frequency. This already gives a hint as to why the blue/yellow and the red/green systems have a much lower resolution than the bright/dark system, as has been observed in psychophysical experiments (Mullen 1985).

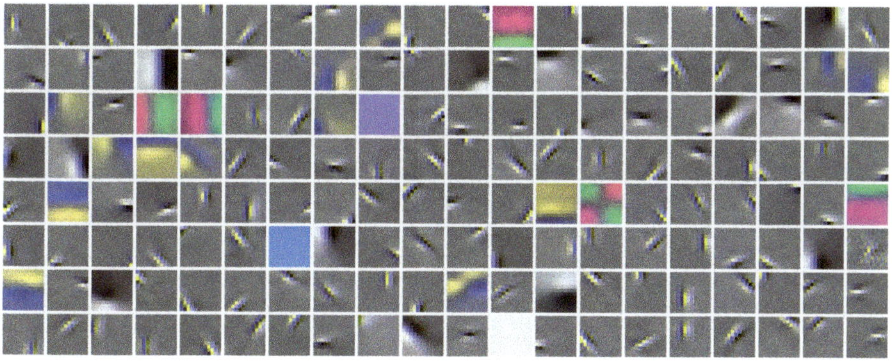

Fig. 15.5 ICA features of color images. Each patch corresponds to one feature A_i. Note that each feature is equally well represented by its negation, i.e. switching each pixel to its opponent color in any one patch is equivalent to changing the sign of $\mathbf{a}_i$ and does not change the ICA model (assuming components with symmetric distributions)

15.1.3 ICA Results and Discussion

The feature vectors A_i estimated by ICA are shown in Fig. 15.5. Examining Fig. 15.5 closely reveals that the features found are very similar to earlier results on grey-scale image data, i.e. the features resemble Gabor-functions. Note that most units are (mainly) achromatic, so they only represent brightness (luminance) variations. This is in agreement with the finding that a large part of the neurons in the primary visual cortex seem to respond equally well to different colored stimuli, i.e. are not selective to color (Hubel and Wiesel 1968; Livingstone and Hubel 1984). In addition, there is a small number of red/green and blue/yellow features. These are also oriented, but of much lower spatial frequency, similar to the grey-scale features of lowest frequency. One could think that the low frequency features together form a "colour" (including brightness) system, and the high-frequency grey-scale features a channel analyzing form. Also, note that the average color (DC-value) of the patches is represented by 3 separate feature vectors, just as the average brightness in an ICA decomposition of grey-scale images is usually separate from the other feature vectors.

We now show typical ICA features plotted in the color-hexagon (Fig. 15.6), as we did with the PCA features. The figure shows a bright/dark feature, a blue-yellow feature, and a red/green feature. There were no "mixtures" of the type seen for PCA; in other words each feature clearly belonged to one of these groups. (Note that the bright/dark features also contained blue/yellow to a quite small degree.)

The dominance of bright/dark features is largely due to the dimension reduction performed while whitening. To test the dependence of the group sizes on the value of n used, we estimated the ICA features for different values of n and counted the group sizes in each case. The results can be seen in Fig. 15.7. Clearly, when n is increased, the proportion of color-selective units increases. However, even in the case of keeping over half of the dimensions of the original space ($n = 250$), the bright/dark features still make up over 60% of all units.

Fig. 15.6 Color content of three ICA filters, projected onto the color hexagon of Fig. 15.3. From left to right: no. 24, 82, and 12

Fig. 15.7 Percentages of achromatic, blue/yellow, and red/green feature vectors for different numbers of retained PCA components (100, 160, 200, and 250). (In each case, the three features giving the mean color have been left out of this count)

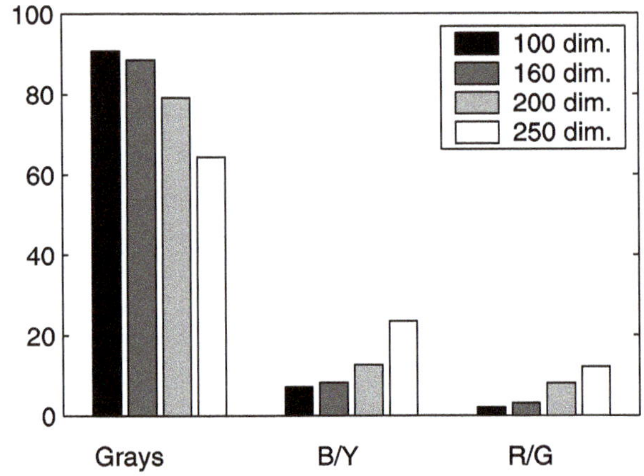

Another thing to note is that each ICA feature is "double-opponent": For blue-yellow features stimulating with a blue spot always gives an opposite sign in the response compared to stimulating with a yellow spot. Red/green and bright/dark features behave similarly. This is in fact a direct consequence of the linear ICA model. It would be impossible to have completely linear filters function in any other way.

Although early results (Livingstone and Hubel 1984) on the chromatic properties of neurons suggested that most color-sensitive cells were unoriented, and exhibited center-surround receptive fields, more recent studies have indicated that there are also oriented color-selective neurons (Ts'o and Gilbert 1988). The fact that our color features are mostly oriented is thus at least in partial agreement with neurophysiological data.

In any case, there is some agreement that most neurons are not selective to chromatic contrast, rather are more concerned about luminance (Hubel and Wiesel 1968; Livingstone and Hubel 1984; Ts'o and Roe 1995). Our basis is in agreement with these findings. In addition, the cytochrome oxidase blobs which have been linked to color processing (Livingstone and Hubel 1984) have also been associated with low spatial frequency tuning (Tootell et al. 1988; Shoham et al. 1997). In other words, color selective cells should be expected to be tuned to lower spatial frequencies. This is also seen in our features.

As stated earlier, we do not pretend that our main image set is representative of all natural environments. To check that the results obtained do not vary wildly with the image set used, we have performed the same experiments on another dataset: single-eye color versions of the 11 stereo images described in Sect. 15.2.1. The found ICA features (not shown) are in most aspects quite similar to that shown in Fig. 15.5: Features are divided into bright/dark, blue/yellow and red/green channels, of which the bright/dark group is the largest, containing Gabor-like filters of mostly higher frequency than the features coding colors. The main differences are that (a) there is a slightly higher proportion of color-coding units, and (b) the opponent colors they code are slightly shifted in color space from those found from our main data. In other words, the qualitative aspects, answering questions such as those proposed in Sect. 15.1.1, are quite similar. However, quantitative differences do exist.

15.2 Stereo Image Experiments

Another interesting extension of the basic grey-scale image ICA model can be made by modeling stereopsis, which means the extraction of depth information from binocular disparity. (Binocular disparity refers to the difference in image location of an object seen by the left and right eyes, resulting from the eyes' horizontal separation.) Now, our artificial neurons are attempting to learn the dependencies of corresponding patches from natural stereo images. The model is shown in Fig. 15.8.

15.2.1 Choice of Data

Again, the choice of data is an important step for us to get realistic results. Different approaches are possible here. In some early work, a binocular correlation function was estimated from actual stereo image data, and subsequently analyzed (Li and Atick 1994). In addition, at least one investigation of receptive field development used artificially generated disparity from monocular images (Shouval et al. 1996).

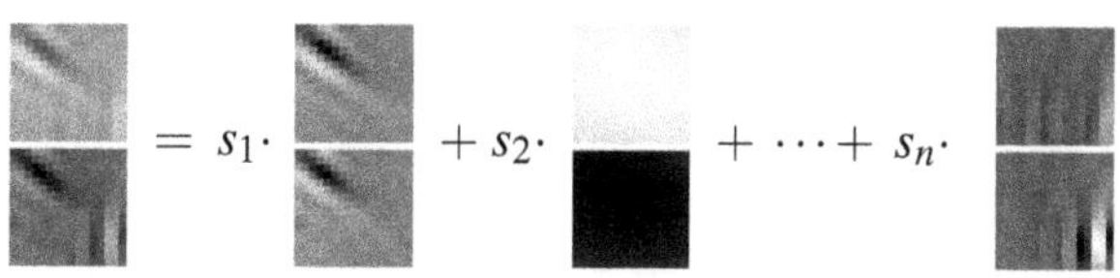

Fig. 15.8 The ICA model for corresponding stereo image patches. The *top row* contains the patches from left-eye image and the *bottom row* corresponding patches from the right-eye image. Just as for grey-scale and color patches, we model the data as a linear combination of feature vectors A_i with independent coefficients s_i

Fig. 15.9 One of the stereo images used in the experiments. The *left image* should be seen with the left eye, and the *right image* with the right eye (so-called uncrossed viewing)

Here, we have chosen to use 11 images from a commercial collection of stereo images of natural scenes; a typical image is given in Fig. 15.9.

To simulate the workings of the eyes, we selected 5 focus points at random from each image and estimated the disparities at these points. We then randomly sampled 16×16-pixel corresponding image patches in an area of 300×300 pixels centered on each focus point, obtaining a total of 50 000 samples. Because of the local fluctuations in disparity (due to the 3D imaging geometry) corresponding image patches often contained similar, but horizontally shifted features; this is of course the basis of stereopsis.

Note that in reality the "sampling" is quite different. Each neuron sees a certain area of the visual field which is relatively constant with respect to the focus point. Thus, a more realistic sampling would be to randomly select 50 000 focus points and from each take corresponding image patches at some given constant positional offset. However, the binocular matching is computationally slow and we thus opted for the easier approach, which should give the same distribution of disparities.

15.2.2 Preprocessing and PCA

The same kind of preprocessing was used in these experiments as for color, in Sect. 15.1. Since each sample consisted of corresponding left and right 16×16-patches our original data was 512-dimensional. First, the mean was removed from each variable, to center the data on the origin. Next, we calculated the covariance matrix of the data, and its eigenvalue decomposition. In order not to waste space,

Fig. 15.10 PCA features of stereo images, i.e. the eigenvectors of the covariance matrix of the data, from left-to-right and top-to-bottom in order of decreasing corresponding eigenvalues. See main text for discussion

we show here (in Fig. 15.10) the principal components for a window size of 8×8 pixels (the result for 16×16 is qualitatively very similar).

The most significant feature is that the principal components are roughly ordered according to spatial frequency, just as in PCA on standard (monocular) image patches. However, in addition early components (low spatial frequency) are more binocular than late ones (high frequency). Also note that binocular components generally consist of features of identical or opposite phases. This is in agreement with the binocular correlation function described in Li and Atick (1994).

As before, we select the first 160 principal components for further analysis by ICA. Again, this is plausible as a coding strategy for neurons, but is mainly done to lower the computational expenses and thus running time and memory consumption. Due to the structure of the covariance matrix, dropping the dimension to 160 is similar to low-pass filtering.

15.2.3 ICA Results and Discussion

Figure 15.11 shows the estimated ICA feature vectors A_i. Each pair of patches represents one feature. First, note that pairs have varying degrees of binocularity. Many of our "model neurons" respond equally well to stimulation from both eyes, but

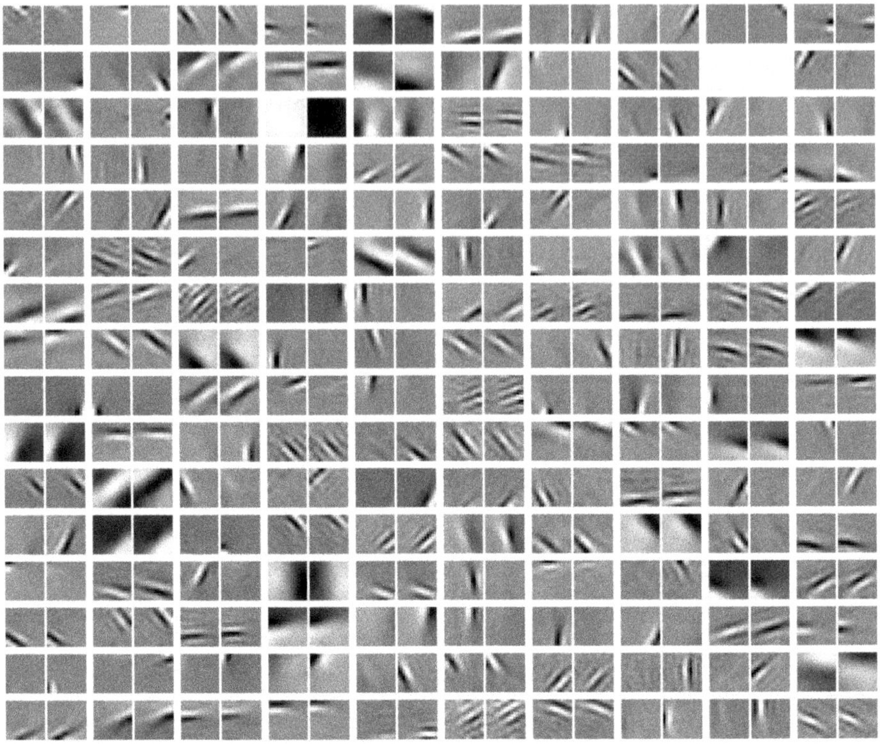

Fig. 15.11 ICA of stereo images. Each pair of patches represents one feature vector A_i. Note the similarity of these features to those obtained from standard image data (Fig. 7.3 on page 161). In addition, these exhibit various degrees of binocularity and varying relative positions and phases

there are also many which respond much better to stimulation of one eye than to stimulation of the other. This is shown quantitatively in Fig. 15.12, which gives an "ocular-dominance" histogram of the features. Ocular dominance thus means whether the neuron prefers input from one of the eyes (monocular) or combines information from both eyes (binocular).

The histogram depends strongly on the area of the sampling around the focus points (which in these experiments was 300×300 pixels). Sampling a smaller area implies that the correlation between the patches is higher and a larger number of features fall into the middle bin of the histogram. In theory, if we chose to sample only exactly at the fixation point, we would obtain (ignoring factors such as occlusion) identical left-right image patches; this would in turn make all feature vectors completely binocular with identical left-right patches, as there would be no signal variance in the other directions of the data space. On the other hand, sampling a larger area leads to a spreading of the histogram towards the edge bins. As the area gets larger, the dependencies between the left and right patches get weaker. In the limit of unrelated left and right windows, all features fall into bins 1 and 7 of the histogram. This was confirmed in experiments (results not shown).

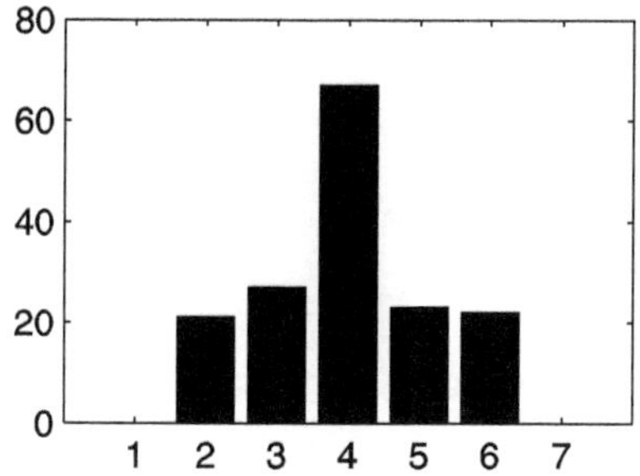

Fig. 15.12 Ocular dominance histogram of the ICA features. For each pair, we calculated the value of $(\|A_i^{\text{left}}\| - \|A_i^{\text{right}}\|)/(\|A_i^{\text{left}}\| + \|A_i^{\text{left}}\|)$, and used the bin boundaries $[-0.85, -0.5, -0.15, 0.15, 0.5, 0.85]$ as suggested in Shouval et al. (1996). Although many units where quite monocular (as can be seen from Fig. 15.11), no units fell into bins 1 or 7. This histogram is quite dependent on the sampling window around fixation points, as discussed in the main text

Taking a closer look at the binocular pairs reveals that for most pairs the left patch is similar to the right patch both in orientation and spatial frequency. The positions of the features inside the patches are close, when not identical. In some pairs, the phases are very similar, while in others they are quite different, even completely opposite. These properties make the features sensitive to different degrees of binocular disparity. Identical left-right receptive fields make the feature most responsive to zero disparity, while receptive fields that are identical except for a phase reversal show strong inhibition (a response smaller than the "base-line" response given by an optimal monocular stimulus) to zero disparity.

To analyze the disparity tuning we first estimated several ICA bases using different random number seeds. We then selected only relatively high frequency, well localized, binocular features which had a clear Gabor filter structure. This was necessary because filters of low spatial frequency were not usually well confined within the patch, and thus cannot be analyzed as complete neural receptive fields. The set of selected feature vectors is shown in Fig. 15.13.

For each stereo pair, we presented an identical stimulus at different disparities to both the left and right parts of the filter corresponding to the pair. For each disparity, the maximum over translations was taken as the response of the pair at that disparity. This gave a disparity tuning curve. For stimuli, we used the feature vectors themselves, first presenting the left patch of the pair to both "eyes", then the right. The tuning curves were usually remarkably similar, and we took the mean of these as the final curve.

We then classified each curve as belonging to one of the types "tuned excitatory", "tuned inhibitory", "near", or "far", which have been identified in physiological experiments (Poggio and Fischer 1977; Fischer and Kruger 1979; LeVay and Voigt 1988). Tuned excitatory units showed a strong peak at zero, usually with smaller inhibition at either side. Tuned inhibitory units on the other hand showed a marked inhibition (canceling) at zero disparity, with excitation at small positive or negative disparities. Features classified as "near" showed a clear positive peak at crossed

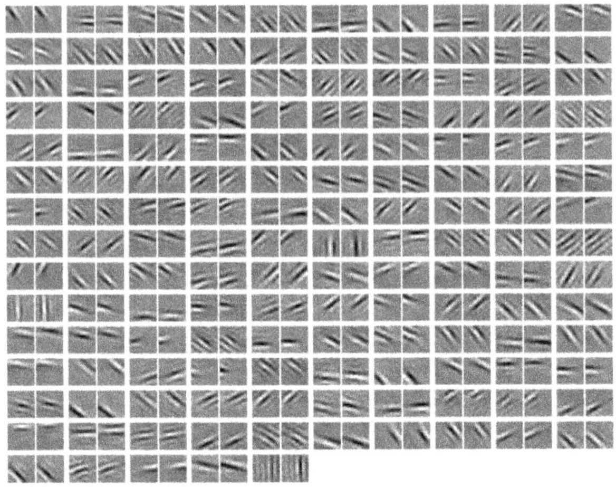

Fig. 15.13 Units selected for disparity tuning analysis. These were selected from bases such as the one in Fig. 15.11 on the basis of binocularity, frequency content, and localization (only well-localized Gabor filters were suitable for further analysis)

(positive) disparity while those grouped as "far" a peak for uncrossed (negative) disparity. Some tuning curves that did not clearly fit any of these classes were grouped into "others".

In Fig. 15.14, we give one example from each class. Shown are the feature vectors and the corresponding tuning curves. It is fairly easy to see how the organization of the patches gives the tuning curves. The tuned excitatory (top) unit has almost identical left-right profiles, and thus shows a strong preference for stimuli at zero disparities. The tuned inhibitory (second) unit has nearly opposite polarity patches which implies strong inhibition at zero disparity. The near (third) unit's right receptive field is slightly shifted (positional offset) to the left compared with the left field, giving it a positive preferred disparity. On the other hand, the far unit (bottom) has an opposite positional offset and thus responds best to negative disparities.

Figure 15.15 shows the relative number of units in the different classes. Note that the most common classes are "tuned excitatory" and "near". One would perhaps have expected a greater dominance of the tuned excitatory over the other groups. The relative number of tuned vs. untuned units probably depends to a great deal on the performance of the disparity estimation algorithm in the sampling procedure. We suspect that with a more sophisticated algorithm (we have used a very simple window-matching technique) one would get a larger number of tuned cells. The clear asymmetry between the "near" and "far" groups is probably due to the much larger range of possible disparities for near than for far stimuli: Disparities for objects closer than fixation can in principle grow arbitrarily large whereas disparities for far objects are limited (Barlow et al. 1967).

It is important to note that completely linear units (simple cells) cannot have very selective disparity tuning. Also, since the disparity tuning curves vary with the stimulus, the concept "disparity tuning curve" is not very well defined (Zhu and Qian 1996). However, disparity tuning is still measurable so long as one keeps in mind that the curve depends on the stimulus. Our tuning curves are "simulations" of experiments where a moving stimulus is swept across the receptive field at

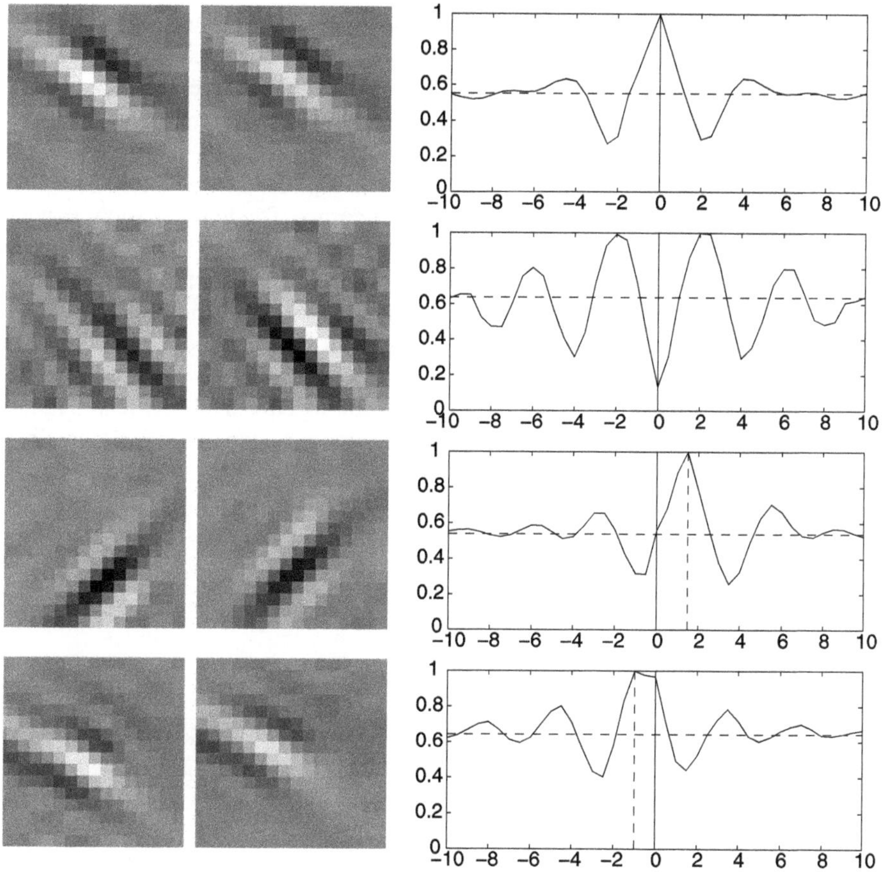

Fig. 15.14 Disparity tuning curves for units belonging to different classes. *Top row*: A "tuned excitatory" unit (no. 4 in Fig. 15.13). *Second row*: a "tuned inhibitory" unit (12). *Third row*: a "near" unit (38). *Bottom row*: a "far" unit (47). Crossed disparity ("near") is labeled positive and uncrossed ("far") negative in the figures. The *horizontal dotted line* gives the "base-line" response (the optimal response to one-eye only) and the *vertical dotted line* the position of maximum deviation from that response

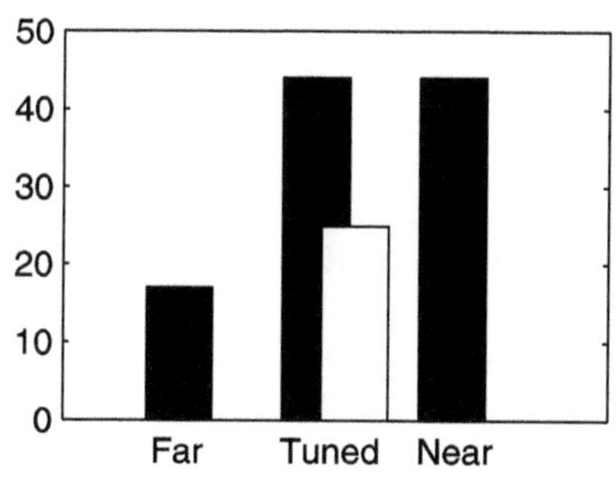

Fig. 15.15 Disparity tuning histogram. The histogram shows the relative amounts of "tuned excitatory" (44), "near" (44), "far" (17) units (in *black*) and "tuned inhibitory" units (25) in *white*. Not shown are those which did not clearly fit into any of these categories (15)

different binocular disparities, and the responses of the neuron in question is measured. As such, it is appropriate to use the estimated feature vectors as input. To obtain stimulus-invariant disparity tuning curves (as well as more complex binocular interactions than those seen here) one would need to model non-linear (complex) cells.

Overall, the properties of the found features correspond quite well to those of receptive fields measured for neurons in the visual cortex. The features show varying degrees of ocular dominance, just as neuronal receptive fields (Hubel and Wiesel 1962). Binocular units have interocularly matched orientations and spatial frequencies, as has been observed for real binocular neurons (Skottun and Freeman 1984). It is easy by visual inspection to see that there exist both interocular position and phase differences, which seems to be the case for receptive fields of cortical neurons (Anzai et al. 1999a). Finally, simulated disparity tuning curves of the found features are also similar to tuning curves measured in physiological experiments (Poggio and Fischer 1977).

15.3 Further References

15.3.1 Color and Stereo Images

Work concerning the second-order statistics of color include Atick et al. (1992), van Hateren (1993), Ruderman et al. (1998). In addition, colored input was used in Barrow et al. (1996) to emerge a topographic map of receptive fields. Again, that work basically concerns only the second-order structure of the data, as the correlation-based learning used relies only on this information. Application of ICA on color images has been reported in Hoyer and Hyvärinen (2000), Wachtler et al. (2001), Doi et al. (2003), Caywood et al. (2004), Wachtler et al. (2007). Related work on LGN neurons can be found in Mante et al. (2005).

In addition to learning chromatic receptive fields, it is also possible to investigate the statistical properties of the chromatic spectra if single pixels (Wachtler et al. 2001). That is, one measures the spectral content of single pixels with a high resolution which gives more than the conventional three dimensions. This can shed light on the optimality of the three-cone dimensionality reduction used in the retina.

Emerging receptive fields from stereo input has been considered in Li and Atick (1994), Shouval et al. (1996), Erwin and Miller (1996, 1998). As with color, most studies have explicitly or implicitly used only second-order statistics (Li and Atick 1994; Erwin 1996, 1998). The exception is Shouval et al. (1996) which used the BCM learning rule (Bienenstock et al. 1982) which is a type of projection pursuit learning closely linked to ICA. The main difference between their work and the one reported in this chapter is that here we use data from actual stereo images whereas they used horizontally shifted (misaligned) data from regular images. In addition, we estimate a complete basis for the data, whereas they studied only single receptive fields.

15.3.2 Other Modalities, Including Audition

Further investigations into the statistical structure of other sensory modalities have been made especially in the context of audition, in which ICA yields interesting receptive fields whether applied on raw audio data (Bell and Sejnowski 1996; Lewicki 2002; Cavaco and Lewicki 2007) or spectrograms (Klein et al. 2003). See also Schwartz et al. (2003) for work related to music perception, and Schwartz and Simoncelli (2001b) for work on divisive normalization for auditory signals.

Further topics which have been addressed using the statistical structure of the ecologically valid environment include visual space (Yang and Purves 2003), somatosensory system (Hafner et al. 2003), and place cells (Lörincz and Buzsáki 2000). Motion in image sequences is considered in Sect. 16.2.

For some work on multimodal integration and natural image statistics, see Hurri (2006), Krüger and Wörgötter (2002) (the latter is on image sequences). An image-processing application combining spatial, temporal, and chromatic information is in Bergner and Drew (2005).

15.4 Conclusion

In this chapter, we have investigated the use of independent component analysis for decomposing natural color and stereo images. ICA applied to color images yields features which resemble Gabor functions, with most features achromatic, and the rest red/green- or blue/yellow-opponent. When ICA is applied on stereo images we obtain feature pairs which exhibit various degrees of ocular dominance and are tuned to various disparities. Thus, ICA seems to be a plausible model also for these modalities and not just grey-scale images.

Chapter 16
Temporal Sequences of Natural Images

Up to this point, this book has been concerned with static natural images. However, in natural environments the scene changes over time. In addition, the observer may move, or the observer may move its eyes. Temporal sequences of natural images, temporal properties of the visual system, and temporal models of processing are the topics of this chapter.

16.1 Natural Image Sequences and Spatiotemporal Filtering

In digital systems, dynamical (time-varying) images are often processed as *image sequences*, which consist of *frames*, each frame being one static image. Figure 16.1 shows an example of an image sequence with lateral camera movement.

Previous chapters have made clear the importance of linear operators as tools and models in image processing. In the case of image sequence data, the fundamental linear operation is *spatiotemporal linear filtering*, which is a straightforward extension of the spatial linear filtering discussed in Chap. 2. Remember that in spatial linear filtering a two-dimensional filter is slid across the image, and the output is formed by computing a weighted sum of the pixels in the area of the filter, with the weights given by the elements of the filter. In spatiotemporal linear filtering, a *three-dimensional* filter is slid across the image sequence, and the output is formed by computing a weighted sum of the pixels in the spatiotemporal area of the filter, with the weights given by the elements of the filter.

Mathematically, let $W(x, y, t)$ denote the filter weights, $I(x, y, t)$ denote the input image sequence, and $O(x, y, t)$ denote the output image sequence. The index t is time. Then linear spatiotemporal filtering is given by

$$O(x, y, t) = \sum_{x_*=-\infty}^{\infty} \sum_{y_*=-\infty}^{\infty} \sum_{t_*=-\infty}^{\infty} W(x_*, y_*, t_*) I(x + x_*, y + y_*, t + t_*), \quad (16.1)$$

where the upper and lower limits of the sums are in practical situations finite. Typically only filters which do not use future time points are used; mathematically, we will denote this *causality* restriction by $W(x, y, t) = 0$ when $t > 0$.

The concepts of frequency-based representations, presented in Sect. 2.2 (page 29) are applicable also in the three-dimensional, spatiotemporal case. An image sequence can be represented as a sum of spatiotemporal sinusoidal components

$$I(x, y, t) = \sum_{\omega_x} \sum_{\omega_y} \sum_{\omega_t} A_{\omega_x,\omega_y,\omega_t} \cos(\omega_x x + \omega_y y + \omega_t t + \psi_{\omega_x,\omega_y,\omega_t}), \quad (16.2)$$

A. Hyvärinen, J. Hurri, P.O. Hoyer, *Natural Image Statistics*,
Computational Imaging and Vision 39,

Fig. 16.1 An example of an image sequence (van Hateren and Ruderman 1998) with 5 frames. Here, time proceeds from left to right

where ω_x and ω_y are spatial frequencies and ω_t is a temporal frequency, $A_{\omega_x,\omega_y,\omega_t}$ is the amplitude associated with the frequency triple, and $\psi_{\omega_x,\omega_y,\omega_t}$ is the phase of the frequency triple. You may want to compare (16.2) with its spatial counterpart, (2.9) on page 35. A spatiotemporal convolution operation is defined by

$$H(x,y,t) * I(x,y,t) = \sum_{x_*=-\infty}^{\infty} \sum_{y_*=-\infty}^{\infty} \sum_{t_*=-\infty}^{\infty} I(x-x_*, y-y_*, t-t*)H(x_*, y_*, t_*), \tag{16.3}$$

where $H(x,y,t)$ is the impulse response, which has a straightforward relationship with the linear filter $W(x,y,t)$

$$H(x,y,t) = W(-x,-y,-t). \tag{16.4}$$

This impulse response has a complex-valued three-dimensional discrete Fourier transform $\tilde{H}(u,v,w)$, the magnitude of which reveals the amplitude response of the filter, and the angle reveals the phase response.

16.2 Temporal and Spatiotemporal Receptive Fields

With the inclusion of time, we get two new kinds of receptive fields: *spatiotemporal* and *temporal*; these are illustrated in Fig. 16.2 in the case of a neuron from the lateral geniculate nucleus. A spatiotemporal receptive field $W(x,y,t)$ (Fig. 16.2a) corresponds to a causal spatiotemporal filter: it defines a linear model that relates the history of all pixels in the image sequence to the output of a neuron. These inherently three-dimensional spatiotemporal receptive fields are often visualized in two dimensions with one spatial dimension and a temporal dimension $W(x,t)$ by taking either a slice at where y is constant ($y = y_{\text{const}}$) or summing over the y-dimension (Fig. 16.2b).[1] A temporal receptive field (Fig. 16.2c) is the time course of a single spatial location in a spatiotemporal receptive field: $W(t) = W(x_{\text{const}}, y_{\text{const}}, t)$. It defines a linear model that relates the history of a single pixel to the output of a neuron.

Spatiotemporal receptive fields are divided into two qualitatively different types based on whether or not they can be described as a cascade of a spatial and a tempo-

[1] When the RF is selective to a certain spatial orientation of the stimulus, this visualization can be improved by rotating the RF spatially so that the preferred orientation becomes the y-axis.

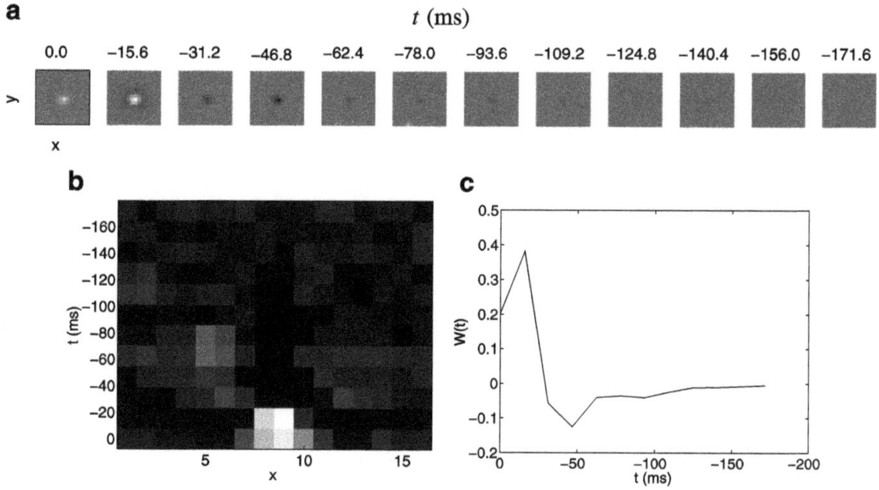

Fig. 16.2 Spatiotemporal and temporal receptive fields of a neuron in the lateral geniculate nucleus (LGN), estimated from measurement data from the neuron. **a** A spatiotemporal receptive field $W(x, y, t)$, the equivalent of a causal linear spatiotemporal filter. **b** A two-dimensional visualization of the RF in **a**, obtained by summing the spatiotemporal RF along the y-axis: $W(x,t) = \sum_y W(x, y, t)$. **c** A temporal receptive field, which is a single time-slice of the spatiotemporal RF: $W(t) = W(x_{\mathrm{const}}, y_{\mathrm{const}}, t)$. For the description of the original measurement data and its source see Dayan and Abbott (2001), Kara et al. (2000)

ral filtering operation. In the case where this is possible, the spatiotemporal filter is called is called *space-time separable*. Let us denote again the output of the filtering by $O(x, y, t)$, the spatial filter by $W_{\mathrm{spat}}(x, y)$ and the temporal filter by $W_{\mathrm{temp}}(t)$. Then the cascade can be combined into a single spatiotemporal filter as follows:

$$\begin{aligned}
O(x, y, t) &= \sum_{x_*=-\infty}^{\infty} \sum_{y_*=-\infty}^{\infty} W_{\mathrm{spat}}(x_*, y_*) \sum_{t_*=-\infty}^{\infty} W_{\mathrm{temp}}(t_*) I(x + x_*, y + y_*, t + t_*) \\
&= \sum_{x_*=-\infty}^{\infty} \sum_{y_*=-\infty}^{\infty} \sum_{t_*=-\infty}^{\infty} \underbrace{W_{\mathrm{spat}}(x_*, y_*) W_{\mathrm{temp}}(t_*)}_{=W(x_*, y_*, t_*)} I(x + x_*, y + y_*, t + t_*) \\
&= \sum_{x_*=-\infty}^{\infty} \sum_{y_*=-\infty}^{\infty} \sum_{t_*=-\infty}^{\infty} W(x_*, y_*, t_*) I(x + x_*, y + y_*, t + t_*). \qquad (16.5)
\end{aligned}$$

Thus, the spatiotemporal filter is obtained as a product of the spatial and temporal parts as

$$W(x, y, t) = W_{\mathrm{spat}}(x, y) W_{\mathrm{temp}}(t). \qquad (16.6)$$

By changing the ordering of the sums in (16.5), it is easy to see that in the space-time separable case, the order in which the spatial and the temporal filtering are done is irrelevant. A spatiotemporal receptive field that is not space-time separable is called *space-time inseparable*.

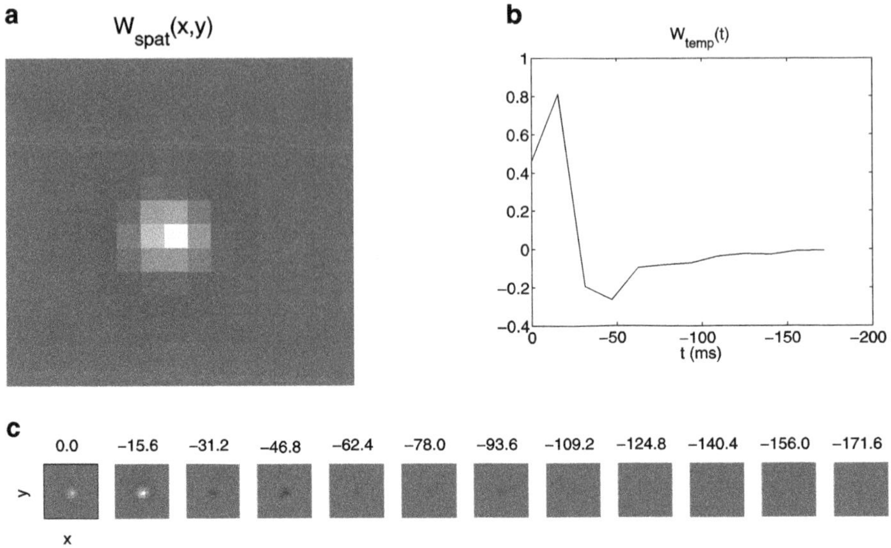

Fig. 16.3 A space-time-separable representation of the spatiotemporal RF of Fig. 16.2. **a**, **b** The optimal spatial RF $W_{\text{spat}}(x, y)$ and temporal RF $W_{\text{temp}}(t)$, estimated using the separability condition $W(x, y, t) = W_{\text{spat}}(x, y)W_{\text{temp}}(t)$. **c** The resulting space-time-separable RF $W_{\text{spat}}(x, y)W_{\text{temp}}(t)$; comparison of this with Fig. 16.2a demonstrates the good match provided by the separable model for this neuron

The spatiotemporal receptive field shown in Fig. 16.2 is approximately space-time separable: Fig. 16.3 shows the decomposition of the receptive field into the spatial part and the temporal part, and the resulting space-time-separable approximation.[2] This suggests that the linear model of the neuron can be divided into a spatial filter and a temporal filter. Intuitively speaking, space-time separability means that the RF does not contain anything that "moves" from one place to another, because the spatial profile is all the time in the same place: only its magnitude (and possibly sign) changes.

16.3 Second-Order Statistics

16.3.1 Average Spatiotemporal Power Spectrum

Now, we begin the investigation of the statistical structure of natural image sequences by characterizing the spatiotemporal correlations between two pixels in

[2]The decomposition has been obtained by minimizing the squared Euclidean distance between the original RF and its space-time-separable version. This can be solved by employing the singular-value decomposition approximation of matrices.

an image sequence. As was discussed in Sect. 5.6 on page 111, a characterization of the average power spectrum is equivalent to an examination of these second-order statistics. Therefore, following Dong and Atick (1995a), we proceed to analyze the average spatiotemporal power spectrum of natural image sequences.

The natural image sequences used as data were a subset of those used in van Hateren and Ruderman (1998). The original data set consisted of 216 monochrome, non-calibrated video clips of 192 seconds each, taken from television broadcasts. More than half of the videos feature wildlife, the rest show various topics such as sports and movies. Sampling frequency in the data is 25 frames per second, and each frame had been block-averaged to a resolution of 128×128 pixels. For our experiments, this dataset was pruned to remove the effect of human-made objects and artifacts. First, many of the videos feature human-made objects, such as houses, furniture, etc. Such videos were removed from the data set, leaving us with 129 videos. Some of these 129 videos had been grabbed from television broadcasts so that there was a wide black bar with height 15 pixels at the top of each image, probably because the original broadcast had been in wide screen format. Our sampling procedure never took samples from this topmost part of the videos.

The results of this section are based on the following procedure. We first took 10 000 samples of size $64 \times 64 \times 64 = \Delta x \times \Delta y \times \Delta t$ from the natural image sequence. We then computed the spatiotemporal power spectrum of each of these samples by computing the squared amplitudes of three-dimensional discrete Fourier transform of the sample. These power spectra were averaged over all of the samples to obtain the average power spectrum $R(\omega_x, \omega_y, \omega_t)$. Image data is often assumed to be approximately rotationally invariant (isotropic, see Sect. 5.7), so a two-dimensional average power spectrum was computed as a function of spatial frequency $\omega_s = \sqrt{\omega_x^2 + \omega_y^2}$ by averaging over all spatial orientations, yielding $R(\omega_s, \omega_t)$.

One way to visualize the resulting two-dimensional function $R(\omega_s, \omega_t)$ is to plot curves of the function while keeping one of the variables fixed. This has been done in Figs. 16.4a and b, keeping ω_t constant in the former and ω_s in the latter. In order to analyze the form of this power spectrum in more detail, one can first fit a space-time separable power spectrum $R_s(\omega_s)R_t(\omega_t)$; the best fit (in terms of least mean square) is visualized in Fig. 16.5 in similar plots as those in Fig. 16.4, but this time plotting curves from both the observed and the best space-time separable power spectrum. As can be seen, at higher frequencies the best separable power spectrum provides a relatively poor match to the observed one.

In order to proceed to a more accurate model of the spatiotemporal power spectrum of natural image sequences, let us reconsider the frequency representation of the s–t-space. Referring to our presentation of the two-dimensional frequency-based representation in Sect. 2.2.2—in particular, see Fig. 2.5 on page 32—let $\boldsymbol{\omega} = [\omega_s\ \omega_t]^T$. The vector $\boldsymbol{\omega}$ has two important properties: direction and magnitude (length). Now consider the direction of the vector. In the case of a two-dimensional spatial frequency-based representation, the direction of the vector $[\omega_x\ \omega_y]^T$ determines the spatial orientation of the sinusoidal in the x–y-space. Analogously, in the

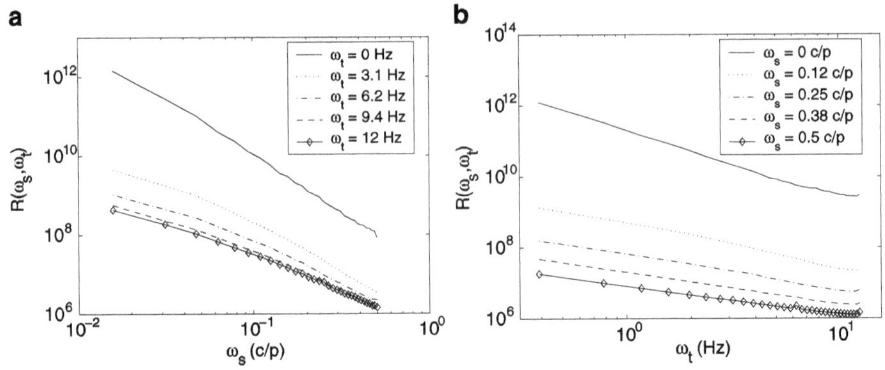

Fig. 16.4 One-dimensional slices of the two-dimensional average spatiotemporal power spectrum $R(\omega_s, \omega_t)$ of natural image sequences. **a** Plots in which ω_t is held constant. **b** Plots in which ω_s is held constant

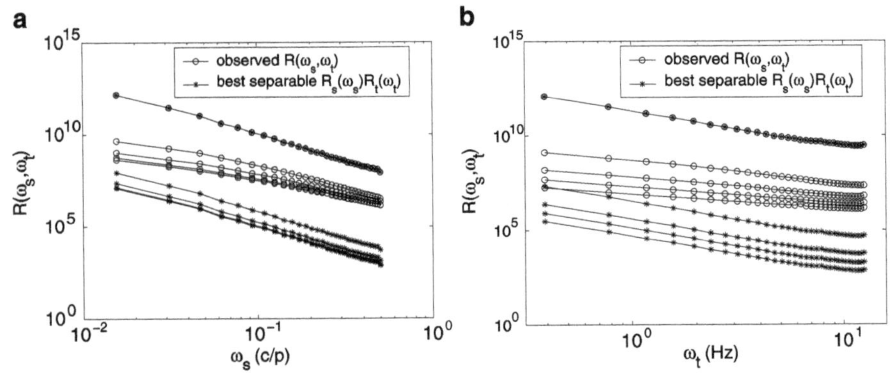

Fig. 16.5 The average spatiotemporal power spectrum of natural image sequences $R(\omega_s, \omega_t)$ is not well approximated by a space-time separable $R_s(\omega_s)R_t(\omega_t)$. These plots show the observed curves plotted in Fig. 16.4 along with plots from the best separable spatiotemporal spectrum (here "best" is defined by minimal least mean square distance). The *uppermost curves* contain both the observed and best separated curves on almost exactly on top of each other, which shows that in the case of the lowest frequencies, the approximation is very good

spatiotemporal case, the direction of the vector $\boldsymbol{\omega} = [\omega_s \ \omega_t]^T$ determines the orientation of the sinusoidal in the s–t-space. We are able to provide a more intuitive interpretation for orientation in the s–t-space: it is the *speed* of the spatial pattern that is moving. Figure 16.6 illustrates this. Points in the (ω_s, ω_t)-space that have the same speed (direction) lie on a line $\omega_t = c\omega_s$, where c is a constant. Therefore, the set of (ω_s, ω_t)-points have the same speed when $\frac{\omega_t}{\omega_s} = \text{constant}$.

It was observed by Dong and Atick (1995b) that the power spectrum has a particularly simple form as a function of spatial frequency ω_s when the speed $\frac{\omega_t}{\omega_s}$ is held constant. Figure 16.7a shows plots of $R(\omega_s, \omega_t)$ as a function of spatial ω_s for different constant values of speed $\frac{\omega_t}{\omega_s}$. As can be seen, in this log–log-plot all the

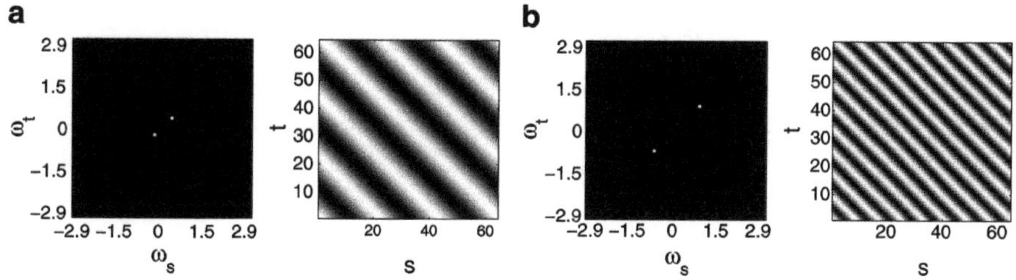

Fig. 16.6 In the frequency-based representation of the s–t-space, the direction of the frequency vector $\boldsymbol{\omega} = [\omega_s \ \omega_t]^T$ is equivalent to the speed of the pixels of a moving spatial grating in the image sequence. This is illustrated here for two different (ω_s, ω_t)-pairs (**a**, **b**), for which the frequency-based representation is shown on the *left*, and the s–t-representation on the *right*. One can see that the pixels in the spatial gratings move with the same speed: for example, looking at the pixel which starts at the corner position $(s, t) = (64, 1)$, it can be seen that in both cases (**a**, **b**), the pixel moves across the whole image when time t runs from 1 to 64, indicating similar speed of movement

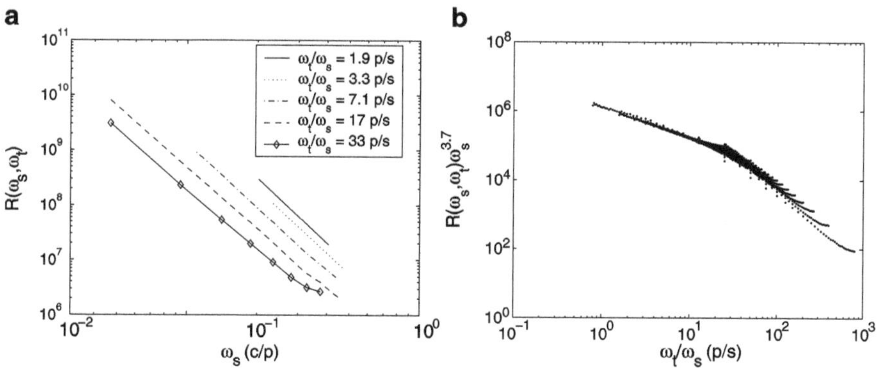

Fig. 16.7 The average spatiotemporal power spectrum $R(\omega_s, \omega_t)$ of natural image sequences can be separated into functions depending on spatial frequency ω_s and speed $\frac{\omega_t}{\omega_s}$. **a** Log–log plots of $R(\omega_s, \omega_t)$ as a function of ω_s are straight lines, suggesting that $R(\omega_s, \omega_t) \approx \omega_s^{-a} f(\frac{\omega_t}{\omega_s})$, where $a > 0$ and $f(\frac{\omega_t}{\omega_s})$ is a function of speed. **b** A plot of $f(\frac{\omega_t}{\omega_s}) \approx \omega_s^{3.7} R(\omega_s, \omega_t)$. See text for details

curves are similar to straight lines with the same slope but different intercepts for different values of $\frac{\omega_t}{\omega_s}$. Denoting the common slope by $-a$, $a > 0$, and the intercepts by $b(\frac{\omega_t}{\omega_s})$, this suggests that

$$\log R(\omega_s, \omega_t) \approx -a \log \omega_s + b\left(\frac{\omega_t}{\omega_s}\right), \tag{16.7}$$

$$R(\omega_s, \omega_t) \approx \omega_s^{-a} \underbrace{\exp\left[b\left(\frac{\omega_t}{\omega_s}\right)\right]}_{=f(\frac{\omega_t}{\omega_s})}, \tag{16.8}$$

$$R(\omega_s, \omega_t) \approx \omega_s^{-a} f\left(\frac{\omega_t}{\omega_s}\right), \tag{16.9}$$

where $f(\cdot)$ is an unknown function of speed. When an estimate of the slope a has been computed (e.g. from the data shown in Fig. 16.7a, an approximate plot of function $f(\cdot)$ can be obtained from

$$f\left(\frac{\omega_t}{\omega_s}\right) \approx \omega_s^{a} R(\omega_s, \omega_t); \tag{16.10}$$

this plot is shown in Fig. 16.7b for $a = 3.7$.

Dong and Atick (1995a) went two steps further in the characterization of $R(\omega_s, \omega_t)$. First, they derived (16.9) from the average power spectrum of static images and a model in which objects move with different velocities at different distances. Second, by assuming a distribution of object velocities they also derived a parametric form for function $f(\cdot)$ which agrees well with the observed $R(\omega_s, \omega_t)$ with reasonable parameter values. See their paper for more details.

The spatiotemporal power spectrum seems to exhibit some anisotropies (see Sect. 5.7, i.e. it is not the same in all orientations). This can be used to explain some psychophysical phenomena (Dakin et al. 2005).

16.3.2 The Temporally Decorrelating Filter

In Sect. 5.9 (page 120), we saw that the removal of linear correlations—that is, whitening—forms the basis of a model that results in the emergence of spatial center-surround receptive fields from natural data. In this section, we apply similar theory to the case of temporal data and temporal receptive fields (see Fig. 16.2c on page 327). We are examining the statistical properties of purely temporal data here, that is, samples consisting of time courses of individual pixels (which are sampled from different spatial locations in the image sequences).

We proceed similarly as in the spatial case. Let $R_\mathrm{i}(\omega_t)$ denote the temporal power spectrum of natural image sequence data (time courses of individual pixels). As in the spatial case, we assume that noise power $R_\mathrm{n}(\omega_t)$ is constant, and given by

$$R_\mathrm{n}(\omega_t) = \frac{R_\mathrm{i}(\omega_{t,c})}{2} \quad \text{for all } \omega_t, \tag{16.11}$$

where $\omega_{t,c}$ is the characteristic frequency at which the data and noise have the same power. As in the spatial case (see (5.49) on page 127), we define the amplitude response of the filter $|\tilde{W}(\omega_t)|$ as the product of the amplitude responses of a whitening filter and a noise-suppressive filter:

$$\left|\tilde{W}(\omega_t)\right| = \frac{1}{\sqrt{R_\mathrm{i}(\omega_t)}} \frac{R_\mathrm{i}(\omega_t) - R_\mathrm{n}(\omega_t)}{R_\mathrm{i}(\omega_t)}. \tag{16.12}$$

As was mentioned in Sect. 5.9, a shortcoming of the decorrelation theory is that it does not predict the phase response of the filter. Here, we use the principle of *minimum energy delay*: the phases are specified so that the energy in the impulse response of the resulting causal filter is delayed the least. The phase response of a minimum energy delay filter is given by the *Hilbert transform* of the logarithm of the amplitude response; see Oppenheim and Schafer (1975) for details. After the amplitude and the phase responses have been defined, the temporal filter itself can be obtained by taking the inverse Fourier transform.

The filter properties that result from the application of (16.11) and (16.12) and the minimum energy delay principle are illustrated in Fig. 16.8 for a characteristic frequency value of $\omega_{s,t} = 7$ Hz (the same value was used in Dong and Atick 1995b). For this experiment, 100 000 signals of spatial size 1×1 pixels and a duration of 64 time points ($\approx$2.5 s) were sampled from the image sequence data of van Hateren and Ruderman (1998). The average temporal power spectrum of these signals was then computed and is shown in Fig. 16.8a. The squared amplitude response of the whitening filter, obtained from (16.12), is shown in Fig. 16.8b. The power spectrum of the filtered data is shown in Fig. 16.8c; it is approximately flat at lower frequencies and drops off at high frequencies because of the higher relative noise power at high frequencies. The resulting filter is shown in Fig. 16.8d; for comparison, a measured temporal receptive field of an LGN neuron is shown in Fig. 16.8e. Please observe the difference in the time scales in Figs. 16.8d and e. Here, the match between the two linear filters is only qualitative; in experimental animals, the latencies of LGN cells seem to vary from tens to hundreds milliseconds (Saul and Humphrey 1990). Similar temporal processing properties are often attributed to retinal ganglion cells (Meister and Berry II 1999), although Dong and Atick (1995a) argue that the temporal frequency response of retinal cells is typically flat when compared with the response of neurons in the LGN.

Dong and Atick (1995a) proceed by showing that when combined with basic neural non-linearities (rectification), the temporally decorrelating filter theory yields response properties that match the timing and phase response properties of LGN neurons. For additional experimental evaluation of the model, see Dan et al. (1996b).

Here, we have used a linear neuron model with a constant filter (static receptive field). In reality, the temporal receptive field of a visual neuron may change, and this adaptation may be related to the short-term changes in stimulus statistics (Hosoya et al. 2005).

16.4 Sparse Coding and ICA of Natural Image Sequences

To analyze the spatiotemporal statistics beyond covariances, the ICA model can be applied directly to natural image sequences. Instead of vectorizing image patches (windows), and using them as data in the ICA model, spatiotemporal image sequence blocks can be vectorized to form the data $\mathbf{x}$. After a spatiotemporal feature detector weight vector $\mathbf{w}$ or, alternatively, a spatiotemporal feature vector $\mathbf{a}$ has been

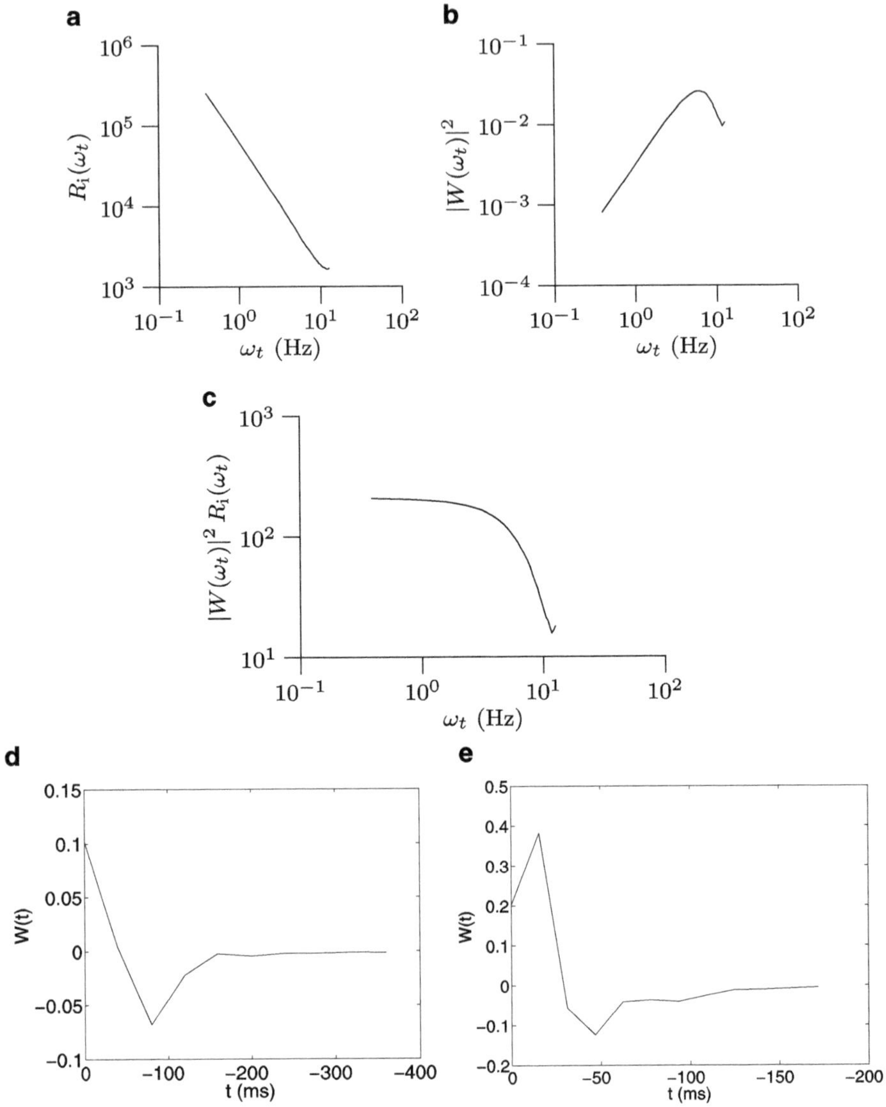

Fig. 16.8 The application of the whitening principle, combined with noise reduction and minimum energy delay phase response, leads to the emergence of filters resembling the temporal receptive fields of neurons in the retina and the LGN. **a** The temporal power spectrum $R_i(\omega_t)$ of natural image sequence data. **b** The squared amplitude response of a whitening filter which suppresses noise: this curve follows the inverse of the data power spectrum at low frequencies, but drops off at high frequencies, because the proportion of noise is larger at high frequencies. **c** The power spectrum of the resulting (filtered) data, showing approximately flat (white) power at low frequencies, and dropping off at high frequencies. **d** The resulting filter which has been obtained from the amplitude response in **b** and by specifying a minimum energy delay phase response; see text for details. **e** For comparison, the temporal receptive field of an LGN neuron. Please note the differences in the time scales in **d** and **e**

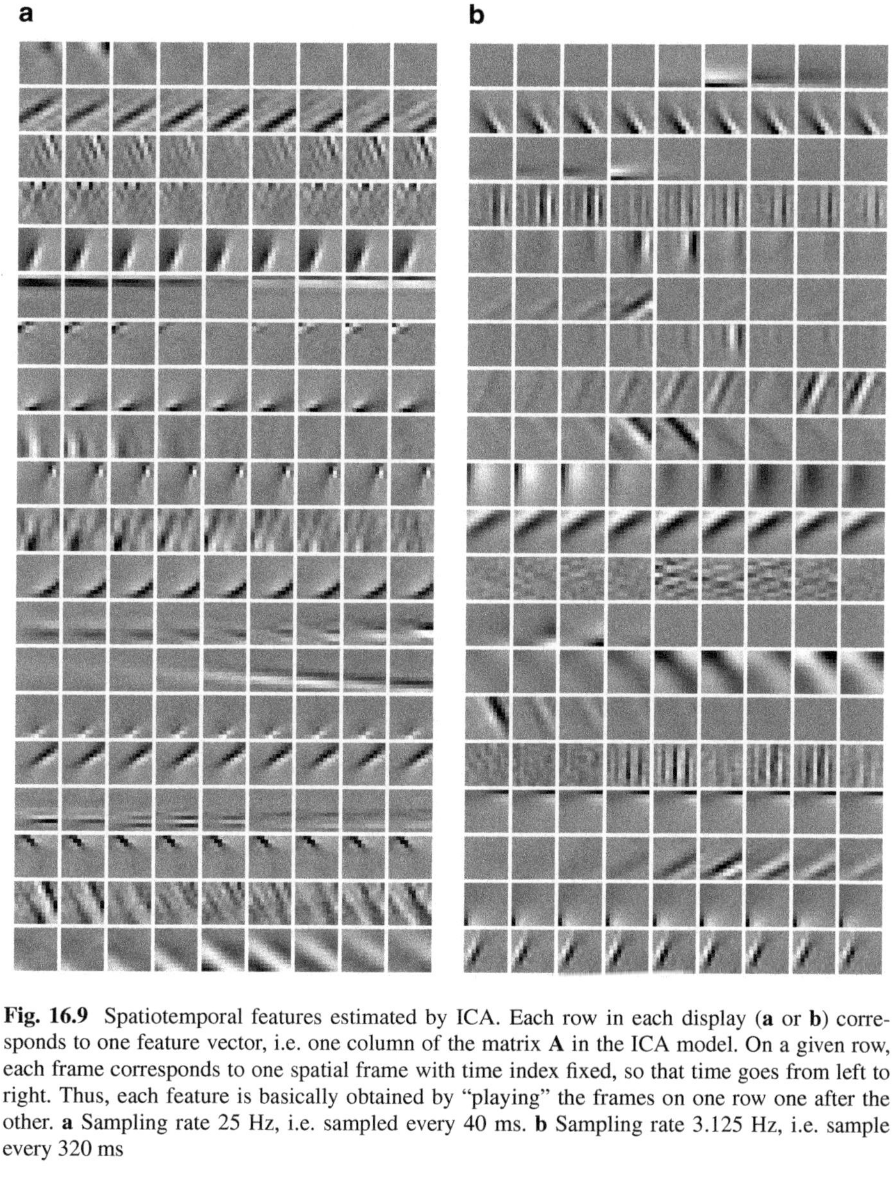

Fig. 16.9 Spatiotemporal features estimated by ICA. Each row in each display (**a** or **b**) corresponds to one feature vector, i.e. one column of the matrix **A** in the ICA model. On a given row, each frame corresponds to one spatial frame with time index fixed, so that time goes from left to right. Thus, each feature is basically obtained by "playing" the frames on one row one after the other. **a** Sampling rate 25 Hz, i.e. sampled every 40 ms. **b** Sampling rate 3.125 Hz, i.e. sample every 320 ms

learned from the data, it can be visualized as an image sequence after "unvectorizing" it, just like in the basic spatial case.

Results of estimating spatiotemporal features by ICA are shown in Fig. 16.9. The data consisted of image sequence blocks of size $(11, 11, 9)$, where the two first values are in pixels and the third value is in time steps. We used two different sampling rates, 25 Hz and 3.125 Hz because that parameter has a visible influence on the results. The number of spatiotemporal patches was 200 000, and the dimension was reduced by approximately 50% by PCA. The data set was the same van Hateren and Ruderman (1998) dataset as used above. FastICA was run in the symmetric mode with non-linearity $g(\alpha) = \tanh(\alpha)$, which corresponds to the familiar log cosh measure of sparseness (see Sect. 18.7).

The estimated features shown in Fig. 16.9 are spatially Gabor-like, some of them are separable and other are not. The results clearly depend on the sampling rate: if the sampling rate is high (a), the features tend to be static, i.e. there is hardly any temporal change. This is intuitively comprehensible: if the time resolution in the data is too high, there is simply not enough time for any changes to occur. When the sampling rate is lower (b), there is much more temporal change in the features.

The results are thus quite well in line with those measured in single-cell recordings in, e.g. DeAngelis et al. (1993a, 1993b, 1995).

Further results on estimating spatiotemporal features vectors obtained by applying FastICA to natural image sequence data can be found at http://hlab.phys.rug.nl/demos/ica/index.html, and the paper (van Hateren and Ruderman 1998).

16.5 Temporal Coherence in Spatial Features

16.5.1 Temporal Coherence and Invariant Representation

Our visual environment has inertia: during a short time interval, the scene we see tends to remain similar in the sense that the same objects persist in our field of vision, the lighting conditions usually change slowly, etc. Could our visual system utilize this property of our environment?

In particular, it has been proposed that those properties which change more quickly are often less important for pattern recognition: The identities of the objects in our visual field change slower than their appearance. For example, when you talk with somebody, you see the same face for a long time, but its appearance undergoes various transformations due to the change in the facial expression and the muscle actions related to speech. So, if you consider those features which change the slowest, they might be directly related to the identity of the interlocutor.

Thus, it has been proposed that a good internal representation for sensory input would be one that changes slowly. The term *temporal coherence* refers to a representation principle in which, when processing temporal input, the representation in the computational system is optimized so that it changes as slowly as possible over time (Hinton 1989; Földiák 1991).

In this section, we will take a look at a model of temporal coherence which results in the emergence of simple-cell-like RFs from natural image sequence data. In the next section, this will be extended to a model that exhibit complex-cell-like behavior and topographical organization of RFs.

16.5.2 Quantifying Temporal Coherence

It has been argued that the neural output in the visual system is characterized temporally as short, intense firing events, or bursts of spikes (Reinagel 2001). Here, we present a model which optimizes a measure of such temporal coherence of activity levels—or energy—and which, when applied to a set of natural image sequences, leads to the emergence of RFs which resemble simple-cell RFs.[3]

We use a set of spatial feature detectors (weight vectors) $\mathbf{w}_1, \ldots, \mathbf{w}_K$ to relate input to output. While it may first sound weird to use purely spatial features with spatiotemporal data, this simplification will make better sense below when we introduce the temporal filtering used in preprocessing; this combination of temporal and spatial features is equivalent to space-time separable spatiotemporal features (see Sect. 16.2, page 327). Let vector $\mathbf{x}(t)$ denote the (preprocessed) input to the system at time t. The output of the kth feature detector at time t, denoted by $s_k(t)$, is given by $s_k(t) = \mathbf{w}_k^T \mathbf{x}(t)$. Let matrix $\mathbf{W} = [\mathbf{w}_1 \; \ldots \; \mathbf{w}_K]^T$ denote a matrix with all the feature detector weights as rows. Then the input-output relationship can be expressed in vector form by $\mathbf{s}(t) = \mathbf{W}\mathbf{x}(t)$, where vector $\mathbf{s}(t) = [s_1(t) \; \ldots \; s_K(t)]^T$.

To proceed to the objective function, we first define a non-linearity $g(\cdot)$ that measures the strength (amplitude) of the feature, and emphasizes large responses over small ones: we require that g is strictly convex, even-symmetric (rectifying), and differentiable. Examples of choices for this non-linearity are $g_1(x) = x^2$, which measures the energy of the response, and $g_2(x) = \log\cosh x$, which is a robustified version of g_1 (less sensitive to outliers). Let the symbol Δt denote a delay in time. *Temporal response strength correlation*, the objective function, is defined by

$$f(\mathbf{W}) = \sum_{k=1}^{K} \sum_{t=1+\Delta t}^{T} g\big(s_k(t)\big) g\big(s_k(t - \Delta t)\big). \tag{16.13}$$

A set of feature detectors which has a large temporal response strength correlation is such that the same features often respond strongly at consecutive time points, outputting large (either positive or negative) values. This means that the same features will respond strongly over short periods of time, thereby expressing temporal coherence of activity levels in the neuronal population.

To keep the outputs of the features bounded we enforce the unit variance constraint on each of the output signals $s_k(t)$, that is, we enforce the constraint $E_t\{s_k^2(t)\} = \mathbf{w}_k^T \mathbf{C}_\mathbf{x} \mathbf{w}_k = 1$ for all k, where $\mathbf{C}_\mathbf{x}$ is the covariance matrix $E_t\{\mathbf{x}(t)\mathbf{x}^T(t)\}$, and E_t means average over t. Additional constraints are needed to keep the feature detectors from converging to the same solution. Standard methods are either to force the set of feature weights to be orthogonal, or to force their outputs to be uncorrelated, from which we choose the latter, as in preceding chapters. This introduces additional constraints $\mathbf{w}_i^T \mathbf{C}_\mathbf{x} \mathbf{w}_j = 0$, $i = 1, \ldots, K$, $j = 1, \ldots, K$, $j \neq i$.

---

[3]This section is based on the article (Hurri and Hyvärinen 2003a) originally published in Neural Computation. Copyright ©2003 MIT Press, used with permission.

These uncorrelatedness constraints limit the number of features K we can find so that if the image data has spatial size $N \times N$ pixels, then $K \leq N^2$. The unit variance constraints and the uncorrelatedness constraints can be expressed by a single matrix equation

$$\mathbf{W}\mathbf{C}_{\mathbf{x}}\mathbf{W}^{\mathrm{T}} = \mathbf{I}. \tag{16.14}$$

Note that if we use a non-linearity $g(x) = x^2$, and $\Delta t = 0$, the objective function becomes $f(\mathbf{W}) = \sum_{k=1}^{K} E_t\{s_k^4(t)\}$. In this case, the optimization of the objective function under the unit variance constraint is equivalent to optimizing the sum of kurtoses of the outputs. As was discussed in Sect. 6.2.1 on page 133, kurtosis is a commonly used measure in sparse coding. Similarly, in the case of non-linearity $g(x) = \log\cosh x$ and $\Delta t = 0$, the objective function can be interpreted as a non-quadratic measure of the non-Gaussianity of features.

Thus, the receptive fields are learned in the model by maximizing the objective function in (16.13) under the constraint in (16.14). The optimization algorithm used for this constrained optimization problem is a variant of the gradient projection method described in Sect. 18.2.4. The optimization approach employs whitening, that is, a temporary change of coordinates, to transform the constraint (16.14) into an orthogonality constraint. Then a gradient projection algorithm employing optimal symmetric orthogonalization can be used. See Hurri and Hyvärinen (2003a) for details.

*16.5.3 Interpretation as Generative Model **

An interpretation of maximization of objective function (16.13) as estimation of a generative model is possible, based on the concept of sources with non-stationary (non-constant) variances (Matsuoka et al. 1995; Pham and Cardoso 2001; Hyvärinen 2001a). The linear generative model for $\mathbf{x}(t)$ is similar to the one in previous chapters:

$$\mathbf{x}(t) = \mathbf{A}\mathbf{s}(t). \tag{16.15}$$

Here, $\mathbf{A} = [\mathbf{a}_1 \; \dots \; \mathbf{a}_K]$ denotes a matrix which relates the image sequence patch $\mathbf{x}(t)$ to the activities of the simple cells, so that each column $\mathbf{a}_k$, $k = 1, \dots, K$, gives the feature that is coded by the corresponding simple cell. The dimension of $\mathbf{x}(t)$ is typically larger than the dimension of $\mathbf{s}(t)$, so that (16.15) is generally not invertible. A one-to-one correspondence between $\mathbf{W}$ and $\mathbf{A}$ can be established by using the pseudo-inverse solution (see Sect. 19.8):

$$\mathbf{A} = \mathbf{W}^{\mathrm{T}}\left(\mathbf{W}\mathbf{W}^{\mathrm{T}}\right)^{-1}. \tag{16.16}$$

The non-stationarity of the variances of sources $\mathbf{s}(t)$ means that their variances change over time, and the variance of a signal is correlated at nearby time points. An example of a signal with non-stationary variance is shown in Fig. 16.10. It can be

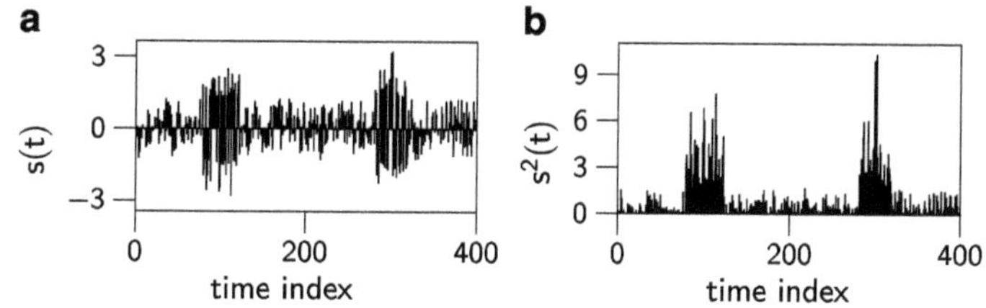

Fig. 16.10 Illustration of non-stationarity of variance. **a** A temporally uncorrelated signal $s(t)$ with non-stationary variance. **b** Plot of $s^2(t)$

shown (Hyvärinen 2001a) that optimization of a cumulant-based criterion, similar to (16.13), can separate independent sources with non-stationary variances. Thus, the maximization of the objective function can also be interpreted as estimation of generative models in which the activity levels of the sources vary over time, and are temporally correlated over time. This situation is analogous to the application of measures of sparseness to estimate linear generative models with independent non-Gaussian sources, i.e. the ICA model treated in Chap. 7.

16.5.4 Experiments on Natural Image Sequences

16.5.4.1 Data and Preprocessing

The natural image data used in the experiments was described in Sect. 16.3.1 (page 329). The final, preprocessed (see below) data set consisted of 200 000 pairs of consecutive 11×11 image patches at the same spatial position, but Δt milliseconds apart from each other. In the main experiment, $\Delta t = 40$ ms; other values were used in the control experiments. However, because of the temporal filtering used in preprocessing, initially 200 000 longer image sequences with a duration of $\Delta t + 400$ ms, and the same spatial size 11×11, were sampled with the same sampling rate.

The preprocessing in the main experiment consisted of three steps: temporal decorrelation, subtraction of local mean, and normalization. The same preprocessing steps were applied in the control experiments; whenever preprocessing was varied in control experiments it is explained separately below. Temporal decorrelation can be motivated in two different ways. First, as was discussed in Sect. 16.3.2 (page 332), it can be motivated biologically as a model of temporal processing in the early visual system. Second, as discussed above, for $\Delta t = 0$ the objective function can be interpreted as a measure of sparseness. Therefore, it is important to rule out the possibility that there is hardly any change in short intervals in video data, since this would imply that our results could be explained in terms of sparse coding or ICA. To make the distinction between temporal response strength correlation and measures of sparseness clear, temporal decorrelation was applied because it enhances temporal changes. Note, however, that this still does not remove all of the static part in the video—this issue is addressed in the control experiments below.

Temporal decorrelation was performed with the temporal filter shown in Fig. 16.8d (page 334). As was already mentioned above, the use of such a temporal filter in conjunction with the learned spatial features makes the overall model spatiotemporal (to be more exact, space-time separable).

16.5.4.2 Results and Analysis

In the main experiment, non-linearity g in objective function (16.13) was chosen to be $g(x) = \log\cosh x$. A set of feature detector weights (rows of $\mathbf{W}$) learned by the model is shown in Fig. 16.11a. The features resemble Gabor functions. They are localized, oriented, and have different scales, and thus have the main properties of simple-cell receptive fields.

To compare the results obtained with this model against those obtained with ICA, we ran both this algorithm and the symmetric version of FastICA with non-linearity tanh 50 times with different initial values and compared the resulting two sets of 6000 ($= 50 \times 120$) features against each other. The results are shown in Fig. 16.12. The measured properties were peak spatial frequency (Figs. 16.12a and b, note logarithmic scale, units cycles/pixel), peak orientation (Figs. 16.12c and d), spatial frequency bandwidth (Figs. 16.12e and f), and orientation bandwidth (Figs. 16.12g and h). Peak orientation and peak frequency are simply the orientation and frequency of the highest value in the Fourier power spectrum. Bandwidths measure the sharpness of the tuning and were computed from the tuning curve as the full width at the point were half the maximum response was attained (full width at half maximum, FWHM); this measure is widely used in vision science. See van Hateren and van der Schaaf (1998) for more details.

Although there are some differences between the two feature sets, the most important observation here is the similarity of the histograms. This supports the idea that ICA/sparse coding and temporal coherence are complementary theories, in that they both result in the emergence of simple-cell-like receptive fields. As for the differences, the results obtained using temporal response strength correlation have a slightly smaller number of high-frequency receptive fields. Also, temporal response strength correlation seems to produce receptive fields that are somewhat more localized with respect to both spatial frequency and orientation.[4]

[4]When these results are compared against the results in van Hateren and van der Schaaf (1998), the most important difference is the peak at zero bandwidth in Figs. 16.12e and f. This difference is probably a consequence of the fact that no dimensionality reduction, anti-aliasing or noise reduction was performed here, which results in the appearance of very small, checkerboard-like receptive fields. This effect is more pronounced in ICA, which also explains the stronger peak at the 45° angle in Fig. 16.12d.

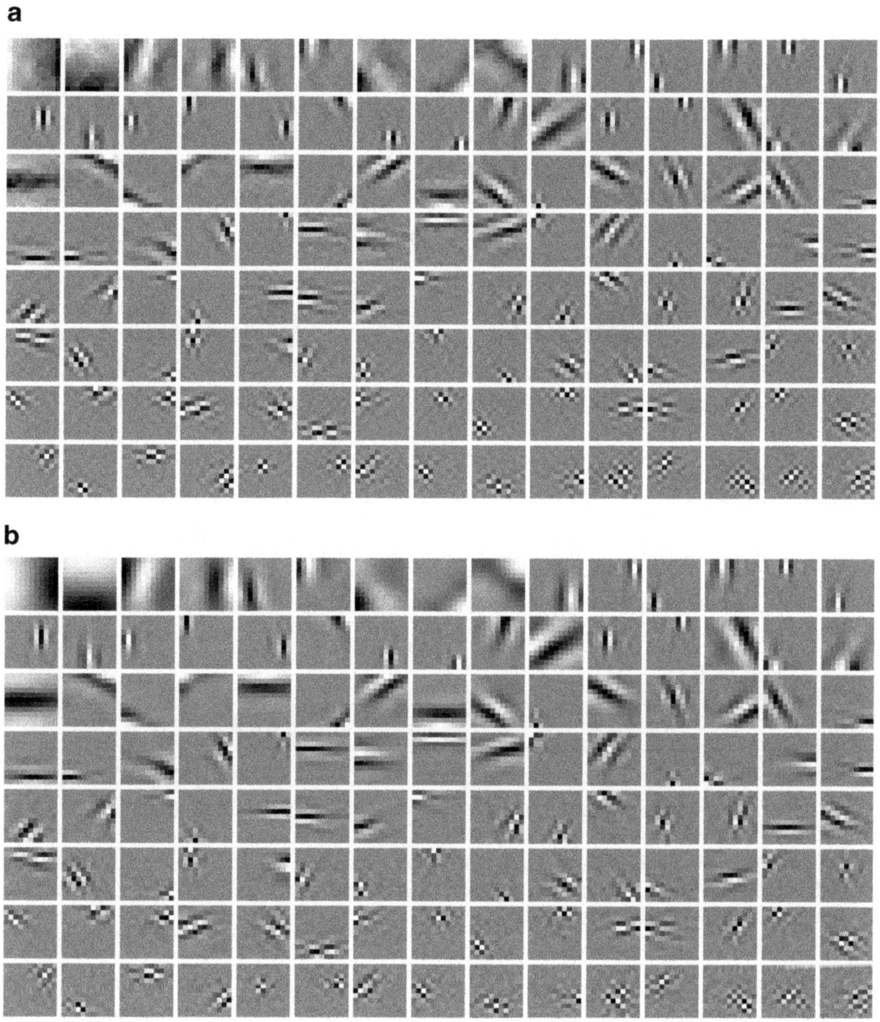

Fig. 16.11 Simple-cell-like filters emerge when temporal response strength correlation is optimized in natural image sequences. **a** Feature weights $\mathbf{w}_k$, $k = 1, \ldots, 120$, which maximize temporal response strength correlation (16.13); here the non-linearity $g(x) = \log\cosh x$. The features have been ordered according to $E_t\{g(s_k(t))g(s_k(t - \Delta t))\}$, that is, according to their "contribution" into the final objective value (features with largest values *top left*). **b** A corresponding set of feature vectors $\mathbf{a}_k$, $k = 1, \ldots, 120$, from a generative-model-based interpretation of the results (see (16.15) and (16.16))

16.5.5 Why Gabor-Like Features Maximize Temporal Coherence

A simplified intuitive illustration of why the outputs of Gabor-like feature have such strong energy correlation over time is shown in Fig. 16.13. Most transformations of objects in the 3D world result in something similar to local translations of lines and

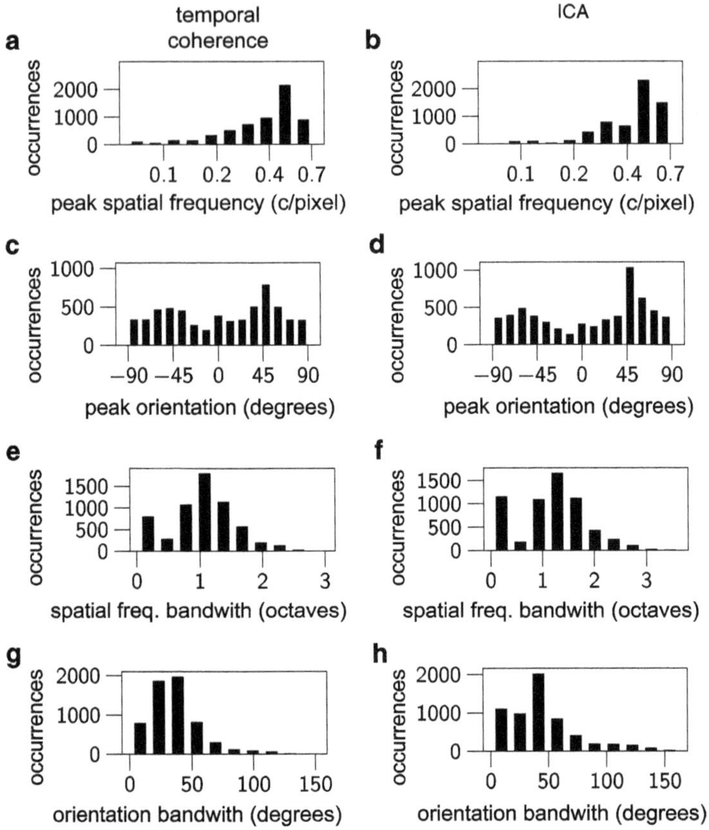

Fig. 16.12 Comparison of properties of receptive fields obtained by optimizing temporal response strength correlation (*left column*, histograms **a**, **c**, **e** and **g**) and estimating ICA (*right column*, histograms **b**, **d**, **f** and **h**). See text for details

edges in image sequences. This is obvious in the case of 3D translations, and is illustrated in Fig. 16.13a for two other types of transformations: rotation and bending. In the case of a local translation, a suitably oriented simple-cell-like RF responds strongly at consecutive time points, but the sign of the response may change. Note that when the output of a feature detector is considered as a continuous signal, the change of sign implies that the signal reaches zero at some intermediate time point, which can lead to a weak measured correlation. Thus, a better model of the dependencies would be to consider dependencies of variances (Matsuoka et al. 1995; Pham and Cardoso 2001), as in the generative-model interpretation described above. However, for simplicity, we consider here the magnitude that is a crude approximation of the underlying variance.

In order to further visualize the correlation of rectified responses at consecutive time points, we will consider the interaction of features in one dimension (orthogonal to the orientation of the feature). This allows us to consider the effect of local

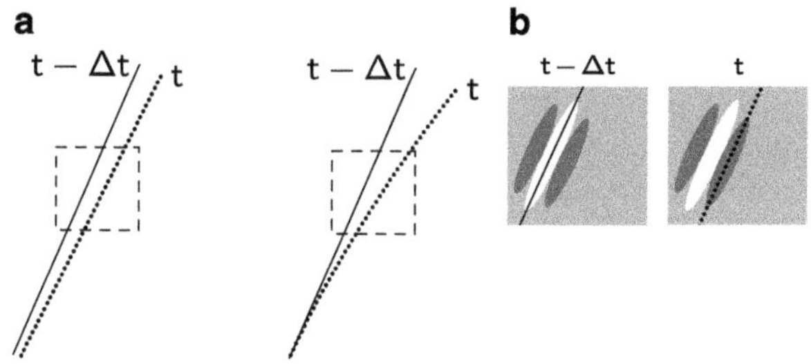

Fig. 16.13 A simplified illustration of temporal activity level dependencies of simple-cell-like features when the input consists of image sequences. **a** Transformations of objects induce local translations of edges and lines in local regions in image sequences: rotation (*left*) and bending (*right*). The *solid line* shows the position/shape of a line in the image sequence at time $t - \Delta t$, and the *dotted line* shows its new position/shape at time t. The *dashed square* indicates the sampling window. **b** Temporal activity level dependencies: in the case of a local translation of an edge or a line, the response of a simple-cell-like features with a suitable position and orientation is strong at consecutive time points, but the sign may change. The figure shows a translating line superimposed on an oriented and localized receptive field at two different time instances (time $t - \Delta t$, *solid line*, *left*; time t, *dotted line*, *right*)

translations in a simplified setting. Figure 16.14 illustrates, in a simplified case, why the temporal response strengths of lines and edges correlate positively as a result of Gabor-like feature structure. Prototypes of two different types of image elements—the profiles of a line and an edge—which both have a zero DC component are shown in the topmost row of the figure. The leftmost column shows the profiles of three different features with unit norm and zero DC component: a Gabor-like feature, a sinusoidal (Fourier basis-like) feature, and an impulse feature. The rest of the figure shows the square rectified responses of the features to the inputs as functions of spatial displacement of the input.

Consider the rectified response of the Gabor-like feature to the line and the edge, that is, the first row of responses in Fig. 16.14. The squared response at time $t - \Delta t$ (spatial displacement zero) is strongly positively correlated with response at time t, even if the line or edge is displaced slightly. This shows how small local translations of basic image elements still yield large values of temporal response strength correlation for Gabor-like features. If you compare the responses of the Gabor-like feature to the responses of the sinusoidal feature—that is, the second row of responses in Fig. 16.14—you can see that the responses to the sinusoidal feature are typically much smaller. This leads to a lower value of our measure of temporal response strength correlation that emphasizes large values. Also, in the third row of responses in Fig. 16.14, we can see that while the response of an impulse feature to an edge correlates quite strongly over small spatial displacements, when the input consists of a line even a very small displacement will take the correlation to almost zero.

Thus, we can see that when considering three important classes of features—features which are maximally localized in space, maximally localized in frequency, or localized in both—the optimal feature is a Gabor-like feature, which is localized both in space and in frequency. If the feature is maximally localized in space, it fails

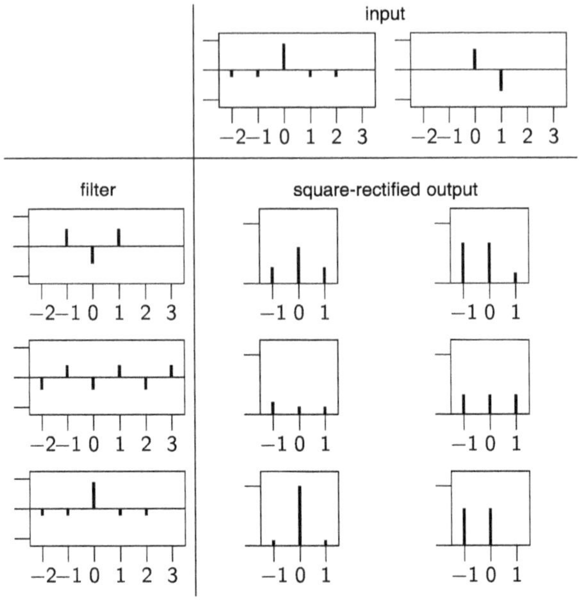

Fig. 16.14 A simplified illustration of why a Gabor-like feature, localized in both space and frequency, yields larger values of temporal response strength correlation than a feature localized only in space or only in frequency. *Top row*: cross sections of a line (*left*) and an edge (*right*) as functions of spatial position. *Leftmost column*: cross sections of three features with unit norm and zero DC component—a Gabor-like feature (*top*), a sinusoidal feature (*middle*), and an impulse feature (*bottom*). The other plots in the figure show the responses of the feature detectors to the inputs as a function of spatial displacement of the input. The Gabor-like feature yields fairly large positively correlated values for both types of input. The sinusoidal feature yields small response values. The impulse feature yields fairly large positively correlated values when the input consists of an edge, but when the input consists of a line even a small displacement yields a correlation of almost zero

to respond over small spatial displacements of very localized image elements. If the feature is maximally localized in frequency, its responses to the localized image features are not strong enough.

Figure 16.15 shows why we need non-linear correlations instead of linear ones: raw output values might correlate either positively or negatively, depending on the displacement. Thus, we see why ordinary linear correlation is not maximized for Gabor-like features, whereas the rectified (non-linear) correlation is.

16.5.6 Control Experiments

To validate the novelty of the results obtained with this model when compared with ICA and sparse coding, and to examine the effect of different factors in the results, a number of control experiments were made. These experiments are summarized here; details can be found in Hurri and Hyvärinen (2003a). The control experiments show that

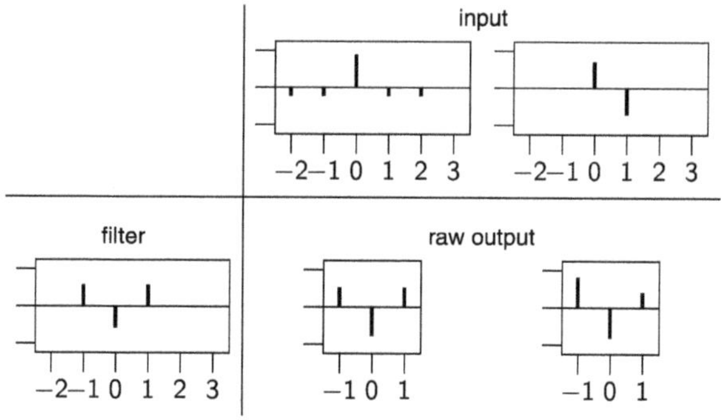

Fig. 16.15 A simplified illustration of why nonlinear correlation is needed for the emergence of the phenomenon. Raw response values of the Gabor-like feature to the line and edge may correlate positively or negatively, depending on the displacement. (See Fig. 16.14 for an explanation of the layout of the figure)

- the results are qualitatively similar when the static part of the video is removed altogether by employing Gram–Schmidt orthogonalization, which strengthens the novelty of this model when compared with static models
- the results are qualitatively similar when no temporal decorrelation is performed
- the results are qualitatively similar when $\Delta t = 120$ ms; when Δ is further increased to $\Delta t = 480$ ms and $\Delta t = 960$ ms, the resulting features start to lose their spatial localization and gradually also their orientation selectivity; finally, when the consecutive windows have no temporal relationship (consecutive windows chosen randomly), the resulting features correspond to noise patterns
- the results are qualitatively similar when observer (camera) movement is compensated by a tracking mechanism in video sampling.

Finally, one further control experiment was made in which the linear correlation $f_\ell(\mathbf{w}_k) = E_t\{s_k(t)s_k(t - \Delta t)\}$ was maximized. The unit variance constraint is used here again, so the problem is equivalent to minimizing $E_t\{(s_k(t) - s_k(t - \Delta t))^2\}$ with the same constraint; we will return to this objective function below in Sect. 16.8. The resulting features resemble Fourier basis vectors, and not simple-cell receptive fields. This shows that non-linear, higher-order correlation is indeed needed for the emergence of simple-cell-like features.

16.6 Spatiotemporal Energy Correlations in Linear Features

16.6.1 Definition of the Model

Temporal response strength correlation, defined in (16.13) on page 337, maximizes the "temporal coherence" in the outputs of individual simple cells. Note that in terms of the generative model described above, the objective functions says nothing about the interdependencies in different $s_k(t)$'s—that is, different cells. Thus, there

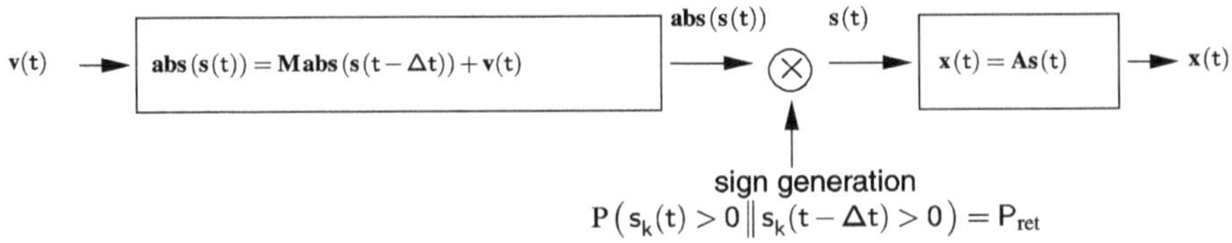

Fig. 16.16 The two layers of the generative model. Let $\mathbf{abs}(\mathbf{s}(t)) = [|s_1(t)| \dots |s_K(t)|]^T$ denote the amplitudes of simple cell responses. In the first layer, the driving noise signal $\mathbf{v}(t)$ generates the amplitudes of simple cell responses via an autoregressive model. The signs of the responses are generated randomly between the first and second layer to yield signed responses $\mathbf{s}(t)$. In the second layer, natural video $\mathbf{x}(t)$ is generated linearly from simple cell responses. In addition to the relations shown here, the generation of $\mathbf{v}(t)$ is affected by $\mathbf{M}\,\mathbf{abs}(\mathbf{s}(t - \Delta t))$ to ensure non-negativity of $\mathbf{abs}(\mathbf{s}(t))$. See text for details

is an implicit assumption of independence in the model, at least if it is interpreted as a probabilistic generative model. In this section, we add another layer to the generative model to extend the theory to simple-cell interactions, and to the level of complex cells.[5]

Like in the many generative models discussed in this book, the output layer of the new model (see Fig. 16.16) is linear, and maps cell responses $\mathbf{s}(t)$ to image features $\mathbf{x}(t)$, but we do not assume that the components of $\mathbf{s}(t)$ are independent. Instead, we model the temporal dependencies between these components in the first layer of our model. Let $\mathbf{abs}(\mathbf{s}(t)) = [|s_1(t)| \dots |s_K(t)|]^T$ denote the activities of the cells, and let $\mathbf{v}(t)$ denote a driving noise signal and $\mathbf{M}$ denote a $K \times K$ matrix; the modeled interdependencies will be "coded" in $\mathbf{M}$. Our model is a *multidimensional first-order autoregressive process*, defined by

$$\mathbf{abs}\big(\mathbf{s}(t)\big) = \mathbf{M}\,\mathbf{abs}\big(\mathbf{s}(t - \Delta t)\big) + \mathbf{v}(t). \tag{16.17}$$

Again, we also need to fix the scale of the latent variables by defining $E_t\{s_k^2(t)\} = 1$ for $k = 1, \dots, K$.

There are dependencies between the driving noise $\mathbf{v}(t)$ and output strengths $\mathbf{abs}(\mathbf{s}(t))$, caused by the non-negativity of $\mathbf{abs}(\mathbf{s}(t))$. To take these dependencies into account, we use the following formalism. Let $\mathbf{u}(t)$ denote a random vector with components which are statistically independent of each other. To ensure the non-negativity of $\mathbf{abs}(\mathbf{s}(t))$, we define

$$\mathbf{v}(t) = \mathbf{max}\big(-\mathbf{M}\,\mathbf{abs}\big(\mathbf{s}(t - \Delta t)\big), \mathbf{u}(t)\big), \tag{16.18}$$

where, for vectors $\mathbf{a}$ and $\mathbf{b}$, $\mathbf{max}(\mathbf{a}, \mathbf{b}) = [\max(a_1, b_1) \ \dots \ \max(a_n, b_n)]^T$. We assume that $\mathbf{u}(t)$ and $\mathbf{abs}(\mathbf{s}(t))$ are uncorrelated. The point in this definition is to make sure that the noise does not drive the absolute values of the $s_K(t)$ negative, which would be absurd.

[5]This section is based on the article (Hurri and Hyvärinen 2003b) originally published in Network: Computation in Neural Systems. Copyright ©2003 Institute of Physics, used with permission.

To make the generative model complete, a mechanism for generating the signs of cell responses $\mathbf{s}(t)$ must be included. We specify that the signs are generated randomly with equal probability for plus or minus after the strengths of the responses have been generated. Note that one consequence of this is that the different $s_k(t)$'s are uncorrelated. In the estimation of the model this uncorrelatedness property is used as a constraint. When this is combined with the unit variance (scale) constraints described above, the resulting set of constraints is the same as in the approach described above in Sect. 16.5 (page 336).

In (16.17), a large positive matrix element $\mathbf{M}(i,j)$, or $\mathbf{M}(j,i)$, indicates that there is strong temporal dependency between the output strengths of cells i and j. Thinking in terms of grouping temporally coherent cells together, matrix $\mathbf{M}$ can be thought of as containing similarities (reciprocals of distances) between different cells. We will use this property below to derive a topography of simple-cell receptive fields from $\mathbf{M}$.

16.6.2 Estimation of the Model

To estimate the model defined above we need to estimate both $\mathbf{M}$ and $\mathbf{A}$. Instead of estimating $\mathbf{A}$ directly, we estimate $\mathbf{W}$ which maps image sequence data to responses

$$\mathbf{s}(t) = \mathbf{W}\mathbf{x}(t), \tag{16.19}$$

and use the pseudo-inverse relationship—that is, (16.16) on page 338—to compute $\mathbf{A}$. In what follows, we first show how to estimate $\mathbf{M}$, given $\mathbf{W}$. We then describe an objective function which can be used to estimate $\mathbf{W}$, given $\mathbf{M}$. Each iteration of the estimation algorithm consists of two steps. During the first step $\mathbf{M}$ is updated, and $\mathbf{W}$ is kept constant; during the second step these roles are reversed.

First, regarding the estimation of $\mathbf{M}$, consider a situation in which $\mathbf{W}$ is kept constant. It can be shown (Hurri and Hyvärinen 2003b) that $\mathbf{M}$ can be estimated by using approximative method of moments, and that the estimate is given by

$$\begin{aligned}\mathbf{M} \approx \beta E_t\left\{\left(\mathbf{abs}(\mathbf{s}(t)) - E_t\{\mathbf{abs}(\mathbf{s}(t))\}\right)\left(\mathbf{abs}(\mathbf{s}(t-\Delta t)) - E_t\{\mathbf{abs}(\mathbf{s}(t))\}\right)^{\mathrm{T}}\right\} \\ \times E_t\left\{\left(\mathbf{abs}(\mathbf{s}(t)) - E_t\{\mathbf{abs}(\mathbf{s}(t))\}\right)\left(\mathbf{abs}(\mathbf{s}(t)) - E_t\{\mathbf{abs}(\mathbf{s}(t))\}\right)^{\mathrm{T}}\right\}^{-1},\end{aligned}$$

where $\beta > 1$. Since this multiplier does not change the relative strengths of the elements of $\mathbf{M}$, and since it has a constant linear effect in the objective function of $\mathbf{W}$ given below, its value does not affect the optimal $\mathbf{W}$, so we can simply set $\beta = 1$ in the optimization. The resulting estimator for $\mathbf{M}$ is the same as the optimal least mean squares linear predictor in the case of unconstrained $\mathbf{v}(t)$.

The estimation of $\mathbf{W}$ is more complicated. A rigorous derivation of an objective function based on well-known estimation principles is very difficult because the statistics involved are non-Gaussian, and the processes have difficult interdependencies. Therefore, instead of deriving an objective function from first principles, we derived an objective function heuristically (Hurri and Hyvärinen 2003b), and verified through simulations that the objective function is capable of estimating the two-layer model. The objective function is a weighted sum of the covariances of feature output strengths at times $t - \Delta t$ and t, defined by

$$f(\mathbf{W}, \mathbf{M}) = \sum_{i=1}^{K} \sum_{j=1}^{K} \mathbf{M}(i, j) \operatorname{cov}\{|s_i(t)|, |s_j(t - \Delta t)|\}. \tag{16.20}$$

In the actual estimation algorithm, **W** is updated by employing a gradient projection approach to the optimization of f in (16.20) under the constraints. The initial value of **W** is selected randomly.

The fact that the algorithm described above is able to estimate the two-layer model has been verified through extensive simulations. These simulations show that matrix **W** can be estimated fairly reliably, and that the relative error of the estimate of matrix **M** also decreases reliably in the estimation, but the remaining error for **M** is larger than in the case of matrix **W**. This difference is probably due to the approximation made in the estimation of **M**; see Hurri and Hyvärinen (2003b). However, the simulations suggest that the error in the estimate of **M** is largely due to a systematic, monotonic, non-linear element-wise bias, which does not affect greatly our interpretation of the elements of **M**, since we are mostly interested in their relative magnitudes. See Hurri and Hyvärinen (2003b) for details. A very closely related model which can be analyzed in detail is in Hyvärinen and Hurri (2004), which shows that a rigorous justification for our objective function above can be found in the case where we use the quadratic function instead of the absolute value function. See also Valpola et al. (2003) for related theoretical work.

16.6.3 Experiments on Natural Images

The estimation algorithm was run on the same data set as for the basic temporal coherence model in Sect. 16.5 to obtain estimates for **M** and **A**. Figure 16.17 shows the resulting feature vectors—that is, columns of **A**. As can be seen, the resulting features are localized, oriented, and have multiple scales, thereby fulfilling the most important defining criteria of simple-cell receptive fields. This suggests that, as far as receptive-field structure is concerned, the method yields receptive fields with similar qualitative properties to those obtained with sparse coding, ICA, or temporal response strength correlation.

What is truly novel in this model is the estimation of matrix **M**, which captures the temporal and spatiotemporal activity-level dependencies between the feature vectors shown in Fig. 16.17. The extracted matrices **A** and **M** can be visualized simultaneously by using the interpretation of **M** as a similarity matrix (see page 347). Figure 16.18 illustrates the feature vectors—that is, columns of **A**—laid out at spatial coordinates derived from **M** in a way explained below. The resulting feature

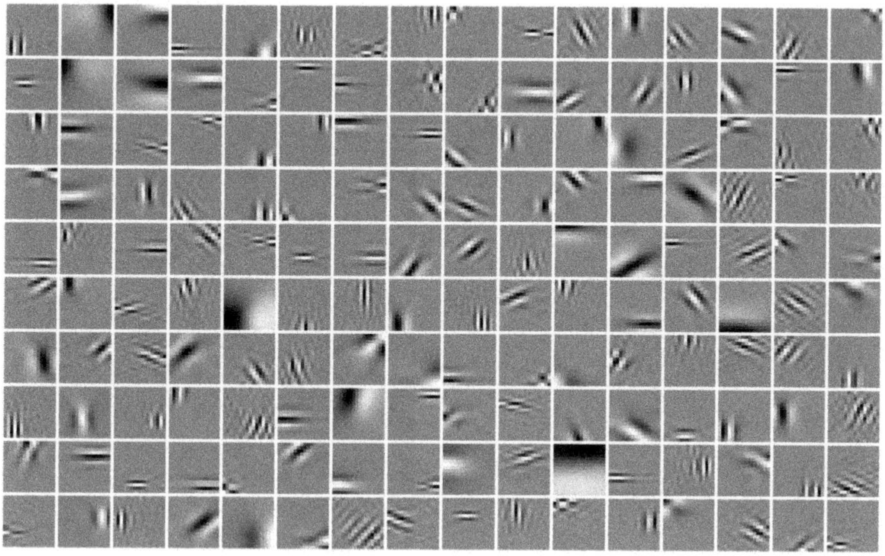

Fig. 16.17 The estimation of the two-layer generative model from natural visual stimuli results in the emergence of localized, oriented receptive fields with multiple scales. The feature vectors (columns of **A**) shown here are in no particular order

vectors are again oriented, localized and multiscale, as in the basic temporal coherence model in Sect. 16.5.

In the resulting planar representation shown in Fig. 16.18, the temporal coherence between the outputs of two cells i and j is reflected in the distance between the corresponding receptive fields: the larger the elements $\mathbf{M}(i, j)$ and $\mathbf{M}(j, i)$ are, the closer the receptive fields are to each other. We can see that local topography emerges in the results: those basis vectors which are close to each other seem to be mostly coding for similarly oriented features at nearby spatial positions. This kind of grouping is characteristic of pooling of simple cell outputs at complex cell level (Palmer 1999). Some global topography also emerges: those basis vectors which code for horizontal features are on the left in the figure, while those that code for vertical features are on the right.

Thus, the estimation of our two-layer model from natural image sequences yields both simple-cell-like receptive fields, and grouping similar to the pooling of simple cell outputs. Linear receptive fields emerge in the second layer (matrix **A**), and cell output grouping emerges in the first layer (matrix **M**). Both of these layers are estimated simultaneously. This is an important property when compared with other statistical models of early vision, because no a priori fixing of either of these layers is needed. The results thus compare with the two-layer models for static images discussed in Sect. 11.8 and Köster and Hyvärinen (2007, 2008). The main difference is that here, **M** describes "lateral" interactions between the features s_k, whereas in Sect. 11.8 considered another stage in hierarchical processing.

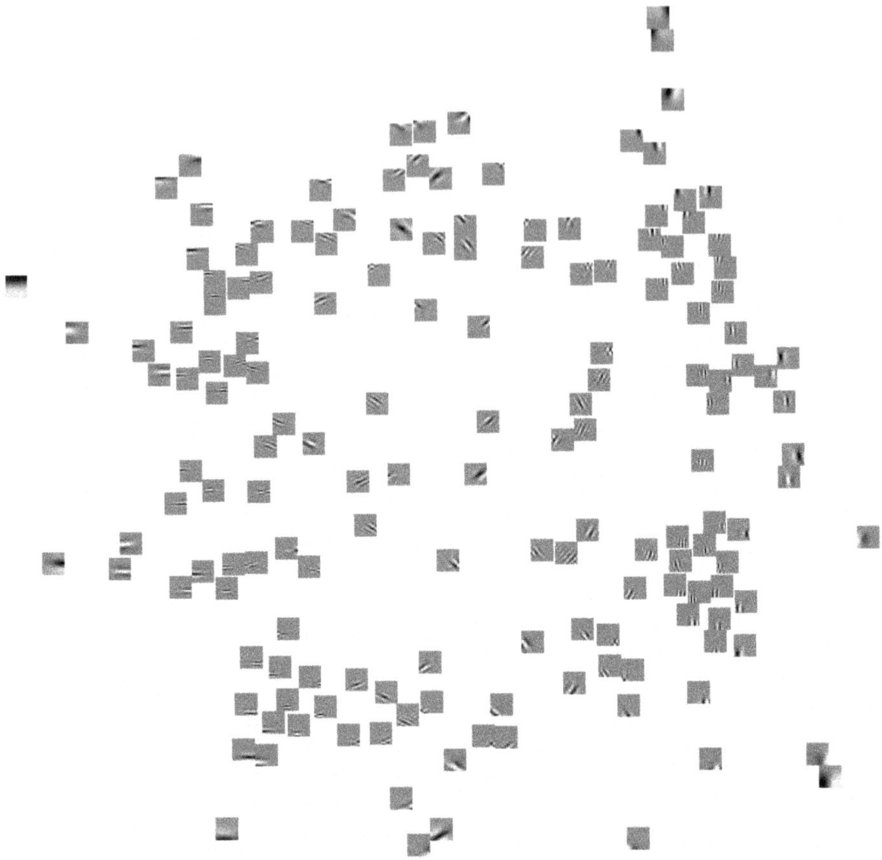

Fig. 16.18 Results of estimating the two-layer generative model from natural image sequences. Features (columns of **A**) plotted at spatial coordinates given by applying multidimensional scaling to **M**. Matrix **M** was first converted to a non-negative similarity matrix $\mathbf{M}_s$ by subtracting $\min_{i,j} \mathbf{M}(i, j)$ from each of its elements, and by setting each of the diagonal elements at value 1. Multidimensional scaling was then applied to $\mathbf{M}_s$ by interpreting entries $\mathbf{M}_s(i, j)$ and $\mathbf{M}_s(j, i)$ as similarity measures between cells i and j. Some of the resulting coordinates were very close to each other, so tight cell clusters were magnified for purposes of visual display. Details are given in Hurri and Hyvärinen (2003b)

16.6.4 Intuitive Explanation of Results

The results shown in Fig. 16.18 suggest that features which prefer similar orientation but different spatial location have spatiotemporal activity dependencies. Why is this the case?

Temporal activity-level dependencies, illustrated in Fig. 16.13, are not the only type of activity-level dependencies in a set of simple-cell-like features. Figure 16.19 illustrates how two *different* cells with similar receptive field profiles—having the same orientation but slightly different positions—respond at consecutive time in-

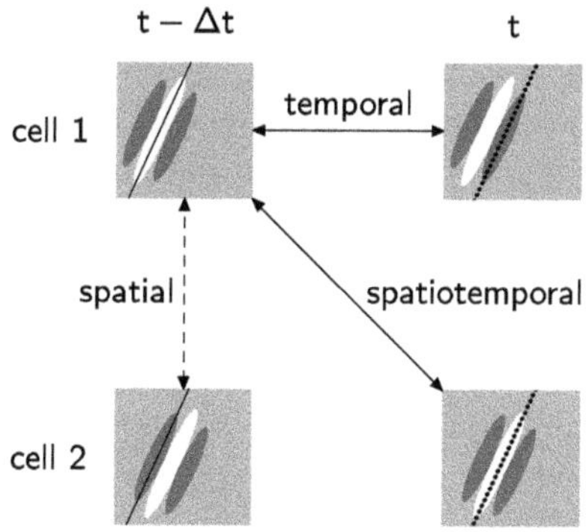

Fig. 16.19 A simplified illustration of static and short-time activity-level dependencies of simple-cell-like receptive fields. For a translating edge or line, the responses of two similar receptive fields with slightly different positions (cell 1, *top row*; cell 2, *bottom row*) are large at nearby time instances (time $t - \Delta t$, *solid line*, *left column*; time t, *dotted line*, *right column*). Each sub-figure shows the translating line superimposed on a receptive field. The magnitudes of the responses of *both* cells are large at *both* time instances. This introduces three types of activity-level dependencies: temporal (in the output of a single cell at nearby time instances), spatial (between two different cells at a single time instance) and spatiotemporal (between two different cells at nearby time instances). The multivariate autoregressive model discussed in this section includes temporal and spatiotemporal activity-level dependencies (marked with *solid lines*). Spatial activity-level dependency (*dashed line*) is an example of energy dependencies modeled in work on static images in Chaps. 9–11

stances when the input is a translating line. The receptive fields are otherwise identical, except that one is a slightly translated version of the other. It can be seen that *both cells* are highly active at *both time instances,* but again, the signs of the outputs vary. This means that in addition to temporal activity dependencies (the activity of a cell is large at time $t - \Delta t$ and time t), there are two other kinds of activity-level dependencies.

Spatial (static) dependencies Both cells are highly active at a single time instance. This kind of dependency is an example of the energy dependencies earlier modeled in static images in Chaps. 9–11.

Spatiotemporal dependencies The activity levels of different cells are also related over time. For example, the activity of cell 1 at time $t - \Delta t$ is related to the activity of cell 2 at time t.

What makes these dependencies important is that they seem to be reflected in the structure of the topography in the primary visual cortex. The results presented in this section suggest that combining temporal activity level dependencies with spatiotemporal dependencies yields both simple-cell-like receptive fields and a set of connections between these receptive fields. These connections can be related to both the way in which complex cells seem to pool simple-cell outputs, and to the topographic organization observed in the primary visual cortex, in the same way as described in Chap. 11. Therefore, the principle of activity level dependencies seems to explain both receptive field structure and their organization.

16.7 Unifying Model of Spatiotemporal Dependencies

In order to motivate the development of a model which unifies a number of statistical properties in natural image sequences, let us summarize the key results on probabilistic modeling of the properties of the neural representation at the simple-cell level.

1. Results obtained using sparse coding or independent component analysis suggest that, on the average, at a single time instant relatively few simple cells are active on the cortex (see Chap. 6); furthermore, each cell is active only rarely.
2. In this chapter, we have described a complementary model, which suggests that simple cells tend to be highly active at consecutive time instants—that is, their outputs are burst-like (see Sect. 16.5).
3. Models on static dependencies between simple-cell-like features, and the relationship between these dependencies and cortical topography, suggest that the active cells tend to be located close to each other on the cortex, as in Chap. 11.
4. As we saw in the preceding section, temporal correlations also lead to topographic properties resembling cortical topography, based on a model which utilizes temporal correlations between the outputs of different features.

These four different principles—sparseness, temporal coherence of activity levels, spatial activity level dependencies, and spatiotemporal activity level dependencies—are not conflicting. That is, none of the principles excludes the existence of another. Perhaps, then, each of these models offers just a limited view to a more complete model of cortical coding at the simple-cell level. In fact, the following description of simple-cell activation is in accordance with all of the principles: when an animal is viewing a natural scene, a relatively small number of patches of cortical area are highly active in the primary visual cortex, and the activity in these areas tends to be sustained for a while. That is, activity is sparse, and contiguous both in space and time. This is the *bubble coding* model (Hyvärinen et al. 2003).

In the bubble coding model, the final generative mapping from latent components to natural image sequence data is linear, like in the previous sections: $\mathbf{x}(t) = \mathbf{A}\mathbf{s}(t)$. The main idea in the bubble coding model is generation of the $\mathbf{s}(t)$ so that they have bubble-like activity. This is accomplished by introducing a bubble-like variance signal for $\mathbf{s}(t)$, as illustrated by an example in Fig. 16.20. The spatiotemporal locations of the variance bubbles are determined by a sparse process $\mathbf{u}(t)$ (Fig. 16.20a). A temporal filter ϕ and spatial pooling function h, both of which are fixed a priori in the model, spread the variance bubbles temporally and spatially (Figs. 16.20b and c). The resulting variance bubbles can also overlap each other, in which case the variance in the overlapping area is obtained as a sum of the variances in each bubble; in Fig. 16.20, however, the variance bubbles are non-overlapping for illustrative purposes. It is also possible that at this point a fixed static non-linearity f is applied to rescale the magnitudes of the variance bubbles. These steps yield the variance signals

$$v_k(t) = f\left(\sum_{\ell} h(k,\ell)\left[\phi(t) * u_\ell(t)\right]\right), \tag{16.21}$$

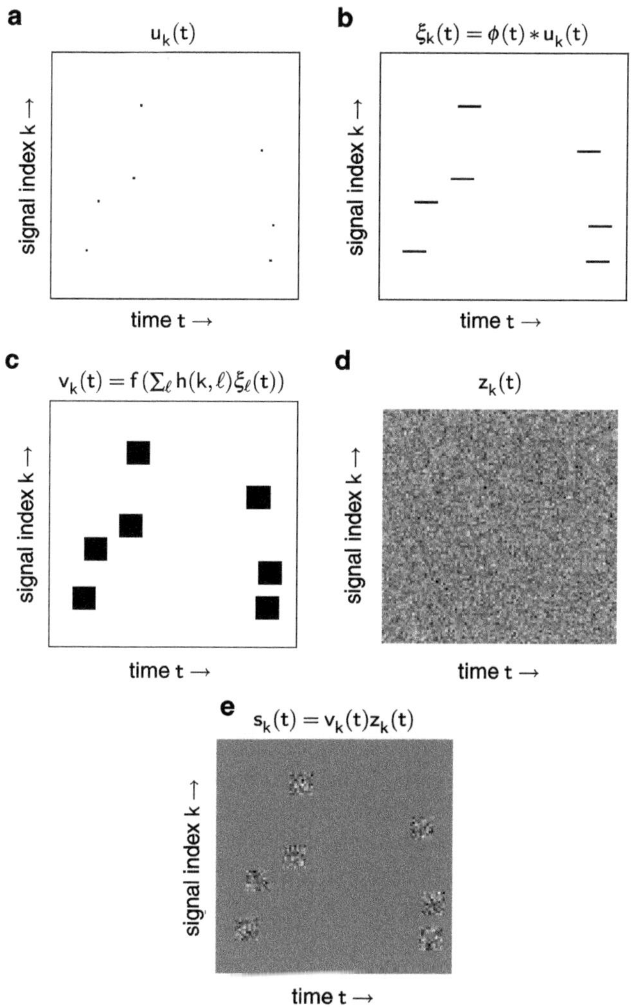

Fig. 16.20 Illustration of the generation of components $s_k(t)$ in the bubble coding model. For simplicity of illustration, we use a one-dimensional topography instead of the more conventional two-dimensional one. **a** The starting point is the set of sparse signals $u_k(t)$. **b** Each sparse signal $u_k(t)$ is filtered with a temporal low-pass filter $\phi(t)$, yielding signals $\phi(t) * u_k(t)$. In this example, the filter $\phi(t)$ simply spreads the impulses uniformly over an interval. **c** In the next step, a neighborhood function $h(k, \ell)$ is applied to spread the bubbles spatially; this is like a spatial low-pass filter. A static non-linearity f may also be applied at this point to rescale the magnitudes of the variance bubbles. This yields variance bubble signals $v_k(t) = f(\sum_\ell h(k, \ell)[\phi(t) * u_\ell(t)])$. In this example, the neighborhood function h is simply 1 close-by and 0 elsewhere, and the static non-linearity f is just the identity mapping $f(\alpha) = \alpha$. **d** Next, we generate Gaussian temporally uncorrelated (white noise) signals $z_k(t)$. **e** Linear components (responses) are defined as products of the Gaussian white noise signals and the spatiotemporally spread bubble signals: $s_k(t) = z_k(t)v_k(t)$. These are transformed linearly by the matrix **A** to give the observed image data (not shown). (Note that in sub-figures **a**–**c**, *white* denotes value zero and *black* denotes value 1, while in sub-figures **d** and **e**, *medium grey* denotes zero, and *black* and *white* denote negative and positive values, respectively)

where $*$ denotes temporal convolution. The burst-like oscillating nature of the components inside the bubbles is introduced through a Gaussian temporally uncorrelated (white noise) process $\mathbf{z}(t)$ (Fig. 16.20d). Thus, the components $s_k(t)$ are generated from the variance bubbles and the noise signals by multiplying the two together (Fig. 16.20e):

$$s_k(t) = v_k(t) z_k(t). \tag{16.22}$$

Note that all three different types of activity level dependencies—temporal, spatial, and spatiotemporal (see Fig. 16.19 on page 351)—are present in the bubble-coding model, as well as sparseness. To complete this generative model, the $s_k(t)$ are finally linearly transformed to the image using a linear transformation, as in almost all models in this book.

An approximative maximum likelihood scheme can be used to estimate the bubble coding model; details can be found in Hyvärinen et al. (2003). Note that because the pooling function h is fixed, it enforces the spatial pooling, while in the two-layer model described in the previous section, this pooling was learned from the data. The temporal smoothing (low-pass) filter ϕ is also fixed in the model.

Figure 16.21 shows the resulting spatial basis vectors, obtained when the bubble coding model was estimated from natural image sequence data. The basis consists of simple-cell-like linear receptive-field models, similar to those obtained by topographic ICA from static images (Fig. 11.4 on page 247), or using the temporal models in Sect. 16.6. The orientation and the location of the feature coded by the vectors change smoothly when moving on the topographic grid. Low-frequency basis vectors are spatially segregated from the other vectors, so there also seems to be some ordering based on preferred spatial frequency. Such an organization with respect to orientation, location, and spatial frequency is similar to the topographic ordering of simple cells in the primary visual cortex, as was discussed in Chap. 11.

One can also estimate spatiotemporal features with this model. An animated example of the resulting spatiotemporal features can be found at www.cs.helsinki.fi/group/nis/animations/bubbleanimation.gif.

The features obtained by the bubble coding models are thus hardly any different from what were obtained by the topographic ICA model, for example. The significance of the model is mainly theoretical in the sense that it gives a unified framework for understanding the different models involved.

16.8 Features with Minimal Average Temporal Change

16.8.1 Slow Feature Analysis

16.8.1.1 Motivation and History

All of the models of temporal coherence discussed above—temporal response strength correlation, the two-layer autoregressive model, and the bubble-coding

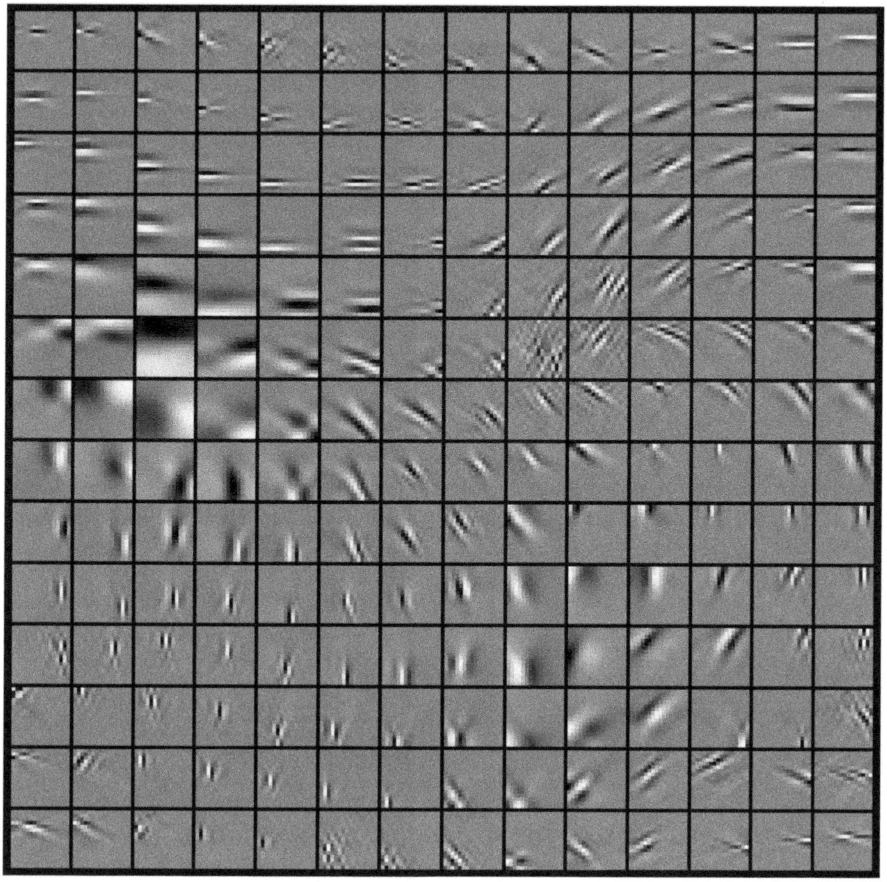

Fig. 16.21 A set of spatial features, estimated from natural image using the bubble coding estimation method, and laid out at spatial coordinates defined by the topographic grid in the bubble coding model. The topographic organization of the features exhibits ordering with respect to orientation, location, and spatial frequency of the vectors, being very similar to that obtained by topographic ICA

model—are based on temporal patterns in output energies (variances) $s^2(t)$. What happens if we just minimize a measure of the temporal change in the outputs of the model neurons? That is, if $s(t)$ is the output of a model at time t, and $\mathbf{w}$ is our vector of model parameters, we could for example minimize the squared difference of the output at close-by time points

$$f_{\mathrm{SFA}}(\mathbf{w}) = E_t\left\{\left(s(t) - s(t - \Delta t)\right)^2\right\}. \tag{16.23}$$

An explicit formalization of this principle was given by Mitchison (1991) who simply described it as "removal of time variation" (see also Hinton 1989; Földiák 1991; Stone 1996). The principle was also used in blind source separation, in which a number of sophisticated methods for its algorithmic implementation have been developed (Tong et al. 1991; Belouchrani et al. 1997; Hyvärinen et al. 2001b). Recently, the principle has been given the name *slow feature analysis* (Wiskott and Sejnowski 2002), thus the subscript SFA in the definition of the objective function.

In order to relate the SFA objective function to the models we have discussed above, let us analyze it in more detail. Expanding the square and taking the expectations of the resulting terms we get

$$\begin{aligned} f_{\mathrm{SFA}}(\mathbf{w}) &= E_t\{s^2(t) - 2s(t)s(t-\Delta t) + s^2(t-\Delta t)\} \\ &= E_t\{s^2(t)\} - 2E_t\{s(t)s(t-\Delta t)\} + \underbrace{E_t\{s^2(t-\Delta t)\}}_{=E_t\{s^2(t)\}} \\ &= 2\big(E_t\{s^2(t)\} - E_t\{s(t)s(t-\Delta t)\}\big) \end{aligned} \tag{16.24}$$

The objective function is non-negative, and a trivial way to minimize it is to compute a zero output $s(t) = 0$ for all t. A standard way to avoid this anomaly is to constrain the "energy" (second moment) of the output signal to unity, that is, define constraint

$$E_t\{s^2(t)\} = 1, \tag{16.25}$$

in which case the objective function become simpler:

$$f_{\mathrm{SFA}}(\mathbf{w}) = 2\big(1 - E_t\{s(t)s(t-\Delta t)\}\big), \tag{16.26}$$

which shows that under the unit energy constraint, SFA is equivalent to maximization of the linear temporal correlation in the output. This is in contrast to the model in Sect. 16.5 (page 336), which was based on maximization of *non-linear* temporal correlation. Also, note that if mean output is zero, that is, if $E_t\{s(t)\} = 0$, then the unit energy constraint is equivalent to the unit variance constraint, since $\mathrm{var}(s(t)) = E_t\{s^2(t)\} - (E_t\{s(t)\})^2$.

16.8.1.2 SFA in a Linear Neuron Model

In Sect. 16.5 (page 336), we mentioned that in a linear neuron model, maximization of linear temporal correlation results in receptive fields which resemble frequency components and not simple-cell-like receptive fields or Gabor functions. In such a linear model, slow feature analysis can be analyzed mathematically in detail. Let $s(t)$ denote the output of the unit at time t :

$$s(t) = \mathbf{w}^{\mathrm{T}}\mathbf{x}(t). \tag{16.27}$$

Assume that the input $\mathbf{x}(t)$ has zero mean ($E_t\{\mathbf{x}(t)\} = \mathbf{0}$), and that we impose the unit variance constraint to avoid the trivial solution $\mathbf{w} = \mathbf{0}$. Then the unit energy constraint also holds, and instead of minimizing the SFA objective function (16.26) we can just maximize linear temporal correlation

$$f_{\text{LTC}}(\mathbf{w}) = E_t\{s(t)s(t-\Delta t)\} = E_t\{\mathbf{w}^T\mathbf{x}(t)\mathbf{x}(t-\Delta t)^T\mathbf{w}\}$$
$$= \mathbf{w}^T E_t\{\mathbf{x}(t)\mathbf{x}(t-\Delta t)^T\}\mathbf{w} \tag{16.28}$$

with the constraint

$$E_t\{s^2(t)\} = 1 \iff \mathbf{w}^T E_t\{\mathbf{x}(t)\mathbf{x}(t)^T\}\mathbf{w} = 1. \tag{16.29}$$

A solution can be derived by adapting the mathematics of PCA described in Sect. 5.8.1. The connection becomes clear if we first whiten the data $\mathbf{x}(t)$ (spatially, i.e. in the same way as in Chap. 5, ignoring the temporal dependencies). For simplicity, we denote the whitened data by $\mathbf{x}(t)$ in the following. For the whitened data, the constraint of unit variance is equivalent to the constraint that $\mathbf{w}$ has unit norm, because $E_t\{\mathbf{x}(t)\mathbf{x}(t)^T\}$ is the identity matrix.

Thus, we have a maximization of a quadratic function under unit norm constraint, just as in PCA. There is a small difference, though: The matrix $E_t\{\mathbf{x}(t)\mathbf{x}(t-\Delta t)^T\}$ defining the quadratic function is not necessarily symmetric. Fortunately, it is not difficult to prove that actually, this maximization is equivalent to maximization using a symmetric version of the matrix:

$$f_{\text{LTC}}(\mathbf{w}) = \mathbf{w}^T\left[\frac{1}{2}E_t\{\mathbf{x}(t)\mathbf{x}(t-\Delta t)^T\} + \frac{1}{2}E_t\{\mathbf{x}(t-\Delta t)\mathbf{x}(t)^T\}\right]\mathbf{w}. \tag{16.30}$$

Thus, the same principle of computing the eigenvalue decomposition applies here (Tong et al. 1991). The optimal vector $\mathbf{w}$ is obtained as the one corresponding to the largest eigenvalue of this matrix. If we want to extract a set of RFs, we can use the following result: assuming that the output of the next selected RF has to be uncorrelated with the outputs of the previously selected ones, then the next maximum is the eigenvector with the next largest eigenvalue.

Figure 16.22 shows the filters that result from such optimization in a linear neuron model. The data set and preprocessing in this experiment was identical to the one in Sect. 16.5. As can be seen, the resulting filters correspond to frequency (Fourier) features, and not the localized RFs in the early visual system.

16.8.2 Quadratic Slow Feature Analysis

As shown above, SFA in the case of a linear neuron model does not produce very interesting results. In contrast, application of the principle in the non-linear case has proven more promising, although results for real natural image sequences have not been reported.

A straightforward and computationally simple way to design non-linear models is the *basis expansion* approach. As a simple example, assume that out original input data is a single scalar x. This data can be expanded by computing the square

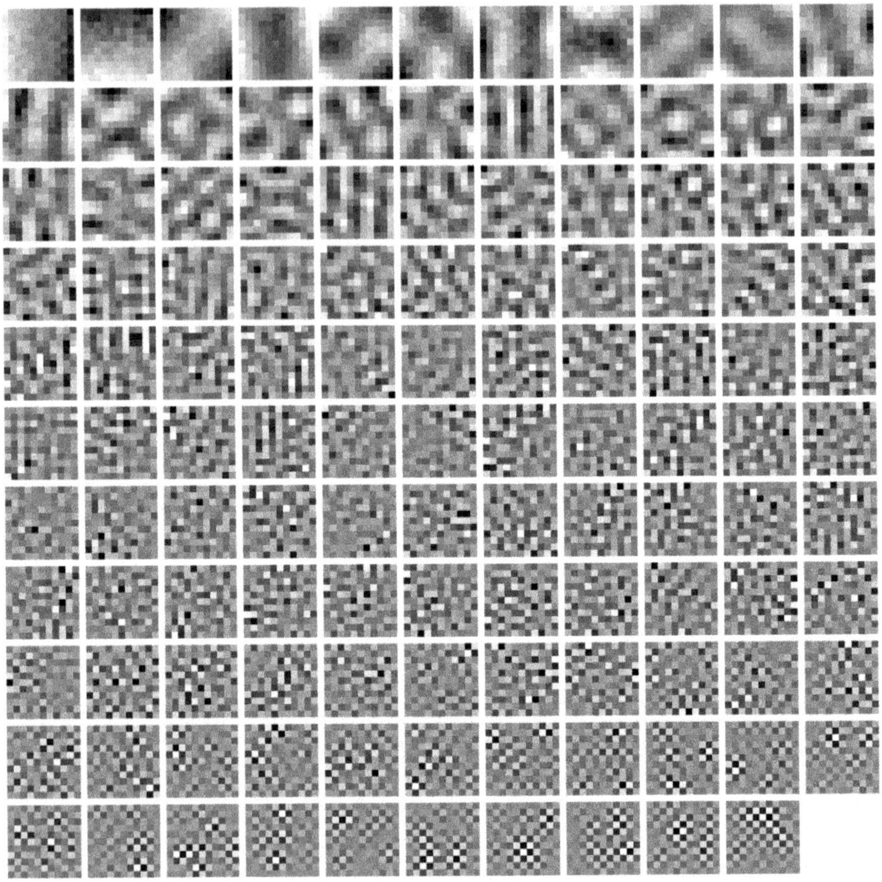

Fig. 16.22 The set of filters which optimize the objective function in slow feature analysis from natural image data in the case of a linear neuron model. As can be seen, the resulting filters do not resemble localized receptive fields of either retina/LGN or V1

of the data point x^2. We can then design a model that is linear in the parameters $\mathbf{a} = (a_1\ a_2)^\mathrm{T}$

$$y = a_1 x + a_2 x^2 = \mathbf{a}^\mathrm{T} \begin{pmatrix} x \\ x^2 \end{pmatrix} \tag{16.31}$$

but obviously non-linear in the data x (here, it is quadratic). A very nice property of this approach is that the analysis developed for the linear case is immediately applicable: all we need to do is to replace the original data with the expanded data. In the SFA case, let $\mathbf{f}(\mathbf{x}(t)) = [f_1(\mathbf{x}(t))\ f_2(\mathbf{x}(t))\ \ldots\ f_M(\mathbf{x}(t))]^\mathrm{T}$ denote a non-linear expansion of the data. Then the output is

$$s(t) = \mathbf{w}^\mathrm{T}\mathbf{f}\big(\mathbf{x}(t)\big), \tag{16.32}$$

and all of the optimization results apply after $\mathbf{x}(t)$ is replaced with $\mathbf{f}(\mathbf{x}(t))$.

The form of basis expansion we are particularly interested in is that of a quadratic basis. Let $\mathbf{x}(t) = [x_1(t)\ x_2(t)\ \cdots\ x_K(t)]^T$ denote our original data. Then a data vector in our quadratic data set $\mathbf{f}(\mathbf{x}(t))$ includes, in addition to the original components $x_1(t), x_2(t), \ldots, x_K(t)$, the products of all pairs of components $x_k(t)x_\ell(t)$, $k = 1, \ldots, K$, $\ell = 1, \ldots, K$; note that this also includes the squares $x_1^2(t), x_2^2(t), \ldots, x_K^2(t)$.

While the computation of the optimum in the case of basis-expanded SFA is straightforward, the interpretation of the results is more difficult: unlike in the case of linear data, the obtained parameters can not simply be interpreted as a template of weights at different positions in the image, because some of the parameters correspond to the quadratic terms $x_k(t)x_\ell(t)$, $k = 1, \ldots, K$, $\ell = 1, \ldots, K$. One way to analyze the learned parameter vectors $\mathbf{w}_1, \ldots, \mathbf{w}_M$ is to compute the input images that elicit maximal and minimal responses, while constraining the norm (energy) of the images to make the problem well posed; that is, if $c > 0$ is a constant

$$\mathbf{x}_{\max} = \arg \max_{\|\mathbf{x}\|=c} \mathbf{w}^T \mathbf{f}(\mathbf{x}), \tag{16.33}$$

$$\mathbf{x}_{\min} = \operatorname*{argmin}_{\|\mathbf{x}\|=c} \mathbf{w}^T \mathbf{f}(\mathbf{x}). \tag{16.34}$$

A procedure for finding $\mathbf{x}_{\max}$ and $\mathbf{x}_{\min}$ is described in Berkes and Wiskott (2007).

Berkes and Wiskott (2005) applied quadratic SFA to *simulated* image sequence data: the image sequences $\mathbf{x}(t)$ were obtained from static natural images by selecting an initial image location, obtaining as $\mathbf{x}(0)$ from the location with random orientation and zoom factor, and then obtaining $\mathbf{x}(t)$, $t > 0$, by applying all of the following transformations at the location of $\mathbf{x}(t-1)$: translation, rotation, and zooming. An important property of the results obtained by Berkes and Wiskott (2005) using SFA with simulated image sequence data is the phase invariance of the quadratic units. This has lead to an association between SFA and complex cells; in fact, Berkes and Wiskott (2005) report a number of observed properties in quadratic SFA models that match those of complex cells. See Fig. 16.23 for a reproduction of those results. However, the results by Berkes and Wiskott were obtained by *simulated* image sequences, whose temporal correlations were basically determined by the experimenters themselves. Thus, they do not really provide a basis for making conclusions about the connection between ecologically valid stimuli and visual processing.

Hashimoto (2003) applied quadratic SFA to *real* natural image sequences. She found that the obtained features were only weakly related to complex cells, and proposed that better results could be found by a *sparse* variant of SFA. This will be treated next.

16.8.3 Sparse Slow Feature Analysis

In sparse SFA, the measure of change is changed to one that emphasizes sparseness. In the original objective in (16.23), it is not necessary to take the squared error.

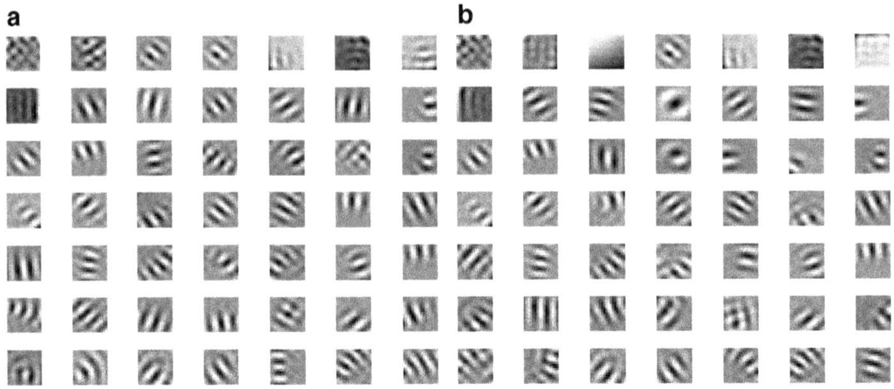

Fig. 16.23 Quadratic SFA of image sequence data *generated from static image data*. **a** Input images $\mathbf{x}_{\max}$ that correspond to maximum output. **b** Input images $\mathbf{x}_{\min}$ that correspond to minimum output. The maximum and minimum input images are at corresponding locations in the lattices. Most of the optimal input images are oriented and bandpass, and also spatially localized to some degree. The maximum and minimum input images of the units have interesting relationships; for example, they may have different preferred orientations, different locations, or one can be orientation-selective while the other is not

The squared error is used here for algebraic and computational simplicity only: it allows us to maximize the objective function using the eigenvalue decomposition. In general, we can consider objective function of the form

$$f_{\mathrm{SSFA}}(\mathbf{w}) = E_t\big\{G\big(s(t) - s(t - \Delta t)\big)\big\}. \tag{16.35}$$

where G is some even-symmetric function. A statistically optimal choice of G is presumably one that corresponds to a sparse pdf because changes in images are usually abrupt, as in edges. We call resulting model Sparse SFA. The statistical justification is based on modeling the data with an autoregressive model in which the driving noise term (innovation process) is super-Gaussian or sparse (Hyvärinen 2001b).

The same sparse model (using a G which is not the square function) can be used in the context of quadratic SFA because quadratic SFA simply means defining the input data in a new way. This leads to the concept of sparse quadratic SFA. It is important not to confuse the two ways in which an SFA model can be quadratic: It can use squared error (i.e. take $G(u) = u^2$), or it can use a quadratic expansion of the input data (using products of the original input variables as new input variables).

Using sparse quadratic SFA, Hashimoto (2003) obtained energy detectors which seem to be much closer to quadrature-phase filter pairs and complex cells than those obtained by ordinary quadratic SFA. Those results were obtained on real natural image sequences, from which ordinary quadratic SFA does not seem to learn very complex-cell like energy detectors.

Thus, we see how sparseness is ubiquitous in natural image statistics modeling, and seems to be necessary even in the context of SFA.

16.9 Conclusion

Different models of temporal coherence have been applied to simulated and natural visual data. The results that emerge from these models agree with neurophysiological observations to varying degrees. Thus, far the principle has mostly been applied to model V1 cells—that is, simple and complex cells. In this chapter we focused on models which have resulted in the emergence of *spatially localized* filters with multiple scales (responding to different frequencies) from *natural* image sequence data. That is, we required that the spatial localization has not been forced in the model, but emerges from learning, as happened in all the sparse coding and ICA-related models treated in this book so far; this is in contrast to some temporal coherence models, in which spatial localization is *enforced* by sampling with a Gaussian weighting window, so that the RFs are then necessarily localized in the center of the patch. Also, we required that the image sequences come from a video camera or a similar device, which is in contrast to some work in which one takes static images and then artificially creates sequence by sampling from them. Further work on temporal coherence, in addition to the work already cited above, include Kohonen (1996), Kohonen et al. (1997), Bray and Martinez (2003), Kayser et al. (2003), Körding et al. (2004), Wyss et al. (2006).

A more philosophical question concerns the priorities between models of static images and image sequences. We have seen models which produce quite similar results in the two cases. For example, simple cell RFs can be learned by sparse coding and ICA from static natural images, or, alternatively, using temporal coherence from natural image sequences. Which model is then more "interesting"? This is certainly a deep question which depends very much of the justification of the assumptions in the models. Yet, one argument can be put forward in general: We should always prefer the simpler model, if both models have the same explanatory power. This is a general principle in modeling, called parsimony, or Occam's razor. In our case, it could be argued that since static images are necessarily simpler than image sequences, we should prefer models which use static images—at least if the models have similar conceptual simplicity. Thus, one could argue that models on image sequences are mainly interesting if they enable learning of aspects which cannot be learned with static images. This may not have been the case with many models we considered in this chapter; however, the principles introduced may very well to lead to discoveries of new properties which cannot be easily, if at all, found in static images.

An important related question concerns learning image transformations (Memisevic and Hinton 2007). One can view image sequences from a viewpoint where each image (frame) is a transformation of the preceding one. This is, in a sense, complementary to the viewpoint of temporal coherence, in which one tries to capture features which are not transformed. It also seems to be closely related to the idea of predictive coding; see Sect. 14.3.

Part V
Conclusion

Chapter 17
Conclusion and Future Prospects

In this chapter, we first provide a short overview of this book. Then we discuss some open questions in the field, as well as alternative approaches to natural image statistics which we did not consider in detail in this book. We conclude with some remarks on possible future developments.

17.1 Short Overview

We started this book in Chap. 1 by motivating the research on natural image statistics from an ecological-statistical viewpoint: The visual system is assumed to be adapted to the statistical structure of natural images because it needs to use Bayesian inference. Next, we prepared the reader by introducing well-known mathematical tools which are needed in natural image statistics models (Part I, i.e. Chaps. 2–4). The rest of the book, up to this chapter, was mainly a succession of different statistical models for natural images.

Part II was dedicated to models using purely linear receptive fields. The first model we considered was principal component analysis in Chap. 5. It is an important model for historical reasons, and also because it provides a preprocessing method (dimension reduction accompanied by whitening) which is used in most subsequent models. However, it does not provide a proper model for receptive fields in the primary visual cortex (V1).

In Chap. 6, the failure of principal component analysis was explained as the failure to consider the sparse, non-Gaussian structure of the data. In fact, natural images have a very sparse structure; the outputs of typical linear filters all have strongly positive kurtosis. Based on this property, we developed a method in which we find the feature weights by maximizing the sparseness of the output when the input to the feature detectors is natural images. Thus, we obtained a fundamental result: sparse coding finds receptive fields which are quite similar to those in V1 simple cells in the sense that they are spatially localized, oriented, and band-pass (localized in Fourier space).

Chapter 7 further elaborated on the linear sparse coding model and brought it firmly into the realm of generative probabilistic models. It also brought the viewpoint of independence: instead of maximizing the sparseness of single features, we can maximize their statistical independence. A most fundamental theoretical results says that these two goals are equivalent for linear features. The resulting model has been named independent component analysis (ICA) in the signal-processing literature. An information-theoretic interpretation of the model was considered Chap. 8, as an alternative to the Bayesian one.

Part III took a clear step forward by introducing non-linear feature detectors. It turns out that independent component analysis is not able to cancel all the depen-

Computational Imaging and Vision 39,

dencies between the components, despite the name of the method. If we measure the dependencies of components given by ICA by computing different kinds of correlations, we see that the squares of the components tend to be strongly correlated (Chap. 9). Such squares are called "energies" for historical reasons. We can model such dependencies by introducing a random variable which controls the variances of all the components at the same time. This enables the reduction of the dependencies based on processing which is quite similar to neurophysiological models of interactions between cells in V1, based on divisive normalization.

In Chap. 10, we used the same kind of energy dependencies to model strongly non-linear features. Here, the non-linearities took the form of computing the squares of linear feature detectors and summing ("pooling") such squares together. Just as with linear features, we can maximize the sparseness of such non-linear features when the input is natural images. The resulting features are quite similar to complex cells in V1. Again, we can build a probabilistic model, independent subspace analysis, based on this maximization of sparseness. Interestingly, the model can also be considered a non-linear version of independent component analysis.

The same idea of maximization of sparseness of energies was extended to model the spatial ("topographic") arrangement of cells in the cortex in Chap. 11. This model is really a simple modification of the complex cells model of the preceding chapter. We order the simple cells or linear feature detectors on a regular grid, which thus defines which cells are close to each other. Then we maximize the sparsenesses of energy detectors which pool the energies of close-by simple cells; we take the sum of such sparsenesses over all grid locations. This leads to a spatial arrangement of linear features which is similar to the one in V1 in the sense that preferred orientations and frequencies of the cells change smoothly when we move on the grid (or cortex); the same applies to the locations of the centers of the features. Because of the close connection to the complex cell model, the pooled energies of close-by cells (i.e. sums of squares of feature detectors which are close to each other on the grid) have the properties of complex cells just like in the preceding chapter.

In Chap. 12, even more complicated non-linear features were learned, although we didn't introduce any new probabilistic models. The trick was to fix the initial feature extraction to computation of energies as in complex cell models. Then we can just estimate a linear model, basic ICA, of the outputs of such non-linear feature detectors. Effectively, we are then estimating a hierarchical three-layer model. The results show that there are strong dependencies between outputs of complex cells which are collinear, even if they are in different frequency bands. Thus, the learned features can be interpreted as short line segments which are, in contrast to the features computed by simple or complex cells, not restricted to a single frequency band (and they are also more elongated).

In Chap. 13, we went back to the basic linear models such as ICA, and introduced two important extensions. First, we considered the case where the number of components is arbitrarily large, which results in what is called an overcomplete basis. Overcomplete bases seem to be important for building a good probabilistic model, although the receptive fields learned are not unlike those learned by basic ICA. This is related to Markov random fields which may allow extension of the models to whole images instead of image patches. The second extension was to consider non-negativity constraints.

To conclude Part III, we showed how the concept of feedback emerges naturally from Bayesian inference in models of natural image statistics (Chap. 14). Feedback can be interpreted as a communication between different feature sets to compute the best estimates of the feature values. Computing the values of the features was straightforward in models considered earlier, but if we assume that there is noise in the system or we have an overcomplete basis, things are much more complicated. The features are interpreted latent (hidden) random variables, and computing the optimal Bayesian estimates of the features is a straightforward application of Bayesian inference. However, computationally it can be quite complicated, and requires updating estimates of some features based on estimates of others; hence the need for feedback from one cortical area to another, or between groups of cells inside the same cortical area. This topic is not yet well developed, but holds great promise to explain the complicated phenomena of cortical feedback which are wide-spread in the brain.

Part IV considered images which are not simple static grey-scale images. For color images and stereo images (mimicking the capture of visual information by the two eyes), ICA gives features which are similar to the corresponding processing in V1, as shown in Chap. 15. For motion (Chap. 16), the same is true, at least to some extent; more interestingly, motion leads to a completely new kind of statistical property, or learning principle. This is temporal coherence or stability, which is based on finding features which change slowly.

17.2 Open, or Frequently Asked, Questions

Next, we consider some questions on the general framework and fundamental assumptions adopted in this book.

17.2.1 What Is the Real Learning Principle in the Brain?

There has been some debate on what is the actual learning principle which the visual cortex "follows", or which it should follow. There are really two questions here: What is the learning principle which the brain *should* follow according to the ecological-statistical approach, and what is the learning principle which best explains the functioning of the brain. Answering the latter question seems impossible considering our modest knowledge of the workings of the visual cortex, but the former question needs some comment because it may seem the existing theory provides several contradictory answers.

In fact, in this book, we saw a few different proposals for the learning principle: sparseness in Chap. 6, independence in Chap. 7, and temporal coherence in Chap. 16. However, in our view, there is no need to argue which one of these is the best since they are all subsumed under the greater principle of describing the statistical structure of natural images as well as possible.

Having a good statistical model of the input is what the visual system needs in order to perform Bayesian inference. Yet, it is true that Bayesian inference may not be the only goal for which the system needs input statistics. Sparse coding, as well as topography, may be useful for reducing metabolic costs (Sects. 6.5 and 11.5). Information theoretic approaches (Chap. 8) assume that the ultimate goal is to store and to transmit the data in noisy channels of limited capacity—the limited capacity being presumably due to metabolic costs.

Our personal viewpoint is that the image analysis and pattern recognition are so immensely difficult tasks that the visual system needs to be optimized to perform them. Metabolic costs may not be such a major factor in the design of the brain. However, we admit that this is, at best, an educated guess, and future research may prove it to be wrong.

In principle, we can compare different models and learning principles as the theory of statistical estimation gives us clear guidelines on how to measure how well a model describes a data set. There may not be a single answer, because one could use, for example, either the likelihood or the score matching distance. However, these different measures of model fit are likely to give very similar answers on which models are good and which are not. In the future, such calculations may shed some light on the optimality of various learning principles.

17.2.2 Nature vs. Nurture

One question which we have barely touched is whether the formation of receptive fields is governed by genes or the input from the environment. One answer to this question is simply that we don't care: the statistical models are modeling the final result of genetic instructions and individual development, and we don't even try to figure out which part has what kind of contribution. The question of nature vs. nurture seems to be highly complex in the case of the visual system, and trying to disentangle the two effects has not produced very conclusive results.

What makes the situation even more complicated in vision research is that there is ample evidence that *pre-natal* experience in the uterus has an important effect on the receptive field properties; see Wong (1999) for a review. In fact, the retinal ganglion cells exhibit spontaneous activity which is characterized by synchronized bursts, and they generate waves of activity that periodically sweep across the retina. If such "traveling waves" are disrupted by experimental manipulations, the development of the visual cortex suffers considerably (Cang et al. 2005).

Spontaneous retinal waves may, in fact, be considered as a primitive form of visual stimuli from which the visual cortex might learn in rather similar ways as it learns from natural images. Application of ICA on such traveling waves can generate something similar to ICA of natural images. Thus, traveling waves may be a method of enabling the rudimentary learning of some basic receptive field properties even before the eyes receive any input. One might speculate that such waves are a clever method, devised by evolution, of simulating some of the most fundamental statistical properties of natural images. Such learning would bridge the gap between nature and nurture, since it is both innate (present at birth) and learned from "stimuli" external to the visual cortex (Albert et al. 2008).

17.2.3 How to Model Whole Images

Our approach was based on the idea of considering images as random vectors. This means, in particular, that we neglect their two-dimensional structure, and the fact that different parts of the image tend to have rather similar statistical regularities. Our approach was motivated by the desire to be sure that the properties we estimate are really about the statistics of images and not due to our assumptions. The downside is that this is computationally a very demanding approach: the number of parameters can be very large even for small image patches, which means that we need large amounts of data and the computational resources needed can be near the limit of what is available—at the time of this writing.

The situation can be greatly simplified if we assume that the dependencies of pixels are just the same regardless of whether the pixels considered are in, say, the upper-left corner of the image, or in the center. We have already considered one approach based on this idea, Markov random fields in Sect. 13.1.7, and wavelet approaches to be considered below in Sect. 17.3.2 are another.

Wavelet theory has been successfully used in many practical engineering tasks to model whole images. A major problem is that it does not really answer the question of what are the statistically optimal receptive fields; the receptive fields are determined largely by mathematical convenience, the desire to imitate V1 receptive fields, or, more recently, the desire to imitate ICA results.

On the other hand, the theory of Markov random fields offers a promising alternative in which we presumably can estimate receptive fields from natural images, as well as obtain a computationally feasible probability model for Bayesian inference. However, at present, the theory is really not developed enough to see whether that promise will be fulfilled.

17.2.4 Are There Clear-Cut Cell Types?

There has been a lot of debate about the categorization of V1 cells into simple and complex cells. Some investigators argue that the cells cannot be meaningfully divided into two classes. They argue that there is a continuum of cell types, meaning that there are many cells which are between the stereotypical simple cells and complex cells.

Consider some quantity (such as phase-variance) which can be measured from cells in the primary visual cortex. The basic point in the debate is whether we can find a quantity such that its distribution is bimodal. This is illustrated in Fig. 17.1. In some authors' view, only such bi-modality can justify classification to simple

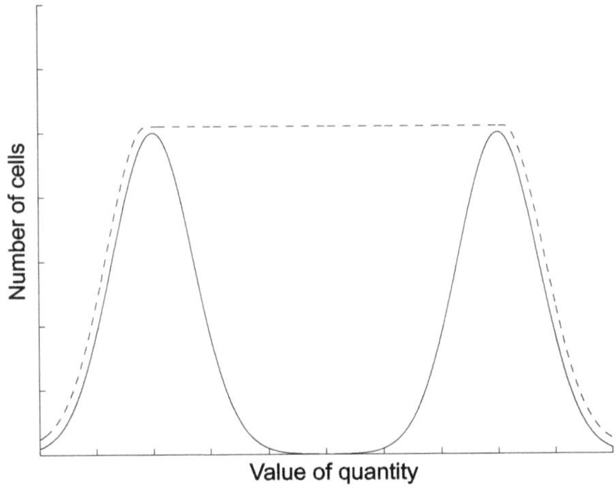

Fig. 17.1 Hypothetical histogram of some quantity for cells in the primary visual cortex. Some authors argue that the histogram should be bimodal (*solid curve*) to justify classification of cells into simple and complex cells. On the other hand, even if the distribution is flat (*dashed curve*), characterizing the cells at the two ends of the distribution may be an interesting approach, especially in computational models which always require some level of abstraction

and complex cells. Thus, there is not much debate on whether there are some cells which fit the classical picture of simple cells, and others which fit the complex cell category. The debate is mainly on whether there is a clear distinction or division between these two classes.

This debate is rather complicated because there are very different dimensions in which one could assume simple and complex cells to be form two classes. One can consider, for example, their response properties as measured by phase-invariance, or some more basic physiological or anatomical quantities in the cells. It has, in fact, been argued that even marked differences in response properties need not imply any fundamental physiological difference which would justify considering two different cell types (Mechler and Ringach 2002).

A related debate is on the validity of the hierarchical model, in which complex cell responses are computed from simple cell responses. It has been argued that complex cell responses might be due to lateral connections in a system with no hierarchical distinction between simple and complex cells (Chance et al. 1999). This dimension, hierarchy vs. lateral connections, might be considered another dimension along which the bi-modality or flatness of the distribution could be considered.

We would argue that this debate is not necessarily very relevant for natural image statistics. Even if the distribution of simple and complex cells is not bimodal with respect to any interesting quantity, it still makes sense to model the two ends of the distribution. This is a useful abstraction even if it neglects the cells in the middle of the distribution. Furthermore, if we have models of the cells at the ends of the "spectrum", it may not be very difficult to combine them into one to provide a more complete model. In any case, mathematical and computational modeling always require some level of abstraction; this includes classification of objects into categories which are not strictly separated in reality.[1]

[1]In fact, when we talk about response properties of a cell, there is always a certain amount of abstraction involves since the response properties change (adapt) depending on various parame-

17.2.5 How Far Can We Go?

So far, the research on natural image statistics has mainly been modeling V1, using the classic distinction of simple and complex cells. Chapter 12 presented an attempt to go beyond these two processing layers. How far is it possible to go with this modeling approach?

A central assumption in natural image statistics research is that learning is *unsupervised*. In the terminology of machine learning, this means learning in which we do not know what is good and what is bad; nor do we know what is the right output of the system in contrast to classic regression methods. Thus, if the system knows that bananas are good (in the sense of increasing some objective function), we are in a domain which is perhaps outside of natural image statistics. So, the question really is: How much of the visual system is involved in processing which applies equally to all stimuli, and does not require knowledge of what the organism needs?

Unsupervised learning may be enough for typical signal-processing tasks such as noise reduction and compression. Noise reduction should be taken here in a very general sense, including operations such as contour completion. More sophisticated tasks which may be possible in an unsupervised setting include segmentation, amodal completion (completion of occluded contours), and various kinds of filling-in of image areas which are not seen due to anatomical restrictions or pathologies.

Certainly, there need not be any clear-cut distinction between processing based on unsupervised learning and the rest. For example, the system might be able to perform a rudimentary segmentation based on generic knowledge of natural image statistics; if that results in recognition of, say, a banana, the prior knowledge about the banana can be used to refine the segmentation. That is, knowledge of the general shapes of objects can be complemented by knowledge about specific objects, the latter being perhaps outside of the domain of natural image statistics.

The greatest obstacle in answering the question in the title of this section is our lack of knowledge of the general functioning of the visual system. We simply don't know enough to make a reasonable estimate on which parts could be modeled by natural image statistics. So, it may be better to leave this question entirely to future research.

17.3 Other Mathematical Models of Images

In this book, data-driven analysis was paramount: We took natural images and analyzed them with models which are as general as possible. A complementary approach is to construct mathematical models of images based on some theoretical assumptions, and then find the best representation. The obvious advantage is that the features can be found in a more elegant mathematical form, although usually

ters. For example, the contrast level may change the classification of a cell to simple or complex (Crowder et al. 2007).

not as a simple formula. The equally obvious disadvantage is that the utility of such a model crucially depends on how realistic the assumptions are. An ancestor of this line of research is Fourier analysis, as well as the more elaborate Gabor analysis, which were discussed in Chap. 2.

In this section, we consider some of the most important models developed using this approach.

17.3.1 Scaling Laws

Most of the mathematical models in this section are related to scaling laws, which are one of the oldest observations of the statistical structure of natural images. Scaling laws basically describe the approximately $1/f^2$ behavior of the power spectrum which we discussed in Sect. 5.6.1. As reviewed in Srivastava et al. (2003), they were first found by television engineers in the 1950s (Deriugin 1956; Kretzmer 1952).

The observed scaling is very closely related to scale-invariance. The idea is to consider how natural image statistics change when you look at natural images at different scales (resolutions). The basic observation, or assumption, is that they don't: natural images look just the same if you zoom in or zoom out. Such scale-invariance is one of the basic motivations of a highly influential theory of signal and image analysis: wavelet theory, which we consider next.

17.3.2 Wavelet Theory

Beginning from the 1980s, the theory of wavelets became very prominent in signal and image processing. Wavelets provide a basis for one-dimensional signals; the basis is typically orthogonal. The key idea is that all the basis vectors (or functions, since the original formulation uses a continuous formalism) are based on a *single* prototype function called the mother wavelet $\phi(x)$. The functions in the wavelet basis are obtained by translations $\phi(x+l)$ and "dilations" (rescalings) $\phi(2^{-s}x)$:

$$\phi_{s,l}(x) = 2^{-s/2}\phi(2^{-s}x - l) \tag{17.1}$$

where s and l are integers that represent scale and translation, respectively. The fundamental property of a wavelet basis is *self-similarity*, which means that the same function is used in different scales without changing it shape. This is motivated by the scale-invariance of natural images. Wavelet theory can be applied on signals sampled with a finite resolution by considering discretized versions of these functions, just like in Fourier analysis we can move from a continuous-time representation to a discretized one.

Much of the excitement around wavelets is based on mathematical analysis which shows that the representation is optimal for several statistical signal-processing tasks such as de-noising (Donoho et al. 1995). However, such theoretical results always assume that the input data comes from a certain theoretical distribution (in this case, from certain function spaces defined using sophisticated functional analysis theory). Another great advantage is the existence of fast algorithms for computing the coefficients in such a basis (Mallat 1989).

This classic formulation of wavelets is for one-dimensional signals, which is a major disadvantage for image analysis. Although it is straightforward to apply a one-dimensional analysis on images by first doing the analysis in, say, the horizontal direction, and then the vertical direction, this is not very satisfactory because then the analysis does not properly contain features of different orientations. In practical image processing, the fact that basic wavelets form an orthogonal basis may also be problematic: It implies that the number of features equals the number of pixels, whereas in engineering applications, an overcomplete basis is usually required. Various bases which are similar to wavelets, but better for images, have therefore been developed. In fact, the theory of multiresolution decompositions of images was one of the original motivations for the general theory of wavelets (Burt and Adelson 1983).

Wavelet-like bases specifically developed for images typically include features of different orientations, as well as some overcompleteness. "Steerable pyramids" are based on steerable filters and, therefore, provide implicitly all possible orientations, see, e.g. Freeman and Adelson (1991), Simoncelli et al. (1992). One of the most recent systems is "curvelets". Curvelets can be shown to provide an optimally sparse representation of edges (Candès et al. 2005), thus providing a basis set which is mathematically well defined and statistically optimal. However, such strong theoretical optimality results only come at the cost of considering edge representation in an artificial setting, and their relevance to natural images remains to be investigated. In any case, such ready-made bases may be very useful in engineering applications.

An interesting hybrid approach is to use a wavelet basis which is *partly* learned (Olshausen et al. 2001; Sallee and Olshausen 2003), thus bridging the wavelet theory and the theory in this book; see also Turiel and Parga (2000).

17.3.3 Physically Inspired Models

Another line of research models the process which generated natural images in the first place. As with wavelet analysis, scale-invariance plays a very important role in these models. In the review by Srivastava et al. (2003), these models were divided into two classes, superposition models and occlusion models.

In the superposition models (Mumford and Gidas 2001; Grenander and Srivastava 2001), it is assumed that the images are a linear sum of many independent "objects". In spirit, the models are not very different from the linear superposition we have encountered ever since the ICA model in Chap. 7. What is different from ICA is that first, the objects come from a predefined model which is not learned, and second, the predefined model is typically richer than the one used in ICA. In fact, the objects can be from a space which defines different sizes, shapes, and textures (Grenander and Srivastava 2001). One of the basic results in this line of research is to show that such superposition models can exhibit both scale-invariance and non-Gaussianity for well-chosen distributions of the sizes of the objects (Mumford and Gidas 2001).

In occlusion models, the objects are not added linearly; they can occlude each other if placed close to each other. An example is the "dead leaves" model, which was originally proposed in mathematical morphology, see Srivastava et al. (2003). It can be shown that scale-invariance can be explained with this class of models as well (Ruderman 1997; Lee et al. 2001).

17.4 Future Work

Modern research in natural statistics essentially started in the mid-1990s with the publication of the seminal sparse coding paper by Olshausen and Field (1996). It coincided with a tremendous increase of interest in independent component analysis (Comon 1994; Bell and Sejnowski 1995; Delfosse and Loubaton 1995; Cardoso and Laheld 1996; Amari et al. 1996; Hyvärinen and Oja 1997) and the highly influential work by Donoho, Johnstone and others on application of wavelets to statistical signal processing (Donoho et al. 1995; Donoho 1995; Donoho and Johnstone 1995). What we have tried to capture in this book is the developments of these ideas in the last 10–15 years.

What might be the next wave in natural image statistics? Multilayer models are seen by many as the Holy Grail, especially if we were able to estimate an arbitrary number of layers, as in classical multilayer perceptrons. Markov random fields may open the way to new successful engineering applications even if their impact on neuroscientific modeling may be modest. Some would argue that image sequences are the key because their structure is much richer than those of static images.

People outside of the mainstream natural image statistics research might put forward arguments in favor of embodiment, i.e. we cannot dissociate information processing from behavior, and possibly not from metabolic needs either. This would mean we need research on robots, or simulated robot-like agents, which interact with their environment. On the other hand, science has often advanced faster when it *has* dissociated one problem from the rest; it may be that using robots makes modeling technically too difficult.

Whatever future research may bring, natural image statistics seems to have consolidated its place as the dominant functional explanation of why V1 receptive fields are as they are. Hopefully, it will lead to new insights on how the rest of the visual system works. Combined with more high-level theories of pattern recognition by Bayesian inference, it has the potential of providing a "grand unified theory" of visual processing in the brain.

Part VI
Appendix:
Supplementary Mathematical Tools

Chapter 18
Optimization Theory and Algorithms

In this book, we have considered features which are defined by some optimality properties, such as maximum sparseness. In this chapter, we briefly explain how those optimal features can be numerically computed. The solutions are based either on general-purpose optimization methods, such as gradient methods, or specific tailor-made methods such as fixed-point algorithms.

18.1 Levels of Modeling

First, it is important to understand the different levels on which we can model vision. A well-known classification is due to Marr (1982), who distinguished between the computational, algorithmic, and implementation levels. In our context, we can actually distinguish even more levels. We can consider, at least, the following different levels:

1. *Abstract principle.* The modeling begins by formulating an abstract optimality principle for learning. For example, we assume that the visual system should have a good model of the statistical properties of the typical input, or that the representation should be sparse to decrease metabolic costs.
2. *Probabilistic model.* Typically, the abstract principle leads to a number of concrete quantitative models. For example, independent component analysis is one model which tries to give a good model of the statistics of typical input.
3. *Objective function.* Based on the probabilistic model, or sometimes directly using the abstract principle, we formulate an objective function which we want to optimize. For example, we formulate the likelihood of a probabilistic model.
4. *Optimization algorithm.* This is the focus of this chapter. The algorithm allows us to find the maximum or minimum of the objective function.
5. *Physical implementation.* This is the detailed physical implementation of the optimization algorithm. The same algorithm can be implemented in different kinds of hardware: a digital computer or a brain, for example. Actually, this level is quite complex and could be further divided into a number of levels: the physical implementation can be described at the level of networks, single cells, or molecules, whereas the detailed implementation of the numerical operations (e.g. matrix multiplication and non-linear scalar functions) is an interesting issue in itself. We will not go into details regarding this level.

Some of the levels may be missing in some cases. For example, in the basic sparse coding approach in Chap. 6, we don't have a probabilistic model: We go directly from the level of principles to the level of objective functions. However, the central idea of this book is that we *should* include the probabilistic modeling level—which was one motivation for going from sparse coding to ICA.

An important choice made in this book is that the level of objective function is *always* present. All our learning was based on optimization of objective functions—which are almost always based on probabilistic modeling principles. In some other approaches, one may go directly from the representational principle to an algorithm. The danger with such an approach is that it may be difficult to understand what such algorithms actually do. Going systematically through the levels of probabilistic modeling, and objective function formulation, we gain a deeper understanding of what the algorithm does based on the theory of statistics. Also, this approach constrains the modeling because we have to respect the rules of probabilistic modeling, and avoids completely ad hoc methods.

Since we always have an objective function, we always need an optimization algorithm. In preceding chapters, we omitted any discussion on how such optimization algorithms can be constructed. One reason for this is that it is possible to use general-purpose optimization methods readily implemented in many scientific computing environments. So, one could numerically optimize the objective functions without knowing anything, or at least not much, on the theory of optimization.

However, it is of course very useful to understand optimization theory when doing natural image statistics modeling for several reasons:

- One can better choose a suitable optimization method, and fine-tune its parameters.
- Some optimization methods have interesting neurophysiological interpretations (in particular, Hebbian learning in Sect. 18.4).
- Some methods have tailor-made optimization methods (FastICA in Sect. 18.7).

That is why in this chapter, we review the optimization theory needed for understanding how to optimize the objective functions obtained in this book.

18.2 Gradient Method

18.2.1 Definition and Meaning of Gradient

The gradient method is the most fundamental method for maximizing a continuous-valued, smooth function in a multi-dimensional space.

We consider the general problem of finding the maximum of a function that takes real values in an n-dimensional real space. Finding the minimum is just finding the maximum of the negative of the function, so the same theory is directly applicable to both cases. We consider here maximization because that is what we needed in preceding chapters. Let us denote the function to be maximized by $f(\mathbf{w})$ where $\mathbf{w} = (w_1, \ldots, w_n)$ is just an n-dimensional vector. The function to be optimized is usually called the objective function.

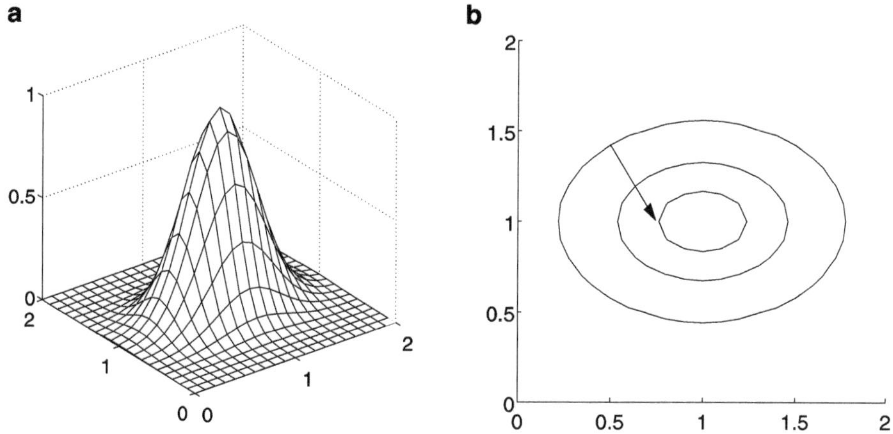

Fig. 18.1 The geometrical meaning of the gradient. Consider the function in (18.3), plotted in **a**. In **b**, the *closed curves* are the sets where the function f has a constant value. The gradient at point (0.5, 1.42) is shown and it is orthogonal to the curve

The gradient of f, denoted by ∇f is defined as the vector of the partial derivatives:

$$\nabla f(\mathbf{w}) = \begin{pmatrix} \frac{\partial f(\mathbf{w})}{\partial w_1} \\ \vdots \\ \frac{\partial f(\mathbf{w})}{\partial w_n} \end{pmatrix}. \tag{18.1}$$

The meaning of the gradient is that it points in the direction where the function grows the fastest. More precisely, suppose we want to find a vector $\mathbf{v}$ which is such that $f(\mathbf{w}+\mathbf{v})$ is as large as possible when we constrain the norm of $\mathbf{v}$ to be fixed and very small. Then the optimal $\mathbf{v}$ is given by a suitably short vector in the direction of the gradient vector. Likewise, the vector that reduces the value of f as much as possible is given by $-\nabla f(\mathbf{w})$, multiplied by a small constant. Thus, the gradient is the direction of "steepest ascent", and $-\nabla f(\mathbf{w})$ is the direction of steepest descent.

Geometrically, the gradient is always orthogonal to the curves in a contour plot of the function (i.e. to the curves that show where f has the same value), pointing in the direction of growing f.

For illustration, let us consider the following function:

$$f(\mathbf{w}) = \exp\bigl(-5(x-1)^2 - 10(y-1)^2\bigr) \tag{18.2}$$

which is, incidentally, like a Gaussian pdf. The function is plotted in Fig. 18.1a. Its maximum is at the point (1, 1). The gradient is equal to

$$\nabla f(\mathbf{w}) = \begin{pmatrix} -10(x-1)\exp(-5(x-1)^2 - 10(y-1)^2) \\ -20(y-1)\exp(-5(x-1)^2 - 10(y-1)^2) \end{pmatrix}. \tag{18.3}$$

Some contours where the function is constant are shown in Fig. 18.1b. Also, the gradient at one point is shown. We can see that taking a small step in the direction of the gradient, one gets closer to the maximizing point. However, if one takes a big enough step, one actually misses the maximizing point, so the step really has to be small.

18.2.2 Gradient and Optimization

The gradient method for finding the maximum of a function consists of repeatedly taking small steps in the direction of the gradient, $\nabla f(\mathbf{w})$, recomputing the gradient at the current point after each step. We have to take small steps because we know that this direction leads to an increase in the value of f only locally — actually, we can be sure of this only when the steps are infinitely small. The direction of the gradient is, of course, different at each point and needs to be computed again in the new point. The method can then be expressed as

$$\mathbf{w} \leftarrow \mathbf{w} + \mu \nabla f(\mathbf{w}) \tag{18.4}$$

where the parameter μ is a small step size, typically much smaller than 1. The iteration in (18.4) is repeated over and over again until the algorithm converges to a point. This can be tested by looking at the change in $\mathbf{w}$ between two subsequent iterations: if it is small enough, we assume the algorithm has converged.

When does such an algorithm converge? Obviously, if it arrives at a point where the gradient it zero, it will not move away from it. This is not surprising because the basic principles of optimization theory tell that at the maximizing points, the gradient is zero; this is a generalization of the elementary calculus result which says that in one dimension, the minima or maxima of a function are obtained at those points where the derivative is zero.

If the gradient method is used for *minimization* of the function, as is more conventional in the literature, the sign of the increment in (18.4) is negative, i.e.

$$\mathbf{w} \leftarrow \mathbf{w} - \mu \nabla f(\mathbf{w}). \tag{18.5}$$

Choosing a good step size parameter μ is crucial. If it is too large, the algorithm will not work at all; if it is too small, the algorithm will be too slow. One method, which we used in the ISA and topographic ICA experiments in this book, is to adapt the step size during the iterations. At each step, we consider the step size used in the previous iteration (say μ_0), and a larger one (say $2\mu_0$) and a smaller one ($\mu_0/2$). Then we compute the value of the objective function that results from using any of these three step sizes in the current step, and choose the step size which gives the largest value for the objective function, and use it as the μ_0 in the next iteration. Such adaptation makes each step a bit slower, but it makes sure that the step sizes are reasonable.

18.2.3 Optimization of Function of Matrix

Many functions we want to maximize are actually functions of a matrix. However, in this context, such matrices are treated just like vectors. That is, an $n \times n$ matrix is treated as an ordinary n^2-dimensional vector. Just like we vectorize image patches and consider them as very long vectors, we consider the parameter matrices as if they had been vectorized. In practice, we don't need to concretely vectorize the parameter matrices we optimize, but that is the underlying idea.

For example, the gradient of the likelihood of the ICA model in (7.15) is given by (see Hyvärinen et al. 2001b for derivation):

$$\sum_{t=1}^{T} g(\mathbf{V}\mathbf{z}_t)\mathbf{z}_t^{\mathrm{T}} + (\mathbf{V}^{-1})^{\mathrm{T}} \tag{18.6}$$

where the latter term is the gradient of $\log|\det \mathbf{V}|$. Here, g is a function of the pdf of the independent components: $g = p_i'/p_i$ where p_i is the pdf of an independent component. Thus, a gradient method for maximizing the likelihood of ICA is given by

$$\mathbf{V} \leftarrow \mathbf{V} + \mu \left[\sum_{t=1}^{T} g(\mathbf{V}\mathbf{z}_t)\mathbf{z}_t^{\mathrm{T}} + (\mathbf{V}^{-1})^{\mathrm{T}} \right] \tag{18.7}$$

where μ is the learning rate, not necessarily constant in time. Actually, in this case it is possible to use a very simple trick to speed up computation. If the gradient is multiplied by $\mathbf{V}^{\mathrm{T}}\mathbf{V}$ from the right, we obtain a simpler version

$$\mathbf{V} \leftarrow \mathbf{V} + \mu \sum_{t=1}^{T} [\mathbf{I} + g(\mathbf{y}_t)\mathbf{y}_t^{\mathrm{T}}]\mathbf{V}, \tag{18.8}$$

where $\mathbf{y}_t = \mathbf{V}\mathbf{z}_t$. This turns out to be a valid method for maximizing likelihood. Simply, the algorithm can be assumed to converge to the same points as the one in (18.7) because $\mathbf{V}^{\mathrm{T}}\mathbf{V}$ is invertible, and thus the points where the change in $\mathbf{V}$ is zero are the same. A more rigorous justification of this natural or relative gradient method is given in Cardoso and Laheld (1996), Amari (1998).

18.2.4 Constrained Optimization

It is often necessary to maximize a function under some constraints. That is, the vector $\mathbf{w}$ is not allowed to take any value in the n-dimensional real space. The most common constraint that we will encounter is that the norm of $\mathbf{w}$ is fixed to be constant, typically equal to one: $\|\mathbf{w}\| = 1$. The set of allowed values is called the constraint set. Some changes are needed in the gradient method to take such constraints into account, but in the cases that we are interested in, the changes are actually quite simple.

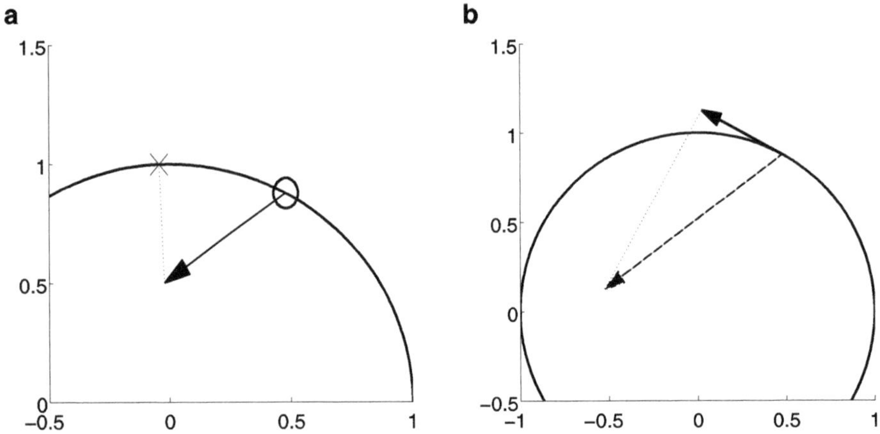

Fig. 18.2 Projection onto the constraint set (**a**), and projection of the gradient (**b**). A function (not shown explicitly) is to be minimized on the unit sphere. **a** Starting at the point marked with "o", a small gradient step is taken, as shown by the *arrow*. Then the point is projected to the closest point on the unit sphere, which is marked by "×". This is one iteration of the method. **b** The gradient (*dashed arrow*) points in a direction in which is it dangerous to take a big step. The projected gradient (*solid arrow*) points in a better direction, which is "tangential" to the constraint set. Then a small step in this projected direction is taken (the step is not shown here)

18.2.4.1 Projecting Back to Constraint Set

The basic idea of the gradient method can be used in the constrained case as well. Only a simple modification is needed: after each iteration of (18.4), we *project* the vector $\mathbf{w}$ onto the constraint set. Projecting means going to the point in the constraint set which is closest. Projection to the constraint set is illustrated in Fig. 18.2a.

In general, computing the projection can be very difficult, but in some special cases, it is a simple operation. For example, if the constraint set consists of vectors with norm equal to one, the projection is performed simply by the division:

$$\mathbf{w} \leftarrow \mathbf{w}/\|\mathbf{w}\|. \tag{18.9}$$

Another common constraint is orthogonality of a matrix. In that case, the projection onto the constraint set is given by

$$\mathbf{W} \leftarrow (\mathbf{W}\mathbf{W}^{\mathrm{T}})^{-1/2}\mathbf{W}. \tag{18.10}$$

Here, we see a rather involved operation: the inverse of the square root of the matrix. We shall not go into details on how it can be computed; suffice it to say that most numerical software can compute it quite efficiently.[1] This operation often called sym-

[1] If you really want to know: the inverse square root $(\mathbf{W}\mathbf{W}^{\mathrm{T}})^{-1/2}$ of the symmetric matrix $\mathbf{W}\mathbf{W}^{\mathrm{T}}$ is obtained from the eigenvalue decomposition of $\mathbf{W}\mathbf{W}^{\mathrm{T}} = \mathbf{E}\,\mathrm{diag}(\lambda_1,\ldots,\lambda_n)\mathbf{E}^{\mathrm{T}}$ as $(\mathbf{W}\mathbf{W}^{\mathrm{T}})^{-1/2} = \mathbf{E}\,\mathrm{diag}(1/\sqrt{\lambda_1},\ldots,1/\sqrt{\lambda_n})\mathbf{E}^{\mathrm{T}}$. It is easy to see that if you multiply this matrix with itself, you get the inverse of the original matrix. See also Sect. 5.9.2.

metric orthogonalization, and it is the way that symmetric decorrelation in sparse coding and other algorithms is usually implemented.

18.2.4.2 Projection of the Gradient

Actually, an even better method is obtained if we first project the gradient onto the "tangent space" of the constraint set, and then take a step in that direction instead of the ordinary gradient direction. What this means is that we compute a direction that is "inside" the constraint set in the sense that infinitely small changes along that direction do not get us out of the constraint set, yet the movement in that direction maximally increases the value of the objective function. This improves the method because then we can usually take larger step sizes and obtain larger increases in the objective function without going in a completely wrong direction as is always the danger when taking large steps. The projection onto the constraint set has to done even in this case. Projection of the gradient is illustrated in Fig. 18.2b.

This is quite useful in the case where we are maximizing with respect to a parameter matrix that is constrained to be orthogonal. The projection can be shown to equal (Edelman et al. 1998):

$$\tilde{\nabla} f(\mathbf{W}) = \nabla f(\mathbf{W}) - \mathbf{W}^T \nabla f(\mathbf{W}) \mathbf{W}^T \tag{18.11}$$

where $\nabla f(\mathbf{W})$ is the ordinary gradient of the function f.

In Sect. 18.5 below, we will see an example of how these ideas of constrained optimization are used in practice.

18.3 Global and Local Maxima

An important distinction is between global and local maxima. Consider the one-dimensional function in Fig. 18.3. The global maximum of the function is at the point $x = 6$; this is the "real" maximum point where the function attains its very largest value. However, there are also two local maxima, at $x = 2$ and $x = 9$. A local maximum is a point in which the function obtains a value which is greater than the values in all neighboring points close-by.

An important point to understand is that the result of a gradient algorithm depends on the *initial point*, that is, the point where the algorithm starts in the first iteration. The algorithm only sees the local behavior of the function, so it will find the closest local maximum. Thus, if in Fig. 18.3, the algorithm is started in the point marked by circles, it will find the global maximum. In contrast, if it is started in one of the points marked by crosses, it will converge to one of the local maxima.

In many cases, we are only interested in the global maximum. Then the behavior of the gradient method can be rather unsatisfactory because it only finds a local optimum. This is actually the case with most optimization algorithms. So, when running optimization algorithms, we have to always keep in mind that an algorithm only gives a local optimum, not the global one.

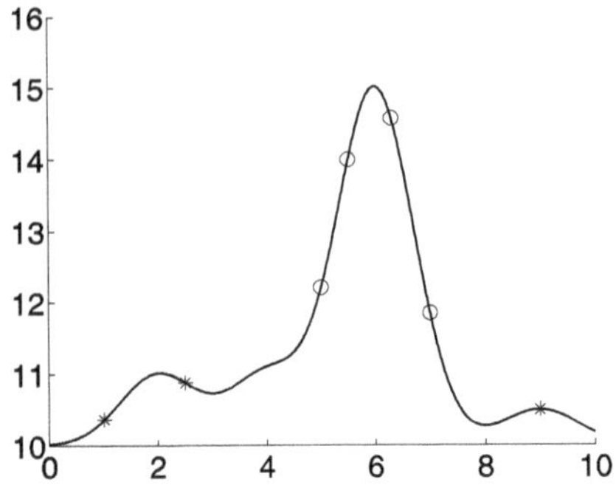

Fig. 18.3 Local vs. global maxima. The function has a global maximum at $x = 6$ and two local maxima at $x = 2$ and $x = 9$. If a gradient algorithm starts near one of the local maxima (e.g. at the points marked by *crosses*), it will get stuck at one of the local maxima and it will not find the global maximum. Only if the algorithm starts sufficiently close to the global maximum (e.g. at the points marked by *circles*), it will find the global maximum

18.4 Hebb's Rule and Gradient Methods

18.4.1 Hebb's Rule

Hebb's rule, or Hebbian learning, is a principle which is central in modern research on learning and memory. It attempts to explain why certain synaptic connections get strengthened as a result of experience, and others don't; this is called *plasticity* in neuroscience, and *learning* is a more cognitive context. Donald Hebb proposed in 1949 that

> When an axon of cell A (...) excites cell B and repeatedly or persistently takes part in firing it, some growth process of metabolic change takes place in one or both cells so that A's efficiency as one of the cells firing B is increased. (Quoted in Kandel et al. (2000).)

This proposal can be readily considered in probabilistic terms: A statistical analysis is about things which happen "repeatedly or persistently".

A basic interpretation of Hebb's rule is in terms of the *covariance* of the firing rates of cells A and B: the change in the synaptic connection should be proportional to that covariance. This is because if the firing rates of A and B are both high at the same time, their covariance is typically large. The covariance interpretation is actually stronger because it would also imply that if both cells are silent (firing rate below average) at the same time, the synaptic connection is strengthened. Even more than that: if one of the cells is typically silent when the other one fires strongly, this has a negative contribution to the covariance, and the synaptic connection should be decreased. Such an extension of Hebb's rule seems to be quite in line with Hebb's original idea (Dayan and Abbott 2001).

Note a difference between this covariance interpretation, in which only the correlation of the firing rates matters, and the original formulation, in which cell A is assumed to "take part in firing [cell B]", i.e. to have a *causal* influence on cell B's firing. This difference may be partly resolved by recent research which shows that the exact timing of the action potentials in cell A and cell B is important, a phenomenon called *spike-timing dependent plasticity* (Dan and Poo 2004), but we will not consider this difference here.

18.4.2 Hebb's Rule and Optimization

Hebb's rule can be readily interpreted as an optimization process, closely related to gradient methods. Consider an objective function of the form which we have extensively used in this book:

$$J(\mathbf{w}) = \sum_t G\left(\sum_{i=1}^n w_i x_i(t)\right). \tag{18.12}$$

To compute the gradient, we use two elementary rules. First, the derivative of a sum is the sum of the derivatives, so we just need to take the derivative of $G(\sum_{i=1}^n w_i x_i(t))$ and take its sum over t. Second, we use the chain rule which gives the derivative of a compound function $f_1(f_2(w))$ as $f_2'(w) f_1'(f_2(w))$. Now, the derivative of $\sum_{i=1}^n w_i x_i(t)$ with respect to w_i is simply $x_i(t)$, and we denote by $g = G'$ the derivative of G. Thus, the partial derivatives are obtained as

$$\frac{\partial J}{\partial w_i} = \sum_t x_i(t) g\left(\sum_{i=1}^n w_i x_i(t)\right). \tag{18.13}$$

So, a gradient method to maximize this function would be of the form

$$w_i \leftarrow w_i + \mu \sum_t x_i(t) g\left(\sum_{i=1}^n w_i x_i(t)\right), \quad \text{for all } i. \tag{18.14}$$

Now, let us interpret the terms in (18.14). Assume that

1. The $x_i(t)$ are the inputs to the ith dendrite of a neuron at time point t.
2. The w_i are the strengths of the synapses at those dendrites.
3. The firing rate of the neuron at time point t is equal to $\sum_{i=1}^n w_i x_i(t)$.
4. The inputs x_i have zero mean, i.e. they describe changes around the mean firing rate.

Further, let us assume that the function g is increasing.

Then the gradient method in (18.14) is quite similar to a Hebbian learning process. Consider the connection strength of one of the synapses i. Then the connection w_i is increased if $x_i(t)$ is repeatedly high at the same time as the firing rate of the neuron in question. In fact, the term multiplied by the learning rate μ is nothing else that the covariance between the input to the ith dendrite and an increasing function of the firing rate of the neuron, as in the covariance-based extension of Hebb's rule.

Such a "learning rule" would be incomplete, however. The reason is that we have to constrain $\mathbf{w}$ somehow, otherwise it might simply go to zero or infinity. In preceding chapters, we usually constrained the norm of $\mathbf{w}$ to be equal to unity. This is quite a valid constraint here as well. So, we assume that in addition to Hebb's rule, some kind of normalization process is operating in cell B.

18.4.3 Stochastic Gradient Methods

The form of Hebb's rule in (18.14) uses the statistics of the data in the sense that it computes the correlation over many observations of $\mathbf{x}$. This is not very realistic in terms of neurobiological modeling. A simple solution for this problem is offered by the theory of stochastic gradient methods (Kushner and Clark 1978).

The idea in a stochastic gradient is simple and very general. Assume we want to maximize some expectation, say $E\{g(\mathbf{w}, \mathbf{x})\}$ where $\mathbf{x}$ is a random vector, and $\mathbf{w}$ is a parameter vector. The gradient method for maximizing this with respect to $\mathbf{w}$ gives

$$\mathbf{w} \leftarrow \mathbf{w} + \mu E\left\{\nabla_{\mathbf{w}} g(\mathbf{w}, \mathbf{x})\right\} \tag{18.15}$$

where the gradient is computed with respect to $\mathbf{w}$, as emphasized by the subscript in the ∇ operator. Note that we have taken the gradient inside the expectation operator, which is valid because expectation is basically a sum, and the derivative of a sum is the sum of the derivatives as noted above.

The stochastic gradient method now proposes that we don't need to compute the expectation before taking the gradient step. For *each observation* $\mathbf{x}$, we can use the gradient iteration given that single observation:

$$\mathbf{w} \leftarrow \mathbf{w} + \mu \nabla_{\mathbf{w}} g(\mathbf{w}, \mathbf{x}). \tag{18.16}$$

So, when given a sample of observations of $\mathbf{x}$, we compute the update in (18.16) for each observation picked in any order. This is reasonable because the update in the gradient will be, *on the average*, equal to the update in the original gradient with the expectation given in (18.15). The step size has to be much smaller, though, because of the large random fluctuations in this "instantaneous" gradient.

So, we can consider Hebbian learning in (18.14) so that the sum over t is omitted, and each incoming observation, i.e. stimulus, is immediately used in learning:

$$w_i \leftarrow w_i + \mu x_i(t) g\left(\sum_{i=1}^{n} w_i x_i(t)\right). \tag{18.17}$$

Such a learning method still performs maximization of the objective function, but is more realistic in terms of neurophysiological modeling: at each time point, the input and output of the neuron make a small change in the synaptic weights.

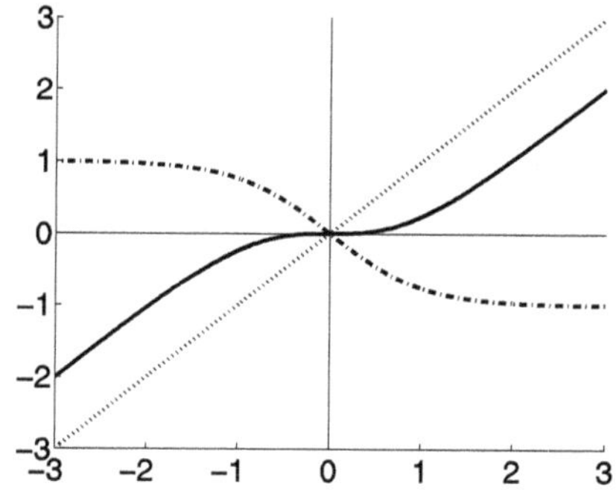

Fig. 18.4 Two non-linearities: g_1 in (18.18), *dash-dotted line*, and g_2 in (18.20), *solid line*. For comparison, the line $x = y$ is given as *dotted line*

18.4.4 Role of the Hebbian Non-linearity

By changing the non-linearity g in the learning rule, and thus the G in the objective function, we see that Hebb's rule is quite flexible and allows different kinds of learning to take place. If we assume that g is *linear* the original function G is quadratic. Then Hebb's rule is actually doing PCA (Oja 1982), since it is simply maximizing the variance of $\mathbf{w}^T\mathbf{x}$ under the constraint that $\mathbf{w}$ has unit norm.

On the other hand, if g is non-linear, G is non-quadratic. Then we can go back to the framework of basic sparse coding in Chap. 6. There, we used the expression $h(s^2)$ instead of $G(s)$ in order to investigate the convexity of h. So, if G is such that it corresponds to a convex h, Hebb's rule can be interpreted as doing sparse coding! The is no contradiction in that almost same rule is able to do both PCA and sparse coding because in Chap. 6 we also assumed that the data is whitened. So, we see that the operation of Hebb's rule depends very much on the preprocessing of the data.

What kind of non-linearities does sparse coding require? Consider the widely-used choice $G_1(s) = -\log\cosh s$. This would give

$$g_1(s) = \quad \tanh s. \tag{18.18}$$

This function (plotted in Fig. 18.4) would be rather odd to use as such in Hebb's rule, because it is decreasing, and the whole idea of Hebb's rule would be inverted. (Actually, such "anti-Hebbian" learning has been observed in some contexts (Bell et al. 1993), and is considered important in some computational models (Földiák 1990).)

However, because the data is whitened, we can find a way of interpreting this maximization as Hebbian learning. The point is that for whitened data, we can add a quadratic term to G, and consider

$$G_2(s) = \frac{1}{2}s^2 - \log\cosh s. \tag{18.19}$$

Since the data is whitened and $\mathbf{w}$ is constrained to unit norm, the expectation of $s^2 = (\mathbf{w}^T\mathbf{x})^2$ is constant, and thus the maximization of G_2 produces just the same result as maximization of G_1. Now, the derivative of G_2 is

$$g_2(s) = s - \tanh s \tag{18.20}$$

which is an increasing function; see Fig. 18.4.

So, using a non-linearity such as g_2, sparse coding does have a meaningful interpretation as a special case of Hebb's rule. The non-linearity g_2 even makes intuitive sense: it is a kind of a thresholding function (actually, a shrinkage function, see Sect. 14.1.2.2), which ignores activations which are small.

18.4.5 Receptive Fields vs. Synaptic Strengths

In the Hebbian learning context, the feature weights W_i are related to synaptic strengths in the visual cells. However, the visual input reaches the cortex only after passing through several neurons in the retina and the thalamus. Thus, the W_i actually model the compound effect of transformations in all those processing stages. How can we then interpret optimization of a function such as $G(\sum_{x,y} W(x,y)I(x,y))$ in terms of Hebb's rule?

In Sect. 5.9, we discussed the idea that the retina and LGN perform something similar to a whitening of the data. Thus, as a rough approximation, we could consider the *canonically preprocessed* data as the input to the visual cortex. Then maximization of a function such as $G(\mathbf{v}^T\mathbf{z})$, where $\mathbf{z}$ is the preprocessed data, is in fact modeling the plasticity of synapses of the cells the primary visual cortex. So, Hebbian learning in that stage can be modeled just as we did above.

18.4.6 The Problem of Feedback

In the Hebbian implementation of ICA and related learning rules, there is one more problem which needs to be solved. This is the implementation of the constraint of orthogonality. The constraint is necessary to prevent the neurons from all learning the same feature. A simple approach would be to consider the minimization of some measure of the covariance of the outputs (assuming the data is whitened as a preprocessing stage):

$$Q(\mathbf{v}_1,\ldots,\mathbf{v}_n) = -M\sum_{j\neq i}\left[E\{s_i s_j\}\right]^2 = -M\sum_{i\neq j}\left[E\{(\mathbf{v}_i^T\mathbf{z})(\mathbf{v}_j^T\mathbf{z})\}\right]^2 \tag{18.21}$$

where M is a large constant (say, $M = 100$). We can add this function as a so-called *penalty* to the measures of sparseness. If we then consider the gradient with respect to $\mathbf{v}_i$, this leads to the addition of a term of the form

$$\nabla_{\mathbf{v}_i} Q = -2M\sum_{j\neq i} E\{\mathbf{z} s_j\}E\{s_i s_j\} \tag{18.22}$$

to the learning rule for $\mathbf{v}_i$. Because M is large, after maximization the sum of the $[E\{s_i s_j\}]^2$ will be very close to zero—corresponding to the case where the s_i are all uncorrelated. Thus, this penalty approximately reinforces the constraint of uncorrelatedness.

The addition of Q to the sparseness measure thus results in the addition of a *feedback* term of the form in (18.22).

18.5 Optimization in Topographic ICA *

As an illustration of the gradient method and constrained optimization, we consider in this section maximization of likelihood of the topographic ICA in (11.5). This section can be skipped by readers not interested in mathematical details.

Because independent subspace analysis is formally a special case of topographic ICA, obtained by a special definition of the neighborhood function, the obtained learning rule is also the gradient method for independent subspace analysis.

First note that we constrain $\mathbf{V}$ to be orthogonal, so $\det \mathbf{V}$ is constant (equal to one), and can be ignored in this optimization. Another simple trick to simplify the problem is to note is that we can ignore the sum over t and just compute the "instantaneous" gradient as in stochastic gradient methods. We can always go back to the sum over t by just summing the gradient over t, because the gradient of a sum is the sum of the gradients. In fact, we can simplify the problem even further by computing the gradient of the likelihood for each term of in the sum over i in the likelihood in (11.5), and taking the sum afterward.

So, the computation of the gradient is essentially reduced to computing the gradient of

$$L_i(\mathbf{v}_1, \ldots, \mathbf{v}_n) = h\left(\sum_{j=1}^{n} \pi(i,j)\left(\mathbf{v}_j^T \mathbf{z}_t\right)^2\right). \tag{18.23}$$

Denote by $\mathbf{v}_k^l$ the lth component of $\mathbf{v}_k$. By the chain rule, applied twice, we obtain

$$\frac{\partial L_i}{\partial \mathbf{v}_k^l} = 2 z_t^l \pi(i,k)\left(\mathbf{v}_k^T \mathbf{z}_t\right) h'\left(\sum_{j=1}^{n} \pi(i,j)\left(\mathbf{v}_j^T \mathbf{z}_t\right)^2\right). \tag{18.24}$$

This can be written in vector form by simply collecting these partial derivatives for all l in a single vector:

$$\nabla_{\mathbf{v}_k} L_i = 2\mathbf{z}_t \pi(i,k)\left(\mathbf{v}_k^T \mathbf{z}_t\right) h'\left(\sum_{j=1}^{n} \pi(i,j)\left(\mathbf{v}_j^T \mathbf{z}_t\right)^2\right). \tag{18.25}$$

(This is not really the whole gradient because it is just the partial derivatives with respect to some of the entries in $\mathbf{V}$, but the notation using ∇ is still often used.)

Since the log-likelihood is simply the sum of the L_i's we obtain

$$\nabla_{\mathbf{v}_k} \log L = \sum_{t=1}^{T} \sum_{i=1}^{n} \nabla_{\mathbf{v}_k} L_i = 2 \sum_{t=1}^{T} \mathbf{z}_t \left(\mathbf{v}_k^{\mathrm{T}} \mathbf{z}_t\right) \sum_{i=1}^{n} \pi(i,k) h' \left(\sum_{j=1}^{n} \pi(i,j) \left(\mathbf{v}_j^{\mathrm{T}} \mathbf{z}_t\right)^2 \right). \tag{18.26}$$

We can omit the constant 2 which does not change the direction of the gradient.

So, the algorithm for maximizing the likelihood in topographic ICA is finally as follows:

1. Compute the gradients in (18.26) for all k. Collect them in a matrix $\nabla_{\mathbf{V}} \log L$ which has the $\nabla \mathbf{v}_k \log L$ as its rows.
2. Compute the projection of this matrix on the tangent space of the constraint space, using the formula in (18.11). Denote the projected matrix as $\tilde{\mathbf{G}}$. (This projection step is optional, but usually it speeds up the algorithm.)
3. Do a gradient step as
$$\mathbf{V} \leftarrow \mathbf{V} + \mu \tilde{\mathbf{G}}. \tag{18.27}$$
4. Orthogonalize the matrix $\mathbf{V}$. For example, this can be done by the formula in (18.10).

To see a connection of such an algorithm with Hebbian learning, consider a gradient update for each $\mathbf{v}_k$ separately. We obtain the gradient learning rule

$$\mathbf{v}_k \leftarrow \mathbf{v}_k + \mu \sum_{t=1}^{T} \mathbf{z}_t \left(\mathbf{v}_k^{\mathrm{T}} \mathbf{z}_t\right) r_t^k \tag{18.28}$$

where

$$r_t^k = \sum_{i=1}^{n} \pi(i,k) h' \left(\sum_{j=1}^{n} \pi(i,j) \left(\mathbf{v}_j^{\mathrm{T}} \mathbf{z}_t\right)^2 \right). \tag{18.29}$$

Equally well, we could use a stochastic gradient method, ignoring the sum over t. In a neural interpretation, the Hebbian learning rule in (18.28) can be considered a "modulated" Hebbian learning, since the ordinary Hebbian learning term $\mathbf{z}_t (\mathbf{v}_k^{\mathrm{T}} \mathbf{z}_t)$ is modulated by the term r_t^k. This term could be considered as top-down feedback, since it is a function of the local energies which could be the outputs of higher-order neurons (complex cells).

18.6 Beyond Basic Gradient Methods *

This section can be skipped by readers not interested in mathematical theory. Here, we briefly describe two further well-known classes of optimization methods. Actually, in our context, these are not very often better than the basic gradient method, so our description is very brief.

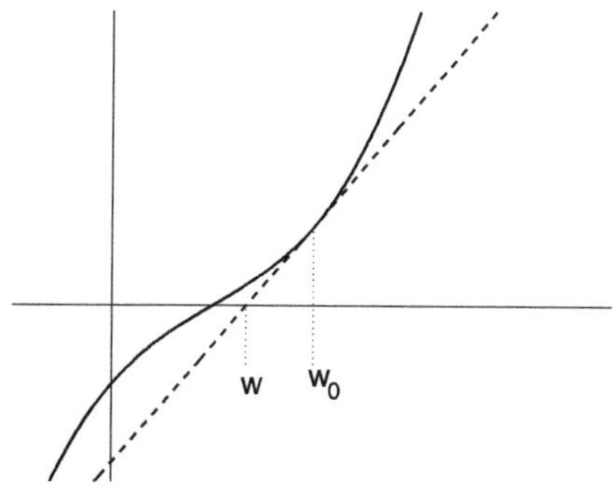

Fig. 18.5 Illustration of Newton's method for solving an equation (which we use in optimization to solve the equation which says that the gradient is zero). The function is linearly approximated by its tangent. The point where the tangent intersects with the x-axis is taken as the next approximation of the point where the function is zero

18.6.1 Newton's Method

As discussed above, the optima of an objective function are found in points where the gradient is zero. So, optimization can be approached as the problem of solving a system of equations given by

$$\begin{aligned} \frac{\partial f}{\partial w_1}(\mathbf{w}) &= 0, \\ &\vdots \\ \frac{\partial f}{\partial w_n}(\mathbf{w}) &= 0. \end{aligned} \tag{18.30}$$

A classic method for solving such a system of equations is Newton's method. It can be used to solve any system of equations, but we consider here the case of the gradient only.

Basically, the idea is to approximate the function linearly using its derivatives. In one dimension, the idea is simply to approximate the graph of the function using its tangent, whose slope is given by the derivative. That is, for a general function g:

$$g(w) \approx g(w_0) + g'(w_0)(w - w_0). \tag{18.31}$$

This very general idea of finding the point where the function attains the value zero is illustrated in Fig. 18.5.

In our case, g corresponds to the gradient, so we use the derivatives of the gradients, which are second partial derivatives of the original function f. Also, we need to use a multidimensional version of this approximation. This gives

$$\nabla f(\mathbf{w}) = \nabla f(\mathbf{w}_0) + \mathbf{H}(\mathbf{w}_0)(\mathbf{w} - \mathbf{w}_0) \tag{18.32}$$

where the function $\mathbf{H}$, called the Hessian matrix, is the matrix of second partial derivatives:

$$\mathbf{H}_{ij} = \frac{\partial^2 f}{\partial w_i \partial w_j}. \tag{18.33}$$

Now, we can at every step of the method, find the new point as the one for which this linear approximation is zero. Thus, we solve

$$\nabla f(\mathbf{w}_0) + \mathbf{H}(\mathbf{w}_0)(\mathbf{w} - \mathbf{w}_0) = \mathbf{0} \tag{18.34}$$

which gives

$$\mathbf{w} = \mathbf{w}_0 - \mathbf{H}(\mathbf{w}_0)^{-1}\big(\nabla f(\mathbf{w}_0)\big). \tag{18.35}$$

This is the idea in the Newton iteration. Starting from a random point, we iteratively update $\mathbf{w}$ according to (18.35), i.e. compute the right-hand side for the current value of $\mathbf{w}$, and take that as the new value of $\mathbf{w}$. Using the same notation as with the gradient methods, we have the iteration

$$\mathbf{w} \leftarrow \mathbf{w} - \mathbf{H}(\mathbf{w})^{-1}\big(\nabla f(\mathbf{w})\big). \tag{18.36}$$

Note that this iteration is related to the gradient method. If the matrix $\mathbf{H}(\mathbf{w}_0)^{-1}$ in (18.35) is replaced by a scalar step size μ, we actually get the gradient method. So, the difference between the methods is threefold:

1. In Newton's method, the direction where $\mathbf{w}$ "moves" is not given by the gradient directly, but the gradient multiplied by the inverse of the Hessian.
2. This "step size" is not always very small: It is directly given by the inverse of the Hessian matrix, and can be quite large.
3. In the gradient method, one can choose between minimization and maximization of the objective function, by choosing the sign in the algorithm (cf. (18.4) and (18.5)). In the Newton method, no such choice is possible. The algorithm just tries to find a local extremum in which the gradient is zero, and this can be either a minimum or a maximum.

The Newton iteration has some advantages and disadvantages compared to the basic gradient method. It usually requires a smaller number of steps to converge. However, the computations needed at each step are much more demanding, because one has to first compute the Hessian matrix, and then compute $\mathbf{H}(\mathbf{w})^{-1}(\nabla f(\mathbf{w}))$ (which is best obtained by solving the linear system $\mathbf{H}(\mathbf{w})\mathbf{v} = \nabla f(\mathbf{w})$).

In practice, however, the main problem with the Newton method is that its behavior can be quite erratic. There is no guarantee any one iteration gives a $\mathbf{w}$ which increases $f(\mathbf{w})$. In fact, a typical empirical observation is that for some functions this does not happen, and the algorithm may completely diverge, i.e. go to arbitrary values of $\mathbf{w}$, eventually reaching infinity. This is because the step size can be arbitrarily large, unlike in the gradient methods. This lack of robustness is why Newton's method is not often used in practice.

As an example of this phenomenon, consider the function

$$f(w) = \exp\left(-\frac{1}{2}w^2\right) \tag{18.37}$$

which has a single maximum as $w = 0$. The first and second derivatives, which are the one-dimensional equivalents of the gradient and the Hessian, can be easily calculated as

$$f'(w) = -w \exp\left(-\frac{1}{2}w^2\right), \tag{18.38}$$

$$f''(w) = \left(w^2 - 1\right) \exp\left(-\frac{1}{2}w^2\right) \tag{18.39}$$

which gives the Newton iteration as

$$w \leftarrow w + \frac{w}{w^2 - 1}. \tag{18.40}$$

Now, assume that we start the iteration at any point where $w > 1$. Then the change $\frac{w}{w^2-1}$ is positive, which means that w is increased and it moves further and further away from zero! In this case, the method fails completely and w goes to infinity without finding the maximum at zero. (In contrast, a gradient method, with a reasonably small step size, would find the maximum.)

However, different variants of the Newton method have proven useful. For example, methods which do something between the gradient method and Newton's method (e.g. the Levenberg–Marquardt algorithm) have proven useful in some applications. In ICA, the FastICA algorithm (see below) uses the basic iteration of Newton's method but with a modification which takes the special structure of the objective function into account.

18.6.2 Conjugate Gradient Methods

Conjugate gradient methods are often considered as the most efficient general-purpose optimization methods. The theory is rather complicated and non-intuitive, so we do not try to explain it in detail.

Conjugate gradient methods try to find a direction which is better than the gradient direction. The idea is illustrated in Fig. 18.6. While the gradient direction is good for a very small step size (actually, it is still the best for an infinitely small step size), it is not very good for a moderately large step size. The conjugate gradient method tries to find a better direction based on information on the gradient directions in previous iterations. In this respect, the method is similar to Newton's method, which also modifies the gradient direction.

In fact, conjugate gradient methods do not just take a step of a fixed size in the direction they have found. An essential ingredient, which is actually necessary for the method to work, is a one-dimension *line search*. This means that once the direction, say $\mathbf{d}$, in which $\mathbf{w}$ should move has been chosen (using the complicated theory of conjugate gradient methods), many different step sizes μ are tried out, and the best one is chosen. In other words, a one-dimensional optimization is performed on the function $h(\mu) = f(\mathbf{w} + \mu\mathbf{d})$, and μ maximizing this function is chosen. (Such line search could also be used in the basic gradient method. However, in the conjugate gradient method it is completely necessary.)

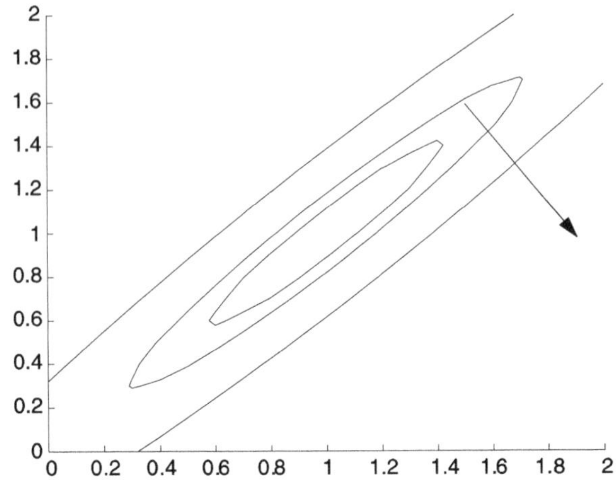

Fig. 18.6 A problem with the gradient method. The gradient direction may be very bad for anything but the very smallest step sizes. Here, the gradient goes rather completely in the wrong direction due to the strongly non-circular (non-symmetric) structure of the objective function. The conjugate gradient method tries to find a better direction

Conjugate gradient methods are thus much more complicated than ordinary gradient methods. This is not a major problem if one uses a scientific computing environment in which the method is already programmed. Sometimes, the method is much more efficient than the ordinary gradient methods, but this is not always the case.

18.7 FastICA, a Fixed-Point Algorithm for ICA

Development of tailor-made algorithms for solving the optimization problems in ICA is a subject of an extensive literature. Here, we explain briefly one popular algorithm for performing the maximization, more information and details can be found in the ICA book (Hyvärinen et al. 2001b).

18.7.1 The FastICA Algorithm

Assume that the data $\mathbf{z}_t$, $t = 1, \ldots, T$, is whitened and has zero mean. The basic form of the FastICA algorithm is as follows:

1. Choose an initial (e.g. random) weight vector $\mathbf{w}$.
2. Let $\mathbf{w} \leftarrow \sum_t \mathbf{z}_t g(\mathbf{w}^T\mathbf{z}_t) - \mathbf{w}\sum_t g'(\mathbf{w}^T\mathbf{z}_t)$.
3. Let $\mathbf{w} \leftarrow \mathbf{w}/\|\mathbf{w}\|$.
4. If not converged, go back to 2.

Note that the sign of $\mathbf{w}$ may change from one iteration to the next; this is in line with the fact that the signs of the components in ICA are not well defined. Thus, the convergence of the algorithm must use a criterion which is immune to this. For example, one might stop the iteration if $|\mathbf{w}^T\mathbf{w}_{\text{old}}|$ is sufficiently close to one, where $\mathbf{w}_{\text{old}}$ is the value of $\mathbf{w}$ at the previous iteration.

To use FastICA for *several* features, the iteration step 2 is applied separately for the weight vector of each unit. After updating all the weight vectors, they are orthogonalized (assuming whitened data). This means projecting the matrix $\mathbf{W}$, which contains the vectors w_i as its rows, on the space of orthogonal matrices, which can be accomplished, for example, by the classical method involving matrix square roots, given in (18.10). See Chap. 6 of Hyvärinen et al. (2001b) for more information on orthogonalization.

18.7.2 Choice of the FastICA Non-linearity

The FastICA algorithm uses a non-linearity, usually denoted by g. This comes from a measure of non-Gaussianity. Non-Gaussianity is measured as $E\{G(s)\}$ for some non-*quadratic* function. The function g is then the derivative of G.

Note that in Chap. 6 we measured non-Gaussianity (or sparseness) as $E\{h(s^2)\}$. Then we have $G(s) = h(s^2)$ which implies $g(s) = 2h'(s^2)s$. So, we must make a clear distinction between the non-linearities h and the functions G and g; they are all different functions but they can be derived from one another.

The choice of the measure of non-Gaussianity, or the non-linearity, is actually quite free in FastICA. We are not restricted to functions such that maximization of G corresponds to maximization of sparseness, or such that G corresponds to the log-pdf of the components. We can use, for example measures of skewness, i.e. the lack of symmetry of the pdf.

In practice, it has been found that $G(s) = \log\cosh s$ works quite well in a variety of domains; it corresponds to the tanh non-linearity as g. (In FastICA, it makes no difference if we take tanh or $-$tanh, the algorithm is immune to the change of sign.)

18.7.3 Mathematics of FastICA *

Here, we present the derivation of the FastICA algorithm, and show its connection to gradient methods. This can be skipped by readers not interested in mathematical details.

18.7.3.1 Derivation of the Fixed-Point Iteration

To begin with, we shall derive the fixed-point algorithm for *one* feature, using an objective function motivated by projection pursuit, see Hyvärinen (1999a) for details. Denote the weight vector corresponding to one feature detector by $\mathbf{w}$, and the canonically preprocessed input by $\mathbf{z}$. The goal is to find the extrema of $E\{G(\mathbf{w}^T\mathbf{z})\}$ for a given non-quadratic function G, under the constraint $E\{(\mathbf{w}^T\mathbf{z})^2\} = 1$. According to the Lagrange conditions (Luenberger 1969), the extrema are obtained at points where

$$E\{\mathbf{z}g(\mathbf{w}^{\mathrm{T}}\mathbf{z})\} - \beta\mathbf{C}\mathbf{w} = 0 \tag{18.41}$$

where $\mathbf{C} = E\{\mathbf{z}\mathbf{z}^{\mathrm{T}}\}$, and β is a constant that can be easily evaluated to give $\beta = E\{\mathbf{w}_0^{\mathrm{T}}\mathbf{z}g(\mathbf{w}_0^{\mathrm{T}}\mathbf{z})\}$, where $\mathbf{w}_0$ is the value of $\mathbf{w}$ at the optimum. Let us try to solve this equation by the classical Newton's method; see Sect. 18.6.1 above. Denoting the function on the left-hand side of (18.41) by F, we obtain its Jacobian matrix, i.e. the matrix of partial derivatives, $JF(\mathbf{w})$ as

$$JF(\mathbf{w}) = E\{\mathbf{z}\mathbf{z}^{\mathrm{T}}g'(\mathbf{w}^{\mathrm{T}}\mathbf{z})\} - \beta\mathbf{C}. \tag{18.42}$$

To simplify the inversion of this matrix, we decide to approximate the first term in (18.42). A reasonable approximation in this context seems to be $E\{\mathbf{z}\mathbf{z}^{\mathrm{T}}g'(\mathbf{w}^{\mathrm{T}}\mathbf{z})\} \approx E\{\mathbf{z}\mathbf{z}^{\mathrm{T}}\}E\{g'(\mathbf{w}^{\mathrm{T}}\mathbf{z})\} = E\{g'(\mathbf{w}^{\mathrm{T}}\mathbf{z})\}\mathbf{C}$. The obtained approximation of the Jacobian matrix can be inverted easily:

$$JF(\mathbf{w})^{-1} \approx \mathbf{C}^{-1}/(E\{g'(\mathbf{w}^{\mathrm{T}}\mathbf{z})\} - \beta). \tag{18.43}$$

We also approximate β using the current value of $\mathbf{w}$ instead of $\mathbf{w}_0$. Thus, we obtain the following approximative Newton iteration:

$$\mathbf{w} \leftarrow \mathbf{w} - [\mathbf{C}^{-1}E\{\mathbf{z}g(\mathbf{w}^{\mathrm{T}}\mathbf{z})\} - \beta\mathbf{w}]/[E\{g'(\mathbf{w}^{\mathrm{T}}\mathbf{z})\} - \beta] \tag{18.44}$$

where $\mathbf{w}^+$ denotes the new value of $\mathbf{w}$, and $\beta = E\{\mathbf{w}^{\mathrm{T}}\mathbf{z}g(\mathbf{w}^{\mathrm{T}}\mathbf{z})\}$. After every step, $\mathbf{w}^+$ is normalized by dividing it by $\sqrt{(\mathbf{w}^+)^{\mathrm{T}}\mathbf{C}\mathbf{w}^+}$ to improve stability. This algorithm can be further algebraically simplified (see Hyvärinen (1999a)) to obtain the original form the fixed-point algorithm:

$$\mathbf{w} \leftarrow \mathbf{C}^{-1}E\{\mathbf{z}g(\mathbf{w}^{\mathrm{T}}\mathbf{z})\} - E\{g'(\mathbf{w}^{\mathrm{T}}\mathbf{z})\}\mathbf{w}. \tag{18.45}$$

These two forms are equivalent. Note that for whitened data, $\mathbf{C}^{-1}$ disappears, giving an extremely simple form of the Newton iteration. In Hyvärinen and Oja (1997), this learning rule was derived as a fixed-point iteration of a gradient method maximizing kurtosis, hence the name of the algorithm. However, image analysis must use the more general form in Hyvärinen (1999a) because of the non-robustness of kurtosis.

18.7.3.2 Connection to Gradient Methods

There is a simple an interesting connection between the FastICA algorithm and gradient algorithms for ICA.

Let us assume that the number of independent components to be estimated equals the number of observed variables, i.e. $n = m$ and $\mathbf{A}$ is square. Denote by $\mathbf{W}$ the estimate of the inverse of $\mathbf{A}$.

Now, consider the preliminary form of the algorithm in (18.44). To avoid the inversion of the covariance matrix, we can approximate it as $\mathbf{C}^{-1} \approx \mathbf{W}^{\mathrm{T}}\mathbf{W}$, since

$\mathbf{C} = \mathbf{A}\mathbf{A}^T$. Thus, collecting the updates for all the rows of $\mathbf{W}$ into a single equation, we obtain the following form of the fixed-point algorithm:

$$\mathbf{W} \leftarrow \mathbf{W} + \mathbf{D}\big[\mathrm{diag}(-\beta_i) + E\{g(\mathbf{y})\mathbf{y}\}\big]\mathbf{W} \tag{18.46}$$

where $\mathbf{y} = \mathbf{W}\mathbf{z}$, $\beta_i = E\{y_i g(y_i)\}$, and $\mathbf{D} = \mathrm{diag}(1/(\beta_i - E\{g'(y_i)\}))$. This can be compared to the natural gradient algorithm for maximization of the likelihood in (18.8). We can see that the algorithms are very closely related. First, the expectation in (18.46) is in practice computed as a sample average as in (18.8). So, the main difference is that in the natural gradient algorithm, the β_i are all set to one, and $\mathbf{D}$ is replaced by identity times the step size μ. So, $\mathbf{D}$ is actually like a step size, although in the form of a matrix here, but it does not affect to the point where the algorithm converges (i.e. the update is zero). So, the only real difference is the β_i. Now, it can be proven that if the g really is the derivative of the log-likelihood, then the β_i are also (for infinite sample) equal to one (Hyvärinen et al. 2001b). In theory, then even this difference vanishes and the algorithms really converge to the same points.

It must be noted that the FastICA algorithm does not maximize sparseness but non-Gaussianity. Thus, in the case of sub-Gaussian features, it may actually be minimizing sparseness; see Sect. 7.9.3.

Chapter 19
Crash Course on Linear Algebra

This chapter explains basic linear algebra on a very elementary level. This is mainly meant as a reminder: The readers hopefully already know this material.

19.1 Vectors

A vector in an n-dimensional real space is an ordered collection of n real numbers. In this book, a vector is typically either the grey-scale values of pixels in an image patch, or the weights in a linear filter or feature detector. The number of pixels is n in the former case, and the number of weights is n in the latter case. We denote images by $I(x, y)$ and the weights of a feature detector typically by $W(x, y)$. It is assumed that the index x takes values from 1 to n_x and the index y takes values from 1 to n_y, where the dimensions fulfill $n = n_x \times n_y$. In all the sums that follow, this is implicitly assumed and not explicitly written to simplify notation.

One of the main points in linear algebra is to provide a notation in which many operations take a simple form. In linear algebra, the vectors such as $I(x, y)$ and $W(x, y)$ are expressed as one-dimensional columns or rows of numbers. Thus, we need to index all the pixels by a single index i that goes from 1 to n. This is obviously possible by scanning the image row by row, or column by column (see Sect. 4.1 for details on such vectorization). It does not make any difference which method is used. A vector is usually expressed in column form as

$$\mathbf{v} = \begin{pmatrix} v_1 \\ v_2 \\ \vdots \\ v_n \end{pmatrix}. \tag{19.1}$$

In this book, the vector containing image data (typically after some preprocessing steps) will be usually denoted by $\mathbf{z}$, and the vector giving the weights of a feature detector by $\mathbf{v}$. In the following, we will use both the vector- and image-based notations side-by-side.

The (Euclidean) *norm* of a vector is defined as

$$\|W(x, y)\| = \sqrt{\sum_{x,y} W(x, y)^2}, \quad \text{or} \quad \|\mathbf{v}\| = \sqrt{\sum_i v_i^2}. \tag{19.2}$$

The norm gives the length (or "size") of a vector. There are also other ways of defining the norm, but the Euclidean one is the most common.

A. Hyvärinen, J. Hurri, P.O. Hoyer, *Natural Image Statistics*,
Computational Imaging and Vision 39,

The *dot-product* (or *inner product* between two vectors is defined as

$$\langle W, I \rangle = \sum_{x,y} W(x, y) I(x, y). \tag{19.3}$$

If W is a feature detector, this could express the value of the feature when the input image is I. It basically computes a *match* between I and W. In vector notation, we use the transpose operator, given by $\mathbf{v}^{\mathrm{T}}$, to express the same operation:

$$\mathbf{v}^{\mathrm{T}} \mathbf{z} = \sum_{i=1}^{n} v_i z_i. \tag{19.4}$$

If the dot-product is zero, the vectors W and I are called *orthogonal*. The dot-product of a vector with itself equals the square of its norm.

19.2 Linear Transformations

A linear transformation is the simplest kind of transformation in an n-dimensional vector space. A vector I is transformed to a vector J by taking weighted sums:

$$J(x, y) = \sum_{x'y'} m(x, y, x', y') I(x', y'), \quad \text{for all } x, y. \tag{19.5}$$

The weights in the sum are different for every point (x, y). The indices x' and y' take all the same values as x and y. Typical linear transformations include smoothing and edge detection.

We can compound linear transformations by taking a linear transformation of J using weights denoted by $n(x, y, x', y')$. This gives the new vector as

$$\begin{aligned} K(x, y) &= \sum_{x''y''} n(x, y, x'', y'') J(x'', y'') \\ &= \sum_{x''y''} n(x, y, x'', y'') \sum_{x'y'} m(x'', y'', x', y') I(x', y') \\ &= \sum_{x'y'} \left(\sum_{x''y''} n(x, y, x'', y'') m(x'', y'', x', y') \right) I(x', y'). \end{aligned} \tag{19.6}$$

Defining

$$p(x, y, x', y') = \sum_{x''y''} n(x, y, x'', y'') m(x'', y'', x', y') \tag{19.7}$$

we see that the compounded transformation is a linear transformation with the weights given by p.

19.3 Matrices

In matrix algebra, linear transformations and linear systems of equations (see below) can be succinctly expressed by products (multiplications). In this book, we avoid using too much linear algebra to keep things as simple as possible. However, it is necessary to understand how matrices are used to express linear transformation, because in some cases, the notation becomes just too complicated, and also because most numerical software takes matrices as input.

A matrix $\mathbf{M}$ of size $n_1 \times n_2$ is a collection of real numbers arranged into n_1 rows and n_2 columns. The single entries are denoted by m_{ij} where i is the row and j is the column. We can convert the weights $m(x, y, x', y')$ expressing a linear transformation by the same scanning process as was done with vectors. Thus,

$$\mathbf{M} = \begin{bmatrix} m_{11} & m_{12} & \dots & m_{1m} \\ \vdots & & & \vdots \\ m_{n1} & m_{n2} & \dots & m_{nm} \end{bmatrix}. \tag{19.8}$$

The linear transformation of a vector $\mathbf{z}$ is then denoted by

$$\mathbf{y} = \mathbf{Mz} \tag{19.9}$$

which is basically a short-cut notation for

$$y_i = \sum_{j=1}^{n_2} m_{ij} z_j, \qquad \text{for all } i. \tag{19.10}$$

This operation is also the definition of the product of a matrix and a vector.

If we concatenate two linear transformations, defining

$$\mathbf{s} = \mathbf{Ny} \tag{19.11}$$

we get another linear transformation. The matrix $\mathbf{P}$ that expresses this linear transformation is obtained by

$$p_{ij} = \sum_{k=1}^{n_1} n_{ik} m_{kj}. \tag{19.12}$$

This is the definition of the product of two matrices: the new matrix $\mathbf{P}$ is denoted by

$$\mathbf{P} = \mathbf{MN}. \tag{19.13}$$

This is the matrix version of (19.7). The definition is quite useful, because it means we can multiply matrices and vectors in any order when we compute $\mathbf{s}$. In fact, we have

$$\mathbf{s} = \mathbf{Ny} = \mathbf{N}(\mathbf{Mz}) = (\mathbf{NM})\mathbf{z}. \tag{19.14}$$

Another important operation with matrices is the transpose. The transpose $\mathbf{M}^T$ of a matrix $\mathbf{M}$ is the matrix where the indices are exchanged: the i, jth entry of $\mathbf{M}^T$ is m_{ji}. A matrix $\mathbf{M}$ is called symmetric if $m_{ij} = m_{ji}$, i.e., if $\mathbf{M}$ equals its transpose.

19.4 Determinant

The determinant answers the question: how are volumes changed when the data space is transformed by the linear transformation m? That is, if I takes values in a cube whose edges are all of length one, what is the volume of the set of the values J in (19.5)? The answer is given by the absolute value of the determinant, denoted by $|\det(\mathbf{M})|$ where $\mathbf{M}$ is the matrix form of m.

Two basic properties of the determinant are very useful.

1. The determinant of a product is the product of the determinants: $\det(\mathbf{MN}) = \det(\mathbf{M})\det(\mathbf{N})$. If you think that the first transformation changes the volume by a factor or 2 and the second by a factor of 3, it is obvious that when you do both transformation, the change in volume is by a factor of $2 \times 3 = 6$.
2. The determinant of a diagonal matrix equals the product of the diagonal elements. If you think is two dimensions, a diagonal matrix simply stretches one coordinate by a factor of, say 2, and the other coordinate by a factor of, say 3, so the volume of a square of area equal to 1 then becomes $2 \times 3 = 6$.

(In Sect. 19.7, we will see a further important result on the determinant of an orthogonal matrix.)

19.5 Inverse

If a linear transformation in (19.5) does not change the dimension of the data, i.e. the number of pixels, the transformation can usually be inverted. That is, (19.5) can usually be solved for I: if we know J and m, we can compute what was the original I. This is the case if the linear transformation is invertible—a technical condition that is almost always true. In this book, we will always assume that a linear transformation is invertible if not otherwise mentioned.

In fact, we can then find a matrix of coefficients $n(x, y)$, so that

$$I(x, y) = \sum_{x'y'} n(x, y, x', y') J(x', y'), \quad \text{for all } x, y. \tag{19.15}$$

This is the inverse transformation of m. In matrix algebra, the coefficient are obtained by computing the inverse of the matrix $\mathbf{M}$, denoted by $\mathbf{M}^{-1}$. So, solving for $\mathbf{y}$ in (19.9) we have

$$\mathbf{y} = \mathbf{M}^{-1}\mathbf{z}. \tag{19.16}$$

A multitude of numerical methods for computing the inverse of the matrix exist.

Note that the determinant of the inverse matrix is simply the inverse of the determinant: $\det(\mathbf{M}^{-1}) = 1/\det(\mathbf{M})$. Logically, if the transformation changes the volume by a factor of 5 (say), then the inverse must change the volume by a factor of $1/5$.

The product of a matrix with its inverse equals the *identity matrix* $\mathbf{I}$:

$$\mathbf{M}\mathbf{M}^{-1} = \mathbf{M}^{-1}\mathbf{M} = \mathbf{I}. \tag{19.17}$$

The identity matrix is a matrix whose diagonal elements are all ones and the off-diagonal elements are all zero. It corresponds to the identity transformation, i.e. a transformation which does not change the vector. This means we have

$$\mathbf{I}\mathbf{z} = \mathbf{z} \tag{19.18}$$

for any $\mathbf{z}$.

19.6 Basis Representations

An important interpretation of the mathematics in the preceding sections is the representation of an image in a basis. Assume we have a number of features $A_i(x, y)$ where i goes from 1 to n. Given an image $I(x, y)$, we want to represent it as a linear sum of these feature vectors:

$$I(x, y) = \sum_{i=1}^{n} A_i(x, y)s_i. \tag{19.19}$$

The s_i are the *coefficients* of the feature vectors A_i. They can be considered as the values of the features in the image I, since they tell "to what extent" the features are in the image. If, for example, $s_1 = 0$, that means that the feature A_1 is not present in the image.

Using vector notation, the basis representation can be given as

$$\mathbf{z} = \sum_{i=1}^{n} \mathbf{a}_i s_i. \tag{19.20}$$

Interestingly, this equation can be further simplified by putting all the s_i into a single vector $\mathbf{s}$, and forming a matrix $\mathbf{A}$ so that the columns of that matrix are the vectors $\mathbf{a}_i$, that is:

$$\mathbf{A} = [\mathbf{a}_1, \mathbf{a}_2, \ldots, \mathbf{a}_n] = \left[\begin{pmatrix} a_{11} \\ \vdots \\ a_{n1} \end{pmatrix} \begin{pmatrix} a_{12} \\ \vdots \\ a_{n2} \end{pmatrix} \cdots \begin{pmatrix} a_{1n} \\ \vdots \\ a_{n3} \end{pmatrix} \right]. \tag{19.21}$$

Then we have equivalently

$$\mathbf{z} = \mathbf{A}\mathbf{s}. \tag{19.22}$$

From this equation, we see how we can apply all the linear algebra machinery to answer the following questions:

- How do we compute the coefficients s_i? This is done by computing the inverse matrix of $\mathbf{A}$ (hoping that one exists), and then multiplying $\mathbf{z}$ with the inverse, since $\mathbf{s} = \mathbf{A}^{-1}\mathbf{z}$.
- When is it possible to represent any $\mathbf{z}$ using the given $\mathbf{a}_i$? This question was already posed in the preceding section. The answer is: if the number of basis vectors equals the dimension of $\mathbf{z}$, the matrix $\mathbf{A}$ is invertible practically always. In such a case, we say that the $\mathbf{a}_i$ (or the A_i) form a basis.

A further important question is: What happens then if the number of vectors $\mathbf{a}_i$ is *smaller* than the dimension of the vector $\mathbf{z}$? Then we cannot represent all the possible $\mathbf{z}$'s using those features. However, we can find the best possible approximation for any $\mathbf{z}$ based on those features, which is treated in Sect. 19.8.

The opposite case is when we have *more* vectors $\mathbf{a}_i$ than the dimension of the data. Then we can represent any vector $\mathbf{z}$ using those features; in fact, there are usually many ways of representing any $\mathbf{z}$, and the coefficients s_i are not uniquely defined. This case is called *overcomplete basis* and treated in Sect. 13.1.

19.7 Orthogonality

A linear transformation is called *orthogonal* if it does not change the norm of the vector. Likewise, a matrix $\mathbf{A}$ is called orthogonal if the corresponding transformation is orthogonal. An equivalent condition for orthogonality is

$$\mathbf{A}^T\mathbf{A} = \mathbf{I}. \tag{19.23}$$

If you think about the meaning of this equation in detail, you will realize that it says two things: the column vectors of the matrix $\mathbf{A}$ are orthogonal, and all normalized to unit norm. This is because the entries in the matrix $\mathbf{A}^T\mathbf{A}$ are the dot-products $\mathbf{a}_i^T\mathbf{a}_j$ between the column vectors of the matrix $\mathbf{A}$.

An orthogonal basis is nothing else than a basis in which the basis vectors are orthogonal and have unit norm; in other words, if we collect the basis vectors into a matrix as in (19.21), that matrix is orthogonal.

Equation (19.23) shows that the inverse of an orthogonal matrix (or an orthogonal transformation) is trivial to compute: we just need to rearrange the entries by taking the transpose. This means that $s_i = \mathbf{a}_i^T\mathbf{z}$, or

$$s_i = \sum_{x,y} A_i(x, y) I(x, y). \tag{19.24}$$

So, in an orthogonal basis, we obtain the coefficients as simple dot-products with the basis vectors. Note that this is *not* true unless the basis is orthogonal.

The compound transformation of two orthogonal transformation is orthogonal. This is natural since if neither of the transformations changes the norm of the image, then doing one transformation after the other does not change the norm either.

The determinant of an orthogonal matrix is equal to plus or minus one. This is because an orthogonal transformation does not change volumes, so the absolute value has to be one. The change in sign is related to reflections. Think of multiplying one-dimensional data by -1: This does not change the "volumes", but "reflects" the data with respect to 0, and corresponds to a determinant of -1.

19.8 Pseudo-Inverse *

Sometimes transformations change the dimension, and the inversion is more complicated. If there are more variables in $\mathbf{y}$ than in $\mathbf{z}$ in (19.9), there are basically more equations than free variables, so there is no solution in general. That is, we cannot find a matrix $\tilde{\mathbf{M}}$ so that for any given $\mathbf{y}$, $\mathbf{z} = \tilde{\mathbf{M}}\mathbf{y}$ is a solution for (19.9). However, in many cases, it is useful to consider an approximative solution: Find a matrix $\mathbf{M}^+$ so that for $\mathbf{z} = \mathbf{M}^+\mathbf{y}$, the error $\|\mathbf{y} - \mathbf{M}\mathbf{z}\|$ is as small as possible. In this case, the optimal "approximative inverse" matrix can be easily computed as:

$$\mathbf{M}^+ = \left(\mathbf{M}^T\mathbf{M}\right)^{-1}\mathbf{M}^T. \tag{19.25}$$

On the other hand, if the matrix $\mathbf{M}$ has fewer rows than columns (fewer variables in $\mathbf{y}$ than in $\mathbf{z}$), there are more free variables than there are constraining equations. Thus, there are many solutions $\mathbf{z}$ for (19.9) for a given $\mathbf{y}$, and we have to choose one of them. One option is to choose the solution that has the smallest Euclidean norm. The matrix that gives this solution as $\mathbf{M}^+\mathbf{y}$ is given by

$$\mathbf{M}^+ = \mathbf{M}^T\left(\mathbf{M}\mathbf{M}^T\right)^{-1}. \tag{19.26}$$

The matrix $\mathbf{M}^+$ in both of these cases is called the (Moore–Penrose) pseudo-inverse of $\mathbf{M}$. (A more sophisticated solution for the latter case, using sparseness, is considered in Sect. 13.1.3.)

Chapter 20
The Discrete Fourier Transform

This chapter is a mathematically sophisticated treatment of the theory of Fourier analysis. It concentrates on the discrete Fourier transform which is the variant used in image analysis practice. It is not necessary to know this material to understand the developments in this book; this is meant as supplementary material.

20.1 Linear Shift-Invariant Systems

Let us consider a system $\mathcal{H}$ operating on one-dimensional input signals $I(x)$. The system is *linear* if for inputs $I_1(x)$ and $I_2(x)$, and scalar α

$$\mathcal{H}\{I_1(x) + I_2(x)\} = \mathcal{H}\{I_1(x)\} + \mathcal{H}\{I_2(x)\}, \tag{20.1}$$

$$\mathcal{H}\{\alpha I_1(x)\} = \alpha\mathcal{H}\{I_1(x)\}; \tag{20.2}$$

similar definitions apply in the two-dimensional case. A system $\mathcal{H}$ is *shift-invariant* if a shift in the input results in a shift of the same size in the output; that is, if $\mathcal{H}\{I(x)\} = O(x)$, then for any integer m

$$\mathcal{H}\{I(x+m)\} = O(x+m); \tag{20.3}$$

or, in the two-dimensional case, for any integers m and n,

$$\mathcal{H}\{I(x+m, y+n)\} = O(x+m, y+n). \tag{20.4}$$

A linear shift-invariant system $\mathcal{H}$ operating on signals (or, in the two-dimensional case, on images) can be implemented by either linear filtering with a filter, or another operation, the *convolution* of the input and the *impulse response* of the system. The impulse response $H(x)$ is the response of the system to an impulse

$$\delta(x) = \begin{cases} 1, & \text{if } x = 0, \\ 0, & \text{otherwise,} \end{cases} \tag{20.5}$$

that is

$$H(x) = \mathcal{H}\{\delta(x)\}. \tag{20.6}$$

By noting that $I(x) = \sum_{k=-\infty}^{\infty} I(k)\delta(x-k)$, and by applying linearity and shift-invariance properties (equations (20.1)–(20.3)) it is easy to show that

$$O(x) = \mathcal{H}\{I(x)\} = \sum_{k=-\infty}^{\infty} I(k)H(x-k) = I(x) * H(x), \tag{20.7}$$

A. Hyvärinen, J. Hurri, P.O. Hoyer, *Natural Image Statistics*,
Computational Imaging and Vision 39,

where the last equality sign defines convolution $*$. Note that convolution is a symmetric operator since by making the change in summation index $\ell = x - k$ (implying $k = x - \ell$)

$$I(x) * H(x) = \sum_{k=-\infty}^{\infty} I(k)H(x-k) = \sum_{\ell=-\infty}^{\infty} H(\ell)I(x-\ell) = H(x) * I(x). \quad (20.8)$$

20.2 One-Dimensional Discrete Fourier Transform

20.2.1 Euler's Formula

For purposes of mathematical convenience, in Fourier analysis, the frequency representation is complex-valued: both the basis images and the weights consist of complex numbers; this is called the representation of an image in the *Fourier space*. The fundamental reason for this is Euler's formula, which states that

$$e^{ai} = \cos a + i \sin a \quad (20.9)$$

where i is the imaginary unit. Thus, a complex exponential contains both the sin and cos function in a way that turns out to be algebraically very convenient. One of the basic reasons for this is that the absolute value of a complex number contains the sum-of-squares operation:

$$|a + bi| = \sqrt{a^2 + b^2} \quad (20.10)$$

which is related to the formula in (2.16) on page 40 which gives the power of a sinusoidal component. We will see below that we can indeed compute the Fourier power as the absolute value (modulus) of some complex numbers.

In fact, we will see that the argument of a complex number on the complex plane is related to the phase in signal processing. The argument of a complex number c is a real number $\phi \in (-\pi, \pi]$ such that

$$c = |c|e^{\phi i}. \quad (20.11)$$

We will use here the signal-processing notation $\angle c$ for the argument.

We will also use the complex conjugate of a complex number $c = a + bi$, denoted by $\bar{c}$, which can be obtained either as $a - bi$ or, equivalently, as $|c|e^{-\phi i}$. Thus, the complex conjugate has the same absolute value, but opposite argument ("phase").

20.2.2 Representation in Complex Exponentials

In signal processing theory, sinusoidals are usually represented in the form of the following complex exponential signal

$$e^{i\omega x} = \cos(\omega x) + i \sin(\omega x), \quad x = 1, \ldots, M. \quad (20.12)$$

A fundamental mathematical reason for this is that these signals are *eigensignals* of linear shift-invariant systems. An eigensignal is a generalization of the concept of an eigenvector in linear algebra (see Sect. 5.8.1). Denote by $H(x)$ the impulse response of a linear shift-invariant system $\mathcal{H}$. Then

$$\mathcal{H}\{e^{i\omega x}\} = H(x) * e^{i\omega x} = \sum_{k=-\infty}^{\infty} H(k)e^{i\omega(x-k)} = e^{i\omega x} \underbrace{\sum_{k=-\infty}^{\infty} H(k)e^{-i\omega k}}_{=\tilde{H}(\omega)}$$
$$= \tilde{H}(\omega)e^{i\omega x}, \tag{20.13}$$

where we have assumed that the sum $\sum_{k=-\infty}^{\infty} H(k)e^{-i\omega k}$ converges, and have denoted this complex number by $\tilde{H}(\omega)$. Equation (20.13) shows that when a complex exponential is input into a linear shift-invariant system, the output is the same complex exponential multiplied by $\tilde{H}(\omega)$; the complex exponential is therefore called an eigensignal of the system.

To illustrate the usefulness of the representation in complex exponentials in analytic calculations, let us derive the response of a linear shift-invariant system to a sinusoidal. This derivation uses the identity

$$\cos(\phi) = \frac{1}{2}\big(e^{i\phi} + e^{-i\phi}\big), \tag{20.14}$$

which can be verified by applying (20.12). Let $\mathcal{H}$ be a linear shift-invariant system, and $A\cos(\omega x + \psi)$ be an input signal; then

$$\begin{aligned}\mathcal{H}\{A\cos(\omega x + \psi)\} &= \frac{A}{2}\mathcal{H}\{e^{i(\omega x+\psi)} + e^{-i(\omega x+\psi)}\}\\ &= \frac{A}{2}\big(e^{i\psi}\mathcal{H}\{e^{i\omega x}\} + e^{-i\psi}\mathcal{H}\{e^{-i\omega x}\}\big)\\ &= \frac{A}{2}\big(e^{i\psi}\tilde{H}(\omega)e^{i\omega x} + e^{-i\psi}\tilde{H}(-\omega)e^{-i\omega x}\big).\end{aligned} \tag{20.15}$$

By the definition of $\tilde{H}(\omega)$ (see (20.13)), $\tilde{H}(-\omega) = \overline{\tilde{H}(\omega)} = |\tilde{H}(\omega)|e^{-i\angle\tilde{H}(\omega)}$. Thus,

$$\begin{aligned}\mathcal{H}\{A\cos(\omega x + \psi)\} &= |\tilde{H}(\omega)|A\frac{1}{2}\big(e^{i(\omega x+\psi+\angle\tilde{H}(\omega))} + e^{-i(\omega x+\psi+\angle\tilde{H}(\omega))}\big)\\ &= \underbrace{|\tilde{H}(\omega)|A}_{\text{amplitude}}\cos\big(\omega x + \underbrace{\psi + \angle\tilde{H}(\omega)}_{\text{phase}}\big).\end{aligned} \tag{20.16}$$

Equation (20.16) is a one-dimensional formal version of the two statements made in Sect. 2.2.3 (page 35):

- When a sinusoidal is input into a linear shift-invariant system, the output is a sinusoidal with the same frequency.
- The change in amplitude and phase depend only on the frequency ω.

Furthermore, the equation contains another important result, namely that both the amplitude and the phase response can be read out from $\tilde{H}(\omega)$: the amplitude response is $|\tilde{H}(\omega)|$, and the phase response $\angle\tilde{H}(\omega)$. This also explains the notation introduced for amplitude and phase responses on page 35.

In order to examine further the use of complex exponentials $e^{i\omega x}$, let us derive the representation of a real-valued signal in terms of these complex-valued signals. Thus, the imaginary parts have to somehow disappear in the final representation. Deriving such a representation from the representation in sinusoidals (equation (2.6), page 31) can be done by introducing *negative frequencies* $\omega < 0$, and using (20.14). Let us denote the coefficient of the complex exponential $e^{i\omega x}$ by $\tilde{I}_*(\omega)$. The representation can be calculated as follows:

$$
\begin{aligned}
I(x) &= \sum_{\omega=0}^{\omega_M} A_\omega \cos(\omega x + \psi_\omega) = \sum_{\omega=0}^{\omega_M} \frac{A_\omega}{2}\left(e^{i(\omega x+\psi_\omega)} + e^{-i(\omega x+\psi_\omega)}\right) \\
&= \sum_{\omega=0}^{\omega_M} \frac{A_\omega}{2}\left(e^{i\psi_\omega} e^{i\omega x} + e^{i(-\psi_\omega)} e^{i(-\omega)x}\right) \\
&= \underbrace{A_0}_{=\tilde{I}_*(0)} + \sum_{\substack{\omega=-\omega_M \\ \omega\neq 0}}^{\omega_M} \underbrace{\frac{A_{|\omega|}}{2} e^{i\,\mathrm{sgn}(\omega)\psi_{|\omega|}}\, e^{i\omega x}}_{\substack{=\tilde{I}_*(\omega) \\ \text{when } \omega\neq 0}} = \sum_{\omega=-\omega_M}^{\omega_M} \tilde{I}_*(\omega) e^{i\omega x}.
\end{aligned} \tag{20.17}
$$

Note the following properties of the coefficients $\tilde{I}_*(\omega)$:

- In general, the coefficients $\tilde{I}_*(\omega)$ are complex-valued, except for $\tilde{I}_*(0)$ which is always real.
- For $\omega \geq 0$, a coefficient $\tilde{I}_*(\omega)$ contains the information about both the amplitude and the phase of the sinusoidal representation—amplitude information is given by the magnitude $|\tilde{I}_*(\omega)|$ and phase information by the angle $\angle\tilde{I}_*(\omega)$:

$$A_\omega = \begin{cases} \tilde{I}_*(0), & \text{if } \omega = 0, \\ 2|\tilde{I}_*(\omega)|, & \text{otherwise}, \end{cases} \tag{20.18}$$

$$\psi_\omega = \begin{cases} \text{undefined}, & \text{if } \omega = 0, \\ \angle\tilde{I}_*(\omega), & \text{otherwise}. \end{cases} \tag{20.19}$$

- A closer look at the derivation (20.17) shows that the magnitude and the angle of the positive and negative frequencies are related to each other as follows:

$$\left|\tilde{I}_*(-\omega)\right| = \left|\tilde{I}_*(\omega)\right|, \tag{20.20}$$

$$\angle\tilde{I}_*(-\omega) = -\angle\tilde{I}_*(\omega). \tag{20.21}$$

  Thus, $\tilde{I}_*(w)$ and $\tilde{I}_*(-w)$ form a complex-conjugate pair. This also means that knowing only the coefficients of the positive frequencies—or only the coefficients of the negative frequencies—is sufficient to reconstruct the whole representation. (This is true only for real-valued signals; if one wants to represent complex-valued signals, the whole set of coefficients is needed. However, such representations are not needed in this book.)

Above it was assumed that a frequency-based representation of a signal $I(x)$ exists:

$$I(x) = \sum_{\omega=0}^{\omega_M} A_\omega \cos(\omega x + \psi_\omega). \tag{20.22}$$

From that, we derived a representation in complex exponentials

$$I(x) = \sum_{\omega=-\omega_M}^{\omega_M} \tilde{I}_*(\omega) e^{i\omega x}, \tag{20.23}$$

$$\tilde{I}_*(\omega) = \begin{cases} A_0 & \text{when } \omega = 0, \\ \frac{A_{|\omega|}}{2} e^{i\,\mathrm{sgn}(\omega)\psi_{|\omega|}} & \text{otherwise.} \end{cases} \tag{20.24}$$

This derivation can be reversed: assuming that a representation in complex exponentials exists—so that the coefficients of the negative and positive frequencies are complex-conjugate pairs—a frequency-based representation also exists:

$$\begin{aligned} I(x) &= \sum_{\omega=-\omega_M}^{\omega_M} \tilde{I}_*(\omega) e^{i\omega x} = \sum_{\omega=-\omega_M}^{\omega_M} \big|\tilde{I}_*(\omega)\big| e^{i(\omega x + \angle \tilde{I}_*(\omega))} \\ &= \tilde{I}_*(0) + \sum_{\omega=\omega_1}^{\omega_M} 2\big|\tilde{I}_*(\omega)\big| \cos\big(\omega x + \angle \tilde{I}_*(\omega)\big). \end{aligned} \tag{20.25}$$

20.2.3 The Discrete Fourier Transform and Its Inverse

Next, we introduce the discrete Fourier transform (DFT) and its inverse, which are the tools that are used in practice to convert signals to their representation in complex exponentials and back. We will first give a definition of the transforms, and then relate the properties of these transforms to the discussion we had above.

The word "discrete" refers here to the fact that the signal (or image) is sampled at a discrete set of points, i.e. the index x is *not* continuous. This is in contrast to the general mathematical definition of the Fourier transform which is defined for functions which take values in a real-valued space. Another point is that the DFT is in a sense closer to what is called the Fourier series in mathematics because the set of frequencies used is discrete as well. Thus, the theory of DFT has a number of differences to the general mathematical definitions used in differential calculus.

The discrete Fourier transformation is used to compute the coefficients of the signal's representation in complex exponentials: this set of coefficients is called the discrete Fourier transform. The inverse discrete Fourier transformation (IDFT) is used to compute the signal from its representation in complex exponentials. Let $I(x)$ be a signal of length N. The discrete Fourier transform pair is defined by

$$\text{DFT:} \quad \tilde{I}(u) = \sum_{x=0}^{N-1} I(x) e^{-i\frac{2\pi x}{N}u}, \quad u = 0, \ldots, N-1. \tag{20.26}$$

$$\text{IDFT:} \quad I(x) = \frac{1}{N} \sum_{u=0}^{N-1} \tilde{I}(u) e^{i\frac{2\pi u}{N}x}, \quad x = 0, \ldots, N-1, \tag{20.27}$$

Notice that the frequencies utilized in the representation in complex exponentials of the IDFT (20.27) are

$$\omega_u = \frac{2\pi u}{N}, \quad u = 0, \ldots, N-1. \tag{20.28}$$

The fact that (20.27) and (20.26) form a valid transform pair—that is, that the IDFT of $\tilde{I}(k)$ is $I(x)$—can be shown as follows. Let $\tilde{I}(u)$ be defined as in (20.26). Then—redefining the sum in (20.26) to be over x_* instead of x to avoid using the same index twice—the IDFT gives

$$\begin{aligned} \frac{1}{N} \sum_{u=0}^{N-1} \tilde{I}(u) e^{i\frac{2\pi u}{N}x} &= \frac{1}{N} \sum_{u=0}^{N-1} \left(\sum_{x_*=0}^{N-1} I(x_*) e^{-i\frac{2\pi x_*}{N}u} \right) e^{i\frac{2\pi u}{N}x} \\ &= \frac{1}{N} \sum_{x_*=0}^{N-1} I(x_*) \left(\sum_{u=0}^{N-1} e^{i\frac{2\pi u(x-x_*)}{N}} \right) \\ &= \frac{1}{N} \sum_{x_*=0}^{N-1} I(x_*) \underbrace{\left[\sum_{u=0}^{N-1} \left(e^{i\frac{2\pi(x-x_*)}{N}}\right)^u \right]}_{\text{term A}}. \end{aligned} \tag{20.29}$$

If $x_* = x$, then term A in (20.29) equals N; when $x_* \neq x$, the value of this geometric sum is

$$\sum_{u=0}^{N-1} \left(e^{i\frac{2\pi(x-x_*)}{N}}\right)^u = \frac{\overbrace{\left(e^{i\frac{2\pi(x-x_*)}{N}}\right)^N}^{=1} - 1}{e^{i\frac{2\pi(x-x_*)}{N}} - 1} = 0. \tag{20.30}$$

Therefore, the IDFT gives

$$\frac{1}{N} \sum_{u=0}^{N-1} \tilde{I}(u) e^{i\frac{2\pi u}{N}x} = \frac{1}{N} I(x) N = I(x). \tag{20.31}$$

We now discuss several of the properties of the discrete Fourier transform pair.

Negative Frequencies and Periodicity in the DFT The representation in complex exponentials in the DFT employs the following frequencies:

$$\omega_u = \frac{2\pi u}{N}, \quad u = 0, \dots, N-1. \tag{20.32}$$

In the previous section, we discussed the use of negative frequencies in representations based on complex exponentials. At first sight, it looks like no negative frequencies are utilized in the DFT. However, the representation used by the DFT (20.27) is *periodic*: as a function of the frequency index u, both the complex exponentials and their coefficients have a period of N. That is, for any integer ℓ, for the complex exponentials, we have

$$e^{i\frac{2\pi(u+\ell N)}{N}x} = e^{i\frac{2\pi u}{N}x} \underbrace{e^{i2\pi\ell x}}_{=1} = e^{i\frac{2\pi u}{N}x}, \tag{20.33}$$

and for the coefficients

$$\begin{aligned}\tilde{I}(u+\ell N) &= \sum_{x=0}^{N-1} I(x) e^{-i\frac{2\pi x}{N}(u+\ell N)} = \sum_{x=0}^{N-1} I(x) e^{-i\frac{2\pi x}{N}u} \underbrace{e^{-i2\pi u\ell}}_{=1} \\ &= \sum_{x=0}^{N-1} I(x) e^{-i\frac{2\pi x}{N}u} = \tilde{I}(u). \end{aligned}\tag{20.34}$$

Therefore, for example, the coefficient $\tilde{I}(N-1)$ corresponding to frequency $\frac{2\pi(N-1)}{N}$ is the same as the coefficient $\tilde{I}(-1)$ corresponding to frequency $\frac{-2\pi}{N}$ would be. In general, the latter half of the DFT can be considered to correspond to the negative frequencies. To be more precise, for a real-valued $I(x)$, the DFT equivalent of the complex-conjugate relationships (20.20) and (20.21) is

$$\begin{aligned}\tilde{I}(N-u) &= \sum_{x=0}^{N-1} I(x) e^{-i\frac{2\pi x}{N}(N-u)} = \sum_{x=0}^{N-1} I(x) \underbrace{e^{-i2\pi x}}_{=1} e^{i\frac{2\pi x}{N}u} \\ &= \overline{\sum_{x=0}^{N-1} I(x) e^{-i\frac{2\pi x}{N}u}} = \overline{\tilde{I}(u)}. \end{aligned}\tag{20.35}$$

This relation also explains why the DFT seems to have "too many numbers" for real-valued signals. It consists of N complex-valued numbers, which seems to contain twice the amount of information as the original signal, which has N real-valued numbers. The reason is that half the information in DFT is redundant, due to the relation in (20.35). For example, if you know all the values of $\tilde{I}(u)$ for u from 0 to $(N-1)/2$ (assuming N is odd), you can compute all the rest by just taking complex conjugates.

Periodicity of the IDFT and the Convolution Theorem The Fourier-transform pair implicitly assumes that the signal $I(x)$ is periodic: applying a derivation similar to (20.34) to the IDFT (20.27) gives

$$I(x+\ell N)=I(x). \tag{20.36}$$

This assumption of periodicity is also important for perhaps the most important mathematical statement about the discrete Fourier transform, namely *the convolution theorem.* Loosely speaking, the convolution theorem states that the Fourier transform of the convolution of two signals is the product of the discrete Fourier transforms of the signals. To be more precise, we have to take border effects into account, i.e. what happens near the beginning and the end of signals, and this is where the periodicity comes into play.

Now, we shall derive the convolution theorem. Let $I(x)$ and $H(x)$ be two signals of the same length N (if they initially have different lengths, one of them can always be extended by "padding" zeros, i.e. adding a zero signal the end). Denote by $\tilde{I}(u)$ and $\tilde{H}(u)$ the Fourier transforms of the signals. Then the product of the Fourier transforms is

$$\begin{aligned}\tilde{H}(u)\tilde{I}(u) &= \left(\sum_{\ell=0}^{N-1} H(\ell)e^{-i2\pi \ell u/N}\right)\left(\sum_{k=0}^{N-1} I(k)e^{-i2\pi k u/N}\right)\\ &= \sum_{\ell=0}^{N-1}\sum_{k=0}^{N-1} H(\ell)I(k)e^{-i2\pi(\ell+k)u/N}.\end{aligned} \tag{20.37}$$

Making a change of index $x=\ell+k$ yields

$$\begin{aligned}\tilde{H}(u)\tilde{I}(u) &= \sum_{\ell=0}^{N-1}\sum_{x=\ell}^{\ell+N-1} H(\ell)I(x-\ell)e^{-i2\pi xu/N}\\ &= \sum_{\ell=0}^{N-1}\left[\sum_{x=\ell}^{N-1} H(\ell)I(x-\ell)e^{-i2\pi xu/N}\right.\\ &\qquad \left. + \underbrace{\sum_{x=N}^{\ell+N-1} H(\ell)I(x-\ell)e^{-i2\pi xu/N}}_{\text{sum A}}\right].\end{aligned} \tag{20.38}$$

If we assume that $I(x)$ is periodic with a period of N, then what has been denoted by sum A in (20.38) can be made simpler: since in that sum $H(\ell)$ is constant and $e^{-i2\pi xu/N}$ is periodic with a period of N, the lower and upper limits in the sum can simply be changed to 0 and $\ell-1$, respectively, yielding

$$\tilde{H}(u)\tilde{I}(u)=\sum_{\ell=0}^{N-1}\left[\sum_{x=\ell}^{N-1} H(\ell)I(x-\ell)e^{-i2\pi xu/N}+\sum_{x=0}^{\ell-1} H(\ell)I(x-\ell)e^{-i2\pi xu/N}\right]$$

$$= \sum_{\ell=0}^{N-1} \sum_{x=0}^{N-1} H(\ell) I(x-\ell) e^{-i2\pi xu/N}$$

$$= \sum_{x=0}^{N-1} \underbrace{\sum_{\ell=0}^{N-1} H(\ell) I(x-\ell)}_{=O(x)} e^{-i2\pi xu/N}$$

$$= \sum_{x=0}^{N-1} O(x) e^{-i2\pi xu/N} = \tilde{O}(u), \tag{20.39}$$

where $\tilde{O}(u)$ is the discrete Fourier transform of $O(x)$. Notice that $O(x)$ is obtained as a convolution of $H(x)$ and $I(x)$ under the assumption of periodicity. That is, we define the values of the signal outside of its actual range by assuming that it is periodic. (If we want to use the basic definition of convolution, we actually have to define the values of the signal up to infinite values of the indices, because the definition assumes that the signals have infinite length.) We call such an operation *cyclic* convolution.

Equation (20.39) proves the cyclic version of the convolution theorem: *the DFT of the cyclic convolution of two signals is the product of the DFTs of the signals.* Note that the assumption of cyclicity is not needed in the general continuous-space version of the convolution theorem; it is a special property of the discrete transform.

In practice, when computing the convolution of two finite-length signals, the definition of cyclic convolution is often not what one wants, because it means that values of the signals near $x = 0$ can have an effect on the values of the convolution near $x = N - 1$. In most cases, one would like to define the convolution so that the effect of finite length is more limited. Usually, this is done by modifying the signals so that the difference between cyclic convolution and other finite-length versions disappear. For example, this can lead to adding ("padding") zeros at the edges. Such zero-padding makes it simple to compute convolutions using DFTs, which is usually much faster than using the definition.[1]

Real- and Complex-Valued DFT Coefficients In general, the coefficients $\tilde{I}(u)$ are complex-valued, except for $\tilde{I}(0)$ which is always real-valued. However, if the signal has an even length so that $\frac{N}{2}$ is an integer, then

$$\tilde{I}\left(\frac{N}{2}\right) = \tilde{I}\left(N - \frac{N}{2}\right) = \overline{\tilde{I}\left(\frac{N}{2}\right)}, \tag{20.40}$$

where in the last step we have applied (20.35). Therefore, when N is even, $\tilde{I}(\frac{N}{2})$ is also real-valued.

[1] See, for example, the MATLAB reference manual entry for the function `conv` for details.

The Sinusoidal Representation from the DFT If N is odd, then starting from (20.27), a derivation similar to (20.25) gives

$$I(x) = \underbrace{\frac{\tilde{I}(0)}{N}}_{=A_0} + \sum_{u=1}^{\frac{N-1}{2}} \underbrace{\frac{2|\tilde{I}(u)|}{N}}_{\substack{=A_u \\ \text{when } u \neq 0}} \cos\Bigg(\underbrace{\frac{2\pi u}{N}}_{\substack{=\omega_u \\ \text{when } u \neq 0}} x + \underbrace{\angle \tilde{I}(u)}_{\substack{=\psi_u \\ \text{when } u \neq 0}} \Bigg). \tag{20.41}$$

If N is even, then

$$I(x) = \underbrace{\frac{\tilde{I}(0)}{N}}_{=A_0} + \sum_{u=1}^{\frac{N}{2}-1} \underbrace{\frac{2|\tilde{I}(u)|}{N}}_{\substack{=A_u \\ \text{when } u \neq 0 \\ \text{and } u \neq \frac{N}{2}}} \cos\Bigg(\underbrace{\frac{2\pi u}{N}}_{\substack{=\omega_u \\ \text{when } u \neq 0 \\ \text{and } u \neq \frac{N}{2}}} x + \underbrace{\angle \tilde{I}(u)}_{\substack{=\psi_u \\ \text{when } u \neq 0 \\ \text{and } u \neq \frac{N}{2}}} \Bigg)$$

$$+ \underbrace{\frac{\tilde{I}(\frac{N}{2})}{N}}_{=A_{\frac{N}{2}}} \cos(\underbrace{\pi}_{=\omega_{\frac{N}{2}}} x). \tag{20.42}$$

Comparing (20.41) to (20.25), we can see that the magnitudes of the DFT coefficients are divided by N to get the amplitudes of the sinusoidals. This corresponds to the $\frac{1}{N}$ coefficient in front of the IDFT (20.27), which is needed so that the DFT and the IDFT form a valid transform pair. However, the placement of this coefficient is ultimately a question of convention: the derivation in (20.29)–(20.31) is still valid if the coefficient $\frac{1}{N}$ would be moved in front of the DFT in (20.26), or even if both the IDFT and DFT equations would have a coefficient of $\frac{1}{\sqrt{N}}$ in front. The convention adopted here in the DFT-IDFT equation pair (equations (20.27) and (20.26)) is the same as in MATLAB.

The Basis is Orthogonal, Perhaps up to Scaling In terms of a basis representation, the calculations in (20.29) show that the complex basis vectors used in DFT are orthogonal to each other. In fact, the dot-product of two basis vectors with frequencies u and u_* is

$$\sum_{x=0}^{N-1} e^{-i\frac{2\pi x}{N}u} e^{i\frac{2\pi x}{N}u_*} = \sum_{x=0}^{N-1} \left(e^{i\frac{2\pi}{N}(u_*-u)}\right)^x \tag{20.43}$$

where we have taken the conjugate of the latter term because that is how the dot-product of complex-valued vectors is defined. Now, this is almost like the "term A" in (20.29) with the roles of u and x exchanged (as well as the signs of u and u_* flipped and the scaling of u changed). So, the calculations given there can be simply adapted to show that for $u \neq u_*$, this dot-product is zero. However, the norms of these basis vectors are not equal to one in this definition. This does not change much because it simply means that the inverse transform rescales the coefficients accordingly. The coefficients in the basis are still obtained by just taking dot-products with the basis vectors (and some rescaling if needed). As pointed out above, different definitions of DFT exist, and in some of them, the basis vectors are normalized to unit norm, so the basis is exactly orthogonal. (In such a definition, it is the convolution theorem which needs a scaling coefficient.)

DFT Can Be Computed by the Fast Fourier Transformation A basic way of computing the DFT would be to use the definition in (20.26). That would mean that we have to do something like N^2 operations because computing each coefficient needs a sum with N terms, and there are N coefficients. A most important algorithm in signal processing is the Fast Fourier Transform (FFT), which computes the DFT using operations which are of the order $N \log N$, based on a recursive formula. This is much faster than N^2 because the logarithm grows very slowly as a function of N. Using FFT, one can compute the DFT for very long signals. Practically all numerical software implementing DFT use some variant of FFT, and usually the function is called `fft`.

20.3 Two- and Three-Dimensional Discrete Fourier Transforms

The two- and three-dimensional discrete Fourier transforms are conceptually similar to the one-dimensional transform. The inverse transform can be thought of as a representation of the image in complex exponentials

$$I(x,y) = \frac{1}{MN} \sum_{u=0}^{M-1} \sum_{v=0}^{N-1} \tilde{I}(u,v) e^{i2\pi(\frac{ux}{M}+\frac{vy}{N})},$$
$$x = 0, \ldots, M-1, \quad y = 0, \ldots, N-1, \tag{20.44}$$

and the coefficients $\tilde{I}(u,v)$ in this representation are determined by the (forward) transform

$$\tilde{I}(u,v) = \sum_{x=0}^{M-1} \sum_{y=0}^{N-1} I(x,y) e^{-i2\pi(\frac{ux}{M}+\frac{vy}{N})},$$
$$u = 0, \ldots, M-1, \quad v = 0, \ldots, N-1. \tag{20.45}$$

The horizontal and vertical frequencies (see Sect. 2.2.2 on page 31) in the representation in complex exponentials (20.44) are

$$\omega_{x,u} = \frac{2\pi u}{M}, \quad u = 0, \ldots, M-1, \tag{20.46}$$

$$\omega_{y,v} = \frac{2\pi v}{N}, \quad v = 0, \ldots, N-1, \tag{20.47}$$

and the amplitude $A_{u,v}$ and phase $\psi_{u,v}$ of the corresponding frequency components

$$A_{u,v}\cos(\omega_{x,u}x+\omega_{y,v}y+\psi_{u,v}), \quad u=0,\ldots,M-1,\ v=0,\ldots,N-1, \tag{20.48}$$

can be "read out" from the magnitude and angle of the complex-valued coefficient $\tilde{I}(u,v)$. In a basis interpretation, the DFT thus uses a basis with different frequencies, phases, *and* orientations.

Computationally, the two-dimensional DFT can be obtained as follows. First, compute a one-dimensional DFT along for each row, i.e. for the one-dimensional slice given by fixing y. For each row, replace the original values $I(x,y)$ by the DFT coefficients. Denote these by $I(u,y)$. Then, just compute a one-dimensional DFT for each column, i.e. for each fixed u. This gives the final two-dimensional DFT $I(u,v)$. Thus, the two-dimensional DFT is obtained by applying the one-dimensional DFT twice; typically, an FFT algorithm is used. The reason why this is possible is the following relation, which can be obtained by simple rearrangement of the terms in the definition in Equation (20.45):

$$\tilde{I}(u,v)=\sum_{y=0}^{M-1}\left[\sum_{x=0}^{N-1}I(x,y)e^{-i2\pi\frac{ux}{N}}\right]e^{-i2\pi\frac{vy}{M}} \tag{20.49}$$

in which the term in brackets is just the one-dimensional DFT for a fixed y.

The three-dimensional discrete Fourier transform pair is defined similarly:

$$I(x,y,t)=\frac{1}{MNT}\sum_{u=0}^{M-1}\sum_{v=0}^{N-1}\sum_{w=0}^{T-1}\tilde{I}(u,v,w)e^{i2\pi(\frac{ux}{M}+\frac{vy}{N}+\frac{wt}{N})},$$
$$x=0,\ldots,M-1,\ y=0,\ldots,N-1,\ t=0,\ldots,T-1, \tag{20.50}$$

$$\tilde{I}(u,v,w)=\sum_{x=0}^{M-1}\sum_{y=0}^{N-1}\sum_{t=0}^{T-1}I(x,y,y)e^{-i2\pi(\frac{ux}{M}+\frac{vy}{N}+\frac{wt}{N})},$$
$$u=0,\ldots,M-1,\ v=0,\ldots,N-1,\ w=0,\ldots,T-1. \tag{20.51}$$

The two- and three-dimensional discrete Fourier transforms enjoy a number of similar properties as the one-dimensional transform. For example, the properties of two-dimensional transform pair include:

- Complex-conjugate symmetry $\tilde{I}(-u,-v)=\overline{I(u,v)}$
- Convolution theorem holds when the convolution is defined as the cyclic variant
- Periodicity of the transform $\tilde{I}(u,v)=\tilde{I}(u+M,v)=\tilde{I}(u,v+N)=\tilde{I}(u+N,v+M)$
- Periodicity of the inverse $I(x,y)=I(x+M,y)=I(x,y+N)=I(x+N,y+M)$

Chapter 21
Estimation of Non-normalized Statistical Models

Statistical models are often based on non-normalized probability densities. That is, the model contains an unknown normalization constant whose computation is too difficult for practical purposes. Such models were encountered, for example, in Sects. 13.1.5 and 13.1.7. Maximum likelihood estimation is not possible without computation of the normalization constant. In this chapter, we show how such models can be estimated using a different estimation method. It is not necessary to know this material to understand the developments in this book; this is meant as supplementary material.

21.1 Non-normalized Statistical Models

To fix the notation, assume we observe a random vector $\mathbf{x} \in \mathbb{R}^n$ which has a probability density function (pdf) denoted by $p_{\mathbf{x}}(.)$. We have a parametrized density model $p(.;\boldsymbol{\theta})$, where $\boldsymbol{\theta}$ is an m-dimensional vector of parameters. We want to estimate the parameter $\boldsymbol{\theta}$ from observations of $\mathbf{x}$, i.e. we want to approximate $p_{\mathbf{x}}(.)$ by $p(.;\hat{\boldsymbol{\theta}})$ for the estimated parameter value $\hat{\boldsymbol{\theta}}$. (To avoid confusion between the random variable and an integrating variable, we use $\boldsymbol{\xi}$ as the integrating variable instead of $\mathbf{x}$ in what follows.)

The problem we consider here is that we only are able to compute the pdf given by the model up to a multiplicative constant $1/Z(\boldsymbol{\theta})$:

$$p(\boldsymbol{\xi};\boldsymbol{\theta}) = \frac{1}{Z(\boldsymbol{\theta})} q(\boldsymbol{\xi};\boldsymbol{\theta}).$$

That is, we do know the functional form of q as an analytical expression (or any form that can be easily computed), but we do *not* know how to easily compute Z which is given by an integral that is often analytically intractable:

$$Z(\boldsymbol{\theta}) = \int_{\boldsymbol{\xi} \in \mathbb{R}^n} q(\boldsymbol{\xi};\boldsymbol{\theta})\, d\boldsymbol{\xi}.$$

In higher dimensions (in fact, for almost any $n > 2$), the numerical computation of this integral is practically impossible as well.

Thus, maximum likelihood estimation cannot be easily performed. One solution is to approximate the normalization constant Z using Monte Carlo methods; see, e.g. Mackay (2003). In this chapter, we discuss a simpler method called score matching.

This chapter is based on (Hyvärinen 2005), first published in Journal of Machine Learning Research. Copyright retained by the author

21.2 Estimation by Score Matching

In the following, we use extensively the gradient of the log-density with respect to the data vector. For simplicity, we call this the score function, although according the conventional definition, it is actually the score function with respect to a hypothetical location parameter (Schervish 1995). For the model density, we denote the score function by $\boldsymbol{\psi}(\boldsymbol{\xi};\boldsymbol{\theta})$:

$$\boldsymbol{\psi}(\boldsymbol{\xi};\boldsymbol{\theta}) = \begin{pmatrix} \frac{\partial \log p(\boldsymbol{\xi};\boldsymbol{\theta})}{\partial \xi_1} \\ \vdots \\ \frac{\partial \log p(\boldsymbol{\xi};\boldsymbol{\theta})}{\partial \xi_n} \end{pmatrix} = \begin{pmatrix} \psi_1(\boldsymbol{\xi};\boldsymbol{\theta}) \\ \vdots \\ \psi_n(\boldsymbol{\xi};\boldsymbol{\theta}) \end{pmatrix} = \nabla_{\boldsymbol{\xi}} \log p(\boldsymbol{\xi};\boldsymbol{\theta}).$$

The point in using the score function is that it does not depend on $Z(\boldsymbol{\theta})$. In fact, we obviously have

$$\boldsymbol{\psi}(\boldsymbol{\xi};\boldsymbol{\theta}) = \nabla_{\boldsymbol{\xi}} \log q(\boldsymbol{\xi};\boldsymbol{\theta}). \tag{21.1}$$

Likewise, we denote by $\boldsymbol{\psi}_{\mathbf{x}}(.) = \nabla_{\boldsymbol{\xi}} \log p_{\mathbf{x}}(.)$ the score function of the distribution of observed data $\mathbf{x}$. This could in principle be estimated by computing the gradient of the logarithm of a non-parametric estimate of the pdf—but we will see below that no such computation is necessary. Note that score functions are mappings from $\mathbb{R}^n$ to $\mathbb{R}^n$.

We now propose that the model is estimated by minimizing the expected squared distance between the model score function $\boldsymbol{\psi}(.;\boldsymbol{\theta})$ and the data score function $\boldsymbol{\psi}_{\mathbf{x}}(.)$. We define this squared distance as

$$J(\boldsymbol{\theta}) = \frac{1}{2}\int_{\boldsymbol{\xi}\in\mathbb{R}^n} p_{\mathbf{x}}(\boldsymbol{\xi}) \left\| \boldsymbol{\psi}(\boldsymbol{\xi};\boldsymbol{\theta}) - \boldsymbol{\psi}_{\mathbf{x}}(\boldsymbol{\xi}) \right\|^2 d\boldsymbol{\xi}. \tag{21.2}$$

Thus, our *score matching* estimator of $\boldsymbol{\theta}$ is given by

$$\hat{\boldsymbol{\theta}} = \arg\min_{\boldsymbol{\theta}} J(\boldsymbol{\theta}).$$

The motivation for this estimator is that the score function can be directly computed from q as in (21.1), and we do not need to compute Z. However, this may still seem to be a very difficult way of estimating $\boldsymbol{\theta}$, since we might have to compute an estimator of the data score function $\boldsymbol{\psi}_{\mathbf{x}}$ from the observed sample, which is basically a non-parametric estimation problem. However, no such non-parametric estimation is needed. This is because we can use a simple trick of partial integration to compute the objective function very easily, as shown by the following theorem.

Theorem 1 *Assume that the model score function $\boldsymbol{\psi}(\boldsymbol{\xi};\boldsymbol{\theta})$ is differentiable, as well as some weak regularity conditions.*[1]

[1]Namely: the data pdf $p_{\mathbf{x}}(\boldsymbol{\xi})$ is differentiable, the expectations $E_{\mathbf{x}}\{\|\boldsymbol{\psi}(\mathbf{x};\boldsymbol{\theta})\|^2\}$ and $E_{\mathbf{x}}\{\|\boldsymbol{\psi}_{\mathbf{x}}(\mathbf{x})\|^2\}$ are finite for any $\boldsymbol{\theta}$, and $p_{\mathbf{x}}(\boldsymbol{\xi})\boldsymbol{\psi}(\boldsymbol{\xi};\boldsymbol{\theta})$ goes to zero for any $\boldsymbol{\theta}$ when $\|\boldsymbol{\xi}\| \to \infty$.

Then the objective function J in (21.2) can be expressed as

$$J(\boldsymbol{\theta}) = \int_{\boldsymbol{\xi} \in \mathbb{R}^n} p_{\mathbf{x}}(\boldsymbol{\xi}) \sum_{i=1}^{n} \left[\partial_i \psi_i(\boldsymbol{\xi}; \boldsymbol{\theta}) + \frac{1}{2} \psi_i(\boldsymbol{\xi}; \boldsymbol{\theta})^2 \right] d\boldsymbol{\xi} + \text{const.} \tag{21.3}$$

where the constant does not depend on $\boldsymbol{\theta}$,

$$\psi_i(\boldsymbol{\xi}; \boldsymbol{\theta}) = \frac{\partial \log q(\boldsymbol{\xi}; \boldsymbol{\theta})}{\partial \xi_i}$$

is the ith element of the model score function, and

$$\partial_i \psi_i(\boldsymbol{\xi}; \boldsymbol{\theta}) = \frac{\partial \psi_i(\boldsymbol{\xi}; \boldsymbol{\theta})}{\partial \xi_i} = \frac{\partial^2 \log q(\boldsymbol{\xi}; \boldsymbol{\theta})}{\partial \xi_i^2}$$

is the partial derivative of the ith element of the model score function with respect to the ith variable.

The proof, given in Hyvärinen (2005), is based on a simple trick of partial integration.

The theorem shows the remarkable fact that the squared distance of the model score function from the data score function can be computed as a simple expectation of certain functions of the non-normalized model pdf. If we have an analytical expression for the non-normalized density function q, these functions are readily obtained by derivation using (21.1) and taking further derivatives.

In practice, we have T observations of the random vector $\mathbf{x}$, denoted by $\mathbf{x}(1), \ldots, \mathbf{x}(T)$. The sample version of J is obviously obtained from (21.3) as

$$\tilde{J}(\boldsymbol{\theta}) = \frac{1}{T} \sum_{t=1}^{T} \sum_{i=1}^{n} \left[\partial_i \psi_i\big(\mathbf{x}(t); \boldsymbol{\theta}\big) + \frac{1}{2} \psi_i\big(\mathbf{x}(t); \boldsymbol{\theta}\big)^2 \right] + \text{const.} \tag{21.4}$$

which is asymptotically equivalent to J due to the law of large numbers. We propose to estimate the model by minimization of $\tilde{J}$ in the case of a real, finite sample.

One may wonder whether it is enough to minimize J to estimate the model, or whether the distance of the score functions can be zero for different parameter values. Obviously, if the model is degenerate in the sense that two different values of $\boldsymbol{\theta}$ give the same pdf, we cannot estimate $\boldsymbol{\theta}$. If we assume that the model is not degenerate, and that $q > 0$ always, we have local consistency as shown by the following theorem and the corollary.

Theorem 2 *Assume the pdf of $\mathbf{x}$ follows the model: $p_{\mathbf{x}}(.) = p(.; \boldsymbol{\theta}^*)$ for some $\boldsymbol{\theta}^*$. Assume further that no other parameter value gives a pdf that is equal*[2] *to $p(.; \boldsymbol{\theta}^*)$,*

[2] In this theorem, equalities of pdf's are to be taken in the sense of equal almost everywhere with respect to the Lebesgue measure.

and that $q(\boldsymbol{\xi};\boldsymbol{\theta}) > 0$ *for all* $\boldsymbol{\xi},\boldsymbol{\theta}$. *Then*

$$J(\boldsymbol{\theta}) = 0 \quad \Leftrightarrow \quad \boldsymbol{\theta} = \boldsymbol{\theta}^*.$$

For a proof, see Hyvärinen (2005).

Corollary 1 *Under the assumptions of the preceding theorems, the score matching estimator obtained by minimization of* $\tilde{J}$ *is consistent, i.e. it converges in probability toward the true value of* $\boldsymbol{\theta}$ *when sample size approaches infinity, assuming that the optimization algorithm is able to find the global minimum.*

The corollary is proven by applying the law of large numbers.[3]

This result of consistency assumes that the global minimum of $\tilde{J}$ is found by the optimization algorithm used in the estimation. In practice, this may not be true, in particular because there may be several local minima. Then the consistency is of local nature, i.e., the estimator is consistent if the optimization iteration is started sufficiently close to the true value.

21.3 Example 1: Multivariate Gaussian Density

As a very simple illustrative example, consider estimation of the parameters of the multivariate Gaussian density:

$$p(\mathbf{x}; \mathbf{M}, \boldsymbol{\mu}) = \frac{1}{Z(\mathbf{M}, \boldsymbol{\mu})} \exp\left(-\frac{1}{2}(\mathbf{x}-\boldsymbol{\mu})^{\mathrm{T}}\mathbf{M}(\mathbf{x}-\boldsymbol{\mu})\right)$$

where $\mathbf{M}$ is a symmetric positive-definite matrix (the inverse of the covariance matrix). Of course, the expression for Z is well known in this case, but this serves as an illustration of the method. As long as there is no chance of confusion, we use $\mathbf{x}$ here as the general n-dimensional vector. Thus, here we have

$$q(\mathbf{x}) = \exp\left(-\frac{1}{2}(\mathbf{x}-\boldsymbol{\mu})^{\mathrm{T}}\mathbf{M}(\mathbf{x}-\boldsymbol{\mu})\right) \tag{21.5}$$

and we obtain

$$\boldsymbol{\psi}(\mathbf{x}; \mathbf{M}, \boldsymbol{\mu}) = -\mathbf{M}(\mathbf{x}-\boldsymbol{\mu})$$

and

$$\partial_i \psi_i(\mathbf{x}; \mathbf{M}, \boldsymbol{\mu}) = -m_{ii}.$$

[3]As sample size approaches infinity, $\tilde{J}$ converges to J (in probability). Thus, the estimator converges to a point where J is globally minimized. By Theorem 2, the global minimum is unique and found at the true parameter value (obviously, J cannot be negative).

Thus, we obtain

$$\tilde{J}(\mathbf{M}, \boldsymbol{\mu}) = \frac{1}{T} \sum_{t=1}^{T} \left[\sum_{i} -m_{ii} + \frac{1}{2} \big(\mathbf{x}(t) - \boldsymbol{\mu}\big)^{\mathrm{T}} \mathbf{M}\mathbf{M} \big(\mathbf{x}(t) - \boldsymbol{\mu}\big) \right]. \tag{21.6}$$

To minimize this with respect to $\boldsymbol{\mu}$, it is enough to compute the gradient

$$\nabla_{\boldsymbol{\mu}} \tilde{J} = \mathbf{M}\mathbf{M}\boldsymbol{\mu} - \mathbf{M}\mathbf{M}\frac{1}{T}\sum_{t=1}^{T} \mathbf{x}(t)$$

which is obviously zero if and only if $\boldsymbol{\mu}$ is the sample average $\frac{1}{T}\sum_{t=1}^{T}\mathbf{x}(t)$. This is truly a minimum because the matrix $\mathbf{MM}$ that defines the quadratic form is positive-definite.

Next, we compute the gradient with respect to $\mathbf{M}$, which gives

$$\nabla_{\mathbf{M}} \tilde{J} = -\mathbf{I} + \mathbf{M}\frac{1}{2T}\sum_{t=1}^{T}\big(\mathbf{x}(t) - \boldsymbol{\mu}\big)\big(\mathbf{x}(t) - \boldsymbol{\mu}\big)^{\mathrm{T}} + \frac{1}{2T}\left[\sum_{t=1}^{T}\big(\mathbf{x}(t) - \boldsymbol{\mu}\big)\big(\mathbf{x}(t) - \boldsymbol{\mu}\big)^{\mathrm{T}}\right]\mathbf{M}$$

which is zero if and only if $\mathbf{M}$ is the inverse of the sample covariance matrix $\frac{1}{T}\sum_{t=1}^{T}(\mathbf{x}(t) - \boldsymbol{\mu})(\mathbf{x}(t) - \boldsymbol{\mu})^{\mathrm{T}}$, which thus gives the score matching estimate.

Interestingly, we see that score matching gives exactly the same estimator as maximum likelihood estimation. In fact, the estimators are identical for any sample (and not just asymptotically). The maximum likelihood estimator is known to be consistent, so the score matching estimator is consistent as well.

This example also gives some intuitive insight into the principle of score matching. Let us consider what happened if we just maximized the non-normalized log-likelihood, i.e., log of q in (21.5). It is maximized when the scale parameters in $\mathbf{M}$ are zero, i.e. the model variances are infinite and the pdf is completely flat. This is because then the model assigns the same probability to all possible values of $\mathbf{x}(t)$, which is equal to 1. In fact, the same applies to the second term in (21.6), which thus seems to be closely connected to maximization of the non-normalized log-likelihood.

Therefore, the first term in (21.3) and (21.6), involving second derivatives of the logarithm of q, seems to act as a kind of a normalization term. Here it is equal to $-\sum_i m_{ii}$. To minimize this, the m_{ii} should be made as large (and positive) as possible. Thus, this term has the opposite effect to the second term. Since the first term is linear and the second term polynomial in $\mathbf{M}$, the minimum of the sum is different from zero.

A similar interpretation applies to the general non-Gaussian case. The second term in (21.3), expectation of the norm of score function, is closely related to maximization of non-normalized likelihood: if the norm of this gradient is zero, then in fact the data point is in a local extremum of the non-normalized log-likelihood. The first term then measures what kind of an extremum this is. If it is a minimum, the first term is positive and the value of J is increased. To minimize J, the first term should be negative, in which case the extremum is a maximum. In fact, the extremum should be as steep a maximum (as opposed to a flat maximum) as possible to minimize J. This counteracts, again, the tendency to assign the same probability to all data points that is often inherent in the maximization of the non-normalized likelihood.

21.4 Example 2: Estimation of Basic ICA Model

Next, we show how score matching can be used in the estimation of the basic ICA model, defined as

$$\log p(\mathbf{x}) = \sum_{k=1}^{n} G\big(\mathbf{w}_k^{\mathrm{T}}\mathbf{x}\big) + Z(\mathbf{w}_1, \dots, \mathbf{w}_n). \tag{21.7}$$

Again, the normalization constant is well known and equal to $\log|\det \mathbf{W}|$ where the matrix $\mathbf{W}$ has the vectors $\mathbf{w}_i$ as rows, but this serves as an illustration of our method.

Here, we choose the distribution of the components s_i to be so-called logistic with

$$G(s) = -2\log\cosh\left(\frac{\pi}{2\sqrt{3}}s\right) - \frac{4\sqrt{3}}{\pi}.$$

This distribution is normalized to unit variance as typical in the theory of ICA. The score function of the model in (21.7) is given by

$$\boldsymbol{\psi}(\mathbf{x}; \mathbf{W}) = \sum_{k=1}^{n} \mathbf{w}_k g\big(\mathbf{w}_k^{\mathrm{T}}\mathbf{x}\big) \tag{21.8}$$

where the scalar non-linear function g is given by

$$g(s) = -\frac{\pi}{\sqrt{3}}\tanh\left(\frac{\pi}{2\sqrt{3}}s\right).$$

The relevant derivatives of the score function are given by

$$\partial_i \psi_i(x) = \sum_{k=1}^{n} w_{ki}^2 g'\big(\mathbf{w}_k^{\mathrm{T}}\mathbf{x}\big)$$

and the sample version of the objective function $\tilde{J}$ is given by

$$\begin{aligned}
\tilde{J} &= \frac{1}{T}\sum_{t=1}^{T}\sum_{i=1}^{n}\left[\sum_{k=1}^{n} w_{ki}^2 g'\big(\mathbf{w}_k^{\mathrm{T}}\mathbf{x}(t)\big) + \frac{1}{2}\sum_{j=1}^{n} w_{ji} g\big(\mathbf{w}_j^{\mathrm{T}}\mathbf{x}(t)\big)\sum_{k=1}^{n} w_{ki} g\big(\mathbf{w}_k^{\mathrm{T}}\mathbf{x}(t)\big)\right] \\
&= \sum_{k=1}^{n}\|\mathbf{w}_k\|^2\frac{1}{T}\sum_{t=1}^{T} g'\big(\mathbf{w}_k^{\mathrm{T}}\mathbf{x}(t)\big) + \frac{1}{2}\sum_{j,k=1}^{n}\mathbf{w}_j^{\mathrm{T}}\mathbf{w}_k\frac{1}{T}\sum_{t=1}^{T} g\big(\mathbf{w}_k^{\mathrm{T}}\mathbf{x}(t)\big) g\big(\mathbf{w}_j^{\mathrm{T}}\mathbf{x}(t)\big).
\end{aligned} \tag{21.9}$$

21.5 Example 3: Estimation of an Overcomplete ICA Model

Finally, we how score matching can be applied in the case of the overcomplete basis model in Sect. 13.1.5. The likelihood is defined almost as in (21.7), but the number of components m is larger than the dimension of the data n, and we introduce some extra parameters. The likelihood is given by

$$\log p(\mathbf{x}) = \sum_{k=1}^{m} \alpha_k G\big(\mathbf{w}_k^T \mathbf{x}\big) + Z(\mathbf{w}_1, \ldots, \mathbf{w}_n, \alpha_1, \ldots, \alpha_n) \tag{21.10}$$

where the vectors $\mathbf{w}_k = (w_{k1}, \ldots, w_{kn})$ are constrained to unit norm (unlike in the preceding example), and the α_k are further parameters. We introduce here the extra parameters α_k to account for different distributions for different projections. Constraining $\alpha_k = 1$ and $m = n$ and allowing the $\mathbf{w}_k$ to have any norm, this becomes the basic ICA model.

We have the score function

$$\boldsymbol{\psi}(\mathbf{x}; \mathbf{W}, \alpha_1, \ldots, \alpha_m) = \sum_{k=1}^{m} \alpha_k \mathbf{w}_k g\big(\mathbf{w}_k^T \mathbf{x}\big).$$

where g is the first derivative of G. Going through similar developments as in the case of the basic ICA model, the sample version of the objective function $\tilde{J}$ can be shown to equal

$$\begin{aligned}\tilde{J} = &\sum_{k=1}^{m} \alpha_k \frac{1}{T} \sum_{t=1}^{T} g'\big(\mathbf{w}_k^T \mathbf{x}(t)\big) \\ &+ \frac{1}{2} \sum_{j,k=1}^{m} \alpha_j \alpha_k \mathbf{w}_j^T \mathbf{w}_k \frac{1}{T} \sum_{t=1}^{T} g\big(\mathbf{w}_k^T \mathbf{x}(t)\big) g\big(\mathbf{w}_j^T \mathbf{x}(t)\big).\end{aligned} \tag{21.11}$$

Minimization of this function thus enables estimation of the overcomplete ICA model using the energy-based formulation. This is how we obtained the results in Fig. 13.1 on page 286.

21.6 Conclusion

Score matching is a simple method to estimate statistical models in the case where the normalization constant is unknown. Although the estimation of the score function is computationally difficult, we showed that the distance of data and model score functions is very easy to compute. The main assumptions in the method are: (1) all the variables are continuous-valued and defined over $\mathbb{R}^n$, (2) the model pdf is smooth enough.

We have seen how the method gives an objective function whose minimization enables estimation of the model. The objective function is typically given as an analytical formula, so any classic optimization method, such as gradient methods, can be used to minimize it.

Two related methods are contrastive divergence (Hinton 2002) and pseudo-likelihood (Besag 1975). The relationships between these methods are considered in Hyvärinen (2007).

References

Albert MV, Schnabel A, Field DJ (2008) Innate visual learning through spontaneous activity patterns. PLoS Comput Biol 4(8)

Albright TD, Stoner GR (2002) Contextual influences on visual processing. Annu Rev Neurosci 25:339–379

Amari SI (1998) Natural gradient works efficiently in learning. Neural Comput 10(2):251–276

Amari SI, Cichocki A, Yang H (1996) A new learning algorithm for blind source separation. In: Advances in neural information processing systems, vol 8. MIT Press, Cambridge, pp 757–763

Angelucci A, Levitt JB, Walton EJS, Hupé JM, Bullier J, Lund JS (2002) Circuits for local and global signal integration in primary visual cortex. J Neurosci 22(19):8633–8646

Anzai A, Ohzawa I, Freeman RD (1999a) Neural mechanisms for encoding binocular disparity: Receptive field position vs. phase. J Neurophysiol 82(2):874–890

Anzai A, Ohzawa I, Freeman RD (1999b) Neural mechanisms for processing binocular information: Simple cells. J Neurophysiol 82(2):891–908

Atick JJ, Redlich AN (1992) What does the retina know about natural scenes? Neural Comput 4(2):196–210

Atick JJ, Li Z, Redlich AN (1992) Understanding retinal color coding from first principles. Neural Comput 4:559–572

Attneave F (1954) Some informational aspects of visual perception. Psychol Rev 61:183–193

Attwell D, Laughlin SB (2001) An energy budget for signaling in the grey matter of the brain. J Cereb Blood Flow Metab 21:1133–1145

Baddeley RJ, Hancock PJB (1991) A statistical analysis of natural images matches psychophysically derived orientation tuning curves. Proc R Soc Ser B 246(1317):219–223

Baddeley RJ, Abbott LF, Booth MC, Sengpiel F, Freeman T, Wakeman EA, Rolls ET (1997) Responses of neurons in primary and inferior temporal visual cortices to natural scenes. Proc R Soc Ser B 264(1389):1775–1783

Bak P, Tang C, Wiesenfeld K (1987) Self-organized criticality: an explanation of $1/f$ noise. Phys Rev Lett 59:381–384

Balasubramaniam V, Kimber D, Berry II MJ (2001) Metabolically efficient information processing. Neural Comput 13:799–815

Barlow HB (1961) Possible principles underlying the transformations of sensory messages. In: Rosenblith WA (ed) Sensory communication. MIT Press, Cambridge, pp 217–234

Barlow HB (1972) Single units and sensation: A neuron doctrine for perceptual psychology? Perception 1:371–394

Barlow HB (2001a) The exploitation of regularities in the environment by the brain. Behav Brain Sci 24(3)

Barlow HB (2001b) Redundancy reduction revisited. Netw Comput Neural Syst 12:241–253

Barlow HB, Blakemore C, Pettigrew JD (1967) The neural mechanism of binocular depth discrimination. J Physiol 193:327–342

Barrow HG, Bray AJ, Budd JML (1996) A self-organized model of 'color blob' formation. Neural Comput 8:1427–1448

Bell AJ, Sejnowski TJ (1995) An information-maximization approach to blind separation and blind deconvolution. Neural Comput 7:1129–1159

Bell AJ, Sejnowski TJ (1996) Learning higher-order structure of a natural sound. Network 7:261–266

Bell AJ, Sejnowski TJ (1997) The 'independent components' of natural scenes are edge filters. Vis Res 37:3327–3338

A. Hyvärinen, J. Hurri, P.O. Hoyer, *Natural Image Statistics*,
Computational Imaging and Vision 39,

Bell CC, Caputi A, Grant K, Serrier J (1993) Storage of a sensory pattern by anti-hebbian synaptic plasticity in an electric fish. Proc Natl Acad Sci USA 90:4650–4654
Belouchrani A, Meraim KA, Cardoso JF, Moulines E (1997) A blind source separation technique based on second order statistics. IEEE Trans Signal Process 45(2):434–444
Bergner S, Drew MS (2005) Spatiotemporal-chromatic structure of natural scenes. In: Proc. IEEE int conf on image processing (ICIP2005), pp II–301–4
Berkes P, Wiskott L (2005) Slow feature analysis yields a rich repertoire of complex cell properties. J Vis 5(6):579–602
Berkes P, Wiskott L (2007) Analysis and interpretation of quadratic models of receptive fields. Nat Protoc 2(2):400–407
Besag J (1975) Statistical analysis of non-lattice data. The Statistician 24:179–195
Bienenstock EL, Cooper LN, Munro PW (1982) Theory for the development of neuron selectivity: orientation specificity and binocular interaction in visual cortex. J Neurosci 2:32–48
Bishop CM (2006) Pattern recognition and machine learning. Springer, Berlin
Blasdel GG (1992) Orientation selectivity, preference, and continuity in monkey striate cortex. J Neurosci 12(8):3139–3161
Bonhoeffer T, Grinvald A (1991) Iso-orientation domains in cat visual cortex are arranged in pinwheel-like patterns. Nature 353:429–431
Bonin V, Mante V, Carandini M (2006) The statistical computation underlying contrast gain control. J Neurosci 26:6346–6353
Boynton G, Hedgé J (2004) Visual cortex: The continuing puzzle of area V2. Current Biol 14:R523–R524
Bray A, Martinez D (2003) Kernel-based extraction of slow features: complex cells learn disparity and translation-invariance from natural images. In: Becker S, Thrun S, Obermayer K (eds) Advances in neural information processing systems, vol 15. The MIT Press, Cambridge, pp 253–260
Brunet JP, Tamayo P, Golub TR, Mesirov JP (2004) Metagenes and molecular pattern discovery using matrix factorization. Proc Natl Acad Sci USA 101:4164–4169
Buchsbaum G, Bloch O (2002) Color categories revealed by non-negative matrix factorization of Munsell color spectra. Vis Res 42:559–563
Burt PJ, Adelson EH (1983) The Laplacian pyramid as a compact image code. IEEE Trans Commun 4:532–540
Burton GJ, Moorehead TR (1987) Color and spatial structure in natural scenes. Appl Opt 26:157–170
Cai D, DeAngelis GC, Freeman RD (1997) Spatiotemporal receptive field organization in the lateral geniculate nucleus of cats and kittens. J Neurophysiol 78(2):1045–1061
Candès EJ, Demanet L, Donoho DL, Ying L (2005) Fast discrete curvelet transforms. Multiscale Model Simul 5:861–899
Cang J, Rentería RC, Kaneko M, Liu X, Copenhagen DR (2005) Development of precise maps in visual cortex requires patterned spontaneous activity in the retina. Neuron 48:797–809
Carandini M (2004) Receptive fields and suppressive fields in the early visual system. In: Gazzaniga MS (ed) The cognitive neurosciences III. MIT Press, Cambridge
Carandini M, Heeger DJ, Movshon JA (1997) Linearity and normalization in simple cells of the macaque primary visual cortex. J Neurosci 17:8621–8644
Carandini M, Heeger DJ, Movshon JA (1999) Linearity and gain control in V1 simple cells. In: Jones EG, Ulinski PS (eds) Models of cortical function, Cerebral cortex, vol 13. Plenum, New York, pp 401–443
Carandini M, Demb JB, Mante V, Tolhurst DJ, Dan Y, Olshausen BA, Gallant JL, Rust NC (2005) Do we know what the early visual system does? J Neurosci 25(46):10,577–10,597
Cardoso JF (1989) Source separation using higher order moments. In: Proc IEEE int conf on acoustics, speech and signal processing (ICASSP'89), Glasgow, UK, pp 2109–2112
Cardoso JF, Laheld BH (1996) Equivariant adaptive source separation. IEEE Trans Signal Process 44(12):3017–3030
Cavaco S, Lewicki MS (2007) Statistical modeling of intrinsic structures in impacts sounds. J Acoust Soc Am 121:3558–3568

Caywood MS, Willmore B, Tolhurst DJ (2004) Independent components of color natural scenes resemble V1 neurons in their spatial and color tuning. J Neurophysiol 91:2859–2873
Chance FS, Nelson SB, Abbott LF (1999) Complex cells as cortically amplified simple cells. Nat Neurosci 2(3):277–282
Chandler DM, Field DJ (2007) Estimates of the information content and dimensionality of natural scenes from proximity distributions. J Opt Soc Am A 24(4):922–941
Chen BL, Hall DH, Chklovskii DB (2006) Wiring optimization can relate neuronal structure and function. Proc Natl Acad Sci USA 103(1):4723–4728
Chen X, Han F, Poo MM, Dan Y (2007) Excitatory and suppressive receptive field subunits in awake monkey primary visual cortex (V1). Proc Natl Acad Sci USA 104:19,120–19,125
Cohen L (1995) Time–frequency analysis. Prentice Hall signal processing series. Prentice Hall, New York
Comon P (1994) Independent component analysis—a new concept? Signal Process 36:287–314
Cover TM, Thomas JA (2006) Elements of information theory, 2nd edn. Wiley, New York
Crowder NA, van Kleef J, Dreher B, Ibbotson MR (2007) Complex cells increase their phase sensitivity at low contrasts and following adaptation. J Neurophysiol 98:1155–1166
Dakin SC, Mareschal I, Bex PJ (2005) An oblique effect for local motion: Psychophysics and natural movie statistics. J Vis 5:878–887
Dan Y, Poo MM (2004) Spike timing-dependent plasticity of neural circuits. Neuron 44:23–30
Dan Y, Atick JJ, Reid RC (1996a) Efficient coding of natural scenes in the lateral geniculate nucleus: Experimental test of a computational theory. J Neurosci 16:3351–3362
Dan Y, Atick JJ, Reid RC (1996b) Efficient coding of natural scenes in the lateral geniculate nucleus: experimental test of a computational theory. J Neurosci 16(10):3351–3362
Davis GW, Naka K (1980) Spatial organization of catfish retinal neurons. I. Single- and random-bar stimulation. J Neurophysiol 43:807–831
Dayan P, Abbott LF (2001) Theoretical neuroscience. The MIT Press, Cambridge
DeAngelis GC, Ohzawa I, Freeman RD (1993a) Spatiotemporal organization of simple-cell receptive fields in the cat's striate cortex. I. General characteristics and postnatal development. J Neurophysiol 69(4):1091–1117
DeAngelis GC, Ohzawa I, Freeman RD (1993b) Spatiotemporal organization of simple-cell receptive fields in the cat's striate cortex. II. Linearity of temporal and spatial summation. J Neurophysiol 69(4):1118–1135
DeAngelis GC, Ohzawa I, Freeman RD (1995) Receptive-field dynamics in the central visual pathways. Trends Neurosci 18(10):451–458
DeAngelis GC, Ghose GM, Ohzawa I, Freeman RD (1999) Functional micro-organization of primary visual cortex: Receptive field analysis of nearby neurons. J Neurosci 19(10):4046–4064
Delfosse N, Loubaton P (1995) Adaptive blind separation of independent sources: a deflation approach. Signal Process 45:59–83
Deriugin NG (1956) The power spectrum and the correlation function of the television signal. Telecommunications 1:1–12
DeValois RL, Albrecht DG, Thorell LG (1982) Spatial frequency selectivity of cells in macaque visual cortex. Vis Res 22:545–559
Doi E, Inui T, Lee TW, Wachtler T, Sejnowski TJ (2003) Spatiochromatic receptive field properties derived from information-theoretic analyses of cone mosaic responses to natural scenes. Neural Comput 15:397–417
Dong DW, Atick JJ (1995a) Statistics of natural time-varying images. Netw Comput Neural Syst 6(3):345–358
Dong DW, Atick JJ (1995b) Temporal decorrelation: a theory of lagged and nonlagged responses in the lateral geniculate nucleus. Netw Comput Neural Syst 6(2):159–178
Donoho DL (1995) De-noising by soft-thresholding. IEEE Trans Inf Theory 41:613–627
Donoho DL, Johnstone IM (1995) Adapting to unknown smoothness via wavelet shrinkage. J Am Stat Assoc 90:1200–1224
Donoho DL, Stodden V (2004) When does non-negative matrix factorization give a correct decomposition into parts? In: Proc NIPS'2003. Advances in neural information processing, vol 16. MIT Press, Cambridge

Donoho DL, Johnstone IM, Kerkyacharian G, Picard D (1995) Wavelet shrinkage: asymptopia? J R Stat Soc Ser B 57:301–337
Dror RO, Willsky AS, Adelson EH (2004) Statistical characterization of real-world illumination. J Vis 4(9):821–837
Durbin R, Mitchison G (1990) A dimension reduction framework for understanding cortical maps. Nature 343:644–647
Edelman A, Arias TA, Smith ST (1998) The geometry of algorithms with orthogonality constraints. SIAM J Matrix Anal Appl 20(2):303–353
Edwards DP, Purpura HP, Kaplan E (1996) Contrast sensitivity and spatial frequency response of primate cortical neurons in and around the cytochromase oxidase blobs. Vis Res 35(11):1501–1523
Elder JH, Goldberg RM (2002) Ecological statistics of gestalt laws for the perceptual organization of contours. J Vis 2(4):324–353
Embrechts P, Maejima M (2000) An introduction to the theory of self-similar stochastic processes. Int J Mod Phys B 14:1399–1420
Emerson RC, Bergen JR, Adelson EH (1992) Directionally selective complex cells and the computation of motion energy in cat visual cortex. Vis Res 32:203–218
Erwin E, Miller KD (1996) Modeling joint development of ocular dominance and orientation maps in primary visual cortex. In: Bower JM (ed) Computational neuroscience: Trends in research 1995. Academic Press, New York, pp 179–184
Erwin E, Miller KD (1998) Correlation-based development of ocularly-matched orientation and ocular dominance maps: determination of required input activities. J Neurosci. 18:5908–5927
Fairhall AL, Lewen GD, Bialek W, de Ruyter van Steveninck RR (2001) Efficiency and ambiguity in an adaptive neural code. Nature 412:787–792
Felsen G, Touryan J, Han F, Dan Y (2005) Cortical sensitivity to visual features in natural scenes. PLoS Biol 3(10):e342
Field DJ (1987) Relations between the statistics of natural images and the response properties of cortical cells. J Opt Soc Am A 4:2379–2394
Field DJ (1994) What is the goal of sensory coding? Neural Comput 6:559–601
Field DJ (1999) Wavelets, vision and the statistics of natural scenes. Philos Trans R Soc A 357(1760):2527–2542
Fischer B, Kruger J (1979) Disparity tuning and binocularity of single neurons in cat visual cortex. Exp Brain Res 35:1–8
Fitzpatrick D (2000) Seeing beyond the receptive field in primary visual cortex. Curr Opin Neurobiol 10(4):438–443
Földiák P (1990) Forming sparse representations by local anti-Hebbian learning. Biol Cybern 64:165–170
Földiák P (1991) Learning invariance from transformation sequences. Neural Comput 3:194–200
Freeman WT, Adelson EH (1991) The design and use of steerable filters. IEEE Trans Pattern Anal Mach Intell 13(9):891–906
Fukushima K (1980) Neocognitron: A self-organizing neural network model for a mechanism of pattern recognition unaffected by shift in position. Biol Cybern 36(4):193–202
Fukushima K, Okada M, Hiroshige K (1994) Neocognitron with dual C-cell layers. Neural Netw 7(1):41–47
Gallant J, Connor C, van Essen D (1998) Neural activity in areas V1, V2 and V4 during free viewing of natural scenes compared to controlled viewing. NeuroReport 9:85–90
Gallant JL, Braun J, van Essen DC (1993) Selectivity for polar, hyperbolic, and Cartesian gratings in macaque visual cortex. Science 259:100–103
Gardner JL, Sun P, Waggoner RA, Ueno K, Tanaka K, Cheng K (2005) Contrast adaptation and representation in human early visual cortex. Neuron 47:607–620
Garrigues PJ, Olshausen BA (2008) Learning horizontal connections in a sparse coding model of natural images. In: Advances in neural information processing systems, vol 20. MIT Press, Cambridge
Gehler P, Welling M (2005) Products of "edge-perts". In: Advances in neural information processing systems, vol 17. MIT Press, Cambridge

Geisler WS, Perry JS, Super BJ, Gallogly DP (2001) Edge co-occurrence in natural images predicts contour grouping performance. Vis Res 41:711–724

Gilbert CD, Wiesel TN (1985) Intrinsic connectivity and receptive field properties in visual cortex. Vis Res 25(3):365–374

Gonzales RC, Woods RE (2002) Digital image processing, 2nd edn. Prentice Hall, New York

Gray MS, Pouget A, Zemel RS, Nowlan SJ, Sejnowski TJ (1998) Reliable disparity estimation through selective integration. Vis Neurosci 15(3):511–528

Grenander U (1976–1981) Lectures in pattern theory I, II, and III: Pattern analysis, pattern synthesis and regular structures. Springer, Berlin

Grenander U, Srivastava A (2001) Probability models for clutter in natural images. IEEE Trans Pattern Anal Mach Intell 23:424–429

Griffin LD (2007) The second order local-image-structure solid. IEEE Trans Pattern Anal Mach Intell 29(8):1355–1366

Griffin LD, Lillholm M, Nielsen M (2004) Natural image profiles are most likely to be step edges. Vis Res 44(4):407–421

Grossberg S, Mingolla E (1985) Neural dynamics of perceptual grouping: textures, boundaries and emergent segmentations. Percept Psychophys 38(2):141–171

Hafner VV, Fend M, Lungarella M, Pfeifer R, König P, Körding KP (2003) Optimal coding for naturally occurring whisker deflections. In: Proc int conf on artificial neural networks (ICANN/ICONIP2002), pp 805–812

Hancock PJB, Baddeley RJ, Smith LS (1992) The principal components of natural images. Netw Comput Neural Syst 3:61–72

Hansel D, van Vreeswijk C (2002) How noise contributes to contrast invariance of orientation tuning in cat visual cortex. J Neurosci 22:5118–5128

Hansen BC, Hess RF (2007) Structural sparseness and spatial phase alignment in natural scenes. J Opt Soc Am A 24(7):1873–1885

Hashimoto W (2003) Quadratic forms in natural images. Netw Comput Neural Syst 14(4):765–788

Heeger D (1992) Normalization of cell responses in cat striate cortex. Vis Neurosci 9:181–198

Hegdé J, Essen DCV (2007) A comparative study of shape representation in macaque visual areas V2 and V4. Cereb Cortex 17(5):1100–1116

Helmholtz H (1867) Handbuch der physiologischen Optik. Dritter Abschnitt. Die Lehre von den Gesichtswahrnehmungen (Physiological optics. Volume III. The theory of the perceptions of vision). Leipzig, Leopold Voss

Henderson JM (2003) Human gaze control during real-world scene perception. Trends Cogn Sci 7(11):498–504

Hérault J, Ans B (1984) Circuits neuronaux à synapses modifiables: décodage de messages composites par apprentissage non supervisé. C R Acad Sci 299(III-13):525–528

Hinton GE (1989) Connectionist learning procedures. Artif Intell 40(1–3):185–234

Hinton GE (2002) Training products of experts by minimizing contrastive divergence. Neural Comput 14(8):1771–1800

Hinton GE, Ghahramani Z (1997) Generative models for discovering sparse distributed representations. Philos Trans R Soc Lond B 352:1177–1190

Hosoya T, Baccus SA, Meister M (2005) Dynamic predictive coding in the retina. Nature 436(10):71–77

Hoyer PO (2004) Non-negative matrix factorization with sparseness constraints. J Mach Learn Res 5:1457–1469

Hoyer PO, Hyvärinen A (2000) Independent component analysis applied to feature extraction from colour and stereo images. Netw Comput Neural Syst 11(3):191–210

Hoyer PO, Hyvärinen A (2002) A multi-layer sparse coding network learns contour coding from natural images. Vis Res 42(12):1593–1605

Hubel DH, Wiesel TN (1962) Receptive fields, binocular interaction and functional architecture in the cat's visual cortex. J Physiol 73:218–226

Hubel DH, Wiesel TN (1963) Receptive fields of cells in striate cortex of very young, visually inexperienced kittens. J Neurophysiol 26:994–1002

Hubel DH, Wiesel TN (1968) Receptive fields and functional architecture of monkey striate cortex. J Physiol (Lond) 195:215–243
Hubel DH, Wiesel TN (1977) Functional architecture of macaque monkey visual cortex (Ferrier lecture). Proc R Soc Lond Ser B 198:1–59
Hübener M, Shoham D, Grinvald A, Bonhoeffer T (1997) Spatial relationships among three columnar systems in cat area 17. J Neurosci 17(23):9270–9284
Hupé JM, James AC, Payne BR, Lomber SG, Girard P, Bullier J (1998) Cortical feedback improves discrimination between figure and background by V1, V2 and V3 neurons. Nature 394:784–787
Hurri J (2006) Learning cue-invariant visual responses. In: Advances in neural information processing systems, vol 18. MIT Press, Cambridge
Hurri J, Hyvärinen A (2003a) Simple-cell-like receptive fields maximize temporal coherence in natural video. Neural Computat 15(3):663–691
Hurri J, Hyvärinen A (2003b) Temporal and spatiotemporal coherence in simple-cell responses: a generative model of natural image sequences. Netw Comput Neural Syst 14(3):527–551
Hyvärinen A (1999a) Fast and robust fixed-point algorithms for independent component analysis. IEEE Trans Neural Netw 10(3):626–634
Hyvärinen A (1999b) Sparse code shrinkage: Denoising of non-Gaussian data by maximum likelihood estimation. Neural Comput 11(7):1739–1768
Hyvärinen A (2001a) Blind source separation by nonstationarity of variance: A cumulant-based approach. IEEE Trans Neural Netw 12(6):1471–1474
Hyvärinen A (2001b) Complexity pursuit: Separating interesting components from time-series. Neural Comput 13(4):883–898
Hyvärinen A (2002) An alternative approach to infomax and independent component analysis. Neurocomputing 44-46(C):1089–1097
Hyvärinen A (2005) Estimation of non-normalized statistical models using score matching. J Mach Learn Res 6:695–709
Hyvärinen A (2007) Connections between score matching, contrastive divergence, and pseudolikelihood for continuous-valued variables. IEEE Trans Neural Netw 18:1529–1531
Hyvärinen A, Hoyer PO (2000) Emergence of phase and shift invariant features by decomposition of natural images into independent feature subspaces. Neural Comput 12(7):1705–1720
Hyvärinen A, Hoyer PO (2001) A two-layer sparse coding model learns simple and complex cell receptive fields and topography from natural images. Vis Res 41(18):2413–2423
Hyvärinen A, Hurri J (2004) Blind separation of sources that have spatiotemporal variance dependencies. Signal Process 84(2):247–254
Hyvärinen A, Inki M (2002) Estimating overcomplete independent component bases from image windows. J Math Imaging Vis 17:139–152
Hyvärinen A, Köster U (2007) Complex cell pooling and the statistics of natural images. Netw Comput Neural Syst 18:81–100
Hyvärinen A, Oja E (1997) A fast fixed-point algorithm for independent component analysis. Neural Comput 9(7):1483–1492
Hyvärinen A, Oja E (2000) Independent component analysis: Algorithms and applications. Neural Netw 13(4-5):411–430
Hyvärinen A, Hoyer PO, Inki M (2001a) Topographic independent component analysis. Neural Comput 13(7):1527–1558
Hyvärinen A, Karhunen J, Oja E (2001b) Independent component analysis. Wiley-Interscience, New York
Hyvärinen A, Hurri J, Väyrynen J (2003) Bubbles: A unifying framework for low-level statistical properties of natural image sequences. J Opt Soc Am A 20(7):1237–1252
Hyvärinen A, Gutmann M, Hoyer PO (2005a) Statistical model of natural stimuli predicts edge-like pooling of spatial frequency channels in V2. BMC Neurosci 6(12)
Hyvärinen A, Hoyer PO, Hurri J, Gutmann M (2005b) Statistical models of images and early vision. In: Proc int conf adaptive knowledge representation and reasoning, Espoo, Finland, pp 1–14

Johnstone IM, Silverman BW (2005) Empirical Bayes selection of wavelet thresholds. Ann Stat 33(4):1700–1752

Jones JP, Palmer LA (1987a) An evaluation of the two-dimensional Gabor filter model of simple receptive fields in cat striate cortex. J Neurophysiol 58:1233–1258

Jones JP, Palmer LA (1987b) The two-dimensional spatial structure of simple receptive fields in cat striate cortex. J Neurophysiol 58:1187–1211

Jones JP, Stepnoski A, Palmer LA (1987) The two-dimensional spectral structure of simple receptive fields in cat striate cortex. J Neurophysiol 58:1212–1232

Jung K, Kim EY (2004) Automatic text extraction for content-based image indexing. In: Advances in knowledge discovery and data mining: proceedings. Lecture notes in artificial intelligence, vol 3056. Springer, Berlin, pp 497–507

Jutten C, Hérault J (1991) Blind separation of sources, part I: An adaptive algorithm based on neuromimetic architecture. Signal Process 24:1–10

Kandel ER, Schwartz JH, Jessell TM (2000) Principles of neural science, 4th edn. McGraw-Hill, New York

Kapadia MK, Ito M, Gilbert CD, Westheimer G (1995) Improvement in visual sensitivity by changes in local context: parallel studies in human observers and in V1 of alert monkeys. Neuron 15(4):843–856

Kapadia MK, Westheimer G, Gilbert CD (2000) Spatial distribution of contextual interactions in primary visual cortex and in visual perception. J Neurophysiol 84:2048–2062

Kara P, Reinagel P, Reid C (2000) Low response variability in simultaneously recorded retinal, thalamic, and cortical neurons. Neuron 27(3):635–646

Karklin Y, Lewicki MS (2005) A hierarchical Bayesian model for learning non-linear statistical regularities in natural signals. Neural Comput 17:397–423

Karklin Y, Lewicki MS (2008) Emergence of complex cell properties by learning to generalize in natural scenes. Nature

Kayser C, Körding K, König P (2003) Learning the nonlinearity of neurons from natural visual stimuli. Neural Comput 15(8):1751–1759

Kersten D, Schrater P (2002) Pattern inference theory: A probabilistic approach to vision. In: Mausfeld R, Heyer D (eds) Perception and the physical world. Wiley, New York

Kim PM, Tidor B (2003) Subsystem identification through dimensionality reduction of large-scale gene expression data. Genome Res 13:1706–1718

Klein DJ, König P, Körding KP (2003) Sparse spectrotemporal coding of sounds. EURASIP J Appl Signal Process 3:659–667

Knill DC, Richards W (eds) (1996) Perception as Bayesian inference. Cambridge University Press, Cambridge

Koenderink JJ, van Doorn AJ (1987) Representation of local geometry in the visual system. Biol Cybern 55:367–375

Kohonen T (1982) Self-organized formation of topologically correct feature maps. Biol Cybern 43(1):56–69

Kohonen T (1996) Emergence of invariant-feature detectors in the adaptive-subspace self-organizing map. Biol Cybern 75:281–291

Kohonen T (2001) Self-organizing maps, 3rd edn. Springer, Berlin

Kohonen T, Kaski S, Lappalainen H (1997) Self-organized formation of various invariant-feature filters in the adaptive-subspace SOM. Neural Comput 9(6):1321–1344

Körding K, Kayser C, Einhäuser W, König P (2004) How are complex cell properties adapted to the statistics of natural stimuli? J Neurophysiol 91(1):206–212

Köster U, Hyvärinen A (2007) A two-layer ICA-like model estimated by score matching. In: Proc int conf on artificial neural networks (ICANN2007), Porto, Portugal, pp 798–807

Köster U, Hyvärinen A (2008) A two-layer model of natural stimuli estimated with score matching. Submitted manuscript

Köster U, Lindgren JT, Gutmann M, Hyvärinen A (2009a) Learning natural image structure with a horizontal product model. In: Proc int conference on independent component analysis and blind signal separation (ICA2009), Paraty, Brazil

Köster U, Lindgren JT, Hyvärinen A (2009b) Estimating Markov random field potentials for natural images. In: Proc int conference on independent component analysis and blind signal separation (ICA2009), Paraty, Brazil
Koulakov AA, Chklovskii DB (2001) Orientation preference patterns in mammalian visual cortex: a wire length minimization approach. Neuron 29:519–527
Kovesi P (1999) Image features from phase congruency. Videre: J Comput Vis Res 1(3)
Kretzmer ER (1952) Statistics of television signals. Bell Syst Tech J 31:751–763
Krieger G, Rentschler I, Hauske G, Schill, K, Zetzsche C (2000) Object and scene analysis by saccadic eye-movements: an investigation with higher-order statistics. Spat Vis 13:201–214
Krüger N (1998) Collinearity and parallelism are statistically significant second order relations of complex cell responses. Neural Process Lett 8:117–129
Krüger N, Wörgötter F (2002) Multi-modal estimation of collinearity and parallelism in natural image sequences. Netw Comput Neural Syst 13:553–576
Kurki I, Hyvärinen A, Laurinen P (2006) Collinear context (and learning) change the profile of the perceptual filter. Vis Res 46(13):2009–2014
Kushner H, Clark D (1978) Stochastic approximation methods for constrained and unconstrained systems. Springer, Berlin
Lamme VAF (1995) The neurophysiology of figure-ground segregation in primary visual cortex. J Neurosci 15:1605–1615
Laughlin S (1981) A simple coding procedure enhances a neuron's information capacity. Z Naturforsch 36 C:910–912
Lee AB, Mumford D, Huang J (2001) Occlusion models for natural images: A statistical study of scale-invariant dead leaves model. Int J Comput Vis 41:35–59
Lee AB, Pedersen KS, Mumford D (2003) The nonlinear statistics of high-contrast patches in natural images. Int J Comput Vis 54:83–103
Lee DD, Seung HS (1999) Learning the parts of objects by non-negative matrix factorization. Nature 401:788–791
Lee DD, Seung HS (2001) Algorithms for non-negative matrix factorization. In: Proc NIPS'2000. Advances in neural information processing, vol 13. MIT Press, Cambridge
Lee TS, Mumford D (2003) Hierarchical Bayesian inference in the visual cortex. J Opt Soc Am A 20(7):1434–1448
Lennie P (2003) The cost of cortical computation. Curr Biol 13:493–497
LeVay S, Voigt T (1988) Ocular dominance and disparity coding in cat visual cortex. Vis Neurosci 1:395–414
Levy WB, Baxter RA (1996) Energy efficient neural codes. Neural Comput 8:531–543
Lewicki MS (2002) Efficient coding of natural sounds. Nat Neurosci 5(4):356–363
Li SZ (2001) Markov random field modeling in image analysis, 2nd edn. Springer, Berlin
Li SZ, Hou X, Zhang H, Cheng Q (2001) Learning spatially localized parts-based representations. In: Proc IEEE conf on computer vision and pattern recognition (CVPR 2001), Hawaii, USA
Li Z (1999) Pre-attentive segmentation in the primary visual cortex. Spat Vis 13(1):25–50
Li Z, Atick JJ (1994) Efficient stereo coding in the multiscale representation. Netw Comput Neural Syst 5:157–174
Lindgren JT, Hyvärinen A (2004) Learning high-level independent components of images through a spectral representation. In: Proc int conf on pattern recognition (ICPR2004). Cambridge, UK, pp 72–75
Lindgren JT, Hyvärinen A (2007) Emergence of conjunctive visual features by quadratic independent component analysis. In: Advances in neural information processing systems, vol 19. MIT Press, Cambridge
Linsker R (1988) Self-organization in a perceptual network. Computer 21:105–117
Liu X, Cheng L (2003) Independent spectral representations of images for recognition. J Opt Soc Am A 20:1271–1282
Livingstone MS, Hubel DH (1984) Anatomy and physiology of a color system in the primate visual cortex. J Neurosci 4:309–356
Lörincz A, Buzsáki G (2000) Two-phase computational model training long-term memories in the entorhinal-hippocampal region. Ann New York Acad Sci 911(1):83–111

Luenberger D (1969) Optimization by vector space methods. Wiley, New York
Lyu S, Simoncelli EP (2007) Statistical modeling of images with fields of Gaussian scale mixtures. In: Advances in neural information processing systems, vol 19. MIT Press, Cambridge
Lyu S, Simoncelli EP (2008) Nonlinear extraction of 'independent components' of elliptically symmetric densities using radial Gaussianization. Tech rep, Courant Inst of Mathematical Sciences, New York University
Mach E (1886) Die Analyse der Empfindungen und das Verhältnis des Physischen zum Psychischen (The analysis of sensations, and the relation of the physical to the psychical). Gustav Fischer, Jena
Mackay DJC (2003) Information theory, inference and learning algorithms. Cambridge University Press, Cambridge
Maldonado PE, Gödecke I, Gray CM, Bonhoeffer T (1997) Orientation selectivity in pinwheel centers in cat striate cortex. Science 276:1551–1555
Malik J, Perona P (1990) Preattentive texture discrimination with early vision mechanisms. J Opt Soc Am A 7(5):923–932
Mallat SG (1989) A theory for multiresolution signal decomposition: The wavelet representation. IEEE Trans Pattern Anal Mach Intell 11:674–693
Malo J, Gutiérrez J (2006) V1 non-linear properties emerge from local-to-global non-linear ICA. Netw Comput Neural Syst 17:85–102
Mandelbrot BB, van Ness JW (1968) Fractional Brownian motions, fractional noises and applications. SIAM Rev 10:422–437
Mante V, Frazor RA, Bonin V, Geisler WS, Carandini M (2005) Independence of luminance and contrast in natural scenes and in the early visual system. Nat Neurosci 8:1690–1697
Mareschal I, Baker CL (1998a) A cortical locus for the processing of contrast-defined contours. Nat Neurosci 1(2):150–154
Mareschal I, Baker CL (1998b) Temporal and spatial response to second-order stimuli in cat area 18. J Neurophysiol 80:2811–2823
Marr D (1982) Vision. Freeman, New York
Martin DR, Fowlkes CC, Malik J (2004) Learning to detect natural image boundaries using local brightness, color, and texture cues. IEEE Trans Pattern Anal Mach Intell 26:530–549
Matsuoka K, Ohya M, Kawamoto M (1995) A neural net for blind separation of nonstationary signals. Neural Netw 8(3):411–419
Mechler F, Ringach DL (2002) On the classification of simple and complex cells. Vis Res 42:1017–1033
Meister M, Berry II MJ (1999) The neural code of the retina. Neuron 22(3):435–450
Mel BW, Ruderman DL, Archie KA (1998) Translation-invariant orientation tuning in visual "complex" cells could derive from intradendritic computations. J Neurosci 18:4325–4334
Memisevic RF, Hinton GE (2007) Unsupervised learning of image transformations. In: Computer vision and pattern recognition
Miller KD, Troyer TW (2002) Neural noise can explain expansive, power-law nonlinearities in neural response functions. J Neurophysiol 87:653–659
Mitchison G (1991) Removing time variation with the anti-Hebbian differential synapse. Neural Comput 3(3):312–320
Mitchison G (1992) Axonal trees and cortical architecture. Trends Neurosci 15:122–126
Mooijaart A (1985) Factor analysis for non-normal variables. Psychometrica 50:323–342
Morrone MC, Burr DC (1988) Feature detection in human vision: a phase-dependent energy model. Proc R Soc Lond Ser B 235:221–224
Moulden B (1994) Collator units: Second-stage orientational filters. In: Higher-order processing in the visual system. Ciba foundation symposium, vol 184. Wiley, Chichester, pp 170–192
Mountcastle VB (1997) The columnar organization of the neocortex. Brain 120:701–722
Mullen KT (1985) The contrast sensitivity of human colour vision to red-green and blue-yellow chromatic gratings. J Physiol 359:381–400
Mumford D (1992) On the computational architecture of the neocortex. II. The role of cortico-cortical loops. Biol Cybern 66(3):241–251

Mumford D (1994) Neuronal architectures for pattern-theoretic problems. In: Koch C, Davis JL (eds) Large-scale neuronal theories of the brain. The MIT Press, Cambridge

Mumford D, Gidas B (2001) Stochastic models for generic images. Q Appl Math 59:85–111

Murray SO, Kersten D, Olshausen BA, Schrater P, Woods D (2002) Shape perception reduces activity in human primary visual cortex. Proc Natl Acad Sci USA 99:15,164–169

Mury AA, Pont SC, Koenderink JJ (2007) Light field constancy within natural scenes. Appl Opt 46:7308–7316

Mussap AJ, Levi DM (1996) Spatial properties of filters underlying vernier acuity revealed by masking: Evidence for collator mechanisms. Vis Res 36(16):2459–2473

Nadal JP, Parga N (1994) Non-linear neurons in the low noise limit: a factorial code maximizes information transfer. Network 5:565–581

Neumann H, Sepp W (1999) Recurrent V1–V2 interaction in early visual boundary processing. Biol Cybern 81:425–444

Nikias C, Mendel J (1993) Signal processing with higher-order spectra. IEEE Signal Process Mag, pp 10–37

Nikias C, Petropulu A (1993) Higher-order spectral analysis—a nonlinear signal processing framework. Prentice Hall, New York

Nykamp DQ, Ringach DL (2002) Full identification of a linear-nonlinear system via cross-correlation analysis. J Vis 2(1):1–11

Ohki K, Chung S, Ch'ng YH, Kara P, Reid RC (2005) Functional imaging with cellular resolution reveals precise micro-architecture in visual cortex. Nature 433:597–603

Oja E (1982) A simplified neuron model as a principal component analyzer. J Math Biol 15:267–273

Olshausen BA (2002) Sparse codes and spikes. In: Rao R, Olshausen B, Lewicki M (eds) Probabilistic models of the brain. MIT Press, Cambridge, pp 257–272

Olshausen BA (2003) Principles of image representation in visual cortex. In: Chalupa LM, Werner JS (eds) The visual neurosciences. MIT Press, Cambridge

Olshausen BA, Field DJ (1996) Emergence of simple-cell receptive field properties by learning a sparse code for natural images. Nature 381:607–609

Olshausen BA, Field DJ (1997) Sparse coding with an overcomplete basis set: A strategy employed by V1? Vis Res 37:3311–3325

Olshausen BA, Field DJ (2004) Sparse coding of sensory inputs. Curr Opin Neurobiol 14:481–487

Olshausen BA, Sallee P, Lewicki MS (2001) Learning sparse image codes using a wavelet pyramid architecture. In: Advances in neural information processing systems, vol 13. MIT Press, Cambridge, pp 887–893

Olzak LA, Wickens TD (1997) Discrimination of complex patterns: orientation information is integrated across spatial scale; spatial-frequency and contrast information are not. Perception 26:1101–1120

Oppenheim A, Schafer R (1975) Digital signal processing. Prentice-Hall, New York

Osindero S, Hinton GE (2008) Modeling image patches with a directed hierarchy of Markov random fields. In: Advances in neural information processing systems, vol 20. MIT Press, Cambridge

Osindero S, Welling M, Hinton GE (2006) Topographic product models applied to natural scene statistics. Neural Comput 18(2):381–414

Paatero P, Tapper U (1994) Positive matrix factorization: A non-negative factor model with optimal utilization of error estimates of data values. Environmetrics 5:111–126

Palmer SE (1999) Vision science—photons to phenomenology. MIT Press, Cambridge

Papoulis A, Pillai SU (2001) Probability, random variables, and stochastic processes, 4th edn. McGraw-Hill, New York

Parra L, Spence CD, Sajda P, Ziehe A, Müller KR (2000) Unmixing hyperspectral data. In: Advances in neural information processing systems, vol 12. MIT Press, Cambridge, pp 942–948

Pasupathy A, Connor CE (1999) Responses to contour features in macaque area V4. J Neurophysiol 82:2490–2502

Pasupathy A, Connor CE (2001) Shape representation in area V4: Position-specific tuning for boundary conformation. J Neurophysiol 86:2505–2519

Pearson K (1892) The grammar of science. Scott, London
Pedersen KS, Lee AB (2002) Toward a full probability model of edges in natural images. In: Proceedings of 7th European conference on computer vision, pp 328–342
Pham DT, Cardoso JF (2001) Blind separation of instantaneous mixtures of non stationary sources. IEEE Trans Signal Process 49(9):1837–1848
Pham DT, Garrat P (1997) Blind separation of mixture of independent sources through a quasi-maximum likelihood approach. IEEE Trans Signal Process 45(7):1712–1725
Pham DT, Garrat P, Jutten C (1992) Separation of a mixture of independent sources through a maximum likelihood approach. In: Proc EUSIPCO, pp 771–774
Poggio GF, Fischer B (1977) Binocular interaction and depth sensitivity in striate and prestriate cortex of behaving rhesus monkey. J Neurophysiol 40:1392–1405
Polat U, Sagi D (1993) Lateral interactions between spatial channels: Suppression and facilitation revealed by lateral masking experiments. Vis Res 33:993–999
Polat U, Tyler CW (1999) What pattern the eye sees best. Vis Res 39(5):887–895
Polat U, Mizobe K, Pettet MW, Kasamatsu T, Norcia AM (1998) Collinear stimuli regulate visual responses depending on cell's contrast threshold. Nature 391:580–584
Pollen D, Ronner S (1983) Visual cortical neurons as localized spatial frequency filters. IEEE Trans Syst Man Cybern 13:907–916
Pollen D, Przybyszewski A, Rubin M, Foote W (2002) Spatial receptive field organization of macaque V4 neurons. Cereb Cortex 12:601–616
Rao RPN, Ballard DH (1999) Predictive coding in the visual cortex: a functional interpretation of some extra-classical receptive field effects. Nat Neurosci 2(1):79–87
Reinagel P (2001) How do visual neurons respond in the real world? Curr Opin Neurobiol 11(4):437–442
Reinagel P, Zador A (1999) Natural scenes at the center of gaze. Netw Comput Neural Syst 10:341–350
Rieke F, Warland D, de van Steveninck R, Bialek W (1997) Spikes: Exploring the neural code. MIT Press, Cambridge
Riesenhuber M, Poggio T (1999) Hierarchical models of object recognition in cortex. Nat Neurosci 2:1019–1025
Ringach DL, Malone BJ (2007) The operating point of the cortex: Neurons as large deviation detectors. J Neurosci 27:7673–7683
Ringach DL, Shapley R (2004) Reverse correlation in neurophysiology. Cogn Sci 28:147–166
Roelfsema PR, Lamme VAF, Spekreijse H, Bosch H (2002) Figure-ground segregation in a recurrent network architecture. J Cogn Neurosci 14:525–537
Romberg J, Choi H, Baraniuk R, Kingsbury N (2000) Hidden Markov tree modelling of complex wavelet transforms. In: Proc IEEE int conf on acoustics, speech and signal processing (ICASSP2000), Istanbul, Turkey, pp 133–136
Romberg J, Choi H, Baraniuk R (2001) Bayesian tree-structured image modeling using wavelet-domain hidden Markov models. IEEE Trans Image Process 10:1056–1068
Roth S, Black MJ (2005) Fields of experts: a framework for learning image priors. In: Proc IEEE int conf computer vision and pattern recognition (CVRP2005), pp 860–867
Ruderman DL (1997) Origins of scaling in natural images. Vis Res 37:3358–3398
Ruderman DL, Bialek W (1994a) Statistics of natural images: scaling in the woods. Phys Rev Lett 73:814–817
Ruderman DL, Bialek W (1994b) Statistics of natural images: Scaling in the woods. Phys Rev Lett 73(6):814–817
Ruderman DL, Cronin TW, Chiao C (1998) Statistics of cone responses to natural images: Implications for visual coding. J Opt Soc Am A 15(8):2036–2045
Rust NC, Schwartz O, Movshon JA, Simoncelli EP (2005) Spatiotemporal elements of macaque V1 receptive fields. Neuron 46:945–956
Sallee P, Olshausen BA (2003) Learning sparse multiscale image representations. In: Advances in neural information processing systems, vol 15. MIT Press, Cambridge, pp 1327–1334
Sanger T (1989) Optimal unsupervised learning in a single-layered linear feedforward network. Neural Netw 2:459–473

Saul AB, Humphrey AL (1990) Spatial and temporal response properties of lagged and nonlagged cells in cat lateral geniculate nucleus. J Neurophysiol 64(1):206–224
Schervish M (1995) Theory of statistics. Springer, Berlin
Schwartz DA, Howe CQ, Purves D (2003) The statistical structure of human speech sounds predicts musical universals. J Neurosci 23:7160–7168
Schwartz O, Simoncelli EP (2001a) Natural signal statistics and sensory gain control. Nat Neurosci 4(8):819–825
Schwartz O, Simoncelli EP (2001b) Natural sound statistics and divisive normalization in the auditory system. In: Advances in neural information processing systems, vol 13. MIT Press, Cambridge
Schwartz O, Sejnowski TJ, Dayan P (2005) Assignment of multiplicative mixtures in natural images. In: Advances in neural information processing systems, vol 17. MIT Press, Cambridge
Shannon CE (1948) A mathematical theory of communication. Bell Syst Techn J 27:379–423
Shoham D, Hübener M, Schulze S, Grinvald A, Bonhoeffer T (1997) Spatio-temporal frequency domains and their relation to cytochrome oxidase staining in cat visual cortex. Nature 385:529–533
Shouval H, Intrator N, Law CC, Cooper LN (1996) Effect of binocular cortical misalignment on ocular dominance and orientation selectivity. Neural Comput 8:1021–1040
Sigman M, Gilbert CD (2000) Learning to find a shape. Nat Neurosci 3(3):264–269
Sigman M, Cecchi GA, Gilbert CD, Magnasco MO (2001) On a common circle: Natural scenes and gestalt rules. Proc Natl Acad Sci USA 98:1935–1940
Silverman MS, Grosof DH, DeValois RL, Elfar SD (1989) Spatial-frequency organization in primate striate cortex. Proc Natl Acad Sci USA 86(2):711–715
Simoncelli EP (2005) Statistical modeling of photographic images. In: Bovik A (ed) Handbook of image and video processing, 2nd edn. Academic Press, San Diego
Simoncelli EP, Adelson EH (1996) Noise removal via Bayesian wavelet coring. In: Proc 3rd IEEE int conf on image processing, Lausanne, Switzerland, pp 379–382
Simoncelli EP, Heeger D (1998) A model of neuronal responses in visual area MT. Vis Res 38(5):743–761
Simoncelli EP, Olshausen BA (2001) Natural image statistics and neural representation. Annu Rev Neurosci 24:1193–216
Simoncelli EP, Freeman WT, Adelson EH, Heeger DJ (1992) Shiftable multiscale transforms. IEEE Trans Inf Theory 38:587–607
Simoncelli EP, Paninski L, Pillow J, Schwartz O (2004) Characterization of neural responses with stochastic stimuli. In: Gazzaniga M (ed) The new cognitive neurosciences, 3rd edn. MIT Press, Cambridge
Skottun BC, Freeman RD (1984) Stimulus specificity of binocular cells in the cat's visual cortex: Ocular dominance and the matching of left and right eyes. Exp Brain Res 56(2):206–216
Smith E, Lewicki MS (2005) Efficient coding of time-relative structure using spikes. Neural Comput 17:19–45
Smith E, Lewicki MS (2006) Efficient auditory coding. Nature 439:978–982
Sonka M, Hlavac V, Boyle R (1998) Image processing, analysis, and machine vision. Brooks/Cole, Pacific Grove
Spanias AS (1994) Speech coding: a tutorial review. Proc IEEE 82(10):1541–1582
Sperling G (1989) Three stages and two systems of visual processing. Spat Vis 4(2/3):183–207
Srivanivasan MV, Laughlin SB, Dubs A (1982) Predictive coding: a fresh view of inhibition in the retina. Proc R Soc Lond B 216:427–459
Srivastava A, Lee AB, Simoncelli EP, Chu SC (2003) On advances in statistical modelling of natural images. J Math Imaging Vis 18:17–33
Stone J (1996) Learning visual parameters using spatiotemporal smoothness constraints. Neural Comput 8(7):1463–1492
Strang G, Nguyen T (1996) Wavelets and filter banks. Wellesley-Cambridge Press, Wellesley
Thomson MGA (1999) Higher-order structure in natural scenes. J Opt Soc Am A 16:1549–1553
Thomson MGA (2001) Beats, kurtosis and visual coding. Netw Comput Neural Syst 12:271–287

Tolhurst DJ, Tadmor Y, Chao T (1992) Amplitude spectra of natural images. Ophthalmic Physiol Opt 12(2):229–232
Tong L, Liu RW, Soon VC, Huang YF (1991) Indeterminacy and identifiability of blind identification. IEEE Trans Circuits Syst 38:499–509
Tootell RBH, Silverman MS, Hamilton SL, Switkes E, Valois RLD (1988) Functional anatomy of macaque striate cortex. V. Spatial frequency. J Neurosci 8:1610–1624
Torralba A, Oliva A (2003) Statistics of natural image categories. Netw Comput Neural Syst 14:391–412
Touryan J, Lau B, Dan Y (2002) Isolation of relevant visual features from random stimuli for cortical complex cells. J Neurosci 22:10,811–10,818
Touryan J, Felsen G, Han F, Dan Y (2005) Spatial structure of complex cell receptive fields measured with natural images. Neuron 45(5):781–791
Ts'o DY, Gilbert CD (1988) The organization of chromatic and spatial interactions in the primate striate cortex. J Neurosci 8(5):1712–1727
Ts'o DY, Roe AW (1995) Functional compartments in visual cortex: Segregation and interaction. In: Gazzaniga MS (ed) The cognitive neurosciences. MIT Press, Cambridge, pp 325–337
Turiel A, Parga N (2000) The multi-fractal structure of contrast changes in natural images: From sharp edges to textures. Neural Comput 12:763–793
Utsugi A (2001) Ensemble of independent factor analyzers with application to natural image analysis. Neural Process Lett 14:49–60
Valpola H, Harva M, Karhunen J (2003) Hierarchical models of variance sources. In: Proceedings of the fourth international symposium on independent component analysis and blind signal separation, pp 83–88
van der Schaaf A, van Hateren JH (1996) Modelling the power spectra of natural images: statistics and information. Vis Res 36:2759–2770
van Hateren JH (1992) Real and optimal neural images in early vision. Nature 360:68–70
van Hateren JH (1993) Spatial, temporal and spectral pre-processing for colour vision. Proc R Soc Ser B 251:61–68
van Hateren JH, Ruderman DL (1998) Independent component analysis of natural image sequences yields spatiotemporal filters similar to simple cells in primary visual cortex. Proc R Soc Ser B 265:2315–2320
van Hateren JH, van der Schaaf A (1998) Independent component filters of natural images compared with simple cells in primary visual cortex. Proc R Soc Ser B 265:359–366
van Vreeswijk C (2001) Whence sparseness? In: Advances in neural information processing systems, vol 13. MIT Press, Cambridge, pp 180–186
Vetterli M, Kovačević J (1995) Wavelets and subband coding. Prentice Hall signal processing series. Prentice Hall, Englewood Cliffs
Vincent B, Baddeley RJ (2003) Synaptic energy efficiency in retinal processing. Netw Comput Neural Syst 43:1283–1290
Vincent B, Baddeley RJ, Troscianko T, Gilchrist ID (2005) Is the early visual system optimised to be energy efficient? Netw Comput Neural Syst 16:175–190
Vinje WE, Gallant JL (2000) Sparse coding and decorrelation in primary visual cortex during natural vision. Science 287(5456):1273–1276
Vinje WE, Gallant JL (2002) Natural stimulation of the nonclassical receptive field increases information transmission efficiency in V1. J Neurosci 22:2904–2915
Wachtler T, Lee TW, Sejnowski TJ (2001) Chromatic structure of natural scenes. J Opt Soc Am A 18(1):65–77
Wachtler T, Doi E, Lee TW, Sejnowski TJ (2007) Cone selectivity derived from the responses of the retinal cone mosaic to natural scenes. J Vis 7(8):1–14
Wainwright MJ, Simoncelli EP, Willsky AS (2001) Random cascades on wavelet trees and their use in analyzing and modeling natural images. Appl Comput Harmon Anal 11:89–123
Wandell BA (1995) Foundations of vision. Sinauer Associates, Sunderland
Weliky M, Fiser J, Hunt RH, Wagner DN (2003) Coding of natural scenes in primary visual cortex. Neuron 37(4):703–718

Wichmann FA, Braun DI, Gegenfurtner KR (2006) Phase noise and the classification of natural images. Vis Res 46:1520–1529

Wilkinson F, James T, Wilson H, Gati J, Menon R, Goodale M (2000) An fMRI study of the selective activation of human extrastriate form vision areas by radial and concentric gratings. Curr Biol 10:1455–8

Williams CB, Hess RF (1998) Relationship between facilitation at threshold and suprathreshold contour integration. J Opt Soc Am A 15(8):2046–2051

Willmore B, Tolhurst DJ (2001) Characterizing the sparseness of neural codes. Netw Comput Neural Syst 12:255–270

Winkler G (2003) Image analysis, random field and Markov chain Monte Carlo methods, 2nd edn. Springer, Berlin

Wiskott L, Sejnowski TJ (2002) Slow feature analysis: Unsupervised learning of invariances. Neural Comput 14(4):715–770

Wong ROL (1999) Retinal waves and visual system development. Ann Rev Neurosci 22:29–47

Wyss R, König P, Verschure PFMJ (2006) A model of the ventral visual system based on temporal stability and local memory. PLoS Biol 4(5):0836–0843

Yang Z, Purves D (2003) A statistical explanation of visual space. Nat Neurosci 6:632 – 640

Yen SC, Baker J, Gray CM (2007) Heterogeneity in the responses of adjacent neurons to natural stimuli in cat striate cortex. J Neurophysiol 97:1326–1341

Yuille A, Kersten D (2006) Vision as Bayesian inference: analysis by synthesis? Trends Cogn Sci 10(7):301–308

Zetzsche C, Krieger G (1999) Nonlinear neurons and high-order statistics: New approaches to human vision and electronic image processing. In: Rogowitz B, Pappas T (eds) Human vision and electronic imaging IV. Proc. SPIE, vol 3644. SPIE, Bellingham, pp 2–33

Zetzsche C, Röhrbein F (2001) Nonlinear and extra-classical receptive field properties and the statistics of natural scenes. Netw Comput Neural Syst 12:331–350

Zetzsche C, Krieger G, Wegmann B (1999) The atoms of vision: Cartesian or polar? J Opt Soc Am A 16:1554–1565

Zhang K, Sejnowski TJ (2000) A universal scaling law between gray matter and white matter of cerebral cortex. Proc Natl Acad Sci USA 97:5621–5626

Zhu SC, Wu ZN, Mumford D (1997) Minimax entropy principle and its application to texture modeling. Neural Comput 9:1627–1660

Zhu Y, Qian N (1996) Binocular receptive field models, disparity tuning, and characteristic disparity. Neural Comput 8:1611–1641

Index

A. Hyvärinen, J. Hurri, P.O. Hoyer, *Natural Image Statistics,*
Computational Imaging and Vision 39,

GPSR Compliance
The European Union's (EU) General Product Safety Regulation (GPSR) is a set of rules that requires consumer products to be safe and our obligations to ensure this.

If you have any concerns about our products, you can contact us on

ProductSafety@springernature.com

In case Publisher is established outside the EU, the EU authorized representative is:

Springer Nature Customer Service Center GmbH
Europaplatz 3
69115 Heidelberg, Germany

www.ingramcontent.com/pod-product-compliance
Ingram Content Group UK Ltd.
Pitfield, Milton Keynes, MK11 3LW, UK
UKHW020712071026
15590UKWH00010B/480

9781849968447